IUTAM Symposium on Creep in Structures

SOLID MECHANICS AND ITS APPLICATIONS
Volume 86

Series Editor: **G.M.L. GLADWELL**
Department of Civil Engineering
University of Waterloo
Waterloo, Ontario, Canada N2L 3GI

Aims and Scope of the Series

The fundamental questions arising in mechanics are: *Why?, How?,* and *How much?*
The aim of this series is to provide lucid accounts written bij authoritative researchers
giving vision and insight in answering these questions on the subject of mechanics as it
relates to solids.

The scope of the series covers the entire spectrum of solid mechanics. Thus it includes
the foundation of mechanics; variational formulations; computational mechanics;
statics, kinematics and dynamics of rigid and elastic bodies: vibrations of solids and
structures; dynamical systems and chaos; the theories of elasticity, plasticity and
viscoelasticity; composite materials; rods, beams, shells and membranes; structural
control and stability; soils, rocks and geomechanics; fracture; tribology; experimental
mechanics; biomechanics and machine design.

The median level of presentation is the first year graduate student. Some texts are monographs defining the current state of the field; others are accessible to final year under-
graduates; but essentially the emphasis is on readability and clarity.

For a list of related mechanics titles, see final pages.

IUTAM Symposium on

Creep in Structures

Proceedings of the IUTAM Symposium
held in Nagoya, Japan,
3–7 April 2000

Edited by

S. MURAKAMI
Nagoya University,
Nagoya, Japan

and

N. OHNO
Nagoya University,
Nagoya, Japan

KLUWER ACADEMIC PUBLISHERS
DORDRECHT / BOSTON / LONDON

A C.I.P. Catalogue record for this book is available from the Library of Congress.

ISBN 0-7923-6737-5

Published by Kluwer Academic Publishers,
P.O. Box 17, 3300 AA Dordrecht, The Netherlands.

Sold and distributed in North, Central and South America
by Kluwer Academic Publishers,
101 Philip Drive, Norwell, MA 02061, U.S.A.

In all other countries, sold and distributed
by Kluwer Academic Publishers,
P.O. Box 322, 3300 AH Dordrecht, The Netherlands.

Printed on acid-free paper

SCIENTIFIC COMMITTEE

S. Murakami, Co-Chair (Nagoya University, Japan)
N. Ohno, Co-Chair (Nagoya University, Japan)
A. Bertram (Otto-von-Göricke-Universität Magdeburg, Germany)
F. Ellyin (University of Alberta, Canada)
E. van der Giessen (Delft University of Technology, The Netherlands)
D. R. Hayhurst (University of Manchester Institute of Science and Technology, UK)
F. Leckie (University of California, Santa Barbara, USA)
J. Lemaitre (Université Paris 6, France)
B. Storåkers (Royal Institute of Technology, Sweden)
M. Życzkowski (Cracow University of Technology, Poland)
Ren Wang (Peking University, P. R. China)

LOCAL ORGANIZING COMMITTEE

S. Murakami, Co-Chair (Nagoya University, Japan)
N. Ohno, Co-Chair (Nagoya University, Japan)
Y. Asada (University of Tokyo, Japan)
T. Inoue (Kyoto University, Japan)
H. Ishikawa (Hokkaido University, Japan)
S. Kubo (Osaka University, Japan)
K. Ohji (Ryukoku University, Japan)
R. Ohtani (Kyoto University, Japan)
M. Sakane (Ritsumeikan University, Japan)
Y. Takahashi (Central Rescarch Institute of Electric Power Industry, Japan)
S. Biwa (Nagoya University, Japan)
M. Mizuno (Nagoya University, Japan)

SPONSORS

PREFACE

The advent of steam turbines and the sudden rise of steam temperature at the beginning of the 20th century gave a great impetus to the start of scientific research on metal creep and high-temperature strength. Then aeronautical and aerospace exploitation in the 1940's and 1950's enlarged the scope of creep research. In this context, the first IUTAM Symposium on "Creep in Structures" was held at Stanford University in July 1960, and about 60 participants from seven countries around the world discussed their recent results on this problem.

Subsequent innovation in science and technology, as in nuclear and new energy technology, new materials, large scale integration of semiconductors etc., has claimed solutions to new and challenging problems in this fundamental field of applied mechanics. In order to discuss the new topics in this discipline, the IUTAM Symposia "Creep in Structures" thereafter have been held every ten years; i.e. the second in 1970 at Gothenburg, Sweden, the third in 1980 at Leicester, U.K. and the fourth in 1990 at Cracow, Poland.

The First (1960) and Second Symposium (1970) were concerned mainly with the phenomenological law of creep and creep analysis of structural elements, whereas the issues of the Third Symposium (1980) shifted toward the problems of creep damage, creep crack growth, practical and effective design methods, etc.

Besides these topics, the Fourth Symposium (1990) highlighted problems related to the internal structural change of materials, e.g., elaboration of constitutive equations for complicated loading conditions, initiation and growth of creep cavities, strain localization in creep fracture, and phase transformation during creep. The materials concerned, furthermore, have been extended to single crystals, composite materials, polymeric materials, ceramics, geological materials, etc.

The research on creep in structures has made great progress also after the Fourth IUTAM Symposium, both in its fundamental aspects and from the application point of view. Research in this field is now expected to make important contributions to the development of the fundamental technology of the 21st century, e.g., environmental and energy enginecring, electronic engineering, aeronautical and aerospace engineering. In view of the importance of consolidating the development of creep research since 1990, and providing a forum to discuss the new horizons in this fundamental field of applied mechanics in the coming century, the Fifth IUTAM Symposium on Creep in Structures was proposed to the General Assembly of IUTAM, and was approved to be held at Nagoya in April 2000.

After the final decision at the General Assembly, according to the Rule and Guidelines of IUTAM symposia, the Scientific Committee of the Symposium carefully selected the candidates for participation in the Symposium, and sent out invitation letters to about 80 of the worlds' best and most active scientists and engineers in relevant fields.

As a result of this invitation, the Fifth IUTAM Symposium on Creep in Structures finally had 91 participants from 15 countries, and 50 innovative papers were presented among which 8 were nominated as General Lectures. Presentation of these papers was divided into 16 Sessions, and 45 and 25 minutes were devoted to each of the general and the ordinary lectures, respectively. The present Proceedings contains the papers presented at the Symposium, together with the Final Program, List and Address of Participants. We believe that this volume has summarized the creep research in the last decade of the 20th century.

Finally, it is our pleasure to express our heartfelt gratitude to all the members of the Scientific Committee and to all the members of the Local Organizing Committee of the Symposium for their generous cooperation, crucial advice and cordial encouragement throughout the planning and preparation. The encouraging support and precious advice of the organizers of the preceding IUTAM Symposia on Creep in Structures, especially from Professor J. Hult (Second Symposium), Professor D. R. Hayhurst (Third Symposium) and Professor M. Zyczkowski (Fourth Symposium) were sincerely appreciated.

The generous support by a number of organizations and foundations in connection with organizing and conducting this Symposium is very gratefully acknowledged. Without their financial support, smooth and successful arrangements of the Symposium would have been virtually impossible. Our special thanks are extended to our colleagues Dr. M. Mizuno, Dr. S. Biwa, Dr. M. Kobayashi and Dr. Y. Sugita for their unfailing cooperation throughout the preparations. We are grateful to our secretaries, Ms. F. Kajiura and Ms. J. Nakatsuka, for performing their considerable secretarial tasks involved in the Symposium. We would like to express our thanks also to the editorial staff of the Kluwer Academic Publishers for their cooperation in publishing this volume.

Sumio Murakami
Nobutada Ohno
July 2000

LIST OF PARTICIPANTS

AINSWORTH, R. A.   British Energy Generation Ltd., Barnett way, Barnwood, Gloucester GL4 3RS, UK

ALTENBACH, H.   FB Ingenieurwissenschaften, Martin-Luther-Universität Halle-Wittenberg, D-06099 Halle, GERMANY

BARRETTE, P. D.   Faculty of Engineering and Applied Science, Memorial University of Newfoundland, St. John's, NF A1B 3X5, CANADA

BENALLAL, A.   LMT, ENS de Cachan/CNRS/Universite Paris 6, 61 Avenue du President Wilson, 94235 Cachan, FRANCE

BIWA, S.   Department of Micro System Engineering, Nagoya University, Furo-cho, Chikusa-ku, Nagoya 464-8603, JAPAN

BOYLE, J. T.   Department of Mechanical Engineering, University of Strathclyde, James Weir Bldg., Montrose St., Glasgow, Scotland, G1 1XJ, UK

BUSSO, E. P.   Department of Mechanical Engineering, Imperial College, Exhibition Road, London SW7 2BX, UK

CHOW, C. L.   Department of Mechanical Engineering, University of Michigan-Dearborn, 4901 Evergreen Road, Dearborn, MI 48128, USA

CHRZANOWSKI, M.   Politechnika Krakowska, ul.Warszawska 24, PL31 155 Krakow, POLAND

DANG VAN, K.   Laboratoire de Mecanique des Solides, Ecole Polytechnique, F-91128 Palaiseau, FRANCE

DUNNE, F. P.   Department of Engineering Science, Oxford University, Parks Road, Oxford OX1 3PJ, UK

DYSON, B. F.   Imperial College, (Home) Granta Lodge, St. George's Road, Weybridge Sorrey KT1 3EP, UK

FUJIKURA, M.   Japan Ultra-High Temperature Materials Research Center, 3-1-8, Higashi-machi, Tajimi-city, Gifu 507-0801, JAPAN

GANCZARSKI, A. W.   Institute of Mechanics and Machine Design, Cracow University of Technology, Jana Pawla II 37, 31-864 Krakow, POLAND

GOLUBOVSKIY, E. R.   All-Russian Institute of Aviation Materials, Radio Street 17, Moscow 107005, RUSSIA

HÄRKEGÅRD, G.   Norwegian University of Science and Technology, NTNU/IMM N-7491 Trondheim, NORWAY

HASEGAWA, S.   Tohoku Electric Power Co., Inc., 7-2-1, Nakayama, Aoba-ku, Sendai, Miyagi 981-0952, JAPAN

HAYHURST, D. R.   Department of Mechanical Engineering, University of Manchester Institute of Science and Technology, P.O. Box 88, Manchester, M60 1QD, UK

HELLMICH, Ch.   Institute for Strength of Materials, Vienna University of Technology, Karlsplatz 13/202, A-1140 Vienna, AUSTRIA

HIROE, T.   Faculty of Engineering, Kumamoto University, Kurokami 2-39-1, Kumamoto 860-8555, JAPAN

HIROTA, M.   Research Laboratory Energy Section, Kyushu Electric Power Co., Inc., 2-1-47 Shiobaru, Minami-ku, Fukuoka 815-8520, JAPAN

HIYOSHI, N.   Department of Mechanical Engineering, Faculty of Science and Engineering, Ritsumeikan University, 1-1-1, Nojihigashi, Kusatsu, Shiga 525-8577, JAPAN

HOMMA, N.   R&D Department, Hokkaido Electric Power Co., Inc., 2-1 Tsuishikari, Ebetsu-shi, Hokkaido 067-0033, JAPAN

HYDE, T. H.   School of Mechanical, Materials and Manufacturing Engineering, University of Nottingham, University Park, Nottingham NG7 2RD, UK

IGARI, T.   Nagasaki R&D Center, Mitsubishi Heavy Ind., Fukahori-machi 5-717-1, Nagasaki 851-0392, JAPAN

INOUE, T.   Department of Energy Conversion Science, Kyoto University, Yoshida-honmachi, Sakyo-ku, Kyoto 606-8501, JAPAN

ISHIKAWA, H.   Division of Mechanical Science, Hokkaido University, N13, W8, Kita-ku, Sapporo 060-8628, JAPAN

ITO, K.   Electric Power Research & Development Center, Chubu Electric Power Co., Inc., 20-1, Kitasekiyama, Ohdaka-cho, Midori-ku, Nagoya 459-8522, JAPAN

ITO, K.   Structure & Strength Department, Research Laboratory, Ishikawajima-Harima Heavy Industry Co., Ltd., 3-1-15 Toyosu, Koto-ku, Tokyo 135-8732, JAPAN

JORDAAN, I. J.   Faculty of Engineering and Applied Science, Memorial
                        University of Newfoundland, St. John's, NF A1B 3X5,
                        CANADA

KATO, Y.   Chubu Plant Service Co. Inc., 11-22 Gohonmatsu-cho,
                        Atsuta-ku, Nagoya 456-8516, JAPAN

KAWAI, M.   Inst. Eng. Mech. & Sys., University of Tsukuba, 1-1-1,
                        Tennoudai, Tsukuba 305-8573, JAPAN

KAWASHIMA, F.   Nagasaki R&D Center, Mitsubishi Heavy Ind.,
                        Fukahori-machi 5-717-1, Nagasaki 851-0392, JAPAN

KITAMURA, T.   Department of Engineering Physics and Mechanics,
                        Graduate School of Engineering, Kyoto University,
                        Yoshida-honmachi, Sakyo-ku, Kyoto 606-8501, JAPAN

KNOWLES, D. M.   Department of Materials Science, University of
                        Cambridge, Pembroke St., Cambridge CB2 30Z, UK

KOHNO, M.   KOBELCO Research Institute Inc., 1-5-5, Takazukadai,
                        Nishi-ku, Kobe, JAPAN

KOWALEWSKI, Z. L.   Institute of Fundamental Technological Research, Polish
                        Academy of Sciences, Swietokrzyska 21, 00-049, Warsaw,
                        POLAND

KREMPL, E.   Rensselaer Polytechnic Institute, ME AE & M - JEC 2001,
                        110 Eight Sreet, Troy, New York 12180-3590, USA

KRUCH, S.   DMSE/LCME, ONERA, BP72-29 Avenue Division
                        Leclerc, 92322 Chatillon Cedex, FRANCE

KUBO, S.   Department of Mechanical Engineering and Systems,
                        Graduate School of Engineering, Osaka University, 2-1,
                        Yamadaoka, Suita, Osaka, 565-0871, JAPAN

LE MAY, I.   Metallurgical Consulting Service Ltd., P.O.Box 5006,
                        Saskatoon, SASK, S7K, 4E3, CANADA

LEMAITRE, J.   Universite Paris 6, L.M.T., 61 Avenue du President
                        Wilson, 94235 Cachan, FRANCE

LIN, J.   School of Man. & Mech. Engineering, University of
                        Birmingham, Edgbaston, Birmingham B15 2TT, UK

LITEWKA, A.   Department of Civil Engineering, Universidade da Beira
                        Interior, Rua Marques d'Avila e Bolama, 6200 Covilha,
                        PORTUGAL

LIU, Y.   Center for Surface Transportation Technology, National
                           Research Council Canada, U-89, Alert Road, Uplands,
                           Ottawa, Ontario   K1A   0R6, CANADA

MARKETZ, W. T.   Institute of Mechanics, Franz-Josef-Strasse 18 A-8700
                           Leoben, AUSTRIA

McLEAN, M.   Department of Materials, Imperial College, Prince Consort
                           Road, London SW7 2BP, UK

MIKKELSEN, L. P.   Department of Building Technology and Structural
                           Engineering, Aalborg University, Sohngaardsholmsvej 57,
                           DK-9000 Aalborg, DENMARK

MIYAZAKI, N.   Department of Materials Process Engineering, Kyushu
                           University, 6-10-1 Hakozaki, Higashi-ku, Fukuoka,
                           812-8581, JAPAN

MIZUNO, M.   Department of Mechanical Engineering, Nagoya
                           University, Furo-cho, Chikusa-ku, Nagoya 464-8603,
                           JAPAN

MURAKAMI, S.   Department of Mechanical Engineering, Nagoya
                           University, Furo-cho, Chikusa-ku, Nagoya 464-8603,
                           JAPAN

NAGAE, Y.   Japan Nuclear Cycle Development Institute, 4002 Narita
                           Oarai-machi, Higashiibaraki-gun, Ibaraki 311-1393,
                           JAPAN

NAKAMURA, K.   Materials Research Section, Technical Research Center,
                           The Kansai Electric Power Company Inc., 11-20 Nakaji,
                           3-chome, Amagasaki, Hyogo 661-0974, JAPAN

NGUYEN, B.-N.   Koiter Institute Delft, Micromechanics of Materials, Delft
                           University of Technology, Mekelweg 2, 2628 CD Delft,
                           THE NETHERLANDS

O'DOWD, N. P.   Department of Mechanical Engineering, Imperial College,
                           London SW7 2BX, UK

OHNO, N.   Department of Micro System Engineering, Nagoya
                           University, Chikusa-ku, Nagoya 464-8603, JAPAN

OHTANI, R.   Department of Engineering Physics and Mechanics,
                           Graduate School of Engineering, Kyoto University,
                           Yoshida-honmachi, Sakyo-ku, Kyoto 606-8501, JAPAN

OKUDA, Y.   Chubu Plant Service Co. Inc., 3-2 Oe-cho, Minato-ku,
                           Nagoya 455-0024, JAPAN

ONCK, P. R.	Delft University of Technology, Mekelweg 2, 2628 CD Delft, THE NETHERLANDS
PONTHOT, J.-P.	University of Liege, LTAS-MC&T, Institut de Mecanique B52/3, Chemin des Chevreuils 1, B4000 Liege, BELGIUM
SAKANE, M.	Department of Mechanical Engineering, Faculty of Science and Engineering, Ritsumeikan University, 1-1-1, Nojihigashi, Kusatsu, Shiga 525-8577, JAPAN
SANOMURA, Y.	Faculty of Engineering, Tamagawa University, 6-1-1, Tamagawa-Gakuen, Machida, Tokyo 194-8610, JAPAN
SASAKI, K.	Division of Mechanical Science, Hokkaido University, N13, W8, Kita-ku, Sapporo 060-8628, JAPAN
SHIMIZU, M.	Kobelco Research Institute Inc., 1-5-5, Takazukadai, Nishi-ku, Kobe, JAPAN
SHINOHARA, N.	Chubu Electric Power Co., Inc., 20-1, Kitasekiyama, Ohdaka-cho, Midori-ku, Nagoya 459-8522, JAPAN
SMITH, D. J.	Department of Mechanical Engineering, University of Bristol, Queen's Building, University Walk, Bristol B58 1TR, UK
SUGITA, Y.	Chubu Electric Power Co., Inc., 20-1, Kitasekiyama, Ohdaka-cho, Midori-ku, Nagoya 459-8522, JAPAN
SUGIYAMA, K.	Chubu Electric Power Co., Inc., 20-1, Kitasekiyama, Ohdaka-cho, Midori-ku, Nagoya 459-8522, JAPAN
SUZUKI, A.	Structure & Strength Department, Research Laboratory, Ishikawajima-Harima Heavy Industry Co., Ltd., 3-1-15 Toyosu, Koto-ku, Tokyo 135-8732, JAPAN
TADA, N.	Department of Engineering Physics and Mechanics, Graduate School of Engineering, Kyoto University, Yoshida-hommachi, Sakyo-ku, Kyoto 606-8501, JAPAN
TAKAHASHI, Y.	Materials Science Department, Central Research Institute of Electric Power Industry, 2-11-1, Iwado Kita, Komae-shi, Tokyo 201-8511, JAPAN
TAKEMASA, F.	Structure & Strength Department, Research Laboratory, Ishikawajima-Harima Heavy Industry Co., Ltd., 3-1-15 Toyosu, Koto-ku, Tokyo 135-8732, JAPAN
TAKIGAWA, Y.	Japan Fine Ceramics Center, 2-4-1 Mutsuno, Atsuta-ku, Nagoya 456-8587, JAPAN

TEZUKA, H.   Materials Engineering Group, Power Engineering R&D
                    Center, The Tokyo Electric Power Company Inc., 4-1
                    Egasaki-cho, Tsurumi-ku, Yokohama 230-8510, JAPAN

VERGER, L.   Laboratoire de Mecanique des Solides, Ecole
                    Polytechnique, F-911128 Palaiseau, FRANCE

WATANABE, K.   Department of Mechanical Engineering, Yamagata
                    University, 4-3-16, Jonan, Yonezawa, Yamagata 992-8510,
                    JAPAN

YAGI, K.   National Research Institute for Metals, 1-2-1 Sengen,
                    Tsukuba, Ibaraki 305-0047, JAPAN

YAGUCHI, M.   Materials Science Department, Central Research Institute
                    of Electric Power Industry, 2-11-1, Iwado Kita, Komae-shi,
                    Tokyo 201-8511, JAPAN

YANG, T.-Q.   Department of Mechanics, Huazhong University of
                    Science and Technology, Wuhan 430074, CHINA

YOSHIDA, F.   Department of Mechanical Engineering, Hiroshima
                    University, 1-4-1, Kagaiyama, Higashi-Hiroshima,
                    739-8527, JAPAN

YUYA, H.   Electric Power Research & Development Center, Chubu
                    Electric Power Co., Inc., 20-1, Kitasekiyama, Ohdaka-cho,
                    Midori-ku, Nagoya 459-8522, JAPAN

PROGRAM

April 3 (Mon.)

9:00-9:30   OPENING ADDRESS
S. Murakami
J. Lemaitre
D. R. Hayhurst

9:30-10:15   GENERAL LECTURE
(Chairpersons: *J. Lemaitre and N. Ohno*)
B. F. Dyson and M. McLean

10:45-12:00   MICROMECHANISMS AND MECHANICAL MODELING 1
(Chairpersons: *J. Lemaitre and N. Ohno*)
W. T. Marketz, A. Chatterjee, F. D. Fischer and H. Clemens
D. M. Knowles and D. W. MacLachlan
E. P. Busso, N. P. O'Dowd and R. J. Dennis

14:00-14:45   GENERAL LECTURE
(Chairpersons: *B. F. Dyson and T. Kitamura*)
P. Onck, B.-N. Nguyen and E. van der Giessen

14:45-15:35   MICROMECHANISMS AND MECHANICAL MODELING 2
(Chairpersons: *B. F. Dyson and T. Kitamura*)
B.-N. Nguyen, P. Onck and E. van der Giessen
N. Tada and R. Ohtani

16:05-17:45   MICROMECHANISMS AND MECHANICAL MODELING 3
(Chairpersons: *I. Le May and K. Yagi*)
T. H. Hyde and W. Sun
A. D. Bettinson, N. P. O'Dowd, K. Nikbin and G. A. Webster
T. Kitamura and T. Shibutani
N. Miyazaki

April 4 (Tue.)

9:00-9:45   GENERAL LECTURE
(Chairpersons: *D. R. Hayhurst and T. Inoue*)
J. P. Sermage, J. Lemaitre and R. Desmorat

9:45-10:35   CONTINUUM DAMAGE MECHANICS 1
              (Chairpersons: *D. R. Hayhurst and T. Inoue*)
              Y. Wei, C. L. Chow, M. K. Neilsen and H. E. Fang
              H. Altenbach

11:05-12:20   CONTINUUM DAMAGE MECHANICS 2
              (Chairpersons: *C. L. Chow and T. Igari*)
              A. Benallal and L. Siad
              S. Murakami, T. Hirano and Y. Liu
              T. Inoue

14:00-14:45   GENERAL LECTURE
              (Chairpersons: *E. Krempl and H. Ishikawa*)
              D. R. Hayhurst

14:45-15:35   CONTINUUM DAMAGE MECHANICS 3
              (Chairpersons: *E. Krempl and H. Ishikawa*)
              A. Litewka and A.C.B. Mesquita
              A. Bodnar and M. Chrzanowski

16:05-17:45   COMPUTATIONAL METHOD IN CREEP AND DAMAGE ANALYSIS
              (Chairpersons: *A. Benallal and N. Miyazaki*)
              L. Adam and J. P. Ponthot
              A. Ganczarski
              Ch. Hellmich, M. Lechner, R. Lackner, J. Macht and H.A. Mang
              A. Epishin, E. Kablov, E. Golubovskiy, I. Svetlov, T. Link, U. Brückner
              and P. Portella

April 5 (Wed.)

9:00-9:45   GENERAL LECTURE
              (Chairpersons: *T. H. Hyde and S. Kubo*)
              H. C. Furtado and I. Le May

9:45-10:35   CREEP DAMAGE AND CREEP CRACK GROWTH 1
              (Chairpersons: *T. H. Hyde and S. Kubo*)
              D. J. Smith, N. S. Walker and S. T. Kimmins
              K. Yagi, F. Abe, K. Kimura and H. Kushima

11:05-11:55   CREEP DAMAGE AND CREEP CRACK GROWTH 2
              (Chairpersons: *D. J. Smith and A. Litewka*)
              S. Kubo, E. Tamura, N. Tagami and K. Ohji
              G. Harkegard and H.-J. Huth

12:00-18:00 *Excursion*

19:00-21:30 *Banquet*

April 6 (Thur.)

9:00-9:45 GENERAL LECTURE
(Chairpersons: *Ky Dang Van and M. Sakane*)
J. T. Boyle and R. Seshadri

9:45-10:35 FRACTURE ASSESSMENT AND DESIGN 1
(Chairpersons: *Ky Dang Van and M. Sakane*)
Y. Takahashi
R. A. Ainsworth, D. W. Dean and P. J. Budden

11:05-12:20 FRACTURE ASSESSMENT AND DESIGN 2
(Chairpersons: *R. A. Ainsworth and Y. Takahashi*)
Ky Dang Van
L. Verger, A. Constantinescu and E. Charkaluk
T. Igari, T. Tokiyoshi and Y. Mizokami

14:00-14:45 GENERAL LECTURE
(Chairpersons: *J. T. Boyle and F. Yoshida*)
E. Krempl and K. Ho

14:45-15:35 MODELING OF CREEP AND RATCHETING 1
(Chairpersons: *J. T. Boyle and F. Yoshida*)
D. R. Hayhurst, J. Lin, Z. L. Kowalewski and B. F. Dyson
H. Ishikawa, K. Sasaki and T. Mayama

16:05-17:45 MODELING OF CREEP AND RATCHETING 2
(Chairpersons: *M. Chrzanowski and M. Kawai*)
Z.L. Kowalewski
M. Sakane and T. Hosokawa
F. Yoshida
Y. Sugita, N. Sinohara and K. Sugiyama

April 7 (Fri.)

9:00-9:45 GENERAL LECTURE
(Chairpersons: *H. Altenbach and Z. L. Kowalewski*)
N. Ohno, T. Ando, T. Miyake and S. Biwa

9:45-10:35 MICROMECHANICS AND COMPOSITE MATERIALS
            (Chairpersons: *H. Altenbach and Z. L. Kowalewski*)
            S. Kruch, J.L. Chaboche and N. Carrere
            J. Carmai and F. P. E. Dunne

11:05-12:20 CREEP OF NONMETALLIC MATERIALS
            (Chairpersons: *L. P. Mikkelsen and Y. Sugita*)
            M. Kawai
            P. D. Barrette and I. J. Jordaan
            T. Hiroe, H. Matsuo, K. Fujiwara and F. Ohashi

14:00-14:50 CREEP AND CREEP DAMAGE IN SHELLS
            (Chairpersons: *E. Busso and S. Biwa*)
            L. P. Mikkelsen
            T.-Q. Yang, X. Zhang, Q. Gang and Q. An

14:50-15:30 CLOSING ADDRESS
            N. Ohno

CONTENTS

D. R. HAYHURST: An opening address 1

B. F. DYSON and M. McLEAN: Micromechanism-quantification for creep constitutive equations 3

W. T. MARKETZ, A. CHATTERJEE, F. D. FISCHER and H. CLEMENS: Creep of γ-TiAl based alloys: experiments and computational modeling 17

D. M. KNOWLES and D. W. MacLACHLAN: Anisotropic creep of single crystal superalloys 31

E. P. BUSSO, N. P. O'DOWD and R. J. DENNIS: A rate dependent formulation for void growth in single crystal materials 41

P. ONCK, B.-N. NGUYEN and E. VAN DER GIESSEN: Microstructural modeling of creep fracture in polycrystalline materials 51

B.-N. NGUYEN, P. ONCK and E. VAN DER GIESSEN: Creep crack growth: from discrete to continuum damage modeling 65

N. TADA and R. OHTANI: Prediction of inner cracking behavior in heat-resistant steel under creep-fatigue condition by means of three-dimensional numerical simulation 75

T. H. HYDE and W. SUN: Creep of welded structures 85

A. D. BETTINSON, N. P. O'DOWD, K. NIKBIN and G. A. WEBSTER: Two parameter characterization of crack tip fields under creep condtions 95

T. KITAMURA and T. SHIBUTANI: Cavity growth induced by electric current andstress in LSI conductor 105

N. MIYAZAKI: Dislocation density simulations for bulk single crystal growth process using dislocation kinetics model 115

J. P. SERMAGE, J. LEMAITRE and R. DESMORAT: Multiaxial creep fatigue under anisothermal conditions 125

Y. WEI, C. L. CHOW, M. K. NEILSEN and H. E. FANG: Constitutive modeling of viscoplastic damage in solder materials 131

H. ALTENBACH: Consideration of stress state influences in the material   141
modeling of creep and damage

A. BENALLAL and L. SIAD: Strain, stress and damage fields in damaged   151
and cracked solids

S. MURAKAMI, T. HIRANO and Y. LIU: Effects of damage on the asymptotic   165
fields of a model I creep crack in steady-state growth

D. R. HAYHURST: Computational continuum damage mechanics: its use   175
in the prediction of creep in structures: past, present and future

A. BODNAR and M. CHRZANOWSKI: Cracking of creeping structures   189
described by means of CDM

L. ADAM and J. P. PONTHOT: A coupled formulation for thermo-   197
viscoplasticity at finite strains: application to hot metal forming

A. GANCZARSKI: Thick axisymmetric plate subjected to thermo-mechanical   207
damage

Ch. HELLMICH, M. LECHNER, R. LACKNER, J. MACHT and H.A. MANG: 217
Creep of shotcrete tunnel shells

A. EPISHIN, E. KABLOV, E. GOLUBOVSKIY, I. SVETLOV, T. LINK,   231
U. BRÜCKNER and P. PORTELLA: Rupture life time prediction and
deformation mechanisms during creep of single-crystal nickel-base
superalloys

H. C. FURTADO and I. LE MAY: Creep damage assessment and void   241
formation in engineering materials

D. J. SMITH, N. S. WALKER and S. T. KIMMINS: Creep damage   257
accumulation and failure in narrow regions of steel welds

K. YAGI, F. ABE, K. KIMURA and H. KUSHIMA: Long-term creep life   267
prediction based on understanding of creep deformation behavior of
ferritic heat resistant steels

S. KUBO, E. TAMURA, N. TAGAMI and K. OHJI: Near-threshold fatigue   277
crack growth in SUS304 steel at elevated temperatures

G. HÄRKEGÅRD and H.-J. HUTH: Approximate viscoplastic notch analysis 287

J. T. BOYLE and R. SESHADRI: The reference stress method in creep design: a thirty year retrospective 297

Y. TAKAHASHI: Study on creep-fatigue life prediction methods based on long-term creep-fatigue tests for austenitic stainless steel 311

R. A. AINSWORTH, D. W. DEAN and P. J. BUDDEN: Developments in creep fracture assessments within the R5 procedures 321

KY DANG VAN: On global approaches to some problems involving plasticity and viscosity effects 331

L. VERGER, A. CONSTANTINESCU and E. CHARKALUK: On the simulation of large viscoplastic structures under anisothermal cyclic loadings 341

T. IGARI, T. TOKIYOSHI and Y. MIZOKAMI: Description of inelastic behavior of perforated plates based on effective stress concept 351

E. KREMPL and K. HO: The overstress model applied to normal and pathological behavior of some engineering alloys 361

D. R. HAYHURST, J. LIN, Z. L. KOWALEWSKI and B. F. DYSON: Creep strain uncertainties associated with testpiece extensometer ridges: their identification and reduction 375

H. ISHIKAWA, K. SASAKI and T. MAYAMA: Equivalence of back stress during plastic and creep deformation 391

Z.L. KOWALEWSKI: Assessment of the multiaxial creep data based on the isochronous creep surface concept 401

M. SAKANE and T. HOSOKAWA: Biaxial and triaxial creep testing of type 304 stainless steel at 923K 411

F. YOSHIDA: Uniaxial/multiaxial creep-ratchetting of several types of steels and its constitutive modelling 419

Y. SUGITA, N. SINOHARA and K. SUGIYAMA: Temperature measurement and lifetime prediction of a high-pressure turbine rotor 429

N. OHNO, T. ANDO, T. MIYAKE and S. BIWA: Effect of matrix creep on fiber stress profiles in unidirectional composites: mesoscopic analysis based on a variational method 439

S. KRUCH, J.L. CHABOCHE and N. CARRERE: Micromechanics based creep damage analysis of unidirectional metal matrix composites 453

J. CARMAI and F. P. E. DUNNE: Micromechanical models for creep in the consolidation of composites 463

M. KAWAI: Off-axis creep behavior of unidirectional polymer matrix composites at high temperature 469

P. D. BARRETTE and I. J. JORDAAN: Creep of ice and microstructural changes under confining pressure 479

T. HIROE, H. MATSUO, K. FUJIWARA and F. OHASHI: Numerical and experimental creep bending behavior of polyethylene beams 489

L. P. MIKKELSEN: A numerical elastic-viscoplastic collapse analysis of circular cylindrical shells under axial compression 499

T.-Q. YANG, X. ZHANG, Q. GANG and Q. AN: Time-dependence of buckling load for a viscoelastic plate under creep condition 509

AUTHOR INDEX 519

AN OPENING ADDRESS

D. R. HAYHURST[*]
Department of Mechanical Engineering,
UMIST, Manchester, M60 1QD, U.K.

As the series of IUTAM Symposia on Creep in Structures enters its 5th decade it is appropriate to begin with some brief notes on the history of IUTAM. In 1899[**] the International Association of Academies was created to initiate and to promote scientific meetings. This association survived World War I. But, in 1917 the International Research Council, known as the IRC, was formed to stimulate exchanges across the borders formed after the war. However, in parallel with the IRC, informal scientific groups were established which became known as International Unions. Subsequently, it was found desirable to establish links between the International Research Council and the International Unions. Hence, in 1931, when the IRC had 40 member states and 8 International Unions existed, the International Council of Scientific Unions, known as ICSU, was formed. In 1946 after World War II, again in an effort to stimulate scientific exchanges across the new borders; the Applied Mechanics community met in Paris and formed IUTAM. It was agreed that it should hold symposia which were: (i) open to invited scientists, (ii) organised to encourage and develop young scientists, (iii) small scale, and (iv) an effective means of establishing personal and international contacts. It is now over one hundred years ago since the formation of the International Association of Academies, that we meet here in Nagoya to develop, and to foster the same ideals.

It was in the late 1950s when the United States space exploration programme was becoming a reality that Nicholas Hoff, Folke Odqvist, Robert Mazet, Yuri Rabotnov, and Shuji Taira, set up the first IUTAM Symposium on Creep in Structures. Amongst the delegates were Jan Hult, Chris Calladine, Michael Zyczkowski, all of whom turned out to be scientific ambassadors, who brought the topic of creep into Europe. But probably most notable, where the efforts of Michael Zyczkowski, member of our organising committee, who took the message of this burgeoning scientific field behind the iron curtain and beyond.

The first Creep in Structures Symposium at Stanford in 1960 was about: deformation, flow, and buckling structures using simple material laws.

---

[*] Member of Scientific Committee
[**] Jan Hult, "1955–1980: A Dynamic Quarter Century for Applied Mechanics", Appl. Mech. Rev., Vol.53, No 1, January 2000

S. Murakami and N. Ohno (eds.), IUTAM Symposium on Creep in Structures, 1–2.
© 2001 *Kluwer Academic Publishers. Printed in the Netherlands.*

The second Creep in Structures Symposium at Gothenburg in 1970 was about: deformation, relaxation, buckling and, the start of damage. Birth was given to the Reference Stress technique and to bounding methods—the cornerstones of the established design assessment Route R5.

The third Creep in Structures Symposium at Leicester in 1980 saw the establishment of computational Continuum Damage Mechanics, constitutive equations, and the formation of cross-disciplinary links with physical metallurgy.

The fourth Creep in Structures Symposium at Cracow in 1990, organised by Michael Zyczkowski, saw the development of: computational analysis, creep-fatigue and ratchetting, a shift from materials towards composites and ceramics, and the introduction of processing.

In the fifth Creep in Structures Symposium here in Nagoya, well: we have all been invited; amongst our delegates are young scientists from something like 12 countries; we are a small conference group of approximately 80; and, I hope that we are about to form long lasting links—both trans-national and cross-disciplinary in the best traditions established, over the last century, by the International Association of Academies, the International Research Council, the International Unions and the International Union of Theoretical and Applied Mechanics.

MICROMECHANISM-QUANTIFICATION FOR CREEP CONSTITUTIVE EQUATIONS

B. F. DYSON and M. McLEAN
Department of Materials
Imperial College of Science, Technology & Medicine
Prince Consort Road
London SW7 2BP
England

Abstract-A uniaxial constitutive equation-set for commercial precipitation-strengthened alloys, consisting of a hyperbolic sine law kinetic creep equation and several microstructure-evolution (damage-rate) equations, has been described. Only a small number of materials- and heat treatment-specific model parameters are required to predict creep curve shapes in these microstructurally unstable alloys. The sinh-law kinetic creep equation is theoretically-derived and so displays several advantages over the ubiquitous but empirical power-law, including explicit predictions of the effects of microstructure on creep rates. Evolution rate equations for three types of microstructural instability (dislocation, particulate and grain boundary cavitation) are quantified generically and are illustrated throughout the paper with a model parameter-set for the nickel-base superalloy IN738LC. Independently variable creep damage terms within an equation-set is extremely parameter-efficient and enables the synthesis of complex materials responses by suitable permutations. For example, creep behaviour of conventionally-cast IN738LC has been predicted from the directionally-solidified alloy parameter-set by reducing the magnitude of creep ductility and activating the cavitation damage term. Similarly, the equation-set has also been used to predict the experimentally-found reciprocity between minimum creep rate/applied stress data and peak stress/applied straining rate data. The (low) inelastic strain to reach minimum creep rate was identical to that required to achieve the corresponding peak-stress because a common *intrinsic* microstructural instability operates in each loading mode. Computed stress-strain trajectories at the lower applied strain rates demonstrate the increasing dominance of damage due to particle-coarsening over that due to dislocation-multiplication. An approximate extension of the uniaxial kinetic law to predict the behaviour of polycrystals under multiaxial stressing is also briefly described and contrasted with the different methodology required for multiaxially-loaded single crystals.

1. Introduction

Finite element analyses of stressed components in power engineering applications require constitutive equations that predict creep of metallic alloys operating in complex

S. Murakami and N. Ohno (eds.), IUTAM Symposium on Creep in Structures, 3–16.
© *2001 Kluwer Academic Publishers. Printed in the Netherlands.*

thermal/mechanical environments. Historically, their development has mainly been the domain of the mechanics community and approached from the perspective of both steady- and cyclic-creep with a strong emphasis on multiaxial loading. Most are based on a power-law kinetic creep equation — for example, Leckie and Hayhurst (1974), Merzer and Bodner (1979), Chaboche and Rousselier (1983), Robinson and Bartolotta (1985), Lemaître and Chaboche (1990). Exponential and hyperbolic sine functions of stress have sometimes also been used, Miller (1997).

A constitutive description of strain rate-dependent behaviour in commercial alloys (which, without exception in the power-engineering area, rely upon microstructures that are thermodynamically unstable) has to take account of several **intrinsic** causes of tertiary creep behaviour under load-control (or intrinsic stress-softening under strain-control). The thermal- or strain-induced characterisation of these intrinsic mechanisms has been an area of active metallurgical research for many years, with a recent emphasis on quantifying the kinetics of evolution of the dislocation, particulate, solid solution and cavitation microstructures. Dyson and McLean (1998) have published a review.

However, damage evolution is only part of the story: a constitutive equation-set must also have a kinetic creep equation that adequately reflects effects on creep resistance of any changes in microstructure. This paper will argue that for precipitation-strengthened alloys, a kinetic creep equation based upon either an empirical or a theoretical power-law is inadequate. An alternative, microstructure-based, functional form (hyperbolic sine) will be described and used to explain differences in polycrystalline and directionally-solidified IN738LC and predict experimentally-found reciprocity between creep and constant strain-rate data. An approximate extension of the uniaxial kinetic law to predict the behaviour of polycrystals under multiaxial stressing will also be briefly described and contrasted with the different methodology required for multiaxially-loaded single crystals.

2. Power-Law Creep Equation: the Problem

Power-law creep equations can be viewed as empirical (Bailey-Norton) or theoretical, with a pedigree that can be traced to active research during the 1950's and 1960's into the physics of *steady state* creep equations of pure metals and simple alloys, Sherby & Burke (1967). Power-laws have retained a pre-eminent position in both the metallurgical and mechanics literatures, in spite of well-known and notable discrepancies between theoretical predictions and experimental creep data. Figure 1 exemplifies the problem in a commercial alloy: data for Nimonic 90 (Harrison & Evans, 1980), a wrought nickel-base superalloy, are plotted at five different temperatures. It is immediately apparent that the stress exponent "n" is not a constant and that its variability with stress depends critically upon temperature. At 700°C for example, "n" is very high throughout the whole stress range investigated and could perhaps even be adequately approximated (at least for research purposes) by a constant value. However, it would not be advisable to extrapolate linearly to service strain rates — 5×10^{-11} s^{-1}. Figure 1 also clearly demonstrates that the stress range over which "n" can be regarded as constant becomes smaller as temperature increases. If the problem were being approached from a pragmatic empirical viewpoint, it would always be possible to generate a function "n(σ,T)", in order to interpolate within the database, but it would be unwise to use this procedure for extrapolation. There is another and more substantial disadvantage in pursuing this route: stress exponents are also known to be a function of accumulated strain during a test at constant stress and temperature. There is thus little virtue in developing a constitutive equation-set from a power-law function where the stress exponent varies with stress, temperature **and** accumulated strain.

When the use of a power-law is approached from the materials-understanding viewpoint, the problems multiply. The seminal attempt to derive an equation for alloys of this type, Ansell and Weertman (1959), assumed climb to be rate controlling but was unable to explain behaviour. Steady-state recovery creep theories such as that of Ansell and Weertman always predict a constant, but *materials-dependent*, activation energy of the order of that for diffusion and a constant, *materials-independent*, stress exponent of the order 3 to 5. Figure 1 illustrates that these predictions are not supported by data from a commercial alloy, Nimonic 90: the activation energy increases with stress level and is always greater than that for diffusion; and the stress exponent varies between

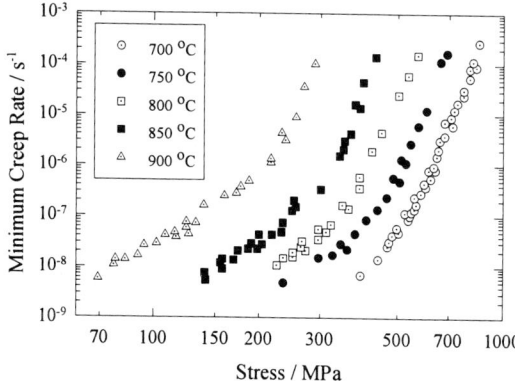

Figure 1. Nimonic 90 Minimum Creep Rate Data

approximately two and twenty. These discrepancies have been reported for many other alloy systems in the metallurgical literature over the past thirty years and several attempts have been made to provide an acceptable explanation. Either a Bailey-Orowan type of equation has been modified, Lagneborg (1968), Threadgill & Wilshire (1972), or an alternative functional form has been derived, Rösler & Arzt (1990). None has been entirely convincing.

There is a further and compelling reason to find an alternative kinetic creep equation: the above equations cannot predict the effects of a progressively changing microstructure on creep resistance, even qualitatively in most cases. This is a major disadvantage when trying to develop a constitutive equation-set for use with complex commercial alloys. In the next section, a microstructure-based kinetic creep equation that meets these requirements will be described and then embedded within a constitutive equation-set in §4 and used to predict behaviour in §5 and §6.

3. A Microstructure-Based Kinetic Creep Equation

Consider a random distribution of precipitate particles with a mean plane-section radius r_s, volume fraction ϕ_p and a total dislocation density ρ_t arranged in a 3-D network. At stresses below the Orowan Stress (which is appropriate for laboratory testing and service applications), dislocations can accumulate only a few thousandths of a percent strain on initial loading before becoming trapped by particles. Thermal activation ensures that the trapping is temporary by enabling dislocations to climb over the particles. This climb/glide mechanism is universally accepted and has been the starting point in the development of a new microstructure-based kinetic creep equation, Dyson (2000). Previous theoretical developments have assumed that the rate-controlling mechanism was either climb, Ansell and Weertman (1959) or a consequence of a strong interaction between dislocations and particles, Rösler & Arzt (1990). By incorporating the geometrically random nature of dislocation-particle intersections into the climb/glide model, Dyson (2000) has followed a different non-deterministic route. It has led to the conclusion that there will always be a small, geometrically-determined, dynamic fraction of the dislocation density which is in a position to "escape" from the climbing network. The probability of a successful "escape" — to be followed by viscous glide until each dislocation is again temporarily trapped at another particle — being a function of stress, temperature and particle dispersion parameters. When the viscous glide rate is fast, the model gives a shear creep rate, $\dot{\gamma}$ which is a very strong function of stress, temperature and inter-particle spacing:

$$\dot{\gamma} = 2\rho\phi_p^{1/2}\left([\pi/4]^{1/2} - \phi_p^{1/2}\right)c_j D_v \sinh\left[\frac{\tau_m b^2 \lambda_p}{kT}\right] \tag{1}$$

In Eq.(1), τ_m is the shear stress acting *within the matrix*; $\lambda_p = 2r_s\left[\left([\pi/4]^{1/2} - \phi_p^{1/2}\right)/\phi_p^{1/2}\right]$ is the square-lattice inter-particle spacing; D_v, the appropriate matrix diffusivity; c_j, the dislocation jog density; ρ, the dislocation density that is potentially able to glide (for example, in an f.c.c. matrix, ρ will be the density of {111}<110> dislocations); k, T, and b are respectively, Bolzmann's constant, absolute temperature and Burger's vector.

There are three interesting points about Eq (1): (i) the pre-sinh term is independent of the state of particle dispersion (so particle-ageing will not affect its magnitude); (ii) the magnitude of the sinh-argument will vary with the state of particle dispersion (λ_p) and so ageing will influence $\dot{\gamma}$; and (iii) as a consequence of (ii), a predetermined amount of ageing prior to creep will influence $\dot{\gamma}$ much more strongly at high stresses than at low.

The uniaxial creep rate, $\dot{\varepsilon}$, in response to a matrix uniaxial stress, σ_m, can be determined from Eq. (1) by using the usual relationships, Dieter (1988); $\dot{\varepsilon} = \dot{\gamma}/\overline{M}$ and $\sigma_m = \overline{M}\tau_m$, where $\overline{M}$ is the Taylor factor:

$$\dot{\varepsilon} = \frac{2\rho\phi_p^{1/2}}{\overline{M}}\left([\pi/4]^{1/2} - \phi_p^{1/2}\right)c_j D_v \sinh\left[\frac{\sigma_m b^2 \lambda_p}{\overline{M}kT}\right] \tag{2}$$

Eq.(2) is identical in form but differs in detail from the one given by Dyson and Osgerby (1993) using Reaction Rate Kinetics.

To progress quantitatively, σ_m in Eq. (2) needs to be related to the applied tensile stress, σ. By treating the particles as elastic, creep will occur only in the matrix and the consequent stress redistribution can be determined using the two-bar analysis of Ion et al (1986). It leads to the following modification of Eq.(2):

$$\dot{\varepsilon} = \frac{2\rho\phi_p^{1/2}}{\overline{M}}\left(1 - \phi_p\right)\left([\pi/4]^{1/2} - \phi_p^{1/2}\right)c_j D_v \sinh\left[\frac{(\sigma - \sigma_k)b^2 \lambda_p}{\overline{M}kT}\right] \tag{3}$$

The evolution equation for the kinematic internal stress, σ_k (the term $(\sigma - \sigma_k)$ being equal to σ_m) is given by

$$\dot{\sigma}_k = \phi_p E \left(1 - \frac{\sigma_k}{\sigma_k^*}\right) \dot{\epsilon} \qquad (4)$$

Young's modulus, E, is assumed to be the same in both matrix and particle. Integrating Eqs.(3) & (4) as a coupled pair gives the primary and secondary creep behaviour in the absence of any instability within the microstructure. The methodology for computing the influence of the latter on strain (or stress) trajectories will be discussed in the following section.

4. Continuum Creep Damage Mechanics

Continuum Damage Mechanics (CDM) provides a convenient framework for dealing with the kinetics of microstructural evolution during time-dependent deformation. The term was coined by Janson and Hult (1977) to quantify *any* progressive loss of load-bearing capacity in stressed materials as inelastic strain accumulated. Following Kachanov (1958) and Robotnov (1960), the loss of load-bearing capacity was represented by an "effective" stress in a purely empirical way. Parallel studies in the metallurgical literature were concerned with the microstructural causes of creep softening and both theoretical and phenomenological rate equations were developed to describe the kinetics of evolution of various microstructural features. Ashby and Dyson (1984) assembled these in an attempt to quantify their effects on creep resistance within the framework of CDM. It was demonstrated that most damage mechanisms could indeed be defined in such a way that their effect on creep resistance was by way of an effective stress. This gives considerable support to CDM, as well as giving metallurgists insight into the (necessarily simplifying) objectives of engineering research. But one important microstructural damage mechanism — softening due to dislocation multiplication, Dyson and McLean (1983) — did not operate as an effective stress and another exception has been described more recently, Kadoya et al (1997). Clearly, empirical CDM lacks the generality required to describe fully the creep behaviour of commercial alloys.

Empirical CDM is a specific case of the internal state variable approach to creep where uniaxial tensile creep for example, is represented by a set of linear differential equations:

$$\boxed{\begin{aligned} \dot{\epsilon} &= \dot{\epsilon}(\sigma, T, H, D_i) \\ \dot{H} &= \dot{H}(\sigma, T, H, D_i) \\ \dot{D}_i &= \dot{D}_i(\sigma, T, H, D_i) \end{aligned}} \qquad (5)$$

The strain rate $\dot{\epsilon}$ in Eq.(5) is a function of the applied uniaxial tensile stress σ, temperature T, an evolving dimensionless hardening parameter H to represent primary creep and a set of evolving dimensionless microstructural damage parameters D_i. Dyson and McLean (1998) have recently reviewed the current knowledge-base on "creep damage" mechanisms (due to microstructure-instabilities), and discussed their application to creep life-prediction within the framework of "physically-based CDM". The most important damage mechanisms in commercial alloys are believed to be associated with the dislocation-, particulate- and grain boundary cavity-microstructures: creep performance of commercial nickel-base superalloys, for example, depends critically upon the (alloy- and heat treatment-specific) rates of evolution of these three microstructural instabilities acting either singly or in parallel. The same three damage mechanisms appear to dominate behaviour in ferritic steels and aluminium alloys but solid solution instabilities may also be an important additional form of damage in certain cases, Kadoya et al (1997).

An explicit form of Eq.(5) for use under uniaxial tensile loading with polycrystalline precipitation-strengthened alloys has been synthesised from Eq.(3) and from the literature knowledge-base on microstructure-damage evolution rates:

$$
\begin{aligned}
\dot{\varepsilon} &= \frac{\dot{\varepsilon}_0}{1-D_d}\sinh\left(\frac{\sigma(1-H)}{\sigma_0(1-D_p)(1-D_c)}\right) \\[2mm]
\dot{H} &= \frac{h'}{\sigma}\left(1-\frac{H}{H^*}\right)\dot{\varepsilon} \\[2mm]
\dot{D}_d &= C(1-D_d)^2\dot{\varepsilon} \\[2mm]
\dot{D}_p &= \frac{K_p}{3}(1-D_p)^4 \\[2mm]
\dot{D}_c &= \frac{1}{3\varepsilon_{f,u}}\dot{\varepsilon}
\end{aligned}
\tag{6}
$$

The "compound" materials parameters, $\dot{\varepsilon}_0$ and σ_0, are related to Eq.(3) by

$$
\dot{\varepsilon}_0 = \frac{2\rho_i\phi_p^{1/2}}{\overline{M}}(1-\phi_p)\left[\pi/4\right]^{1/2}-\phi_p^{1/2}c_jD_v
\tag{7}
$$

$$
\sigma_0 = \left[\frac{\overline{M}kT}{b^2\lambda_{p,i}}\right]
\tag{8}
$$

where ρ_i and $\lambda_{p,i}$ are initial values.

The second term in Eq.(6) contains a dimensionless parameter, $H = \sigma_k/\sigma$ where $\sigma(1-H)$ is the stress acting within the matrix; it is equivalent to Eq.(4) when stress transients are ignored. The three following equations refer to each of the microstructural instabilities: subscript "d" refers to dislocation; "p" to particle; and "c" to cavitation. The evolution expression for $\dot{D}_c$ is that for continuous nucleation of cavitated grain boundary facets where the cavities are assumed to grow under creep constraint. Each damage term is defined to range from zero to unity: $D_d = 1-(\rho_i/\rho)$; $D_p = 1-(\lambda_{p,i}/\lambda)$; and $D_c = 1-(A_{c,i}/A_c)$ where ρ, λ_p, and A_c are respectively, the dislocation density, average inter-particle spacing and area fraction of cavitated grain boundary facets, with "i" signifying initial values. As mentioned above, damage due to instability in the dislocation microstructure does not enter as an "effective" stress in the expression for creep rate. These instabilities manifest themselves as a premature tertiary stage in a creep test and as a premature stress-softening stage in a constant strain-rate test. The term "premature" being with respect to any form of mechanical instability associated with the external loading system and testpiece geometry (e.g. necking in steady-load tension).

Seven model parameters are required to predict *isothermal* strain/time (stress/time) behaviour under constant/variable load/stress or constant/variable straining-rate respectively: $\dot{\varepsilon}_0$, σ_0, h', H^*, K_p, C and $\varepsilon_{f,u}$, the cavitation-induced ductility. This increases to ten parameters when *non-isothermal* behaviour is being considered since $\dot{\varepsilon}_0$, σ_0, and K_p are all functions of temperature. Thus the temperature-independent model parameters are:

$$h' = E\phi_p \tag{9}$$

$$H^* = \frac{2\phi_p}{1 + 2\phi_p} \tag{10}$$

$$C = \frac{\phi_p \rho^{1/2}}{r_s \rho_i} \tag{11}$$

where Eq.(9) was derived from the two-bar model of primary creep due to Ion et al (1986). Dyson (2000) derived Eq.(10) from a plasticity constraint placed upon the upper-bound level of stress redistribution around elastically-stressed particles and Eq.(11) was derived from Eq.(3) using the dislocation multiplication model of Kocks et al (1975).

The temperature-dependence of parameters $\dot{\varepsilon}_0$, & K_p can each be accounted for by an appropriate activation energy:

$$\dot{\varepsilon}_0 = \dot{\varepsilon}_0' \exp - \left(\frac{Q_{j/d}}{RT}\right) \tag{12}$$

$$K_p = K_p' \exp\left(-\frac{Q_p}{RT}\right) \tag{13}$$

where $\dot{\varepsilon}_0' = \frac{2\rho_i \phi_p^{1/2}}{M}\left(1 - \phi_p\right)\left([\pi/4]^{1/2} - \phi_p^{1/2}\right)c_{j,0} D_{v,0}$ and K_p has been derived from standard particle-coarsening theory, controlled by volume diffusion, Greenwood (1956). $K_p' = k_p / \lambda_{p,i}^3$ where k_p is the rate constant in particle-coarsening theory. Only a phenomenological relationship exists between σ_0 and temperature, Dyson (2000), based on the fact that the volume fraction of particles decreases (λ_p increases) with temperature to reach zero at the appropriate solvus:

$$\sigma_0 = \sigma_{0,m}\left[1 - \exp\left(-\frac{\Delta H}{RT_s}\left(\frac{T_s}{T} - 1\right)\right)\right] \tag{14}$$

where $\sigma_{0,m}$ is the maximum (at T=0) value of σ_0, ΔH is an enthalpy of solution and T_s is the particle solvus temperature (at which $\sigma_0 = 0$).

5. Modelling the Nickel-Base Superalloy IN738LC

5.1. MODEL PARAMETER-SET FOR DIRECTIONALLY-SOLIDIFIED IN738LC

Initial estimates, Table 1, of most of the ten model parameters for the directionally-solidified (d.s.) nickel-base superalloy IN738LC were made using Eqs.(7-13) with

$\rho_i = 10^{10} m^{-2}$; $\phi_p = 0.5$ $2r_s = 5\times10^{-7} m$; $E = 1.6\times10^5 MPa$; $b = 2.4\times10^{-10} m$; $\overline{M} = 3$;

$c_{ji}=1$; and $\lambda_{p,i} = 1.27\times10^{-7} m$. The parameters for particle-coarsening are those of a similar alloy, Nimonic 115, and were taken from the literature. Iterative numerical integration of Eq.(6) (with the cavitation term inactive) was performed with Table 1 input as initial values until adequate fits were found to the data in Figs.2 & 3 and used to construct the solid lines. Comparison of the two sets of parameters is encouraging, particularly with regard to C, h', and $H^{\cdot}$. It should be noted that the theory behind

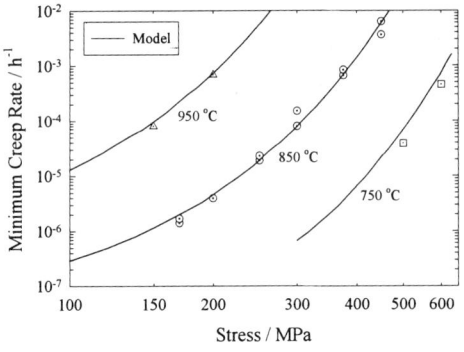

Figure 2. Directionally-solidified IN738LC minimum creep rate data & model.

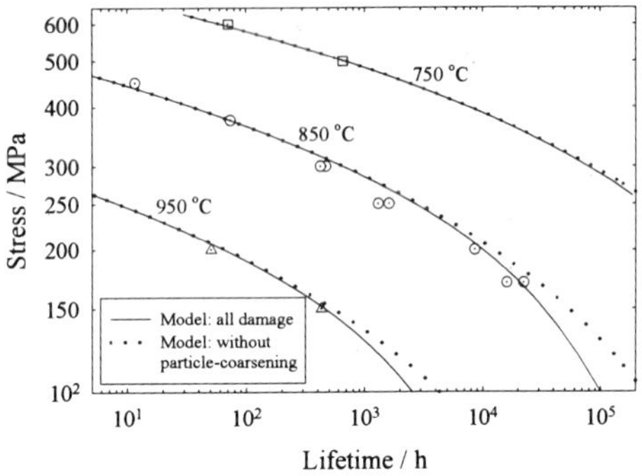

Figure 3. Directionally-solidified IN738LC lifetime data & model.

Eq.(11) predicts that C should vary with dislocation density whereas it has always been assumed constant in the numerical modelling. The value quoted in Table 1 is at t=0 and is thus a lower limit. The large difference between initial and final values for $\sigma_{0,m}$ is disappointing and may be a consequence of Eq.(14) being phenomenological.

TABLE 1. Initial model parameter-set for IN738LC d.s.

$\dot{\varepsilon}'_0$ (s^{-1})	$Q_{d/j}$ (kJ/mol)	$\sigma_{0,m}$ (MPa)	ΔH (kJ/mol)	T_s (K)	h' (MPa)	$H^{\cdot}$	C	K'_p (s^{-1})	Q_p (kJ/mol)	$\varepsilon_{f,u}$
$<4\times10^4$	>280	6	?	≈ 1400	8×10^4	0.5	>21	8×10^6	305.62	0.2

TABLE 2. Model parameter-set for IN738LC d.s. fitted to the database in Figs 2&3.

$\dot{\varepsilon}'_0$ (s^{-1})	$Q_{d/j}$ (kJ/mol)	$\sigma_{0,m}$ (MPa)	ΔH (kJ/mol)	T_s (K)	h' (MPa)	$H^{\cdot}$	C	K'_p (s^{-1})	Q_p (kJ/mol)	$\varepsilon_{f,u}$
2.5×10^3	311.25	33	49.83	1394	6×10^4	0.42	30	10^6	305.62	0.2

Figure 3 illustrates the importance of particle-coarsening on lifetimes at lower stresses and higher temperatures. The dotted curves were computed after de-activating the particle-coarsening term in Eq.(6). Particle-coarsening does not appreciably affect the predicted minimum creep rates investigated in Fig.2. The most important point to note about particle-coarsening is that it is a thermally-induced mechanism and so, to maximise a lifetime, the integrated total thermal exposure should be minimised rather than the creep strain accumulated. In structures therefore, it is potentially damaging to reduce the mechanical load while still retaining the temperature level.

5.2. PREDICTION OF CONVENTIONALLY-CAST IN738 CREEP BEHAVIOUR FROM DIRECTIONALLY-SOLIDIFIED MODEL PARAMETER-SET

IN738LC is also used commercially in conventionally-cast (c.c.) form with a grain size of approximately 1mm. Creep rupture occurs because of the nucleation and growth of grain boundary cavities which requires the cavitation term in Eq.(6) to be active. Numerical computations have been performed over a range of stresses at a single temperature of 850°C , where experimental creep and rupture data are available for comparison. The only change made to the parameter-set Table 2 was to substitute the lower-bound creep ductility of 5%, acquired from the data given in Fig.4, for the 20% used for the d.s. alloy.

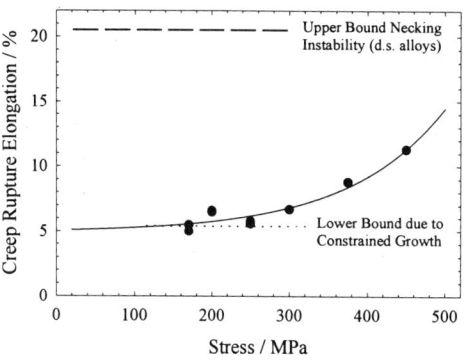

Figure 4. Creep ductility as a function of stress for IN738LC c.c. at 850°C

Experimental and computed data for minimum creep rates and lifetimes of both c.c. and d.s. are plotted in Figs 5 & 6. Experimental minimum creep rate data are seen to be virtually indistinguishable, although there is a tendency for the c.c. data to be slightly faster than d.s. as the stress decreases. This is also mirrored in the computed m.c.r. data but, for clarity, only d.s. data are plotted. Lifetime data in Fig.6 present a different

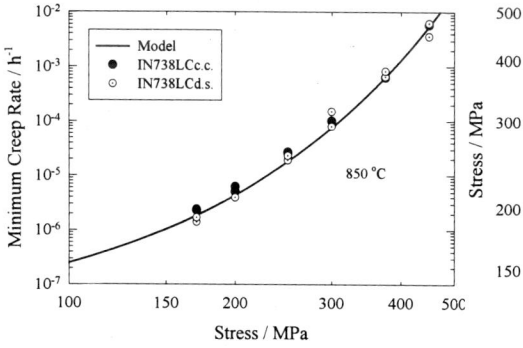

Figure 5. Comparison of c.c. and d.s. IN738LC minimum creep rate data with model predictions.

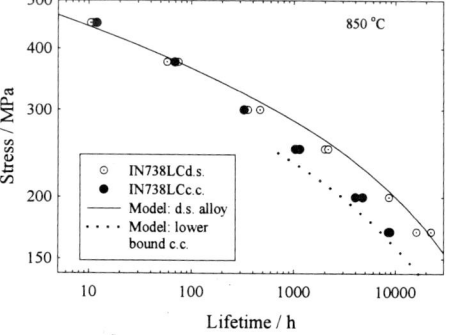

Figure 6. Comparison of d.s. and c.c. IN738LC lifetime data with model predictions

story: at low stresses, c.c. lifetimes are less than half those of the d.s. alloy. Beyond 300 MPa, lifetimes converge. Also plotted in Fig.6 are the computed c.c. and d.s. lifetimes. Lower-bound lifetimes are very close to the two lowest stress experimental points and the conclusion drawn is that the assumption of cavities growing under creep constraint is only justified at the lower stresses. This is consistent with the ductility data in Fig.4 increasing at stresses greater than about 250-300 MPa.

Figure 7 is very instructive because it shows that equation-set (6) along with the parameter-set in Table 2 mimic the essential characteristics of the creep curves found

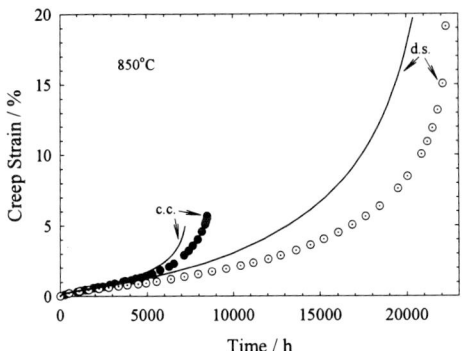

for c.c. and d.s. IN738LC (and c.c. and d.s. alloys in general) within the lower-bound region. Primary and secondary (computed and experimental) creep rates in each variant are virtually identical but diverge markedly in the tertiary stage. This is because the rapid cavitation-rate in the c.c. alloy has begun to dominate tertiary creep behaviour, whereas the more benign dislocation-multiplication dominates d.s. tertiary behaviour. Figure 7 also demonstrates the necessity for having an interactive cavitation term. The alternative method of predicting c.c. lifetimes — using a reduced ductility but

Figure 7. Comparison of experimental and model creep curves for c.c. & d.s. IN738LC at 850° C and 170 MPa.

not re-activating the cavitation damage term — would have given far longer lives, more consistent with lifetimes found in the unconstrained region (stresses>300 MPa).

6. Application to Creep and Constant Straining Rate

Equation-set (6) can also be integrated to predict stress/strain behaviour under a constant applied straining rate. Limited experimental data suggest that there is reciprocity between minimum creep rate/applied stress data and peak stress/applied straining rate data. Figure 8 is an example taken from the work of Osgerby and Dyson (1993) using the wrought nickel-base superalloy Nimonic 101 (IN597). The minimum

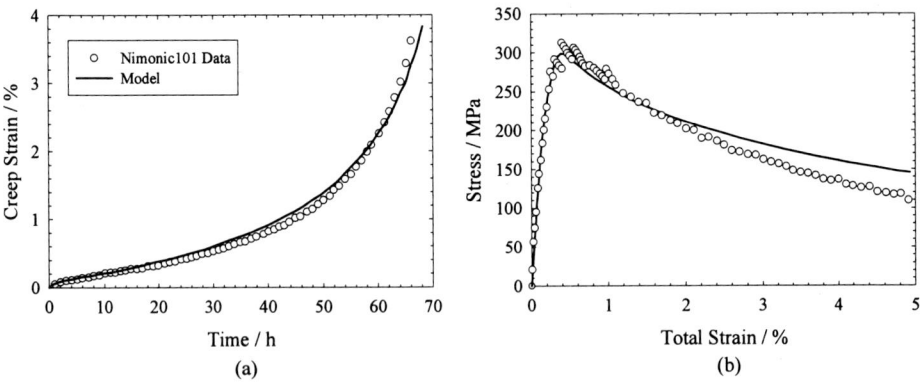

Figure 8. (a) Creep data and model for Nimonic 101 at 300 MPa / 800°C; and (b) corresponding stress-strain experimental and computed data generated at the minimum creep rate found in (a).

creep rate of 1.4×10^{-4} h^{-1} under an initial stress of 300 MPa occurs at about 0.3% inelastic strain; the peak stress of approximately 300 MPa under a straining rate of 1.4×10^{-4} h^{-1} (2.8×10^{-8} s^{-1}) occurs also at about 0.3% (0.005 total –(300/1.6x10^5) elastic). A parameter-set for a simple power-law strain-damaging model was derived from experimental creep data and used to predict the stress/strain behaviour in Fig.8(b). The model was adjusted to fit the creep data in Fig.8(a) and, to a first approximation, fits the stress/strain curve in Fig.8(b). It is significant that deviations between experimental and computed data in both loading modes are self consistent: an inadequate strain acceleration in the later stages of tertiary manifesting itself as an insufficient stress-softening.

The parameter-set in Table 2 has been used to generate stress/strain curves for IN738LCds over a range of applied straining rates, Fig.9. They confirm the behaviour

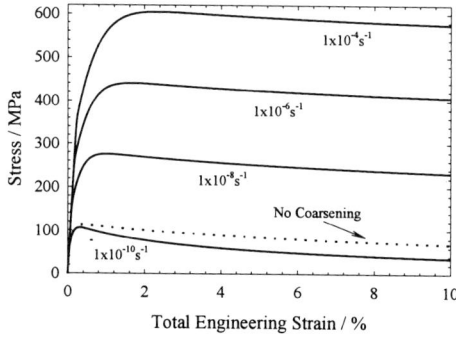

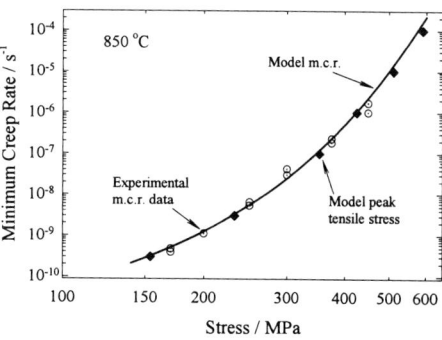

Figure 9. Predicted stress-strain curves at various strain rates for IN738LC d.s. using the parameter-set in Table 2.

Figure 10. Illustrating the predicted reciprocity between minimum creep rate/applied stress and peak stress/applied straining rate.

found for Nimonic 101 except that the post-peak softening is less strong in this alloy than in Nimonic 101. Particle-coarsening has little significant effect on softening rates at this temperature until the straining rate falls below about 10^{-8} s^{-1}. It becomes very significant at a strain rate of 10^{-10} s^{-1}. Fig.10 demonstrates that the reciprocity between minimum creep rate/applied stress data and peak stress/applied straining rate data is predicted to occur at least within the strain rate range 10^{-4} s^{-1} to 10^{-10} s^{-1}. Inelastic strains at peak stress/minimum creep rate again correspond with values ranging from about 0.1% at low applied stresses/straining rates to 2% at high stresses/straining rates.

7. Extension to Multiaxially-Loaded Single- and Poly-Crystals

7.1. POLY-CRYSTAL MULTIAXIAL MODELLING

A multiaxial version of the uniaxial equation-set given by Eq.(6) has been suggested by Kowalewski et al (1994), based on the work of Othman et al (1994), and is reproduced in Eq.(15). Some implicit and explicit assumptions were made during the derivation of Eq.(15). The normalised internal stress term, H, representing the extent of stress redistribution around elastic particles was regarded as a scalar whereas it is almost certainly a tensor, at least within a single crystal. It is possible that the scalar representation is adequate in polycrystals but this has not been assessed experimentally or theoretically. The cavitation damage evolution rate term is phenomenological and there is a need for more experimental verification. The reasoning leading to the cavitation damage term entering the strain rate equation as a modified von Mises stress probably restricts usage to axisymmetric loading. In view of the above comments, multiaxial modelling of unstable microstructures using the microstructure-based Eq.(15) should be undertaken with due diligence and is clearly an area requiring more theoretical and experimental input.

$$\dot{\varepsilon}_{ij} = \frac{3S_{ij}}{2\sigma_e} \frac{\dot{\varepsilon}_0}{1-D_d} \sinh\left(\frac{\sigma_e(1-H)}{\sigma_0(1-D_p)(1-D_c)}\right)$$

$$\dot{H} = \frac{1}{\sigma_e}\left(1-\frac{H}{H^*}\right)\dot{\varepsilon}_e$$

$$\dot{D}_d = C(1-D_d)^2\dot{\varepsilon}_e \qquad\qquad (15)$$

$$\dot{D}_p = \frac{K_p}{3}(1-D_p)^4$$

$$\dot{D}_c = \frac{1}{3\varepsilon_{f,u}}\left(\frac{\sigma_1}{\sigma_e}\right)^2\dot{\varepsilon}_e$$

7.2. SINGLE-CRYSTAL MULTIAXIAL MODELLING

A number of empirical models have been proposed to account for the anisotropic creep behaviour of single crystal superalloys by summing the shear strains on a limited number of allowed slip systems - usually taken to be $\{111\}<\bar{1}\bar{1}0>$ and $\{001\}<110>$. Equation 1 can in principle be modified to represent the rate of shear deformation in a specific crystallographic slip system and so follow the route used by McLean and co-workers - Ghosh et al (1990), Pan et al (1997), Basoalto et al (2000). These analyses used a linearised primary model based upon the work of Ion et al (1986), with the damage being due only to dislocation-multiplication. The latter can easily be justified since grain boundary cavitation clearly does not occur in single crystal superalloys and coarsening of the large γ' particles found in modern high particle volume fraction alloys is slow and appears to have a relatively small effect on creep rates. In the model of McLean and co-workers, deformation on each slip system is represented by:

$$\dot{\gamma}^k = \dot{\gamma}_i^k(1-S^k)(1+\omega^k)$$

$$S^k = H^k\dot{\gamma}_i^k\left(1-\frac{S^k}{S_{ss}^k}\right) \qquad\qquad (16)$$

$$\dot{\omega}^k = \beta^k\dot{\gamma}^k$$

where $\dot{\gamma}_i^k$, H^k, S_{ss}^k and β^k are constants; S^k and ω^k are state variables representing respectively, hardening due to stress redistribution around the particles and softening due to dislocation-multiplication; and $\dot{\gamma}_i^k$ is given by an equation similar to Eq. (1) with $\dot{\gamma}_i^k$ determined by the magnitude of the resolved shear stress on slip system "k".

The strain tensor is then computed by summing all of the shear contributions:

$$\varepsilon_{ij} = \sum_{k=1}^{N}\gamma^k b_i^k n_j^k$$

where n_i, b_i are vectors normal to the slip plane and parallel to the slip direction respectively. This leads to both a change in orientation of an arbitrary direction from $[x_1 x_2 x_3]$ to $[X_1 X_2 X_3]$:

$$\begin{bmatrix} X_1 \\ X_2 \\ X_3 \end{bmatrix} = \begin{bmatrix} 1+\varepsilon_{11} & \varepsilon_{12} & \varepsilon_{13} \\ \varepsilon_{21} & 1+\varepsilon_{22} & \varepsilon_{23} \\ \varepsilon_{31} & \varepsilon_{32} & 1+\varepsilon_{33} \end{bmatrix} \cdot \begin{bmatrix} x_1 \\ x_2 \\ x_3 \end{bmatrix}$$

and a linear strain in that direction of $(\overline{X} - \overline{x})/\overline{x}$.

The extension of uniaxial isotropic creep models to multiaxial loading generally assume that deformation takes place by shear. However, when von Mises stresses are used, this implicitly assumes that shear always occurs in the direction of the maximum resolved shear stress and that an isotropic material can deform on any plane on which the resolved shear stress is maximum. This is not the case here. For a single crystal, the resolved shear stress on any possible slip system can be calculated and the rate of shear deformation determined. The contributions from all allowed slip systems can then been summed as described above to predict strains in arbitrary directions, shape changes of the crystal and changes in crystal orientation.

This approach simplifies the mathematics of representing multiaxial creep relative to those previously used for isotropic materials. The creep law applies to one dimensional shear deformation on each possible slip system and the state variables representing damage and stress redistribution are both scalars. The anisotropy is introduced through the crystallographic symmetry operations. This implicitly assumes that there is no interaction between the damage on different slip systems and the model may have to be modified in the future to account for such effects. Similarly, stress redistribution in these alloys has been approximated by a scalar and current work using Eq.(1) is addressing this deficiency. Nevertheless, the relative simplicity of the McLean et al anisotropic model makes it particularly appropriate for incorporation into finite element codes to simulate component behaviour.

8. Conclusions

1) The ubiquitous and empirical power-law kinetic creep equation has little to offer when constructing a constitutive equation-set for commercial precipitation-strengthened alloys.

2) In contrast, a theoretically-derived sinh-law kinetic creep equation, when embedded within a constitutive equation-set having evolving quantified microstructural damage parameters displays several advantages:

    *   damage parameters are quantified separately and therefore their relative effects on creep rates can be "investigated" computationally;

    *   interactive effects of an evolving multi-component microstructure on creep resistance (leading to tertiary creep under applied loading or stress-softening under applied straining) can be predicted; an impossible task when using conventional empirical and single-damage constitutive laws;

    *   the number of (material- and heat treatment-specific) model parameters required for creep strain- and life-prediction is far smaller than reported in the literature for empirical models — for example, 17 are required for the θ Projection Method, Evans et al (1982) compared with 11 here — and this

parameter-efficiency should become even more apparent under complex thermal/mechanical loading.

3) The polycrystalline multiaxial model derived from this equation-set needs further theoretical development and experimental verification.

4) A current multiaxial model for single crystals considers shear deformation to be represented by scalar state variables, with anisotropy introduced solely through crystallographic symmetry operations. Anisotropy due to tensorial hardening during stress redistribution needs to be addressed and is under active consideration.

Acknowledgements Data for conventionally-cast IN738LC were supplied from the NPL creep databank by Mr M S Loveday. The authors thank EPSRC for support of this work through Grant Number GR/M93123.

References

Ansell, G. S. and Weertman, J. (1959) *Trans AIME,* **215,** 838-843.

Ashby, M F and Dyson, B F 1984 in *Advances in Fracture Research 1,* (ed. S.R. Valluri et al), Pergamon Press, Oxford, 3-30.

Basoalto, H C Ghosh, R N Ardakani, M G Shollock, B A & McLean, M (2000) in *Proceedings of Ninth International Symposium on Superalloys,* ed. T.Pollock et al., Seven Springs, PA, TMS (in press)

Chaboche, J L & Rousselier, G (1983) *J Press. Vess. Tech.,* **105,** 153-159.

Dieter, G E (1988) *Mechanical Metallurgy* McGraw-Hill Book Company (UK) Ltd p189.

Dyson, B F and McLean, M (1983) *Acta Metall.* **31,** 17-27.

Dyson, B F & Osgerby, S: (1993) *NPL Report DMA (A116).*

Dyson, B F & McLean, M (1998) in *Microstructural Stability of Creep Resistant Alloys for High Temperature Plant Applications,* edited by A Strang et al., Institute of Materials, London, 371-394.

Dyson, B F (2000) to be published.

Evans, R W Parker, J D & Wilshire, B: (1982) in *Recent Advances in Creep and Fracture of Engineering Materials and Structures,* edited by B Wilshire & D R J Owen, Pineridge Press, Swansea, UK, 135-184.

Ghosh, R N Curtis, R V & McLean, M (1990) *Acta Metall.,* **38,** 1977-1922.

Greenwood, G W: (1956) *Acta Metall.,* **4,** 243-248.

Harrison, G F & Evans, W J (1980) *Engineering Aspects of Creep* **1,** Inst. of Mech. Eng. London, 69-76.

Ion, J C Barbosa, A Ashby, M F Dyson, B F and McLean, M (1986) *The Modelling of Creep for Engineering Design* - I. DMA A115.

Janson, J & Hult, J: (1977) *J de Mécanique Appliquée* **1,** 69-84.

Kachanov, L M: (1958) *Izv. Ak. Nauk SSSR Otdel. Tekh. Nauk* **8,** 26-31.

Kocks, U F Argon, A S & Ashby, M F (1975) *Progress in Materials Science* Pergamon Press Oxford p90.

Kadoya, Y Nishimura, N Dyson, B F and McLean, M: (1997) in *Creep & Fracture of Engineering Materials & Structures,* edited by J C Earthman and F A Mohamed, TMS, Warrendale, USA, 343-352.

Kowalewski, Z L Hayhurst, D R & Dyson, B F (1994) *Journal of Strain Analysis* **29,** 309-315.

Lagneborg, R (1968) *J Materials Science,* **3,** 596-602.

Leckie, F A & Hayhurst, D R (1974) *Proc. Roy. Soc. Lond.,* **A340,** 323-347.

Lemaître, J & Chaboche, J L (1990) *Mechanics of Materials,* Cambridge University Press.

Merzer, A & Bodner, S R (1979) *ASME J. Eng. Mater. Tech.* **101,** 388-397.

Miller, A K (1997) in *Creep and Fracture of Engineering Materials and Structures,* edited by J C Earthman & F A Mohamed, TMS Warrendale Pa, 159-70.

Osgerby, S & Dyson, B F (1993) in *Creep and Fracture of Engineering Materials and Structures* Eds. B Wilshire and R W Evans. The Inst. of Metals, London, 53-61.

Othman, A M Dyson, B F Hayhurst D R and Lin, J: (1994) *Acta Metall. & Mater.* **42,** 597-611.

Pan, L-M, Shollock, B A & McLean, M (1997) *Proc R Soc Lond* **A453,** 1689-1715.

Robinson, D N & Bartolotta, P A (1985) *NASA Report CR 174836.*

Rabotnov, Y N: (1969) *Proc. XII IUTAM Congress,* Stamford, eds. Hetenyi & Vincenti, Springer, 137-141.

Rösler, J. and Arzt, E. (1990) **38,** 671-690.

Sherby, O D & Burke, P M (1967) *Progress in Materials Science,* **13,** 325-386.

Threadgill, P. L. and Wilshire, B. (1972) in *Creep Strength in Steel and High Temperature Alloys* ISI: Sheffield: UK 8-14.

CREEP OF γ-TIAL BASED ALLOYS – EXPERIMENTS AND COMPUTATIONAL MODELING

W.T. MARKETZ
Institut fuer Mechanik, Montanuniversitaet Leoben,
Franz-Josef-Strasse 18,A-8700 Leoben, Austria
A. CHATTERJEE
Max-Planck-Institut fuer Metallforschung, Seestrasse 92,
D-70174 Stuttgart, Germany
F.D. FISCHER
Institut fuer Mechanik, Montanuniversitaet Leoben,
Franz-Josef-Strasse 18,A-8700 Leoben, Austria
H. CLEMENS
Institut für Metallkunde, Universitaet Stuttgart, Seestrasse 71,
D-70174 Stuttgart, Germany

1. Introduction

Intermetallic γ-TiAl based materials are qualified to become an important material for advanced applications especially in aeroengine and aerospace industries (Kim, 1994). Research and development have progressed significantly within the last few years and led to comprehensive understanding of fundamental correlations between alloy composition and microstructure, processing behavior and mechanical properties. It is well known that the mechanical properties of γ-TiAl alloys depend strongly on microstructure, which in turn is influenced by the alloy chemistry and the applied heat treatments. A designed fully lamellar (DFL) microstructure which consists of colonies of parallel γ-TiAl (tetragonal face centered $L1_0$ structure) and α_2-Ti_3Al (ordered hexagonal DO_{19} structure) laths with a colony size in the range of 150-200μm possesses superior creep resistance.

Recent studies (Parthasarathy et al. 1998, Crofts et al. 1996, Maruyama et al. 1997) have shown that interface spacing has a major influence on the creep behavior. This may allow the conclusion that the interfaces γ/γ and α_2/γ must play a role in limiting creep flow or providing creep strength. In this paper differently spaced DFL microstructures were adjusted in order to investigate their influence on creep. Fine grained sheet material exhibiting a nominal composition of Ti-46.5at%Al-4at%(Cr,Nb,Ta,B) was used and short term creep tests were carried out in air at 700°C and 800°C under a load stress of 175MPa constant in time. The interface spacing was varied in the range of 1.2 μm to 0.14 μm by altering the cooling rates from 1 K/min to 200K/min. A first approach in modeling the steady state creep deformation of the fully lamellar material in question is presented. A power law description for diffusion

S. Murakami and N. Ohno (eds.), IUTAM Symposium on Creep in Structures, 17–30.
© 2001 *Kluwer Academic Publishers. Printed in the Netherlands.*

controlled dislocation creep is proposed and a structure factor is introduced which depends on the lamellar orientation with respect to the loading axis as well as on the mean lamellar interface spacing.

2. Material and Experimental

In this study the investigated sheet material with a nominal composition of Ti-46.5at% Al-4at%(Cr,Nb,Ta,B) was fabricated at Plansee AG (Reutte, Austria) by a powder metallurgical process followed by rolling. A detailed description of the rolling process is reported by Clemens et al. 1997. A subsequent heat treatment at 1000°C for 2h delivered the starting material exhibiting a primary annealed (PA) microstructure. The PA microstructure predominantly consists of equiaxed γ-TiAl grains with an average grain diameter of 15-20μm and approximately 5vol% of α_2Ti$_3$Al phase located at grain boundaries and triple junctions of γ-grains. Creep test specimens with an overall length of 50 mm and a gauge area of 30 mm x 3 mm (thickness 1 mm) were cut from the PA sheet material by spark erosion parallel to the rolling direction. In order to obtain a so-called designed fully lamellar (DFL) microstructure the specimens were heat-treated at 1350°C, which corresponds to a temperature above the α-transus temperature of the alloy. Differently spaced lamellae were obtained by varying the cooling rates. Figure 1 demonstrates the course of the heat treatments schematically.

After a certain hold time at 1350°C the samples were cooled with 15 K/min below the α-transus temperature. This step was performed to prevent possible grain growth within the α-transus field, thus to achieve comparable colony sizes. Subsequently, the specimens were cooled to 1000°C applying different cooling rates. This results in the formation of different mean interface spacings. After reaching 1000°C the specimens were cooled to room temperature by furnace cooling. The highest cooling rate (200 K/min) was obtained by floating with argon.

The creep samples were tested in tension in air at 700°C and 800°C and a constant load, which corresponded to a constant load stress of 175 MPa. In order to investigate the microstructure developing during deformation within the secondary creep regime by transmission electron microscopy (TEM), the tests were interrupted at a strain of approximately 2.5 %. To preserve the creep microstructure the specimens were cooled down to room temperature under applied load.

Optical microscopy under polarized light was used to determine the mean colony size by using the linear intercept method. The mean interface spacing of "edge on" tilted colonies was measured by TEM. At least 200 to 400 laths were considered, and no distinction was made between α_2/γ and γ/γ interfaces.

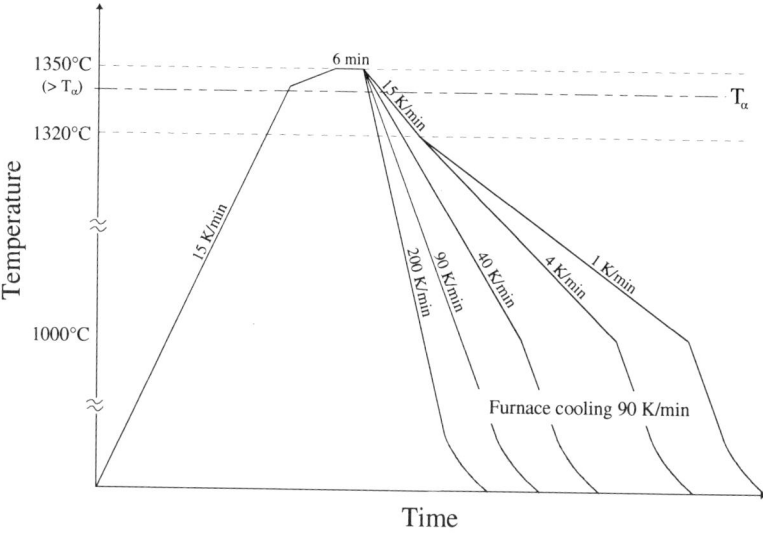

Figure 1. Schematic drawing of the heat treatment resulting in microstructures with different lamellar spacing but comparable colony size.

3. Experimental Results

3.1 MICROSTRUCTURE

The heat-treatments mentioned above resulted in lamellar microstructures with a mean colony size of 130 µm. The light optical micrographs of three representative specimens which were cooled with 1 K/min, 40 K/min and 200 K/min, respectively, are shown in Figure 2. While the colony size of all specimens remained similar, the mean interface spacing decreased from 1.2 µm to 0.14 µm with decreasing cooling rate. Metallographic examinations of differently cooled specimens have shown an almost undisturbed fully lamellar microstructure for cooling rates in the range of 4 K/min and 200 K/min. Typical defects such as primary γ-grains at colony boundaries occur with cooling rates < 4 K/min. Cooling rates higher than 200 K/min lead to the appearance of Widmannstätten-like features (Chatterjee et al. 1999).

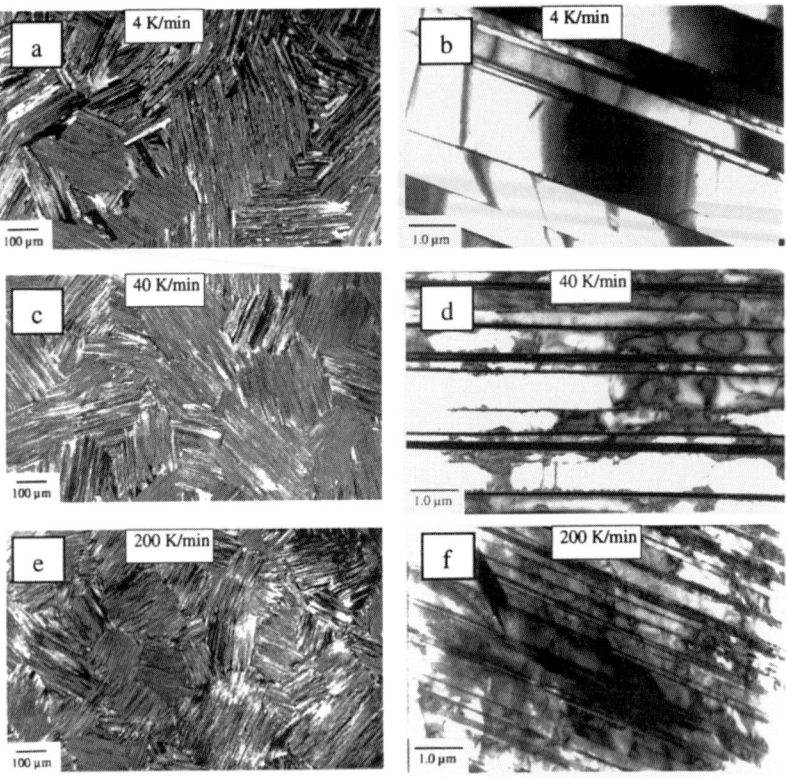

Figure 2. Colony size (light optical micrographs) and lamellar spacing (TEM micrographs) obtained in Ti-46.5at%Al-4at%(Cr,Nb,Ta,B) sheet specimens which were annealed at 1350°C for 6 minutes and subsequently cooled with different cooling rates: (a,b) 4 K/min, (c,d) 40 K/min, (e,f) 200 K/min.

The lamellar microstructure of specimens cooled within 4 - 200 K/min consists of regular patterns of alternate α_2 and γ lamellae, as shown in Figure 2. The mean interface spacing, including α_2/γ and γ/γ interfaces, was found to decrease with increasing cooling rate. The correlation between the mean interface spacing λ [μm] and the cooling rate R [K/min] is depicted in Figure 3 and can be expressed by a fit according to Equation 1:

$$\lambda = 1.18 \, R^{-0.39} \tag{1}$$

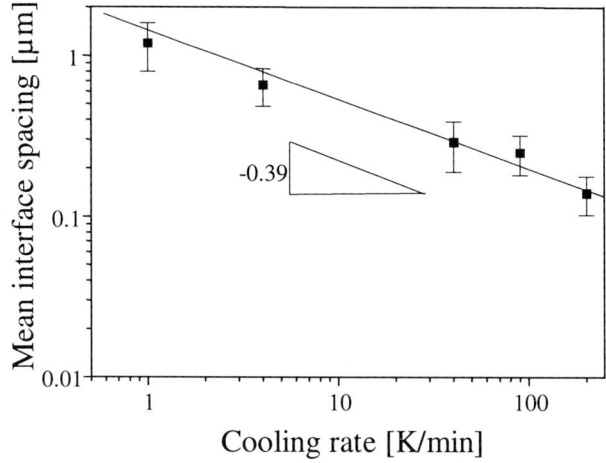

Figure 3. Dependence of the mean interface spacing λ on cooling rate R obtained for Ti-46.5at% Al-4at%(Cr,Nb,Ta,B) sheet material.

3.2 CREEP BEHAVIOR

Heat-treated creep specimens with different mean interface spacings but comparable colony size were tested at 700°C and 800°C applying a constant load stress of 175 MPa with the loading axis parallel to the rolling direction. The creep test results obtained for differently spaced DFL microstructures at 800°C are summarized in Figure 4. All creep tests were carried out to the secondary creep regime. The minimum creep rate decreases significantly with decreasing interface spacing. The correlation between the minimum creep rate $\dot{\varepsilon}_{min}$ [s^{-1}] and the mean interface spacing λ [μm] at the creep conditions described above can be expressed as:

$$\dot{\varepsilon}_{min} (700°C) = 1.59*10^{-9} \lambda^{0.68} \qquad (2)$$

$$\dot{\varepsilon}_{min} (800°C) = 1.03*10^{-7} \lambda^{0.66} \qquad (3)$$

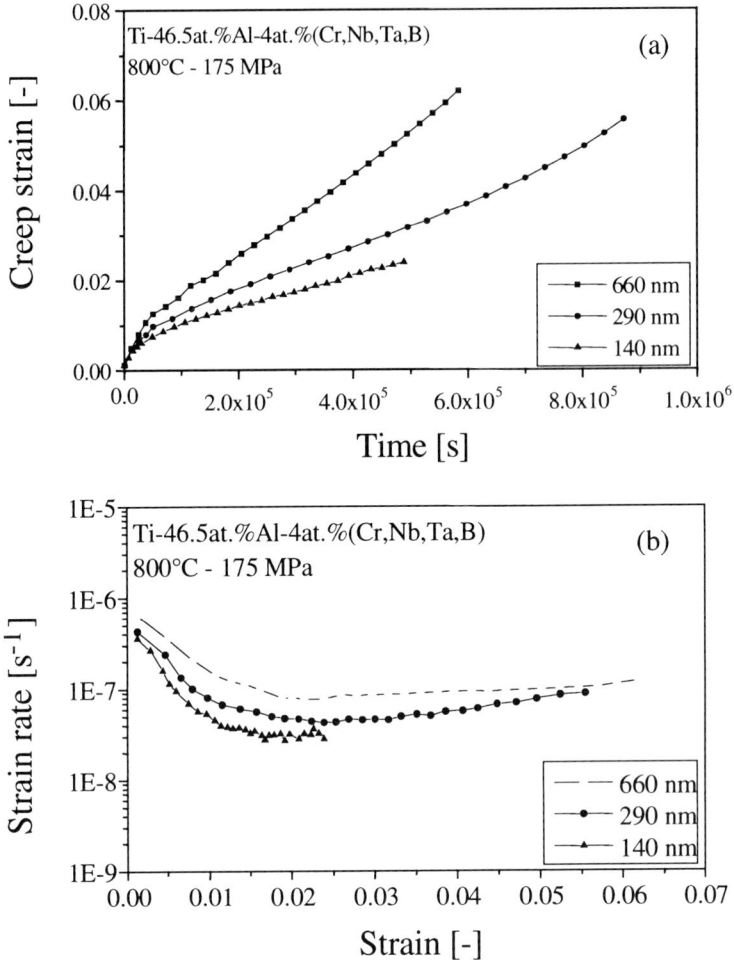

Figure 4. Creep behavior of Ti-46.5at% Al-4at%(Cr,Nb,Ta,B) sheet material tested at
800°C and 175 MPa. (a) strain ε vs. time, (b) strain rate ἐ vs. strain ε.

An activation energy of Q=350kJ/mol, which is independent of the mean lamellar
spacing, has been determined within the temperature range of 700°C to 800°C and a
stress of 175 MPa.

3.3 DISCUSSION OF EXPERIMENTAL RESULTS

The experimental results point out the significant effect of lamellar spacing in fully lamellar microstructures on the creep behavior of the alloy at the investigated temperatures of 700°C and 800°C under a constant load stress of 175 MPa in terms of minimum creep strain rate. The enhanced creep resistance in terms of decreasing minimum creep rates with the decrease of the lamellar spacing observed in the present study is in accordance with the results reported by Parthasarathy et al. (1998) and Maruyama et al. (1997). A similar correlation between secondary strain rate and mean interface spacing (Equation 2 and 3) is also reported by Parthasarathy et al. (1998). However, Parthasarathy argued that substructure invariant models discussed for dispersion strengthened materials as well as power law creep models are inapplicable to fully lamellar alloys.

In general, two mechanisms are suggested to be responsible for the improvement of creep behavior due to reduced mean interface spacing as shown in Figure 4a. The first mechanism is the impediment of dislocation glide by a reduced interface spacing. The second mechanism is based on the assumption that the lamellar interfaces restrict dislocation motion parallel to α_2/γ lamellae causing a bowing out of dislocation segments (e.g. see also Wang et al. 1995). TEM investigations conducted on fully lamellar Ti-46.5at%Al-4at%(Cr,Nb,Ta,B) sheet material confirm both mechanisms. These observations have shown frequent interceptions across γ-lamellae due to the formation of small angle grain boundaries and dislocation pile ups as well as the formation of subgrain boundaries within γ-lamellae. Furthermore, the formation and emission of dislocation loops as well as the bowing out of dislocations segments between the interfaces were observed. Similar results were found by Chen et al (1999), who recently reported dislocation bowing between lamellar interfaces of the soft oriented grains, resulting in the assumption that the strain rate of soft grains obeys a power law relation.

4. Micromechanical Modeling

As indicated in the previous chapter there are contradictory statements reported in the literature as to whether a power law description of creep of fully lamellar TiAl-materials is applicable. Therefore, the purpose of our computational modeling is a first approach to investigate the application of a power law creep model to describe the creep behavior of DFL γ-TiAl materials.

For many pure metals and alloys an established, largely phenomenological, relationship exists between the steady state strain rate, $\dot{\varepsilon}_{ss}$, and stress, σ:

$$\dot{\varepsilon}_{ss} = A \cdot \exp\left\{-\frac{Q}{R \cdot T}\right\} \cdot \sigma^n \tag{4}$$

where A is a structure factor, Q is the activation energy for creep, R is the gas constant, T is the temperature and n denotes the stress exponent.

It seems obvious that such a general relation like type (4) should fail in the description of the creep behavior of fully lamellar microstructures consisting of differently oriented colonies of α_2- and γ-laths. Furthermore, the creep behavior of the hexagonal α_2-Ti$_3$Al phase differs from the creep behavior of the tetragonal face centered γ-phase (Bartholomeusz et al. 1993). The effects of the lamellar structure can best be studied with so called polysynthetically twinned (PST) crystals of TiAl. These are single-crystals with lamellar structure (Fujiwara et al. 1990). A detailed investigation of the creep behavior of PST TiAl at 877°C and 200 MPa has been reported by Wegmann et al. (2000). It has been shown that PST orientations with their lamellae parallel or perpendicular to the loading axis (hard mode orientations) exhibit longer creep lives and lower minimum creep rates than soft oriented PST specimens. The hard PST orientation deforms mainly by deformation modes oblique to the lamellar interfaces, whereas the soft orientation deforms mainly by slip parallel to the lamellar interfaces. The effect of the lamellar orientation on the creep resistance is discussed on the basis of the resolved shear stress for ordinary glide dislocations. Due to the fact that the γ-phase is considerably weaker than the α_2-phase at high temperatures the creep resistance is expected to be controlled by the deformation of the γ-constituent (Wegmann et al., 2000). A detailed information on hard- and soft mode deformation of PST TiAl single-crystals is given by Inui et al. (1995).

4.1 MODELING OF THE CREEP BEHAVIOR OF SINGLE-CRYSTAL PST TIAL

Based on the above mentioned experimental results and Equation (4) we suggest to modify the latter relation in order to take the lamellar orientation as well as the lamellar interface spacing of the γ-phase into account. The DFL TiAl investigated does not show an extended steady state creep region but only a minimum creep strain rate. Consequently, the notation $\dot{\varepsilon}_{ss}$ in Equation (4) changes to $\dot{\varepsilon}_{min}$ in the following. Equation (5) gives a new relation between the minimum creep strain rate $\dot{\varepsilon}_{min}$, temperature T, activation energy Q_γ and applied stress introducing a structure factor B (ϕ,λ) depending on the angle ϕ between lamellae orientation and load axis and on the mean interface spacing of the lamellae, respectively. Therefore, the creep behavior of the γ-TiAl constituent can be expressed as follows:

$$\dot{\varepsilon}^\gamma{}_{min} = B(\phi,\lambda) \cdot \exp\left\{-\frac{Q_\gamma}{R \cdot T}\right\} \cdot \sigma^{n_\gamma} \qquad (5)$$

The creep behavior of the α_2-phase is described by a relation according to Equation (6):

$$\dot{\varepsilon}^{\alpha}{}_{min} = C \cdot \exp\left\{-\frac{Q_{\gamma}}{R \cdot T}\right\} \cdot \sigma^{n_{\alpha}} \tag{6}$$

The input data for our modeling are taken from literature available for single α_2-Ti$_3$Al phase and single γ-TiAl phase without additional alloying elements (He et al. 1997, Schafrik 1977, Bartholomeusz et al. 1993, El-Souni et al. 1993). The elastic constants, the activation energy and the stress exponent for both phases at 800°C are given in Table I. For a first approach the structure factor C for the α_2-phase is assumed to be independent of lamellae orientation and lamellae thickness. This assumtion seems to be justified since the volume content of the α_2-phase is lower than 10vol%. The value is calculated from experiments by Bartholomeusz et al. 1993 to be C=5.49*10^{14} [GPa^{-n}s^{-1}].

Table I. Elastic constants, activation energy and stress exponents for γ-TiAl- and α_2-Ti$_3$Al single phase materials (Schafrik 1977, Bartholomeusz et al. 1993, He et al. 1997, El-Souni et al. 1993).

Material	E (800°C)	ν (800°C)	Activation energy Q	Stress exponent n
γ-TiAl	102.0 GPa	0.28	380 kJ/mol	7.2
α_2-Ti$_3$Al	150.0 GPa	0.247	352kJ/mol	5.5

The activation energy for γ-TiAl is taken from creep experiments conducted on fine-grained Ti-46.5at% Al-4at%(Cr,Nb,Ta,B) sheet with near gamma microstructure. The stress exponent of n=7.2 for γ-TiAl indicates that the creep process is mainly determined by dislocation glide and dislocation climb. Simulations with the orthotropic elasticity tensor for TiAl and Ti$_3$Al did not show any significant difference in the results compared to cases, where isotropic elasticity in each constituting phase is assumed.

4.1.1 Determination of the structure factor B(ϕ,λ) and C
To determine the structure factor B (ϕ,λ) we fit our model to the experimental data obtained by Wegmann et al. (2000). The nominal composition of the PST TiAl crystal used in their study was binary Ti-48at%Al.

The arrangement of the ordered γ-domains between the α_2-lamellae can be seen in Figure 5. A three-dimensional finite element (FE) model is chosen for the simulations of PST crystals in this work. The representative volume element (RVE) represents the microstructure and the behavior of the whole specimen. The orientation angle ϕ is determined by the length/height ratio of the unit cell. An amount of 10vol% of α_2-Ti$_3$Al and all six ordered domains of the γ-phase are considered. The selected orientations of the lamellae relative to the load axis are taken from Wegmann et al. (2000) as well as the dimensions of the specimen. A periodic arrangement of a set of unit cells represents a great number of parallel α_2-Ti$_3$Al and γ-TiAl lamellae by applying periodic boundary conditions which ensure periodicity.

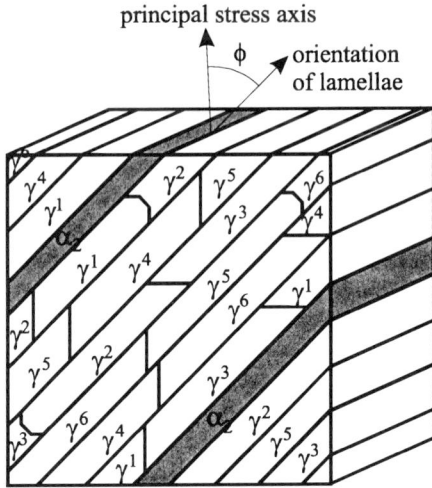

Figure 5. Three-dimensional unit cell representing the lamellar microstructure of a PST crystal formed of six ordered domains of γ-TiAl and 10 vol% of α_2-Ti$_3$Al. (Schlögl and Fischer 1997)

In Figure 6 the dependence of the structure factor B on the orientation of the lamellae at a fixed mean interface spacing of $\lambda=520$ nm is depicted (B $\rightarrow$ B($\phi,\lambda=520$nm) . For our further simulations we assume that the relationship of B(ϕ) obtained for Ti-48at%Al PST crystals is the same as in the polycrystalline Ti-46.5at%Al-4at%(Cr,Nb,Ta,B) material with DFL microstructure that we investigated.

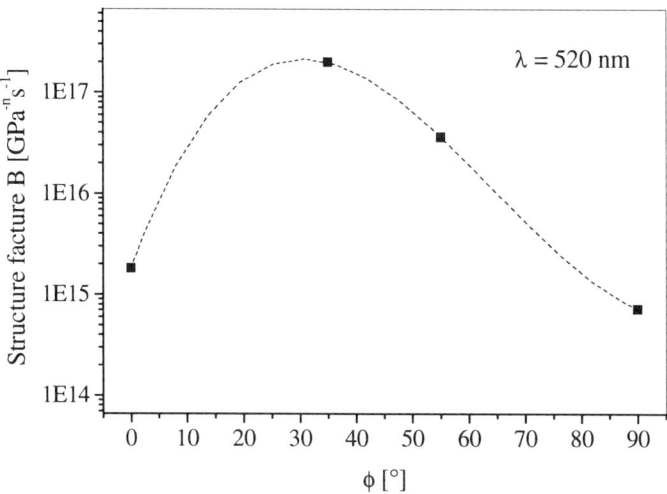

Figure 6. Dependence of the structure factor B on the orientation angle φ between the lamellae and the loading axis. The dashed line indicates a polynomial fit to B obtained for four orientations φ.

4.2 MODELING OF THE CREEP BEHAVIOR OF POLYCRYSTALLINE TIAL WITH FULLY LAMELLAR MICROSTRUCTURE

In order to simulate the creep behavior of a DFL polycrystal with different mean interface spacings the dependence of the structure factor B on the interface spacing at a fixed orientation angle φ is needed additionally. From the experimental results (Figure 4) and Equation (2) and (3) we infer the following relation:

$$\frac{B(\phi,\lambda)}{\lambda^w} = \text{constant} \tag{7}$$

with w given in Equation (2) for 700°C and Equation (3) 800°C.

The major difficulty inherently associated with the modeling of a lamellar polycrystal is to find a RVE just large enough to represent the entire material sufficiently well, and, on the other hand, as small as possible in order to keep the computational effort low. In our case the microstructure is assumed to show periodic patterns with different oriented lamellar colonies. The developed RVE for the DFL microstructure exhibits of six hexagonal grains with different orientation of the lamellae and is depicted in Figure 7. The finite element mesh consists of 2085 6-node bilinear quadrilateral generalized plane strain elements. Periodic boundary conditions are applied to ensure periodicity. Detailed information on periodic boundary conditions has been provided by Antretter (1998).

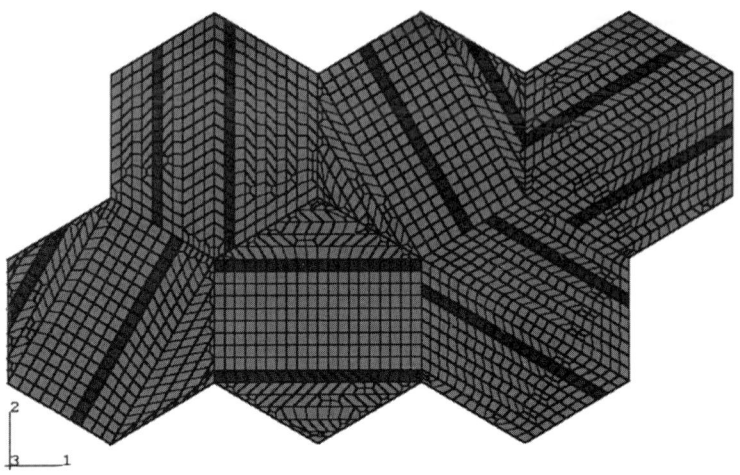

Figure 7. Finite element mesh for the DFL microstructure. The α_2-laths are indicated by the darker shading. The RVE applies periodic boundary conditions to ensure periodicity.

Using Equations (5), (6) and (7) to describe creep properties for the individual phases, we can compute the global creep deformation behavior of the DFL microstructure applying our FE-model. Several orientations and mean lamellar interface spacings were studied for the DFL microstructure. Figure 8 shows the dependence of the minimum creep rate $\dot{\varepsilon}_{min}$ on the mean lamellar interface spacing at T=700°C and T=800°C under a load stress of σ=175 MPa obtained experimentally and by simulation, respectively.

In our micromechanical model, the DFL microstructure is composed of a number of differently oriented lamellar single crystals. We introduced a structure factor B which depends on the lamellar orientation ϕ and the mean lamellar interface spacing λ. $B(\phi,\lambda)$ was determined for each colony orientation as well as for different lamellar spacings from creep data obtained on PST single crystals. The FE simulations matches quite well with the experimental data for at temperatures of 700°C and 800°C. However, further research is necessary to extent the finite element model for other creep mechanisms occurring under different creep conditions. One limiting factor that still remains is the availability of experimental data of the single constituent phases, such as the stress exponent at lower temperatures. Finally, the role of additional alloying elements , i.e. Cr, Nb, Ta in our investigated material has to be studied and considered in the model. Moreover, the dependence of the exponent w in Equation (7) on temperature, prevailing creep mechanism and content of alloying elements has to be studied.

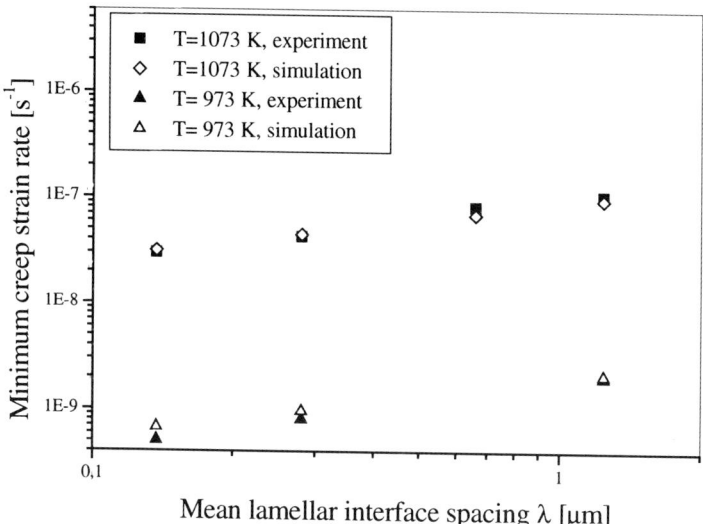

Figure 8. Minimum creep rate $\dot{\varepsilon}_{min}$ as a function of the mean lamellar interface spacing λ at temperatures of 700°C and 800°C and a constant load of 175 MPa. Calculated data (open symbols) and experimental data (solid symbols) obtained on Ti-46.5at% Al-4at%(Cr,Nb,Ta,B) material with DFL microstructure (colony size~130μm)

5. Summary and Outlook

Depending on the cooling rate the average interface spacing of a Ti-46.5at% Al-4at%(Cr,Nb,Ta,B) material, as measured by TEM, varied between 0.14 μm and 1.2 μm for fully lamellar microstructures with comparable colony sizes. Creep tests conducted at 700°C and 800°C under a constant load stress of 175MPa indicated that the minimum creep rate decreases monotonically with decreasing interface spacing. Creep tests performed on this sheet material with DFL microstructure did not show any grain boundary sliding and the main creep deformation mechanism under the given conditions appeared to be diffusion controlled dislocation creep. A first approach in computational modeling simulation confirmed that a power law creep model can describe the creep behavior of DFL microstructures if a structure factor depending on the lamellar orientations and the mean lamellar interface spacing is introduced. However, further research is necessary to extent the finite element model for other creep mechanisms occurring under different creep conditions.

6. Acknowledgement

The authors appreciate the funding by FWF and OeNB under the project number 491/P12418-TEC. In addition we would like to thank Plansee AG (Reutte, Austria) for providing sheet material. Special thanks to T. Antretter for very helpful discussions.

7. References

Antretter, T. (1998), *Micromechanical modeling of high speed steel,* Fortschritt-Berichte VDI, **18**,VDI Verlag, Düsseldorf.

Bartholomeusz, M.F., Yang, Q. and Wert, J.A. (1993), Creep deformation of a two- phase TiAl/Ti₃Al lamellar alloy and the individual TiAl and Ti₃Al constituent phases, *Scripta Met.Mat.* **29**, 389-394.

Chatterjee, A., Bolay, U., Sattler, U. and Clemens, H. (1999a), Adjustment of differently spaced fully lamellar microstructures in a γ-TiAl based alloy and their creep behavior, *Proc. Euromat 1999*, in print.

Chen, W.R., Triantafillou, J., Beddoes, J., Zhao, L., (1999), Effect of fully lamellar   morphology on creep of a near gamma TiAl intermetallic, *Intermetallics*, 7, 171-178.

Clemens, H. Glatz, W., Eberhardt, N. Martinz, H.P. and Knabl, W. (1997), Processing, properties and application of gamma titanium aluminide sheet and foil materials, *Mat. Res. Soc. Symp. Proc.* **460**, 29-43.

Crofts, P.D., Bowen, P. and Jones, I.P. (1996), The effect of lamella thickness on the creep behaviour of Ti-48Al-1Nb-2Mn, *Scripta Mat.* **35** (12), 1391-1396.

Es-Souni, M., Bartels, A., Wagner, R. (1993), Creep behaviour of a near γ-TiAl alloy Ti-48Al-2Cr: Effect of microstructure, in Darolia, R., Lewandowski, J.J., Liu, C.T., Martin, P.L., Miracle, D.B. and Nathal, M.V. (eds.) *Structural Intermetallics*, TMS, Warrendale, 335-343.

Fujiwara, T., Nakamura, A., Hosomi, M., Nishitani, S.R., Shirai, Y. and Yamaguchi, M. (1990), Deformation of polysynthetically twined crystals of TiAl with a nearly stoichiometric composition, *Phil. Mag. A*, **61**, 591-606.

Inui, H., Kishida, K., Misaki, M., Kobayashi, M., Shirai, Y. and Yamaguchi, M. (1995), Temperature dependence of yield stress, tensile elongation and deformation structures in polysynthetically twinned crystals of Ti-Al, *Phil. Mag. A*, **72** , 1609-1631.

Kim, Y.-W. (1994), Ordered Intermetallic Alloys, Part III: Gamma Titanium Aluminides, *JOM*, **46**, 30-40.

Maruyama, K., Yamamoto, R., Nakakuki, H. and Fujitsuna, N. (1997), Effects of lamellar spacing, volume fraction and grain size on the creep strength of fully lamellar TiAl alloys, *Mat. Sci.and Eng.* **A239-240**, 419-428.

Parthasarathy, T.A., Keller, M. and Mendiratta, M.G. (1998), The effect of lamellar lath spacing on the creep behavior of Ti-47at%Al, *Scripta Mat.* **37**, 1025-1031.

Parthasarathy, T.A., Mendiratta, M.G. and Dimiduk, D.M. (1997), Observations on the creep behavior of polycrystalline TiAl: Identification of critical effects, *Scripta Mat.* **37**, 315-321.

Schloegl, S.M.; Fischer, F.D.(1997), The role of slip and twinning in the deformation behaviour of polysynthetically twinned crystals of TiAl: a micromechanical model, *Phil. Mag. A*, **75**, 621-636.

Wang, J.N., Schwartz, A.J., Nieh, T.G., Liu, C.T., Sikka, V.K. and Clemens, D. (1995), Creep of a fine grained, fully lamellar, two phase TiAl alloy at 760°, in Kim, Y-W., Wagner, R. and Yamaguchi, M. (eds.) *Gamma Titanium Aluminides*, TMS, Warrendale, 949-957.

Wegmann, G., Suda, T. and Maruyama, K. (2000), Creep deformation of polysynthetically twinned (PST) Ti-48mol%Al, *Key Engineering Materials*, **170-174**, 709-716.

ANISOTROPIC CREEP OF SINGLE CRYSTAL SUPERALLOYS

D.M. Knowles and D.W. MacLachlan
University of Cambridge, Department of Materials Science,
Pembroke St., Cambridge, CB2 3QZ UK

1. Introduction

Single crystal Ni-base superalloys are an integral feature of modern gas turbine engines. Their excellent high temperature tensile properties are a result of a number of factors including an absence of grain boundaries and a matrix which comprises a high volume fraction of cuboidal γ' in a solid solution strengthened γ matrix. Such a combination of a two-phase structure embedded within a single crystal leads to complex anisotropic creep properties. Accurate modelling of creep anisotropy in these materials is crucial for prediction of stress redistribution and life. In this paper, a physically based, isothermal slip system model, which incorporates interaction between slip system types is introduced. Its predictive capabilities for both uniaxial and multiaxial loading at 1223K are discussed using a creep test matrix for calibration and validation purposes.

2. Materials and Methods

The materials used in this study were SRR99 and CMSX-4. The CMSX-4 bars were seeded to give a number of different axis orientations, figure 1. SRR99 bars were all grown in the conventional [100] direction. The alloy underwent a standard heat treatment to minimise residual dendritic microsegregation and maximise phase stability. Conventional creep specimens were machined from CMSX-4 and uniaxial tests were performed in air on a constant load creep machine with self aligning joints. The test temperature was controlled by platinum/13% rhodium-platinum thermocouples to ±2 K. Creep strain was measured by averaging two high temperature extensometers with linear displacement voltage transducers (LDVTs). Biaxial testing was undertaken on both CMSX-4 and SRR99 with their principal axial directions being [001]. A tension-torsion servohydraulic was used to apply load. The specimens had external and internal diameters of 3 and 4 mm respectively. Tensile and torsional strains were measured remotely from the gauge length.

Foils taken at a distance greater than 3 mm from the fracture surface were prepared for TEM examination. Lattice rotations that occurred along the specimen gauge length during creep deformation were determined using electron backscattered diffraction analysis (EBSP).

S. Murakami and N. Ohno (eds.), IUTAM Symposium on Creep in Structures, 31–40.
© 2001 *Kluwer Academic Publishers. Printed in the Netherlands.*

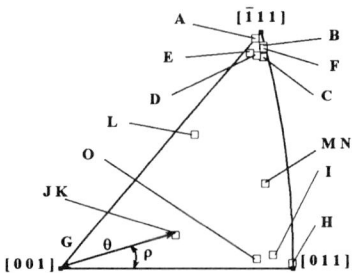

Figure 1: Standard [001] stereographic projection showing the axis orientations of the uniaxial loaded specimens.

3. Mechanisms

For tests towards the [001] pole networks of $a/2<110>$ dislocations were observed in the γ-phase and around the γ′-precipitates in the early stages of life. For brevity {111}<101> and {111}<112> slip will from here on be referred to as type I and type II slip respectively. For specimens tested towards the [111] pole deformation twinning was observed. Figure 2(a) shows the resultant microstructure in specimen M. This mechanism is typical of the deformation occurring in specimens orientated towards [111].

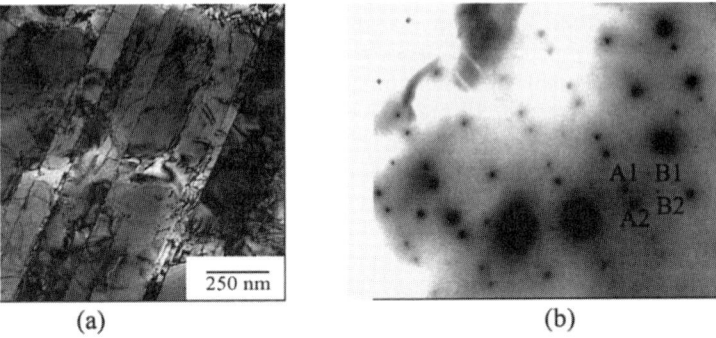

(a)   (b)

Figure 2: (a) TEM micrograph of CMSX-4 specimen M tested at 350 MPa. (b) Electron diffraction pattern for (a), twin diffraction spots from the γ-phase, (A1 and A2) and from the γ′-phase, (B1 and B2) are shown.

As with slip deformation, the shearing of a crystal resulting from twinning causes plastic deformation. The magnitude of the twinning shears shown in figure 2(a) results in the reorientation of parts of the crystal. The twins extend through the γ/γ′ microstructure twinning both the γ and γ′ phases. In a $L1_2$ structure this requires the passage of partial dislocations twice the magnitude seen in FCC structures, giving a shear $s=\sqrt{2}$. Such a shear also twins the FCC lattice, but is not generally observed for energetic reasons. As the twins observed in CMSX-4 are continuous through both phases it would appear that they are the result of the passage of $a/3<112>$ dislocations through the whole structure. The amount of longitudinal plastic strain resulting from deformation twinning can be quite significant. For specimen M in this study, for, instance it contributes 0.8 % to the strain for every one vol.% of twinned lattice.

Figure 3 shows the lattice rotations resulting from creep testing of three different initial orientations along with that expected from slip if only the most highly stressed type II system was activated. For specimen J, tested at a stress of 250 MPa, no twin deformation was observed, the rotation shown is reasonably consistent with a/2<101> dislocation activity observed in the γ-phase. For specimen H, tested with an initial stress of 350 MPa the rotation is towards the [111] pole consistent with the operation of a type II slip mechanism. For specimen M the lattice rotation generally proceeds towards the [211] pole. The failure strain for this specimen was 35 %. The volume fraction of twins in this specimen is approximately 15-20 %, which equates to a longitudinal strain of approximately 12–15 %. Twining is clearly a significant deformation mechanism in this orientation but may not be responsible for the total axial strain observed.

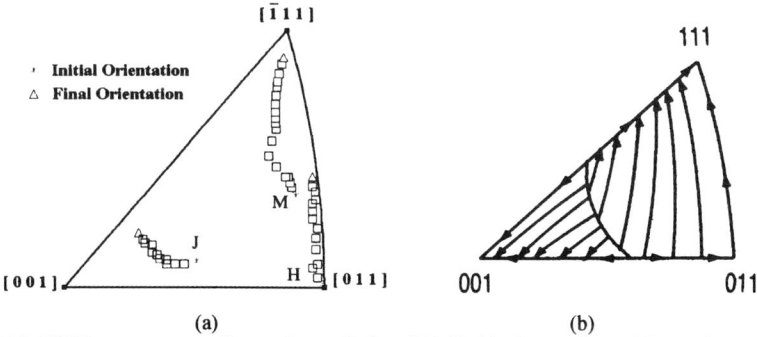

(a)   (b)

Figure 3(a): EBSP measurements for specimens H, J, and M (b) ideal crystallographic rotations that would occur if deformation resulted from slip on the most highly stressed type II system.

4. Discussion

A significant problem in the use of slip system models to predict anisotropic creep at relatively high temperatures has been their inability to explain the low creep resistance of the [111] orientation relative to the [001] orientation. If deformation is assumed to occur in both orientations by the operation of type I slip, then based on the relative Schmid factors for the two orientations, resistance in the [001] orientation should be orders of magnitude stronger. The test results suggest that a different mechanism is operating in the [111] orientation. Previous slip system approaches have addressed this issue by claiming that cube slip is responsible for the apparent lower resistance in the [111] orientation [2,3]. There is, however very little evidence for such cube slip in single crystal superalloys [4] and its use has been called into question. Experimental evidence shows that type II slip systems may be operative in orientations towards the [111] pole. The rotations measured (figure 3) are similar to those that would occur if the primary type II system was active. TEM analysis has shown that a major mechanism of deformation is by twinning in this orientation, which is derived from type II slip and is consistent with incorporating such a mechanism into a macroscopic model. The total global strain experienced by the specimen does not come from deformation twinning and additional strain is due to conventional type I or type II slip. The latter having been observed at higher stress levels. In either case the unexpectedly high creep rates towards the <111> pole may be ascribed to the operation of an additional deformation mechanism.

If type II systems are considered to be operative towards the [111] orientation a similar problem is encountered as for modelling with the type I systems mentioned previously; the Schmid factors for type II slip are much higher at the [001] pole than they are for the [111] pole. Creep deformation on type I slip systems however, is confined to the continuous matrix phase for a significant portion of a creep test and does not usually occur within the precipitates before the onset of tertiary creep. It is well known that during creep, type I dislocations percolate throughout the matrix, which has a very low initial dislocation density [1]. As this process proceeds interfacial dislocations are laid down at the precipitate boundaries. This situation is in contrast to that which occurs when type II dislocations are active. Type II dislocations pass through both matrix and precipitate. For type II dislocations to fully restore the $L1_2$ lattice the burgers vector must be of magnitude a<112>. This vector may dissociate in various ways as it shears both matrix and precipitate, but there is strong evidence which suggests that such a resultant dislocation can pass through and completely restore the structure, leaving no deformation or dislocation debris in its wake [5].

In order for significant strain to be associated with specific slip system activity the dislocations need to be able to travel through large portions of the lattice. Type I dislocations reside chiefly within the interconnected γ-channels and although slip is hindered by the γ'-precipitates, climb and cross slip processes provide mechanisms for continued progress through the structure. For deformation via type II slip the dislocations must be able to pass through both phases. This may occur by dislocation pairs expanding or contracting in the two phases, but it necessarily requires that they can proceed through the precipitate interfaces and γ channels relatively unhindered. If this is not the case their activity will be confined to individual precipitates leading to insignificant strain levels.

In summary the passage of type II dislocations through the lattice will lead to little disruption of the structure which affects type I activity, the same cannot be said of the converse case. If type I slip systems are highly activated their progress through the material will leave a wake of dislocation networks at the precipitate interfaces. If conditions are such that type II dislocations are activated they will interact with these networks on entering or leaving the precipitates leading to the rapid formation of jogs and tangles which will render them sessile. It is suggested therefore that type II activity will have little influence on type I slip systems. However if conditions are such that the type I slip systems are stressed significantly the subsequent dislocation network which ensues in the γ channels may sufficient to inhibit significant type II deformation leading to rapid hardening [10].

5. Modelling

The reasoning behind the orientation dependence of creep can be explained in the following manner. In the [001] orientation deformation occurs rapidly on type I slip systems due to the high Schmid factors they experience, causing the formation of a network of dislocations throughout the matrix, located primarily at the interfaces. Such a network prevents precipitate shear by type II dislocations and hence restricts activity of type II systems. As the orientation changes from [001] towards [111] the activity on type I slip systems decreases preventing hardening and thus allowing long term activation of the type II systems.

The model used in this analysis is based on finite deformation single crystal plasticity theory [6]. Deformation is analysed using the FE package ABAQUS through use of a user material subroutine[7]. The full specimen is simulated because during creep in unstable crystallographic orientations elastic rotation causes bending of the specimen resulting in non-uniform stresses and an elliptical cross section. Such features can be incorporated into the creep analysis using a solid model. The equations used to describe the shear strain rates on the different slip systems are given by [7]:

$$\dot{\gamma}_a = E_{101} \times \left(\frac{\tau_a}{\tau_{C101} - \tau_a} \right)^{u_{101}} \quad \ldots \quad \alpha=1 \text{ to } 12 \text{ (Type I)} \tag{1}$$

$$\dot{\gamma}_a = E_{112} \times \left(\frac{\tau_a}{\tau_{C112} - \tau_a} \right)^{u_{112}} \quad \ldots \quad \alpha=13 \text{ to } 24 \text{ (Type II)} \tag{2}$$

α is the slip system number, E and u are material constants describing shear strain rate and $\tau_{C101/C112}$ is the critical resolved shear stress determined from tensile tests for the separate slip systems. Whilst it is straightforward to conclude that shear strain should accumulate on specific slip systems it is not evident how damage should evolve as a function of stress or strain. Uniaxial rupture life cannot be correlated with stress based parameters such as the stress tensor or Von Mises equivalent stress as they predict a rupture life independent of orientation. Rupture life does not correlate directly with Schmid factors on active slip systems [7]. Furthermore, as will be shown later, biaxial rupture life does not correlate with Von Mises effective stress. Damage in single crystal superalloys can be viewed as occurring via a number of mechanisms. Flow in the matrix causes load redistribution which generates local stresses. Precipitate shear in itself can lead to local softening and crack initiation either from oxygen spikes or casting porosity is commonly observed late in life. It is clear that slip on the different slip systems types will contribute in different ways to the development of these processes. The effect of different types of shear on damaging the material have therefore been de-coupled. As the stress based parameters mention above are inappropriate for correlating strain rate and rupture life, damage accumulation is taken as a function of the macroscopic strain rate tensor derived from microscopic slip on the different slip systems:

$$\dot{\omega}_{101} = C_{101} \times \left[\frac{\dot{\varepsilon}_{101}}{E_{101}} \right]^{v_{101}/u_{101}} \qquad \dot{\omega}_{112} = C_{112} \times \left[\frac{\dot{\varepsilon}_{112}}{E_{112}} \right]^{v_{112}/u_{112}} \tag{3}$$

$$\dot{\varepsilon}_{112} = \sum_a \dot{\gamma}_a \mathbf{b}_a \mathbf{n}_a \ldots \alpha=13 \text{ to } 24 \qquad \dot{\varepsilon}_{101} = \sum_a \dot{\gamma}_a \mathbf{b}_a \mathbf{n}_a \ldots \alpha=1 \text{ to } 12 \tag{4}$$

$$\dot{\omega} = \dot{\omega}_{101} + \dot{\omega}_{112} \tag{5}$$

C and v are further material parameters and $\mathbf{b}$ and $\mathbf{n}$ are the slip direction normal and slip plane normal respectively. The subsequent effect of global damage on stress and strain rates is factored into the model in the conventional damage mechanics method using an effective stress, equation 6 from which the resolved shear stresses in equations 1 and 2 are calculated.

$$\sigma = \frac{\sigma_0}{1 - \omega} \tag{6}$$

Using strain rate rather than stress to determine a global damage state allows the combined effects of load and crystallographic orientation to be factored into the rate of damage accumulation. The influence of resolved shear stresses on the various slip systems is therefore incorporated into the equations as a global rather than a local effect. Two tensor measures of damage are used which describe how slip on different types of slip system effect the material. Thus the effects of different origins of shear strain on the material is incorporated into the model at the macroscopic level.

If the model is to be used consistently to describe strain evolution in a range of orientations, including the [001] and [111] poles, it is necessary to utilise equations describing the hardening mechanisms discussed earlier. In classical single crystal plasticity, interaction between different slip systems is introduced through the hardening matrix [6]. In this analysis the equations and especially the hardening matrix take a slightly different form from those conventionally used [6] due to the different nature of the problem. Firstly, there are two different intrinsic types of slip system, between which hardening will be different, hence the hardening matrix is not symmetric. Secondly the attainment of stable interfacial dislocation networks surrounding precipitates is a very rapid process [1,8] which rapidly hardens deformation with respect to type II slip. The more common linear dependence of hardening rate on shear strain rate does not describe this high sensitivity of hardening between the slip systems and is replaced by a hyperbolic tangent function. Given these factors the equation set developed to describe the deformation and interaction process is:

$$\dot{\gamma}_\alpha = \frac{E}{1+g_\alpha} \times \left(\frac{\tau_\alpha}{\tau_C - \tau_\alpha}\right)^u \qquad\qquad \dot{g}_\alpha = \sum_\beta h_{\alpha\beta} f(\dot{\gamma}_\beta) \qquad\qquad (7)$$

$$f(\dot{\gamma}_\beta) = \left\{1 + \tanh\left[a_0 \ \log\left(\frac{\dot{\gamma}_\beta}{\dot{\gamma}_{ref}}\right)\right]\right\} \qquad\qquad h_{\alpha\beta} = H_{\alpha\beta} \sec h^2(\mu\gamma) \qquad\qquad (8)$$

In these equations μ, a_0 and $\dot{\gamma}_{ref}$ are material constants. g characterises the current strain hardened state of the crystal. E, u and τ_c are constants for type I and type II slip as previously described. The rate of material hardening is specified by the evolution equations for $\dot{g}_\alpha$. As the effect of hardening is restricted to type I systems on type II systems γ in equation 8 is the sum of accumulated shear strain on all type I slip systems. The constant $H_{\alpha\beta}$ represents an initial hardening rate matrix:

$$H_{\alpha\beta} = \begin{array}{c|ccccc} & 1 & . & 12 & 13 & . & 24 \\ \hline 1 & 0 & . & 0 & 0 & . & 0 \\ . & . & . & . & . & . & . \\ 12 & 0 & . & 0 & 0 & . & 0 \\ 13 & H_{13,1} & . & H_{13,12} & 0 & . & 0 \\ . & . & . & . & . & . & . \\ 24 & H_{24,1} & . & H_{24,12} & 0 & . & 0 \end{array} \qquad (9)$$

Where $H_{ij}=H_{kl}$ for $13 \le i,k \le 24$ and $j,l \le 12$. The procedure used to calibrate the model described thus far has been to fit equations 1-6 describing strain and damage associated with shear on the type I and II slip systems to data in the [001] and [111] orientations respectively. This is first done for the type I systems and a correction is then made for the small amount of strain occurring on the type I systems in the [111] orientation before fitting the type II systems. The hardening equations are then introduced and the degree of type I on type II hardening is set so that it is sufficient to negate the effect of type II slip in the [001] orientation and also to optimise the prediction of further test data in unstable orientations. The overall amount of hardening is controlled through the initial hardening rate and the rate of saturation of hardening with accumulated shear strain.

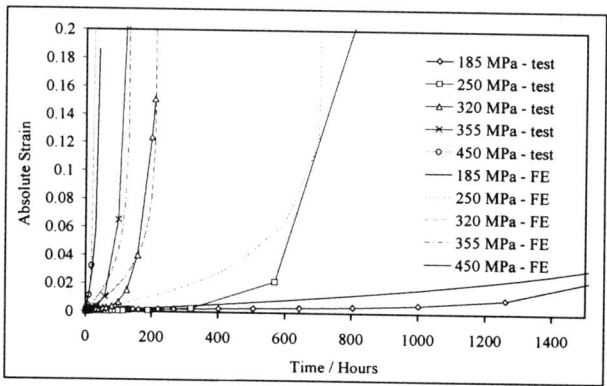

Figure 4: <100> creep data and model simulations at 950°C.

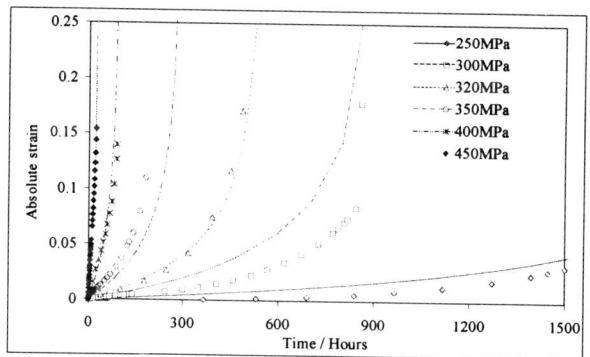

Figure 5: [111] creep test data and model simulations950°C

Figures 4-6 illustrate the results of creep simulations using the model presented here. Figures 4 and 5 show available data in the [100] and [111] orientations whilst figure 6 contains data from unstable orientations inside the stereographic triangle. It can be seen that the desired accuracy of the fits in stable orientations such as [111] and [100] is maintained within the typical scatter levels encountered from such tests. For the unstable orientations the accuracy of fit for a wide range of initial tensile axes and stresses is generally acceptable. Rupture lives are predicted extremely well, but in some instances particularly towards the [110] pole secondary creep rates are lower than the model would suggest. It appears in these regimes that the internal variable of damage is evolving at a different rate from that seen in some other orientations. The reasons for this are not immediately clear and this is an area of ongoing investigation, although further optimisation may improve the fit.

In its current form the model represents a macroscopic approach to the simulation of a two phase structure. This can be viewed as being accounted for by the introduction of separate deformation mechanisms. An area not addressed currently is anisotropy associated with the microstructure itself. Shear stresses parallel or perpendicular to the matrix channels are likely to lead to different levels of constraint and this clearly needs to be taken into account for a more complete description of the deformation process. It has not yet proved possible to integrate the slip system model with microstructural features, but this is an area which further work needs to consider in order to fully simulate the material response.

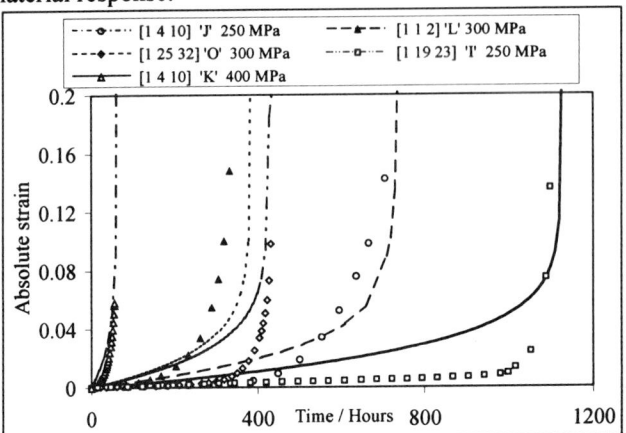

Figure 6: Creep test data and model simulations in various unstable orientations.

Biaxial creep testing can be characterised by two parameters, the equivalent stress and the stress angle [9], given by:

$$\sigma_{eq} = \left[\sigma^2 + \left(\sqrt{3}\tau\right)^2\right]^{\frac{1}{2}} \qquad \theta = tan^{-1}\left(\frac{\sigma}{\sqrt{3}\tau}\right) \tag{11}$$

In these equations σ_{eq} is the equivalent stress, θ is the stress angle, σ in normal stress and τ is shear stress. For CMSX-4 tests were done at an equivalent stress of 360 MPa and stress angles of 90, 30 and 0 degrees. Additional tests were also performed on another single crystal alloy SRR99 at an equivalent stress of 270 MPa and at stress angles of 90, 53, 30 and 0 degrees. For a given equivalent stress the results show that the materials are least resistant in tension, the resistance increases up to a certain combination of tension and torsion and then decreases back to pure torsion which has a resistance slightly higher than pure tension (figure 8). Comparable results have been found previously for the biaxial behaviour of a similar single crystal superalloy [9].

When considering the behaviour of single crystals subjected to biaxial tension / torsion creep there are two primary factors which need to be taken into account. The first is the change in cumulative shear strain rate on the relevant slip systems as the stress angle is changed progressively from tension to torsion, the second is the change in shear strain rate, for a given stress angle, as one rotates around the surface of the thin cylinder. With respect to the latter the shear strain rate remains the same for pure tension but changes around the circumference if a torsional stress is present. The circumferential angle around the surface of the test cylinder, starting at a point in a cubic crystallographic direction is denoted [ϕ] Figure 7 shows the predicted variation with θ of the combined shear strain rate derived from all slip systems, for each of the type I and type II slip systems, and for different values of the parameter ϕ. These profiles are obviously dependent on the stress exponents u_{101} and u_{112}. For the values these exponents take of 3.2 and 3 respectively and it can be seen that for those values there is little effect of ϕ. The cumulative shear strain rate on both types of slip system, decreases with decreasing θ, for any value of ϕ. This would suggests that creep resistance in torsion should be stronger than that in tension, however as mentioned previously the resistance is highest for some combination of tension and torsion (stress angle approximately 30 degrees).

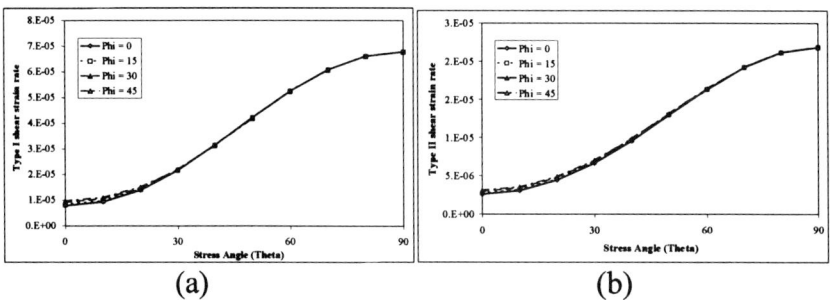

(a)   (b)

Figure 7: Plot of the cumulative shear strain rate, on the type I, figure (a), and type II, figure (b), systems as a function of the stress angle for different values of the parameter ϕ.

The situation in progressing from tension to torsion is similar to that which occurs in uniaxial creep when the tensile orientation changes from [001] to [111] – the type I Schmid factors decrease but the strength does not increase accordingly. In terms of modelling the behaviour the same phenomenology can be applied to the two situations. In tension, deformation along [001] occurs on type I slip systems and the rate of such deformation is sufficient to harden the type II slip systems. As the stress angle is reduced, and the torsional stress component increased, the creep rate initially decreases due to the decreasing type I shear strain rate. With further reduction in stress angle the shear strain rate on type I slip systems becomes too low to permit hardening of the type II slip systems, which become active. As the stress angle is reduced to zero degrees the type II slip systems experience less hardening and the creep resistance decreases.

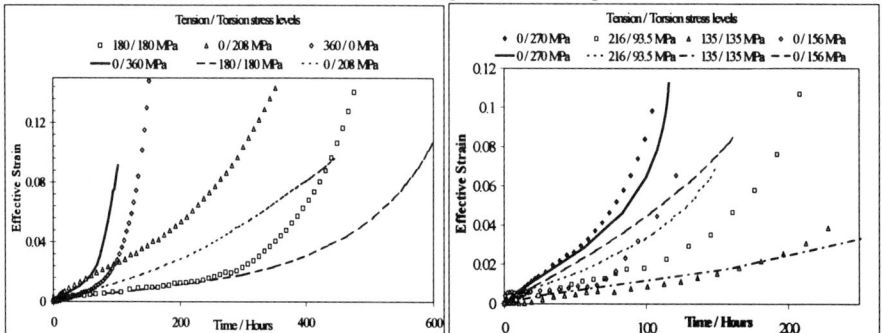

Figure 8: Comparison of biaxial test behaviour and model simulations. The effective strain is plotted against time for different stress angles and constant effective stresses, symbols are test data and lines are model results. (a) shows CMSX-4 data at an effective stress of 360 MPa, (b) shows SRR99 test data at an effective stress of 270 MPa.

The hardening functions and parameters obtained from fitting the model to uniaxial data at 1223 K have been found sufficient to give reasonable predictions of creep behaviour under combined tension / torsion testing. The comparison between model simulations and available test data for SRR99 and CMSX-4 are shown in figure 7.

6. Conclusions

A slip system model has been developed to analyse creep of CMSX-4 and SRR99 at 1223 K. The model incorporates type I and type II slip, experimental evidence indicated that deformation near the [001] pole is due primarily to the former and deformation near the [111] pole is due to the latter.

Microstructural evidence suggests the formation of a dislocation network in the matrix causes hardening of the type II slip systems by preventing them from entering and leaving the precipitates. This phenomena has been accounted for in the model through the use of hardening equations. This allows a consistent description of creep in the [001] and [111] poles and also describes creep in unstable orientations in the centre of the stereographic triangle to a good degree of accuracy.

Thin cylinder biaxial creep behaviour has been predicted using the uniaxial model developed at 1223 K. From tests the material is weakest in pure tension, resistance increases to a maximum during combined tension and torsion after which it decreases to pure torsion, which is approximately a factor of two stronger than pure tension. The reasons for this success are due to the similarity in relative activation of type I and type II slip systems for the stress angle changing from tension to torsion to that which occurs when the uniaxial stress orientation moves from [100] to [111].

7. Acknowledgements

The authors would also like to thank Rolls-Royce plc, DERA and EPSRC for funding this work.

8. References

1. T. M. Pollock, A. S. Argon, "Creep Resistance of CMSX-3 Nickel Base Superalloy Single Crystals," Acta metall. mater., 40 (1) (1992), 1-30.
2. R. N. Ghosh, R. V. Curtis, M. McLean, "Creep Deformation of Single Crystal Superalloys - Modelling the Crystallographic Anisotropy," Acta metall. mater., 38 (10) (1990), 1977-1992.
3. P. E. McHugh, R. Mohrmann, "Modelling of Creep in a Ni base superalloy using a single crystal plasticity model," Computational Materials Science, 9 (1997), 134.
4. M. Kolbe, A. Dlouhy, G. Eggeler, "Dislocation reactions at γ/γ' interfaces during shear creep deformation in the macroscopic crystallographic shear system (001)[110] of CMSX-6," Materials Science & Engineering A: Structural Materials: Properties, Microstructure and Processing, A246 (1998) 133-142.
5. G.R. Leverant and B.H. Kear, "The Mechanism of Creep in Gamma Prime Precipitation-Hardened Nickel-Base Alloys at Intermediate Tempratures" Metall. Trans. A, 1, (1970) 49-498.
6. D. Peirce, R. J. Asaro, A. Needleman, "An Analysis of Non-uniform and Localised Deformation in Ductile Single Crystals," Acta Metall., 30 (1982), 1087-1119.
7. D. W. MacLachlan et al, "Constitutive Modelling of Anisotropic Creep Deformation in Single Crystal Blade Alloys SRR99 and CMSX-4", Int. J. Plasticity, in press.
8. D. W. MacLachlan and D. M. Knowles, "Creep Behaviour Modelling of Singel Crystal Superalloy CMSX-4", Met Trans A, in press.
9. N. Ohno, T. Takeuchi, "Anisotropy in Multiaxial Creep of Nickel-Based Single-Crystal Superalloy CMSX-2: Experiments and Indentification of Active Slip Systems," JSME International Journal, Series A, 37 (2) (1994).
10. K.Kakehi, T. Sakaki, J.M. gui and Y. Misaki, " The influence of orientation and heat treatment on creep strngth of single crystal superalloys", Creep and Fracture of Engineering Structures, Ed . B. Wilshire, Swansea, Institute of Materials (1993), 221-230.

A RATE DEPENDENT FORMULATION FOR VOID GROWTH IN SINGLE CRYSTAL MATERIALS

E.P. BUSSO, N.P. O'DOWD and R.J. DENNIS
Department of Mechanical Engineering
Imperial College
London SW7 2BX, UNITED KINGDOM

Abstract

The presence of casting defects within Ni-base single crystal superalloy components in gas turbine engines is known to lead to the nucleation of microcracks during service. In this work, a micromechanics-based formulation is proposed to describe the growth of initially spherical defects within single crystals under different multiaxial stress states. The framework provides an explicit link between the mesoscopic (at the level of the voids) and the macroscopic (at the component level) length scales. A recently proposed finite strain rate-dependent crystallographic theory is used to describe the constitutive behaviour of the single crystal. The effects of material anisotropy, stress state, temperature and interaction with a free surface on the growth rate of a single void are quantified from detailed finite element analyses of a representative material volume containing a void. Based on these results, a meso-mechanics formulation of void growth for single crystal materials is presented.

1. Introduction

Gas turbine blades are generally subjected to severe thermo-mechanical loading conditions which often lead to cracks initiating in regions where inelastic deformation is highly localized. In Nickel-base single crystal superalloys, whose microstructures consist of hard (γ') precipitates dispersed in a soft (γ) matrix, a major contributing factor to the localisation of the inelastic deformation is the heterogeneity of the material microstructure, in particular inclusions and solidification related porosities. Such porosities are typically found in inter-dendritic regions and are widely spaced. The aim of this work is to develop assessment tools for the life prediction of single-crystal superalloy components operating under typical service conditions. An integrated meso-macro constitutive framework has been de-

41

S. Murakami and N. Ohno (eds.), IUTAM Symposium on Creep in Structures, 41–50.

veloped which enables the presence of 10–20 μm diameter casting defects at the mesoscale to be accounted for at the macroscopic level.

2. Macroscopic Crystallographic Formulation

A rate dependent crystallographic formulation has recently been proposed to describe the average macroscopic stress-strain behaviour of the super-alloy single crystal (SC) CMSX4 (Busso et al., 2000). It relies on the multiplicative decomposition of the total deformation gradient, $\mathbf{F}$, into an inelastic component, $\mathbf{F}^p$, associated with pure slip while the lattice remains undistorted and unrotated, and an elastic component, $\mathbf{F}^e$, which accounts for the elastic stretching and rigid-body rotations (Asaro and Rice, 1977). The flow and evolution equations for the active slip system are based on those proposed by Busso & McClintock (1996). The flow rule relies on a stress-dependent activation energy expressed in terms of two internal state variables per slip system, α: a macroscopically average slip resistance S^α, and a back or internal stress B^α. Thus,

$$\dot{\gamma}^\alpha = \dot{\gamma}_o \exp\left[-\frac{F_o}{k\theta}\left\langle 1 - \left\langle \frac{|\tau^\alpha - B^\alpha| - S^\alpha \mu/\mu_0}{\hat{\tau}_0 \mu/\mu_0} \right\rangle^p \right\rangle^q\right] sgn(\tau^\alpha)\,, \quad (1)$$

where τ^α is the resolved shear stress, θ the absolute temperature, μ, μ_0 the shear moduli at θ and 0 K, respectively, and F_0, $\hat{\tau}_o$, p, q and $\dot{\gamma}_0$ are material parameters.

The evolutionary behaviour of the overall slip resistance is given by,

$$\dot{S}^\alpha = \sum_{\beta=1}^{n_\alpha} \delta_S^{\alpha\beta}\left[h_s - d_D(S^\beta - S^\beta_{S0})\right]|\dot{\gamma}^\beta|\,, \quad (2)$$

where, S^β_{S0} is the initial value of S^β, d_D is a dynamic recovery parameter, and n_α the total number of slip systems. In this SC formulation, an explicit link between the γ' precipitate population at the microscale and the behaviour of the homogeneous equivalent material at the macroscale is incorporated. This link is introduced through the hardening function h_s in Eq. 2, which depends on the characteristics of the current precipitate population. Here,

$$h_s = \hat{h}_s\left\{l/l_m, V_f, a/b, \tilde{\epsilon}^{in}\right\}\,, \quad (3)$$

where l/l_m is the precipitate size normalised by a reference mean value, V_f the precipitate volume fraction, a/b its aspect ratio, and $\tilde{\epsilon}^{in}$ the accumulated uniaxial equivalent inelastic strain at the macroscale. Equation 3 has been calibrated from FE analyses of periodic unit cells at the microscale containing the individual precipitates (Meissonnier *et al.*, 1999). In Eq. 2, $\delta_G^{\alpha\beta}$ is the latent hardening or interaction function,

$$\delta_G^{\alpha\beta} = \omega_1 + (1 - \omega_2)\delta_{\alpha\beta}. \qquad (4)$$

where $\delta_{\alpha\beta}$ is the Kroneker delta so that $\omega_1 = \omega_2 = 0$ corresponds to self-hardening, and $\omega_1 = \omega_2 = 1$ to Taylor hardening.

The back stress evolves according to ,

$$\dot{B}^\alpha = h_B \dot{\gamma}^\alpha - r_D B^\alpha |\dot{\gamma}^\alpha|, \qquad (5)$$

where h_B is the hardening coefficient, and r_D a dynamic recovery function expressed in terms of the current overall deformation resistance (Busso and McClintock, 1996),

$$r_D = \frac{h_B \mu_o}{S^{(\alpha)}} \left\{ \frac{\mu_0'}{f_c \lambda} - \mu \right\} . \qquad (6)$$

Here, f_c and λ are statistical factors, and μ_0', μ_0 and μ known shear moduli.

2.1. Numerical Integration Procedure

The above crystallographic formulation has been implemented numerically into the finite element (FE) method using a large strain fully implicit algorithm (Busso *et al.*, 2000). The commercial FE code ABAQUS (1999) was used to perform the FE computations. The implementation required the development of a fully implicit finite strain material subroutine to update, at each integration point, the stresses, solution-dependent variables and material jacobian.

The incremental procedure starts with an estimate of the deformation gradient at the end of a generic time increment, $\mathbf{F}_{n+1}$, and the known initial conditions at the beginning, namely $\mathbf{F}_n$, $\boldsymbol{\sigma}_n$, and the solution-dependent variables associated with the inelastic deformation and state, $\{\mathbf{F}_n^p, S_n^\alpha, B_n^\alpha\}$. An implicit time-integration procedure with a single level of iteration is then used to solve the simultaneous incremental non-linear equations associated with the crystallographic formulation. The solution of the incremental problem leads to the update of the inelastic deformation gradient, slip resistances, and back stresses $\{\mathbf{F}_{n+1}^p, S_{n+1}^\alpha, B_{n+1}^\alpha\}$, and the new lattice orientation, $\{\mathbf{F}_{n+1}^e \mathbf{m}^\alpha, \mathbf{n}^\alpha \mathbf{F}_{n+1}^{e-1}\}$, where $\mathbf{m}^\alpha$ and $\mathbf{n}^\alpha$ are the slip system unit vectors defining the slip direction and the slip plane normal, respectively. Furthermore, the jacobian required to iteratively solve the global equilibrium equations, $\partial\sigma_{n+1}/\partial\epsilon_{n+1}$, was analytically determined. For details, see Busso *et al.* (2000).

The constitutive SC formulation, which accounts for both octahedral ($\{111\} < 110 >$) and cubic ($\{110\} < 100 >$) slip, is shown to predict accurately the behaviour of CMSX4 under creep, monotonic, cyclic and thermomechanical loading conditions within the 650°C - 1025°C temperature range at crystallographic orientations. In Figs. 1 and 2, examples of the model predictive capabilies are given.

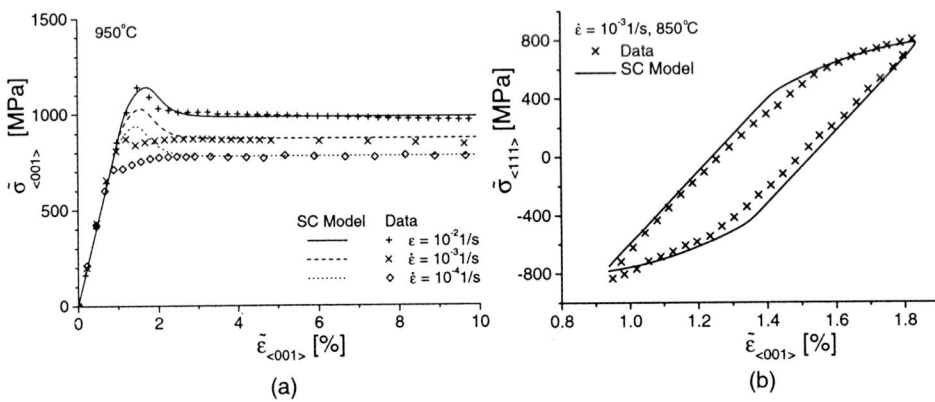

Fig. 1 Comparison between measured and predicted (a) $< 001 >$ monotonic response at different strain rates and 950°C , and (b) $< 111 >$ steady state cyclic response at 850°C

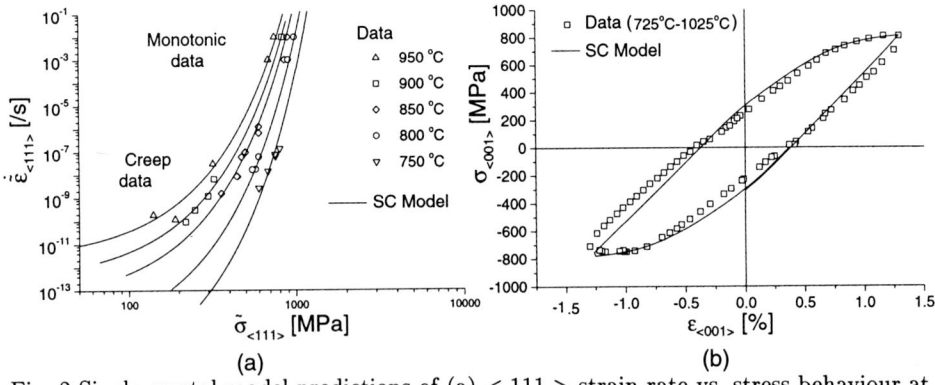

Fig. 2 Single crystal model predictions of (a) $< 111 >$ strain rate vs. stress behaviour at different temperatures, and (b) $< 001 >$ thermo-mechanical load with out-of-phase ± 1.2% strain and 725-1025°C temperature histories

3. Mechanistic Study of Void Growth

The macroscopic crystallographic formulation was used to investigate the deformation of a representative material volume containing an initially spherical void of approximately 10–20 μm diameter. FE analyses were used to determine the functional dependence of the void growth rates on material anisotropy, stress state, temperature and interaction with a free surface. In that way, it is possible to determine the local conditions which give rise to microcracks initiating from the surface of embedded casting defects. Note that generally no crack-like defect are found in superalloy components prior to service. A typical cracked porosity observed in CMSX4 after a 900 hour $< 001 >$ creep test at 950°C is shown in Fig. 3.

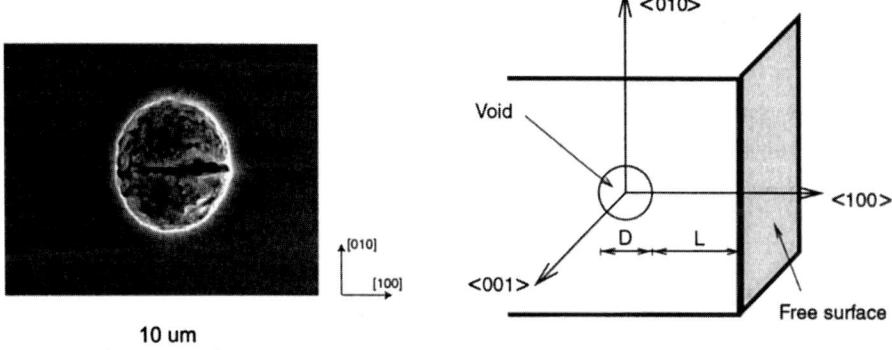

Fig. 3 Typical cracked porosity observed after 931 hrs at 950°C when loaded under creep	Fig. 4 Definition of the length scale L/D for a voided material volume where one of its boundaries is a free surface

The relevant length scales which define the interaction between a single void with a free surface are indicated in Fig. 4. Here, the free surface is assumed to be parallel to the $< 100 >$ crystallographic cubic plane, D is the void diameter, and L the distance between the void and the free surface. Note that due to the fact that these types of voids are widely spaced, interaction between individual voids is not considered here. Therefore, for a fixed $\gamma - \gamma'$ size and volume fraction, the ratio L/D uniquely defines the length scale of the problem.

A typical finite mesh, made up of 4000 3D linear isoparametric elements, is shown in Fig. 5. A contour plot is given in Fig. 6, showing the distribution of the accumulated uniaxial equivalent inelastic strain distribution in a unit cell containing a void with $L/D = 1$. The localisation of the deformation around the void can be clearly seen — note that the maximum accumulated inelastic strain in the vicinity of the void (i.e. 45%) is approximately five times greater than the far field value.

Representative FE results are shown by the symbols in Fig. 6. Here, the void growth rate $\dot{V}$ is normalised by the product of the void volume and the far field uniaxial equivalent strain rate in the material. The results show that the rate of growth of casting defects in an infinite single crystal medium is strongly dependent on the applied triaxiality and the applied stresses.

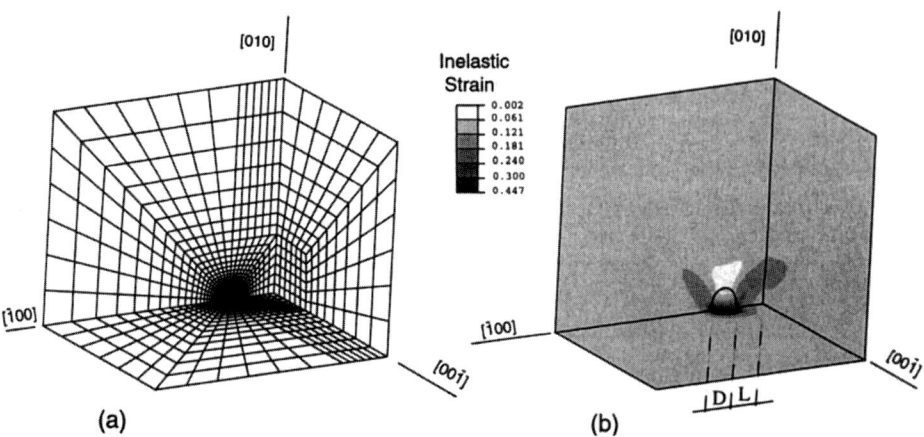

Fig. 5 (a) Typical finite element model for the void-free surface interaction studies, and (b) contour plot of accumulated inelastic strain around a void ($L/D = 1$) after 830 hours at 850°C under a uniaxial far-field < 010 > applied stress of 600 MPa

Furthermore, it was found that (see Fig. 6(b)), for the acceleration of the defect growth rate as the result of its proximity to a free-surface to be non-negligible, the void needs to be within two diameters of the free surface.

A framework has been proposed (Busso et al., 2000b) to describe the growth rate of embedded casting defects within a superalloy single crystal at a given temperature, L/D ratio, relative lattice orientation, and applied multiaxial stress state. The structure of the proposed formulation follows that of Budiansky et al. (1982) for isotropic materials, where a normalised measure of the void growth rate is assumed to have a power dependency on triaxiality.

The contributions to the total volumetric growth rate is assumed to arise independently from each of the two active slip system families; diffusional flow effects were implicitly accounted for in the constitutive model's flow rule. Then,

$$\dot{V} = \left(a_1 \dot{V}_{\text{oct}}^r + a_2 \dot{V}_{\text{cub}}^r\right)^{\frac{1}{r}} \tag{7}$$

where $\dot{V}_{\text{oct}}$ and $\dot{V}_{\text{cub}}$ are the contributions to the overall rate of void growth arising from slip in the octahedral ($\{111\} < 110 >$) and cubic ($\{110\} < 100 >$) slip systems, respectively, and r, a_1 and a_2 are anisotropy-related parameters calibrated numerically from the finite element void calculations (Busso *et al.*, 2000b). Note that the splitting of the void growth contributions into octahedral and cubic parts allows the generalisation of the formulation to an arbitrary orientation.

Individual expressions for $\dot{V}_{\text{oct}}$ and $\dot{V}_{\text{cub}}$ are,

$$\frac{\dot{V}_{\text{oct}}}{V\dot{\epsilon}_{\text{oct}}} = \hat{a}_{\text{oct}}(n, L/D) \left|\frac{\sigma_m}{\sigma_e}\right|^{w_{\text{oct}}}, \tag{8}$$

$$\frac{\dot{V}_{\text{cub}}}{V\dot{\epsilon}_{\text{cub}}} = \hat{a}_{\text{cub}}(n, L/D) \left|\frac{\sigma_m}{\tilde{\sigma}_e}\right|^{w_{\text{cub}}}. \tag{9}$$

Here, V is the void volume, $\dot{\epsilon}_{\text{oct}}$ and $\dot{\epsilon}_{\text{cub}}$ are the uniaxial equivalent inelastic strain rates due to octahedral and cubic slip for a void-free material subjected to the equivalent stress $\tilde{\sigma}_e$, respectively. Also,

$$n = \hat{n}(\theta) = \delta(\log \dot{\tilde{\epsilon}})/\delta(\log \sigma_e), \tag{10}$$

is the material strain rate-sensitivity at the temperature θ, which can be determined directly from the single crystal constitutive law, and

$$w_i = \hat{w}_i(n), \tag{11}$$

$$a_i = \hat{a}_i(\dot{n}, L/D), \tag{12}$$

(for $i = \text{cub, oct}$), are functions calibrated from an extensive parametric study. Further details of the particular form of the above functions in Eqs. 7–11 will be provided in Busso *et al.* (200b).

A comparison between the predictions given by Eq. 7 for different triaxialities, L/D ratios, and applied equivalent stresses, and the accurate FE reference solutions are given in Figs. 6 (*a*) and (*b*).

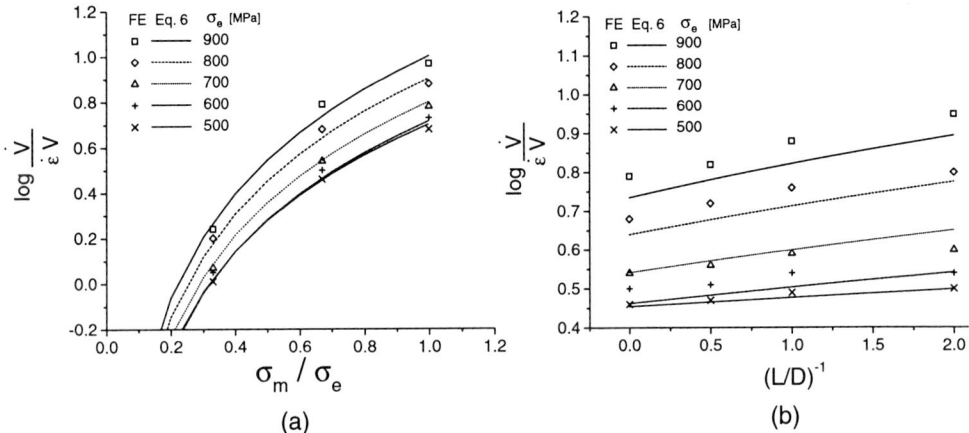

Fig. 6 Effect of (a) mean triaxiality normalised by the equivalent stress, $\sigma_m/\tilde{\sigma}_e$, and (b) the void diameter relative to the distance from the free surface, D/L, on the void growth rate $\dot{V}$

4. Integrated Damage-Deformation Constitutive Framework and Life Prediction Methodology

A life assessment methodology combining the SC model and the void growth formulation has been developed. The approach assumes that the void growth is controlled by the local stresses but that, due to the discrete nature and small size of the voids, the macroscopic stresses are unaffected by the voids.

A failure criterion for the initiation of a microcrack from the surface of the void was defined in terms of a critical inelastic strain in the vicinity of the void. This critical value can in turn be linked to a critical void volume. The failure criterion was calibrated from failed blunt notched specimens tested under creep loading in conjunction with FE analyses. The diameters of cracked pores in the notch centreline of failed specimens were recorded and the local triaxiality computed from 3D FE analyses of the test specimens. A unit cell FE analysis of a void subjected to a similar triaxiality was then performed till it reached the critical volume at which the cracks were observed to initiate. The resulting maximum accumulated inelastic strain near the void surface was then used as a critical magnitude to post-process all other unit cell analyses, and the void volumes at which this critical failure strain was reached, recorded. The results are summarised in Fig. 7. Here, the different symbols at the same relative triaxiality level correspond to different magnitudes of the applied stress. These results reveal that the

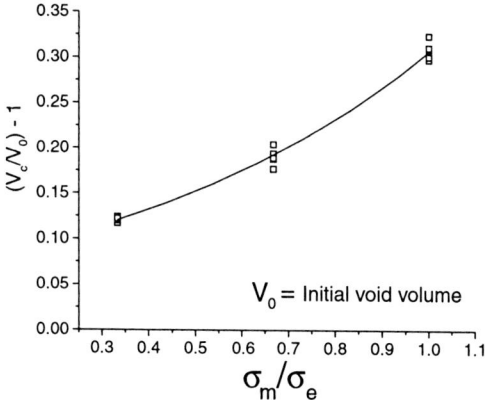

Fig. 7 Effect of triaxiality on the critical void volume for microcrack nucleation

critical void volume is strongly dependent on triaxiality and weakly on the magnitude of the macroscopic stress.

In a structural integrity analysis, the void growth rates can be calculated directly from Eq. 7 and the failure condition identified from Fig. 7. This procedure has been incorporated into the user-defined SC material subroutine to enable macroscopic failure predictions in actual components to be made.

5. Acknowledgments

Financial support for this work by the EPSRC (UK) through grant GR/K73688 and the provision of the CMSX4 test data by ALSTOM (UK) are gratefully acknowledged. The ABAQUS program was provided under academic license by HKS Inc, Providence, Rhode Island.

6. References

1. Busso, E.P., Dennis, R. and O'Dowd, N.P. (2000). A Crystallographic rate-Dependent Formulation for Superalloy Single Crystals with Microstructural Length Scales. Submitted for publication.
2. Asaro, R. J. and Rice, J. R. (1977). Strain Localization in Ductile Single Crystals. *J. Mech. Phys. Solids.* **25**, 309–338.
3. Busso, E. P. and McClintock, F. A. (1996), A Dislocation Mechanics-Based Crystallographic Model of a B2-Type Intermetallic Alloy, *Int. J. Plasticity*, **12**, 1-28.
4. Meissonnier, F. T., Busso, E. P. and O'Dowd, N. P. (1999), Finite Element Implementation of a Non-Local Visco-Plastic Crystallographic Formulations. Accepted for publication in the *Int. J. Plasticity.*.

5. ABAQUS (1999), HKS Inc., Providence, Rhode Island, USA.

6. Busso, E.P., Dennis, R. and O'Dowd, N.P., (2000b), A Micromechanics Study of Void Growth in Single Crystal Materials. Submitted for publication.

7. Budiansky, B., Hutchinson, J.W. and Slutsky, S., (1982), Void Growth and Collapse in Viscous Solids. In *Mechanics of Solids: Rodney Hill 60th Anniversary*, V. 13.

MICROSTRUCTURAL MODELING OF CREEP FRACTURE IN POLYCRYSTALLINE MATERIALS

P.R. ONCK, B.-N. NGUYEN & E. VAN DER GIESSEN
Delft University of Technology, Koiter Institute Delft
Mekelweg 2, 2628 CD Delft, The Netherlands

Abstract

This paper is concerned with a recent microstructural approach to model creep crack growth. The model spans three different length scales, from the scale of individual cavities, through the grain scale up to the macroscopic scale of cracks in components and test specimens. In order to study the initial stages of creep crack growth, we consider a near-tip process window in which a large number of grains is represented discretely. This window is surrounded by a standard continuum. Macroscopic specimen dimensions and loading configuration are communicated to this near-tip region by applying boundary conditions in accordance with higher-order asymptotic stress fields for power-law creeping materials. The paper presents an overview of some recent results obtained with this type of modeling.

1. Introduction

High-temperature failure of polycrystalline metals (see Riedel, 1987, for an overview) spans a wide range of length scales, as illustrated in Fig. 1. At the largest, i.e. macroscopic, scale we consider a component or test specimen containing a crack (Fig. 1a), while the smallest relevant scale is that of the key failure mechanism, i.e. the nucleation and growth of small cavities along the grain boundaries (Fig. 1e). The intermediate scales determine how these elemental mechanisms lead to growth of the macroscopic crack. At the second smallest length scale (Fig. 1d), we observe the individual grains in the material and the distribution of cavitation damage along its grain boundaries. This is also the scale where two other key mechanisms are operating, namely creep of the grains themselves and sliding of adjacent grains relative to each other. Coalescence of the cavities after sufficient growth leads to microcracks along the grain facets, and at the next larger length scale we are concerned with the distribution of these microcracks near the tip of the crack (Fig. 1c). Growth of the crack at this scale occurs by the linking-up of facet microcracks with the main crack. Zooming out further (Fig. 1b) brings us to the size scale at which we are no longer able to distinguish the grain microstructure of the material; what remains is that we can identify a zone in the neighborhood of the crack tip in which damage occurs. This damage zone is surrounded by a zone of material that is not damaged and which only creeps (Fig. 1b).

S. Murakami and N. Ohno (eds.), IUTAM Symposium on Creep in Structures, 51–64.
© 2001 *Kluwer Academic Publishers. Printed in the Netherlands.*

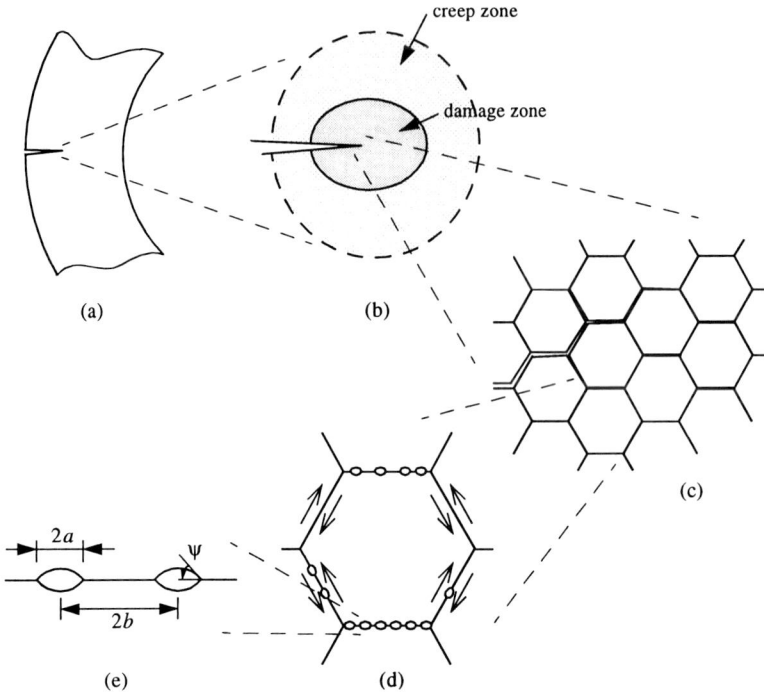

Figure 1. Size scales involved in creep fracture: (a) macroscopic crack; (b) crack-tip neighbourhood with creep and damage zones; (c) mesoscopic near-tip region inside damage zone; (d) individual grains and grain boundary damage; (e) microscopic grain boundary cavities.

A thorough understanding of creep fracture requires that all relevant length scales are bridged. This poses an enormous challenge, since there are various competing mechanisms involved, each having its own characteristic time scale and dependence on stress and microstructure. Approaches in the literature have either started from the macroscopic scale (Figs. 1a,b) or from the opposite microscopic scales (Figs. 1d,e). The more engineering, macroscopic approaches include non-linear fracture mechanics and continuum damage modeling (e.g. Hayhurst *et al.*, 1984; Tvergaard, 1986). On the other hand, micromechanical studies have focused on the basic damage processes of cavity growth (e.g. Needleman and Rice, 1980) and grain boundary sliding (e.g. Ghahremani, 1980), and their combination (e.g. Van der Giessen and Tvergaard, 1991). The missing link is at the size scale depicted in Fig. 1c.

Onck and Van der Giessen (1998, 1999) have proposed a computational procedure to forge this missing link. The approach adopts a two-dimensional microstructural model of the material by individually representing a large number of grains surrounding the propagating crack. Cavitation and sliding along all grain boundaries in this aggregate is considered, and is described by a set of constitutive equations based on micromechanical studies at the smaller scales (Figs. 1d,e). A small-scale damage formulation is used, so that loading, specimen geometry and crack length are fully communicated through the boundary conditions, according to higher-order, mode I near-tip fields in creeping materials.

Such a model allows to investigate how the crack growth process (including the linking-up mechanism, crack growth rate and direction) is affected by *microstructural* features such as the grain boundary diffusivity, cavity nucleation, grain size and initial crack tip radius. Moreover, due to the small-scale damage assumption, also the effect of *macroscopic* features, such as loading, crack length and specimen type can be analyzed through the near-tip field parameters. The objective of the present paper is to review some of these issues. The results are compiled from recent papers that deal with isolated aspects of the problem. Here they are pulled together to provide a more coherent picture.

The paper is structured as follows. Section 2 opens with a formulation of the small-scale damage problem, followed by a description of the microstructural model, featuring grain elements and grain boundary elements, together with an overview of the mesoscopic constitutive equations for grain deformation and for grain boundary cavitation and sliding (Section 3). In Section 4 all loading and material parameters entering the problem are formulated in dimensionless form. Finally, in Section 5 results are presented concerning the effect of a number of microstructural as well as macroscopic (loading) parameters on the crack growth process.

2. Small-scale damage formulation

We consider a region near the tip of a pre-existing macroscopic crack in a particular specimen that is subjected to mode I loading. We assume that it is contained well within the creep zone shown in Fig. 1b but is much larger than the damage zone, i.e. we consider small-scale damage conditions. Outside the damage zone, the material deforms elastically and by dislocation creep. In this region (see Fig. 2a), it is appropriate to represent the polycrystalline material by a continuum giving a description averaged over many grains. Inside the anticipated damage zone or process zone (Fig. 2b), all grains making up the aggregate are represented individually, assuming that they have the same hexagonal shape. Symmetry is assumed with respect to the crack plane, allowing to analyze only half the near-tip domain. Almost all results to be presented are for an initially sharp crack (as shown in Fig. 2), but also an initially blunt crack with a crack tip radius equal to $12\sqrt{3}R_0$ ($2R_0$ is the grain boundary facet length) is considered.

Under steady-state conditions, the material sufficiently far away from the damage zone deforms by creep only. Assuming an isotropic response, the creep rate is given by the Norton power-law

$$\dot{\epsilon}_e^C = B\sigma_e^n, \tag{1}$$

with B the creep parameter, n the creep exponent and σ_e the Mises stress. At the macroscopic scale, creep fracture mechanics has often relied on the application of the Hoff analogy by which the HRR-solution for power-law creeping materials can

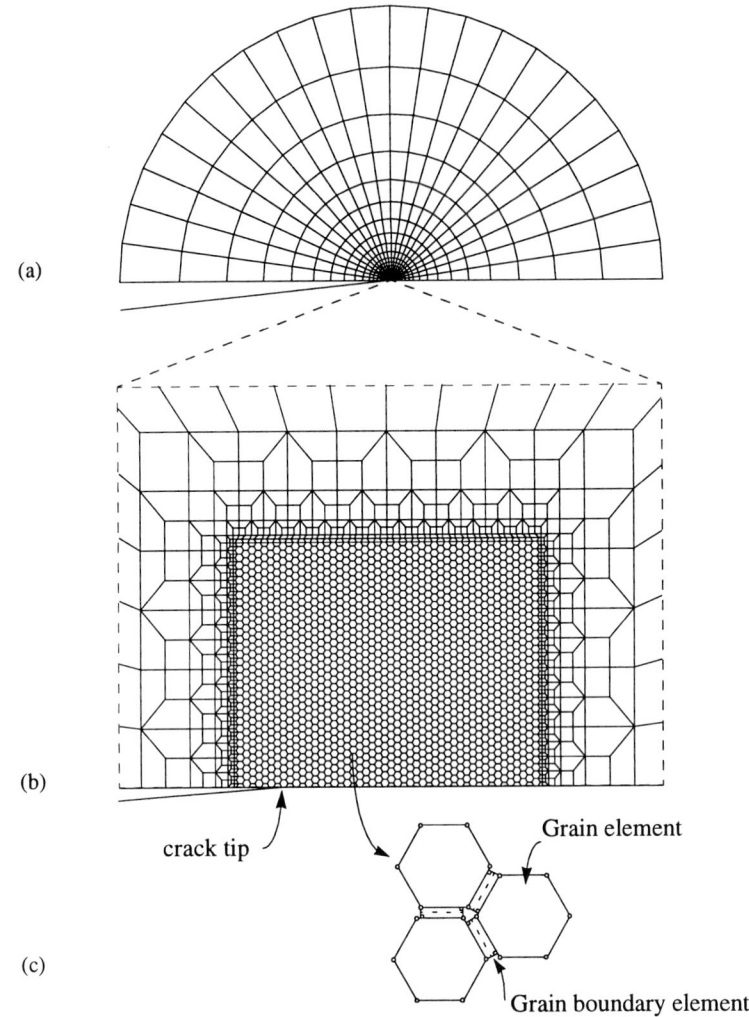

Figure 2. Finite element mesh used for small-scale damage analysis. (a) Semi-circular domain. (b) Crack tip region, showing the process window which consists of (c) grain elements and grain boundary elements. Cavitation damage is allowed to occur along all grain boundaries in the process window.

be obtained by replacing the J-integral by the C^*-integral,

$$\sigma_{ij}(r,\theta) = \left(\frac{C^*}{BI_n r}\right)^{1/(n+1)} \tilde{\sigma}_{ij}^{(0)}(\theta,n), \tag{2}$$

with (r,θ) polar coordinates measured from the crack tip and I_n a dimensionless function dependent on n. At increasing distance away from the crack tip ($r = 0$), the actual fields in, for example, a test specimen will differ from the HRR fields.

An improved description can be provided by including higher-order terms in the asymptotic expansion. The higher-order terms have recently been established by Nguyen *et al.* (2000a) through an asymptotic analysis similar to that of Yang *et al.* (1993) for elastoplastic bodies. After truncation of the series expansion after the third term, the stress field is then obtained as (Nguyen *et al.*, 2000a)

$$\sigma_{ij}(r,\theta) = \left(\frac{C^*}{BI_n\Lambda}\right)^{\frac{1}{n+1}} [\left(\frac{r}{\Lambda}\right)^{\frac{-1}{n+1}} \tilde{\sigma}_{ij}^{(0)}(\theta,n) +$$
$$A_2^* \left(\frac{r}{\Lambda}\right)^{p} \tilde{\sigma}_{ij}^{(1)}(\theta,n) + A_2^{*2} \left(\frac{r}{\Lambda}\right)^{q} \tilde{\sigma}_{ij}^{(2)}(\theta,n)], \qquad (3)$$

where $\Lambda = C^*/B\sigma_\infty^{n+1}$ is a material and loading-dependent length scale which straightforwardly follows from dimensional considerations. The stress σ_∞ is a measure for the applied loading. In (3) p, q as well as $\tilde{\sigma}_{ij}^{(1)}(\theta,n)$ and $\tilde{\sigma}_{ij}^{(2)}(\theta,n)$ are the same exponents and dimensionless functions as determined for elastoplastic materials in (Yang *et al.*, 1993) and (Chao and Zhang, 1997), respectively. The nondimensional functions $\tilde{\sigma}_{ij}^{(0)}(\theta,n)$ in the leading term (see also eq. (2)) are given by Shih (1983). The second and third terms in (3) are controlled by the same parameter A_2^*, which is independent of specimen size and load level, and a function only of n, the specimen shape and loading configuration (Nguyen *et al.*, 2000a). When $A_2^* = 0$, (3) reduces to the HRR solution (2). Clearly, (3) constitutes a three-parameter characterization of the steady creep fields in terms of C^*, A_2^* and σ_∞. The fact that the corresponding elastoplastic fields of Yang *et al.* (1993) only have two parameters J and A_2 is caused by the fact that the role of σ_∞ is then taken over by the yield stress, which is absent in steadily creeping materials, cf. (1).

3. Microstructural model

The problem outlined above is analyzed using an incremental, finite strain, finite element model involving two discretizations: one for the material inside the process zone and one outside this zone. Details of the procedure may be found in (Onck and Van der Giessen, 1999); it suffices here to only reiterate the key ingredients.

The region in Fig. 2a outside the damage zone is described by a standard continuum and discretized with standard continuum finite elements. The continuum constitutive equations are assumed to represent the average behaviour of an aggregate of grains. Hence, isotropic elasticity and dislocation creep, as prescribed by (1), are assumed. Each grain in the process zone (see Fig. 2b) is represented by a single, so-called grain element, while the grain boundary facets are treated by special-purpose grain-boundary elements (see Fig. 2c). The grain elements used here are, in effect, super-elements with six nodes per element (Onck and Van der Giessen, 1997). Each grain element accounts for the elastic and creep behaviour of an individual grain. For simplicity, we neglect any grain anisotropy, so that the constitutive equations for the grain material are also of the form (1).

The grain-boundary elements are designed to incorporate the relevant mechanisms that take place inside the grain boundaries, i.e. grain boundary sliding and

grain boundary cavitation. The viscous grain boundary sliding is described by a
Newton viscous law with a viscosity η_B expressed in terms of the strain-rate like
parameter $\dot{\epsilon}_B$:

$$\dot{\epsilon}_B = \left(\frac{w}{d_0} \frac{B^{-1/n}}{\eta_B} \right)^{n/(n-1)}, \tag{4}$$

with w the thickness of the boundary and d_0 the effective grain size, $d_0 = 3.64 R_0$.
The cavitation process is governed by the nucleation, growth and coalescence of
cavities. The average separation between two adjacent grains is specified by $\delta_c = V/(\pi b^2)$ where V is the cavity volume, determined by the cavity radius a, and the
cavity half-spacing b. By a smeared-out approach where a continuous variation of
δ_c is used instead of its discrete distribution, the rate of change of δ_c,

$$\dot{\delta}_c = \frac{\dot{V}}{\pi b^2} - \frac{2V}{\pi b^2} \frac{\dot{b}}{b}, \tag{5}$$

is determined by the volumetric growth rate $\dot{V}$ and the rate of change of b. Based
on the results of micromechanical studies of void growth by coupled grain boundary
diffusion and creep by Needleman and Rice (1980), Sham and Needleman (1983)
and Tvergaard (1984), $\dot{V}$ is given by

$$\dot{V} = \dot{V}_{\text{dif}}(\mathcal{D}, \sigma_n, a, b, L) + \dot{V}_{\text{cr}}(\sigma_m, \sigma_e, a), \tag{6}$$

where $\dot{V}_{\text{dif}}$ and $\dot{V}_{\text{cr}}$ are the contributions by diffusion (with grain boundary diffusion parameter $\mathcal{D}$) and creep, respectively. Here, σ_n is the average stress normal
to the grain boundary, and σ_m and σ_e are the average mean and Mises stresses, respectively, remote from the cavities. The coupling between diffusive and creep con-
tributions to cavity growth is governed by the length parameter $L = [\mathcal{D}\sigma_e/\dot{\epsilon}_e^C]^{1/3}$
(Needleman and Rice, 1980). Cavity growth is dominated by diffusion for rela-
tively large values of L, while creep gives a larger contribution for smaller values
of L. The volumetric growth rate $\dot{V}$ is used to compute the rate of change of
the cavity size as $\dot{a} = \dot{V}/(4\pi a^2 h(\psi))$ with $h(\psi) = [(1 + \cos\psi)^{-1} - \frac{1}{2}\cos\psi]/\sin\psi$,
assuming that the cavity tip angle ψ is maintained by rapid surface diffusion.

The variation of cavity spacing b during the damage process is expressed as a
function of the change of the cavity density N through cavity nucleation:

$$\frac{\dot{b}}{b} = -\frac{1}{2} \frac{\dot{N}}{N}. \tag{7}$$

Nucleation is taken to be governed by (Onck and Van der Giessen, 1999)

$$\dot{N} = F_n \, (\sigma_n/\Sigma_0)^2 \, \dot{\epsilon}_e^C, \qquad \text{for } \sigma_n > 0 \qquad \text{and } S > S_{\text{thr}}. \tag{8}$$

Here, F_n is a material parameter for the nucleation activity, Σ_0 a reference stress
(introduced solely for normalization purposes), and S is the cumulative quantity
$S = (\sigma_n/\Sigma_0)^2\epsilon_e^C$. The threshold value S_{thr} for cavity nucleation to occur is taken
to be $S_{\text{thr}} = N_I/F_n$ with N_I the initial cavity density.

During the computation, the expressions for $\dot{a}$ and $\dot{b}$ are integrated to give
the current cavity radius a and spacing b. When the ratio a/b approaches unity,

coalescence of cavities occurs; here, we use $a/b = 0.7$ to signal coalescence. Once this has occurred, the grain boundary facet has lost its stress-carrying capacity and the grain-boundary element henceforth represents a facet microcrack.

4. Nondimensional parameters

As we shall present all results in nondimensional form, all parameters that govern the problem will also be specified nondimensionally. All lengths are scaled with the initial half-width of the grain facets, R_0, while cavity densities are normalized by $N_R = 1/(\pi R_0^2)$. Stresses are normalized with the reference stress Σ which is taken to be the magnitude of the HRR stress field at a distance R_0 from the tip: $\Sigma = (C^*/BI_nR_0)^{1/(n+1)}$. The reference time $t_{R0} = 1/\dot{E}_e^C$, based on $\dot{E}_e^C = B\Sigma^n$, is used to normalize time.

All cases to be presented have used $n = 5$ and $\nu = 0.3$, and are for a crack loading C^* specified through $\Sigma/E = 0.9 \times 10^{-3}$, so that elastic deformations remain small. The grain boundary viscosity η_B is specified in terms of the ratio $\dot{E}_e^C/\dot{\epsilon}_B$. Here, we have used $\dot{E}_e^C/\dot{\epsilon}_B = 10$. Free sliding has been discussed by Onck and Van der Giessen (1999). The density and size of freshly nucleated cavities is taken according to $N_I/N_R = 40$ and $a_I/R_0 = 0.67 \times 10^{-3}$, respectively, while $N_{max} = 100N_R$ in all cases. The two main parameters that govern nucleation and diffusive growth are F_n and $\mathcal{D}$ (Σ_0 in (8) is arbitrarily chosen equal to Σ), and they are specified in terms of F_n/N_R and L_R/R_0, respectively, with L_R defined as $L_R = (\mathcal{D}\Sigma/\dot{E}_e^C)^{1/3}$. A relatively large value of L_R/R_0 (here, $L_R/R_0 = 0.1$), implies that cavity growth occurs predominantly by diffusion, with creep strains remaining small so as to lead to 'brittle' fracture. For L_R/R_0 smaller (here, $L_R/R_0 = 0.032$), creep deformations increase (while cavity growth may still be diffusion dominated) and fracture is more 'ductile'. In the present study, most calculations are shown for $L_R/R_0 = 0.032$, while also results are shown for $L_R/R_0 = 0.1$, leading to more brittle, creep-constrained crack growth (Dyson, 1976). The value of the nucleation activity F_n is taken to be coupled to that of $\mathcal{D}$ so that the maximum density N_{max} is approximately reached at the moment of cavity coalescence.

5. Results

The non-dimensional parameters associated with ductile fracture as specified in the previous section, define a reference case. Results will be shown of this reference case, after which we will study the effect of (i) the parameter L_R/R_0 ($L_R/R_0 = 0.1$, leading to more brittle, creep-constrained crack growth), (ii) increasing the grain size (by changing R_0), (iii) randomizing the distribution of F_n, (iv) a blunted initial crack tip geometry, and (v) higher-order near-tip fields by analyzing nonzero values of A_2^* in (3).

We start out with the reference case. Figure 3 shows snapshots of the grain boundary cavitation state and the Mises effective stress distribution near the crack tip in the process window (see Fig. 2b) at two instances during early crack extension. The cavitation state is shown by plotting the value of a/b perpendicular to each facet and with the ordinate along the facet. Wherever microcracking has

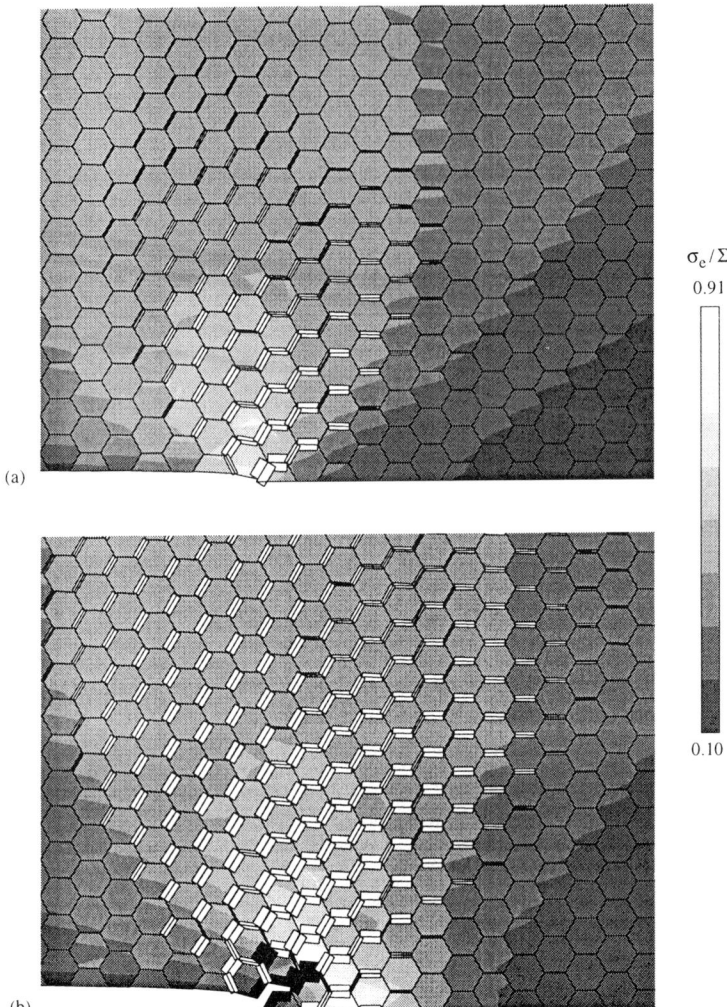

σ_e/Σ

0.91

0.10

Figure 3. Grain boundary cavitation state and Mises effective stress distribution in the process window (see Fig. b) for $L_R/R_0 = 0.032$ at two instances during early crack extension. Values of a/b are plotted along and on either side of the facets. Microcracked facets, where $a/b = 0.7$, are highlighted in black. (a) $t/t_{R0} = 0.31$; (b) $t/t_{R0} = 0.64$. [From (Onck and Van der Giessen, 1999)]

occurred due to cavity coalescence at $a/b = 0.7$, this is highlighted in black. Figure 3a corresponds to $t/t_{R0} = 0.31$. At this stage, damage has not changed the stress state very drastically, so that still the initial HRR effective stress contours above the crack tip can be observed. Cavity nucleation has taken place on facets above the crack, in a region where creep deformation is dominant. In this region, two families of cavitating facets can be distinguished: one group of horizontal

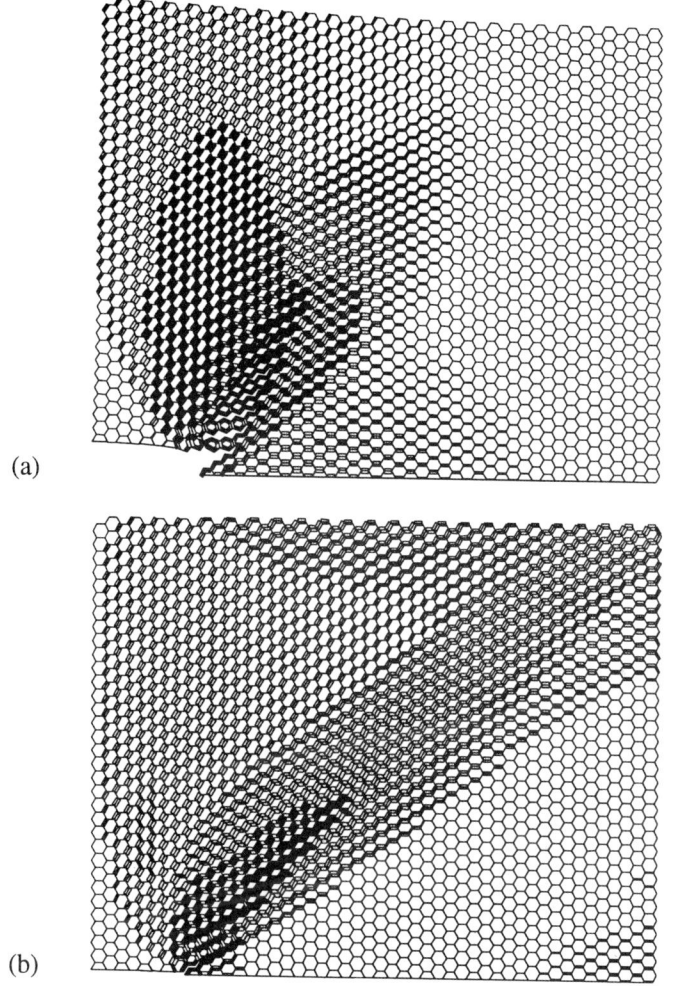

(a)

(b)

Figure 4. Grain boundary cavitation state in the process window for the sharp crack analysis. (a) $L_R/R_0 = 0.032$, $t/t_{R0} = 1.86$ (b) $L_R/R_0 = 0.1$, $t/t_{R0} = 0.14$ [From (Onck and Van der Giessen, 1999)].

facets above and in front of the crack tip and one group of facets inclined at $-30°$, located above and behind the crack tip. This agrees with the character of the initial HRR-field, thus confirming that nucleation, according to (8), occurs first in a region where creep activity is dominant and at facets where normal stresses are highest. In Fig. 3b the stress and damage state is plotted for $t/t_{R0} = 0.64$. Cavity coalescence near the crack tip has led to several stress-free microcracks, which have linked up with the main crack. The new crack-tip position is clearly marked by the stress peak.

To follow the subsequent crack extension, we have zoomed out somewhat in Fig.

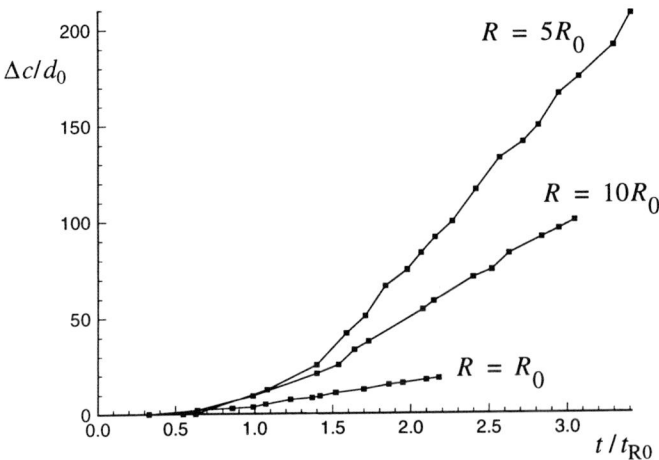

Figure 5. Crack growth increment vs. time, normalized with respect to the reference case $R = R_0$. [From (Onck and Van der Giessen, 1998b)].

4a, so that the complete process window (cf. Fig. 2b) can be distinguished. The crack propagates roughly along a 60° direction with the crack plane. Multiple microcracks have developed above as well as in front of the crack tip. Again, two families can be clearly distinguished: horizontal microcracks, more dominant in front of the crack, and inclined microcracks above the crack. It is seen that the actual crack growth takes place in the zone where both families meet. In this zone, regions of linked microcracks can be observed, which have a length of approximately three grain diameters and are oriented at 30° with the crack front. The actual 60° macrocrack propagates by the joining of the 30° regions of linked microcracks.

(i). The next case corresponds to a larger value of the grain boundary diffusivity $\mathcal{D}$, leading to $L_R/R_0 = 0.1$. Figure 4b corresponds to $t/t_{R0} = 0.14$. The same two families of cavitating facets can be distinguished as in the previous case, but comparison with Fig. 4a reveals that microcracking has occurred in a localized band along the 50° direction with the crack plane. Clearly, fracture is more brittle. This can be directly related to the higher value of the diffusion coefficient $\mathcal{D}$, making diffusional cavity growth much faster than creep deformation. For these parameters we indeed find that grain boundary damage development at the crack tip is constrained by creep of the surrounding material (Dyson, 1976). This causes the near-tip stresses to be redistributed away from the region where cavitation takes place. In the more ductile case for $L_R/R_0 = 0.032$ stress redistribution occurs only as a consequence of microcracking (see e.g. Fig. 3b).

(ii). One of the attractive features of the microstructural model is that the grain size is an essential ingredient in the theory. This allows an investigation into how the grain size affects the crack growth process. Experimental creep crack growth results show that the brittleness and crack growth rate increase with grain size for fixed specimen dimensions (e.g. Gooch *et al.*, 1977; Zhu *et al.*,

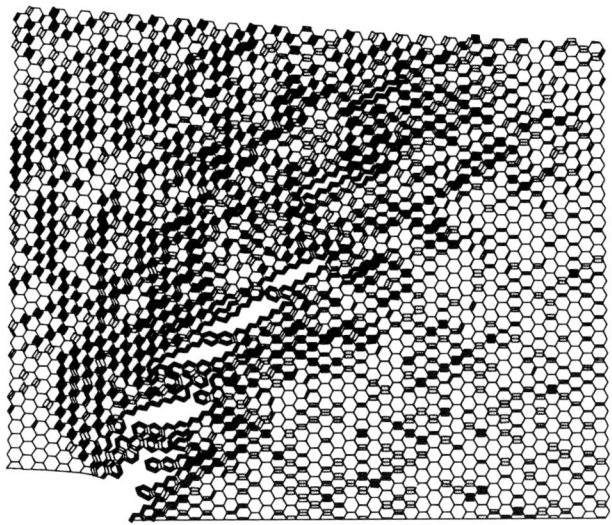

Figure 6. Grain boundary cavitation state for $L_R/R_0 = 0.032$ at $t/t_{R0} = 2.73$ with a random distribution of F_n. [From (Onck and Van der Giessen, 1999)].

1993). To investigate this, we kept the parameters C^*, n, a_I, N_I, E, B, η_B/w, $\mathcal{D}$ and N_{max} unchanged and analyzed two cases with larger grain sizes: $R = 5R_0$ and $R = 10R_0$. To ensure N_{max} was reached at the moment of coalescence, the nucleation parameter F_n in (8) was tuned to the grain size. For details, the reader is referred to Onck and Van der Giessen (1998b). Figure 5 shows the normalized crack extension against normalized time, showing an increase in crack growth rate, $\dot{c}$, with grain size, in accordance with experimental observations. Based on simple dimensional arguments, the crack growth rate was found to depend on grain size according to $\dot{c} \propto R^{n/n+1}$ (Onck and Van der Giessen, 1998b), which fits the numerical results of Fig. 5 quite well. In addition, the crack tip opening rates, $\dot{u}^{ctod}$, were identical for all cases, leading to an increase of the 'brittleness factor', $\dot{c}/\dot{u}^{ctod}$ (Gooch *et al.*, 1977), with grain size.

(iii). Experiments show that the nucleation activity is higher on some grain boundary facets than on others, while also the moment at which nucleation takes place varies among grain boundaries. These experimental observations are incorporated by randomly assigning the nucleation activity F_n to the facets in the process window according to a Gaussian distribution. Figure 6 shows that damage is distributed more diffusely and that isolated cavitating facets have developed away from the crack tip. Crack growth occurs under an angle of approximately 55° and again by the joining of regions of linked microcracks along the 30° direction. It was reported by Onck and Van der Giessen (1999) that both the crack growth rate and direction remain almost unaffected by a random distribution of F_n.

In micrographs of polished sections of failed specimens, the damage zone is often associated with the region in which microcracks are present. A comparison with such micrographs can be made more directly by plotting only the deformed geometry with the crack opening and microcracked regions painted black. Figure

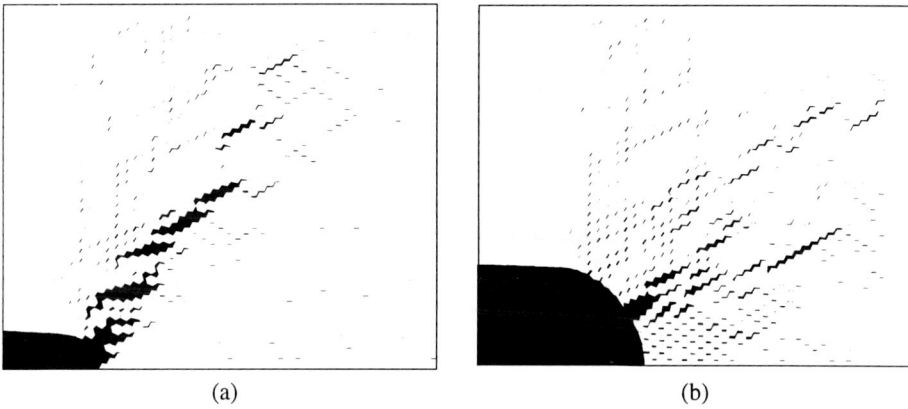

(a)   (b)

Figure 7. 'Micrographs', where the crack opening and microcracked regions are painted black. (a) Sharp crack tip at $t/t_{R0} = 2.73$, corresponding to Fig. 6. (b) Blunted crack tip at $t/t_{R0} = 2.77$. [From (Onck and Van der Giessen, 1998a)].

7a is an example of such a representation corresponding to Fig. 6. This 'micrograph' clearly depicts the opened up regions of linked microcracks as well as the horizontal and inclined microcracks surrounding the actual extended crack.

(iv). Creep crack growth experiments show a large variety of crack growth patterns, depending, among others, on crack tip geometry (e.g. Ozmat *et al.*, 1991). To gain some understanding of the observed phenomena, we analyzed crack growth from an initially blunted crack tip, with a crack tip radius of $12\sqrt{3}R_0$. Figure 7b shows the resulting 'micrograph' for the same random distribution of F_n as Figure 7a. Comparison with Figure 7a yields that the crack growth direction has shifted from $55°$ to approximately $30°$ degrees and that damage development takes place in a much more diffuse region, more in front of the crack tip. This is a result of the fact that the blunted geometry has redistributed the effective stress (driving nucleation through the creep strain) more uniformly around the crack tip and more towards the crack plane.

(v). If we focus on the actual opening stresses at physically-relevant distances from the crack tip (e.g. by performing finite element calculations), we often observe that they are lower than the HRR stresses (see (2)). This is commonly called the 'constraint effect', which depends on the loading configuration and specimen type. These constraint effects can be accounted for by including higher-order terms in the asymptotic series expansion of the near-tip stress fields, see (3). These fields contain three important parameters. The first one is C^*, which sets the load level and the remaining ones, A_2^* and σ_∞ account for the constraint effect. Practical values of A_2^* are found to lie in the interval $(-2, 0)$, corresponding to moderately deep to shallow crack situations (Nguyen *et al.*, 2000a). Figure 8 shows the crack extension vs. time for fixed values of C^* and σ_∞ ($\sigma_\infty/E = 1.7 \times 10^{-4}$), while A_2^* is varied between 0 and -2. The results show that the crack growth rate decreases with decreasing values of A_2^*. Nguyen *et al.* (2000b) also analyzed a variation of σ_∞ for fixed values of C^* and A_2^* ($A_2^* = -2$). The crack growth rate was observed to decrease with increasing σ_∞.

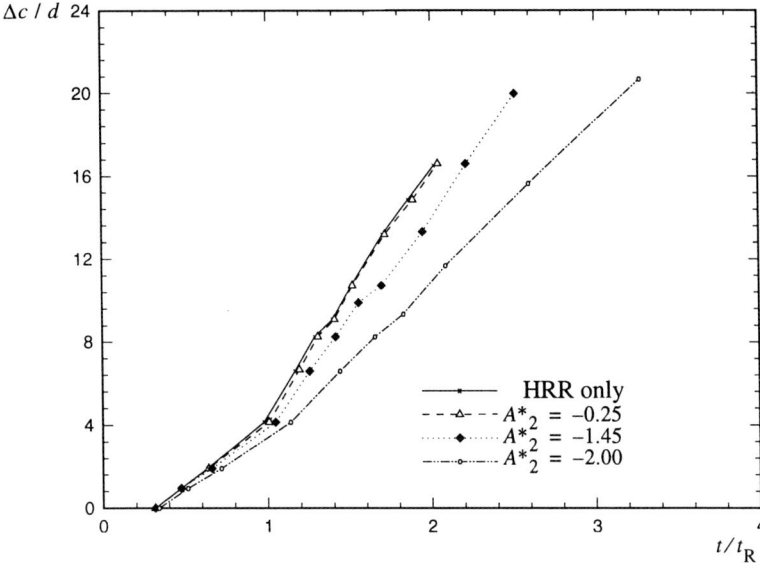

Figure 8. Crack advance Δc normalized by average grain size d_0 vs. normalized time for different values of A_2^*. [From (Nguyen *et al.*, 2000b)].

6. Concluding remarks

The microstructural modelling of creep crack growth presented here has been able to reproduce many features that have been observed experimentally either by metallographic investigations or by macroscopic creep testing (see Onck and Van der Giessen, 1998a,b, 1999; Nguyen *et al.*, 2000b). It has provided a numerical tool to investigate the complex and highly coupled mechanisms that are involved in creep fracture and thereby to improve our understanding of the phenomenon and its dependence on material characteristics.

It should also be noted, however, that there are several limitations. First of all, the studies so far have been entirely two-dimensional. The three-dimensionality of grains will definitely give rise to deviations from the predictions discussion here. However, in principle, the idea can be applied also in three dimensions. There are only practical limitations, such as computing power, that has prevented us from pursuing this.

Another limitation is the uncertainty in the nucleation law (8). Even though experimental studies have found a similar stress dependence at the macroscopic level (e.g. Dyson and McLean, 1977), it has not been verified experimentally at the grain boundary level nor is there any good theoretical origin. Yet, as mentioned above, several features of the predictions are related to this particular form. Several variations of (8) have been investigated by Onck and Van der Giessen (1999), and the results are indeed sensitive to the assumptions. Until the moment that an improved nucleation law becomes available, little more seems possible than to be aware of this.

Acknowledgement

The research of Dr. P.R. Onck was made possible by a fellowship of the Royal Netherlands Academy of Arts and Sciences.

References

Chao, Y.J. and Zhang, L. (1997), Tables of plane strain crack tip fields: HRR and higher order terms. ME-Report 97-1, University of South Carolina.

Dyson, B.F. (1976), Constraints on diffusional cavity growth rates, *Metal Sci.* 10, 349–353.

Dyson, B.F. and McLean, D. (1977), Creep of Nimonic 80A in torsion and tension. *Metal Sci.* 11, 37–45.

Ghahremani, F. (1980), Effect of grain boundary sliding on steady creep of polycrystals. *Int. J. Solids Struct.* 16, 847–862.

Gooch D.J., Haigh J.R., King B.L. (1977), Relationship between engineering and metallurgical factors in creep crack growth. *Met. Sci.* 11, 545–550.

Hayhurst, D.R., P.R. Brown and C.J. Morrison (1984), The role of continuum damage in creep crack growth, *Phil. Trans. Roy. Soc. Lond.* A311, 131–158.

Needleman, A. and J.R. Rice (1980), Plastic creep flow effects in the diffusive cavitation of grain boundaries, *Acta Metall.* 28, 1315–1332.

Nguyen, B.-N., P.R. Onck and E. Van der Giessen (2000a), On Higher-Order Crack-Tip Fields in Creeping Solids, *J. Appl. Mech.* in print.

Nguyen, B.-N., P.R. Onck and E. Van der Giessen (2000b), Crack-tip Constraint Effects on Creep Fracture, *Engng Fract. Mech.* 65, 467–490.

Onck P.R. and Van der Giessen E. (1997), Microstructurally-based Modelling of intergranular creep fracture using grain-elements, *Mech. Mater.* 26, 109–126.

Onck, P.R. and Van der Giessen, E. (1998a), Microstructural modelling of creep crack growth from a blunted crack. *Int. J. Frac.* 92, 373–399.

Onck, P.R. and Van der Giessen, E. (1998b), Numerical simulation of grain-size effects on creep crack growth by means of grain elements. *J. Phys. IV France,* 8, 285–292.

Onck, P.R. and Van der Giessen, E. (1999), Growth of an initially sharp crack by grain boundary cavitation. *J. Mech. Phys. Solids,* 47, 99–139.

Ozmat, B., Argon, A.S.and Parks, D.M. (1991), Growth modes of cracks in creeping type 304 stainless steel. *Mech. Mater.* 11, 1–17.

Riedel, H. (1987), Fracture at high temperatures. Materials Research and Engineering Series, Springer-Verlag.

Sham, T.-L and A. Needleman (1983), Effects of triaxial stressing on creep cavitation of grain boundaries, *Acta Metall.* 31, 919–926.

Shih, C.F. (1983), Tables of HRR Singular Field Quantities, *Report MRL E-147, Materials Research Laboratory, Brown University, Providence.*

Tvergaard, V. (1984), On the creep constrained diffusive cavitation of grain boundary facets, *J. Mech. Phys. Solids* 32, 373–393.

Tvergaard, V. (1986), Analysis of creep crack growth by grain boundary cavitation, *Int. J. Fract.* 31, 183–209.

Van der Giessen, E. and V. Tvergaard (1991), A creep rupture model accounting for cavitation at sliding grain boundaries, *Int. J. Fract.* 48, 153–178.

Yang, S., Y.J. Chao and M.A. Sutton (1993), Higher Order Asymptotic Crack Tip Fields in a Power-law Hardening Material, *Engng Fract. Mech., Vol. 45,* 1-20.

Zhu S.M., Wang F.G., Zhu S.J. (1993), Grain size dependence of creep crack growth in Ni-Cr austenitic steels, *Mater. Trans. JIM* 34, 450–454.

CREEP CRACK GROWTH: FROM DISCRETE TO CONTINUUM DAMAGE MODELING

B.-N. NGUYEN, P.R. ONCK & E. VAN DER GIESSEN
Delft University of Technology, Koiter Institute Delft
Mekelweg 2, 2628 CD Delft, The Netherlands

1. Introduction

Creep crack growth in specimens and structural components has intensively been studied by many authors in the past decades. Depending on the scale adopted for the formulation of the problem, the related approach can be quite different. At the scale of individual cavities and those of the grain and grain boundaries, there is the micromechanical approach which allows to model the intergranular creep fracture process resulting from nucleation, growth and coalescence of cavities leading to microcracks as well as the grain boundary viscous sliding (e.g. Rice, 1981; Tvergaard, 1984). Later, the behavior of a polycrystalline aggregate containing many grains whose boundaries suffer from creep cavitation and viscous sliding have been investigated in the work of Van der Giessen and Tvergaard (1994). All these micromechanical approaches allow to study the above-mentioned damage mechanisms as well as their effect on the overall behavior and the lifetime. More recently, an attempt to link these mechanisms to the behavior of a macroscopic crack has been performed by Onck and Van der Giessen (1999) by means of a small-scale damage microstructural approach. Their studies have revealed how cavitation damage and microcracking evolve around the crack tip and how the macroscopic crack grows by linking-up with surrounding microcracks, both for typical brittle and more ductile fracture conditions.

On the other hand, at the macroscopic level, a continuum damage mechanics (CDM) approach is often adopted. In the usual purely phenomenological approach, the damage mechanisms are ignored and just their effect on the overall response is accounted for through a damage variable which is scalar or tensorial in nature. Besides these micro- and macromechanical approaches, there is also a mixed approach in which the concept of a continuum damage parameter is used while its evolution law is based on micromechanical modeling (see e.g. Tvergaard, 1986).

Until now, there seems to be no link between micro- and macromechanical approaches. At the same time, CDM approaches seem of practical value for engineering purposes, but they suffer from mesh sensitivity problems, whereas micromechanical approaches present feasibility difficulties to simulate crack propagation at the macroscopic level. This paper is concerned with a first step to remedy the missing link, which is believed to be due to the lack of an inherent material characteristic length scale in CDM models. For two-dimensional situations, we will explore a novel mixed approach and derive a nonlocal formulation of creep damage making use of the gradients of the stresses and damage vari-

S. Murakami and N. Ohno (eds.), IUTAM Symposium on Creep in Structures, 65–74.
© 2001 *Kluwer Academic Publishers. Printed in the Netherlands.*

ables. Starting from previous micromechanical modeling, a heuristic homogenization approach is adopted to derive a theory for the macroscopic response. For this reason, we will refer to it as the "homogenized microstructural model" (HMM).

The basic philosophy is illustrated in Fig. 1, which summarizes the various scales involved in creep rupture. We consider a polycrystalline material formed by regular hexagonal grains which can be deformed by creep strain and damage under a multiaxial stress state (Fig. 1a). At the grain level (Fig. 1b), one can observe the occurrence of grain boundary cavitation and microcracking caused by coalescence of these cavities. By going down to the grain boundary scale (Fig. 1c), one distinguishes individual cavities of average radius a and distributed with an average half-spacing of b. For the modeling, illustrated in the right-hand side of the figure, we replace the discrete distribution of individual cavities on a grain boundary by a continuous one characterized by an homogenized set of cavitation parameters a and b, leading to an average separation between neighboring grains (Fig. 1d). After discussion of the governing equations at this level (Sec. 2), we proceed by returning to the grain level (Fig. 1e) for which we will advance the HMM procedure (Sec. 3) that provides a macroscopic continuum description (Fig. 1f) of creep, grain boundary cavitation as well as grain boundary sliding (Sec. 4). The adequacy of the HMM model is then investigated (Sec. 5) by comparing its predictions under homogeneous macroscopic stress states with two truely microstructural models of different degrees of complexity.

2. Damage evolution laws

The cavitation process is governed by the nucleation, growth and coalescence of cavities. The presence of a distribution of grain boundary voids with radius a and half-spacing b (see Fig. 1c) gives rise to an average separation between the adjacent grains, specified by $\delta_c = V/(\pi b^2)$ where V is the cavity volume, determined by the cavity radius and its precise shape. In a smeared-out approach where a continuous variation of δ_c is used (Fig. 1d) instead of its discrete distribution (Fig. 1c), the rate of change of δ_c,

$$\dot{\delta}_c = \delta_c(\dot{V}/V - 2\dot{b}/b), \tag{1}$$

is determined by the volumetric growth rate $\dot{V}$ and the rate of change of b. Based on the results of micromechanical studies of void growth by coupled grain boundary diffusion and creep by e.g. Needleman and Rice (1980) and Tvergaard (1984), a set of approximate yet accurate expressions has been devised for $\dot{V}$ of the form

$$\dot{V} = \dot{V}_{\text{dif}}(\mathcal{D}, \sigma_{\text{n}}, a, b, L) + \dot{V}_{\text{cr}}(\sigma_{\text{m}}, \sigma_{\text{e}}, a), \tag{2}$$

where $\dot{V}_{\text{dif}}$ and $\dot{V}_{\text{cr}}$ are the contributions by diffusion (with grain boundary diffusion parameter $\mathcal{D}$) and creep, respectively. Here, σ_{n} is the average stress normal to the grain boundary, and σ_{m} and σ_{e} are the average mean and Mises stresses, respectively, remote from the cavities. The creep contribution is based on the assumption that the grains behave according to the Norton creep law

$$\dot{\epsilon}_{\text{e}}^C = B\sigma_{\text{e}}^n \tag{3}$$

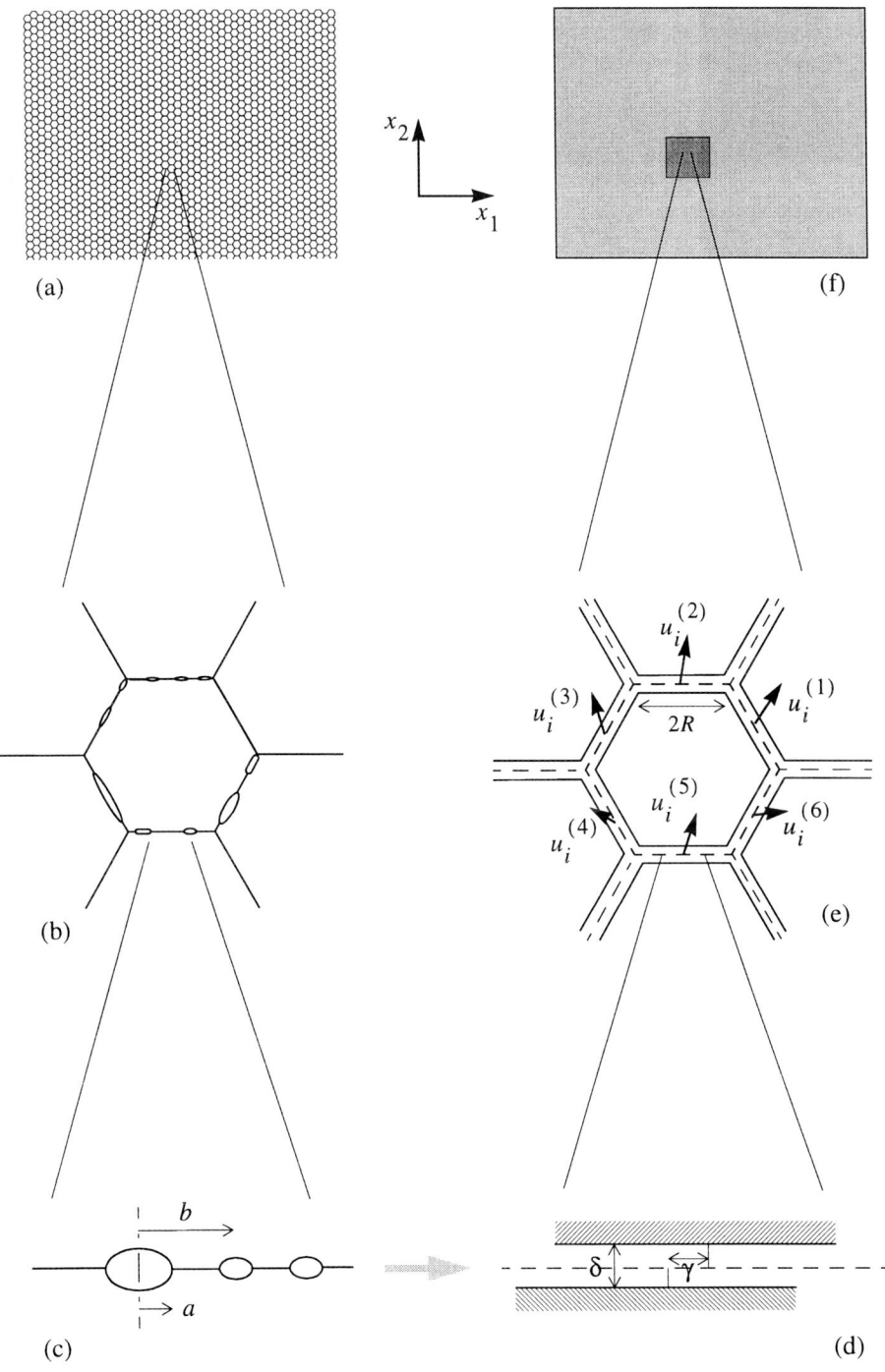

Figure 1. Creep damage from discrete to continuum point of view. See text for explanation.

with B and n being the creep parameter and exponent, respectively. The coupling between diffusive and creep contributions to cavity growth is governed by the length parameter (Needleman and Rice, 1980)

$$L = \left[\mathcal{D}\sigma_e/\dot{\epsilon}_e^C\right]^{1/3}. \tag{4}$$

Cavity growth is dominated by diffusion for relatively large values of L, while creep gives a larger contribution for smaller values of L. The volumetric growth rate $\dot{V}$ is used to compute the rate of change of the cavity size $\dot{a} \propto \dot{V}/(4\pi a^2)$.

The variation of cavity spacing b during the damage process is due to the change of the cavity density N by cavity nucleation, i.e. $\dot{b}/b = -\frac{1}{2}\dot{N}/N$. The same nucleation law is used here as introduced in (Van der Giessen and Tvergaard, 1994):

$$\dot{N} = F_n \left(\sigma_n/\Sigma_0\right)^2 \dot{\epsilon}_e^C, \qquad \text{for } \sigma_n > 0, \tag{5}$$

where F_n specifies the nucleation activity while Σ_0 is a reference stress used for the normalization of σ_n.

Introducing equations (2–5) into (1), we conclude that the separation rate is governed by a (constitutive) function as follows:

$$\dot{\delta}_c = \dot{\delta}_c(a, b, \sigma_n, \sigma_m, \sigma_e). \tag{6}$$

3. The Homogenized Microstructural Model (HMM)

We consider a two-dimensional polycrystalline material comprising perfectly hexagonal grains (See Fig. 1a), and start out by assuming that it is subjected to a state of macroscopically homogeneous (plane strain) deformations. The perfect hexagonal geometry of the grains allows us to define a representative volume which corresponds to a single grain of the aggregate (Fig. 1b,e). Each grain deforms elastically and by power-law creep according to (3). Macroscopically, we observe additional deformations due to the average separation between grains, described by δ_c according to (6), and the slip γ due to grain boundary sliding (to be discussed in the next section). All these are in general nonuniform fields over the grain or its facets (cf. Van der Giessen and Tvergaard, 1994), but for the sake of the homogenization we will here work from the average uniform quantities.

As a consequence, we can identify uniform displacement rate fields $\dot{u}_i^{(k)}$ acting on its $k = 1, \ldots, 6$ boundaries (see Fig. 1e). Denoting with $n_i^{(k)}$ and $t_i^{(k)}$ the unit vectors normal and parallel to the boundary facet k, respectively ($n_i^{(k)} t_i^{(k)} = 0$), Nguyen et al. (2000) use standard homogenization arguments to derive the following expression for the resulting macroscopic creep strain rate

$$\dot{E}_{ij}^C = \frac{R}{S} \left(\sum_{k=1}^{6} [\dot{u}_i^{c\,(k)} n_j^{(k)} + \dot{u}_j^{c\,(k)} n_i^{(k)}] \right.$$

$$\left. + \sum_{k=1}^{6} \frac{1}{2}\dot{\gamma}[t_i^{(k)} n_j^{(k)} + t_j^{(k)} n_i^{(k)}] + \sum_{k=1}^{6} \dot{\delta}_c^{(k)} n_i^{(k)} n_j^{(k)} \right) \tag{7}$$

in which the terms on the right-hand side represent the contributions of creep, grain boundary sliding and cavitation damage, respectively. In (7), R is the half-length of a facet (see Fig. 1e) and the grain area is $S = 6\sqrt{3}R^2$.

As revealed by detailed micromechanical investigations (e.g. Ghahremani, 1980), grain boundary sliding enhances the creep rate of an aggregate of grains. This effect is usually incorporated at the macroscale through a stress enhancement factor called f, as discussed in more detail in the next section. Therefore, the first and second terms of (7) can be unified so that the expression of the macroscopic creep strain rate becomes

$$\dot{E}^C_{ij} = \frac{3}{2} \dot{E}^C_e \frac{S_{ij}}{\Sigma_e} f^n + \frac{R}{S} \sum_{k=1}^{6} \dot{\delta}^{(k)}_c n^{(k)}_i n^{(k)}_j, \tag{8}$$

in which the first term is the well-known expression for sliding-enhanced power-law creep in terms of the macroscopic stress deviator S_{ij} and the corresponding Mises stress Σ_e. With the separation rate on each facet being determined by the current damage and stress state at the facet according to (6), the contributions $\dot{\delta}^{(k)}_c$ in the last term on the right-hand side of this expression is controlled by the values $a^{(k)}, b^{(k)}, \sigma^{(k)}_n, \sigma^{(k)}_m, \sigma^{(k)}_e$ at facet k.

As explained in full detail in (Nguyen *et al.*, 2000), the facet damage values $a^{(k)}$ and $b^{(k)}$ are obtained in two steps. First of all, we introduce fields for a and b at the macrolevel for each of the three families of facets $(1, 4)$, $(2, 5)$ and $(3, 6)$ (cf. Fig. 1e). Next, the value at the specific facet is obtained from a series expansion of the field variable in the corresponding family. The expansion includes terms up to the first order, thus involving the gradients of a or b on the specific family of facets. Similarly, the values of $\sigma^{(k)}_n, \sigma^{(k)}_m$, $\sigma^{(k)}_e$ are determined by the stress state $\Sigma^{(k)}_{ij}$ at the facet; we start out from the standard definitions, which for the normal stress $\sigma^{(k)}_n$ reads

$$\sigma^{(k)}_n = n^{(k)}_i \Sigma^{(k)}_{ij} n^{(k)}_j, \tag{9}$$

but modify the normal and effective stresses to account for grain boundary sliding (see Sec. 4). The facet stress state itself, $\Sigma^{(k)}_{ij}$, is obtained from a series expansion of the macroscopic stress field Σ_{ij} up to the first order, involving its gradients. With all of these ingredients, the entire right-hand side in (8) is specified at each instant, and the constitutive theory is complete.

4. Grain boundary sliding

At elevated temperatures, grains can slide relative to each other in a linear viscous manner. This gives a contribution to the overall creep rate that is incorporated in the HMM model through the concept of a stress enhancement factor f. The value of f depends on the grain boundary viscosity η_B, but also on the creep rate inside the grains. Ghahremani (1980) rigorously showed that for a two-dimensional aggregate of hexagonal grains as considered here (Fig. 1b, e), f is a function of n and of the ratio between the pure creep rate $\dot{E}^C_e$ and the strain-rate parameter

$$\dot{\epsilon}^B = \left(\frac{w}{d} \frac{B^{-1/n}}{\eta_B} \right)^{n/(n-1)}, \tag{10}$$

where w is the thickness of the grain boundary and d is related to the grain size ($d \approx 2\sqrt{3}R$). For any n, the value of f thus varies smoothly from $f = 1$ without grain boundary sliding ($\dot{E}_e^C/\dot{\epsilon}^B \to \infty$) to a finite value ($f \approx 1.2$ for $n = 5$) under free sliding ($\dot{E}_e^C/\dot{\epsilon}^B = 0$). This dependence can be expressed in a fully equivalent manner in terms of the ratio between Σ_e and the stress measure $\sigma_B = (\dot{\epsilon}^B/B)^{1/n}$. The latter function has been calculated and parameterized in a simple expression for $n = 5$ by Nguyen et al. (2000).

Grain boundary sliding also affects the stress state inside the grain and the normal stresses on the facets (e.g. Tvergaard, 1984; Van der Giessen, Tvergaard, 1994)). In the absence of sliding, the average facet normal stress, σ_n^{nsl} is simply the resolved normal stress according to (9). For completely free sliding and in the absence of any damage, the average normal stress follows directly from equilibrium (Rice, 1981),

$$\sigma_n^{fsl} = 2n_i^{(k)}\Sigma_{ij}^{(k)}n_j^{(k)} - \Sigma_m \,. \tag{11}$$

Nguyen et al. now assume that σ_n in presence of both cavitation and viscous sliding for a given value of σ_B/σ_e is determined by interpolation between its values at the no-sliding and free-sliding limits given in (9) and (11), respectively:

$$\sigma_n = \alpha(\sigma_B/\sigma_e)\sigma_n^{fsl} + [1 - \alpha(\sigma_B/\sigma_e)]\sigma_n^{nsl}, \tag{12}$$

where α is the interpolation parameter dependent on σ_B/σ_e. The latter function is also computed in this reference. Finally, the effective stress inside a grain becomes nonuniform in the presence of viscous sliding. The effect of this on the macroscopic creep rate is already incorporated through the enhancement factor f. However, it also affects the local effective stress near the grain boundary cavities and therefore their growth rate by creep. This is incorporated in an approximate manner by introducing an enhancement factor g for the effective stress in the cavity growth relations, in particular eq. (6).

5. Comparison of HMM with true microstructural models

In order to test the accuracy of the continuum HMM model, we compare its predicted damage evolution with the predictions of two more detailed models. We will do so for a macroscopically homogeneous state of stress, and therefore of damage, so that we need not be concerned with gradient effects. This means that we apply the HMM model locally in a material point of the continuum (Fig. 1f). The finest-scale model we shall confront the results with is essentially the type of model presented by Van der Giessen and Tvergaard (1994) in which all fields, inside the grains as well as on the grain boundaries, are resolved in great detail by means of a fine finite element discretization. Analyses with this model will be referred to as the "detailed analyses". The HMM results will also be compared with results obtained using Onck and Van der Giessen's (1997) grain elements. In this technique, each grain is still represented individually but at a lower resolution, i.e. by a single (super) element. The calculations with both reference models will involve a unit cell calculation, similar to those in (Van der Giessen and Tvergaard, 1994; Onck and Van der Giessen, 1997), under macroscopically constant stresses, as illustrated in the inset of Fig. 2.

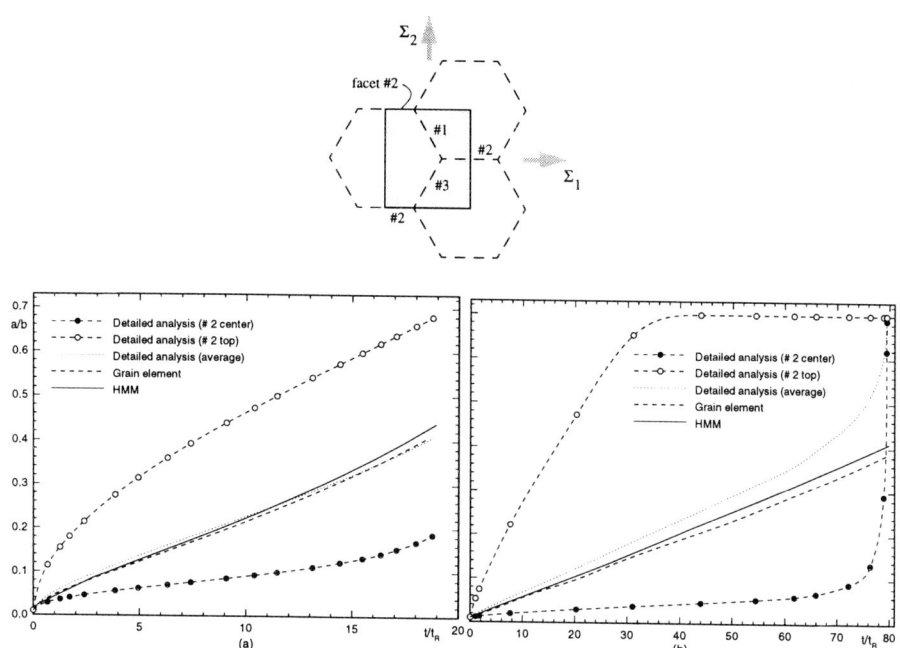

Figure 2. Evolution of damage parameter a/b on facets #2 (see inset) according to various models under uniaxial macroscopic stress ($\Sigma_1 = 0$) and with free grain boundary sliding for (a) $(a/L)_{\mathrm{I}} = 0.025$, (b) $(a/L)_{\mathrm{I}} = 0.1$.

In every case analyzed, all grain facets are assumed to have an initial density $N_{\mathrm{I}} = 1/\pi R^2$ of small grain boundary cavities as specified by $(a/b)_{\mathrm{I}} = 0.01$. Nucleation of new cavities takes place continuously according to (5) with the nucleation activity parameter F_{n} taken equal to $100 N_{\mathrm{I}}$. These values are the same as used in some of the cases studied by Onck and Van der Giessen (1997) in order to facilitate comparison. A number of different cases will be discussed which differ in: (i) the relative contributions of diffusion and creep to cavity growth; (ii) the grain boundary viscosity; (iii) the triaxiality of the macroscopic stress state. The latter is controlled by the ratio of the principal stresses Σ_2 and Σ_1, while keeping the Mises equivalent stress Σ_{e} at a constant level of $0.5 \times 10^{-3} E$ (E being Young's modulus). The grain boundary diffusion contribution to cavity growth is specified in terms of the value of the macroscopic L according to (4), i.e. in terms of the macroscopic Σ_{e} and the corresponding macroscopic power-law creep rate $\dot{E}_{\mathrm{e}}^{\mathrm{C}}$. The creep exponent is taken to be $n = 5$. All times are normalized by the reference time $t_{\mathrm{R}} = \Sigma_{\mathrm{e}}/(E\dot{E}_{\mathrm{e}}^{\mathrm{C}})$.

Figure 2 first shows the comparison of the predicted damage evolution under uniaxial stress ($\Sigma_1 = 0$) in the absence of any grain boundary viscosity ($\dot{E}_{\mathrm{e}}^{\mathrm{C}}/\dot{\epsilon}^{\mathrm{B}} = 0$). Two cases are shown, one corresponding to a value of L so that $(a/L)_{\mathrm{I}} = 0.025$ (Fig 2a: diffusion relatively fast), the other such that $(a/L)_{\mathrm{I}} = 0.1$ (Fig 2b: creep contribution to cavity growth more important). Only damage on the facets that are transverse to the stress state,

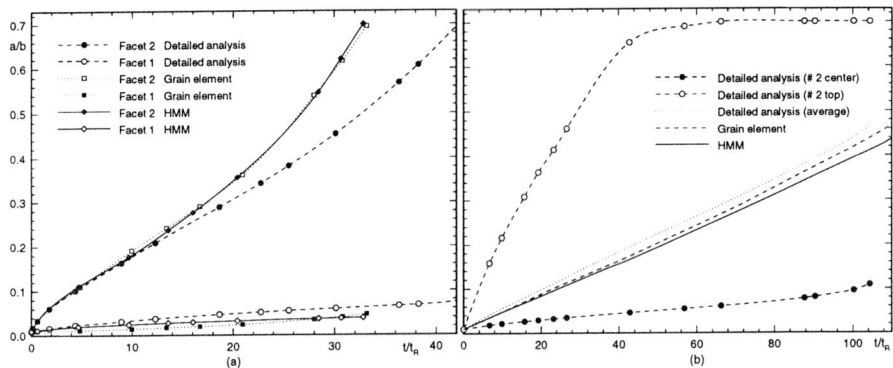

Figure 3. Evolution of damage parameter a/b according to various models under uniaxial macroscopic stress ($\Sigma_1 = 0$) and with viscous grain boundary sliding for (a) $(a/L)_I = 0.025$, (b) $(a/L)_I = 0.1$.

i.e. the facets #2 ($k = 2, 4$ in Fig. 1e), is shown; there is hardly any damage on the other families of facets in this case. Due to the free grain boundary sliding on facets #1 and #3, the damage on the #2 facets is revealed to be very non-uniform by the detailed calculations, which is exemplified in Fig. 2 by the results at the center of facet #2 as well as near the triple point (#2 top). This is even more so when creep is significant (Fig. 2b). Neither the grain elements nor the HMM continuum model capture these variations, but their predictions agree well with the average damage accumulation for both values of L. Note that these results already suggest that the corrections on creep rate and facet stresses due to free grain boundary sliding are approximate yet quite adequate.

Figure 3 shows a similar comparison but now in the presence of viscous sliding, such that $\dot{E}_e^C / \dot{\epsilon}^B = 0.1$. In this case, the damage on facet #2 is quite uniform for $(a/L)_I = 0.025$, so that only average values are shown in Fig. 3a. We observe now that the HMM tends to overestimate the damage development compared to the detailed analysis, but in the same way as the grain element solution. Damage development on the other facets #1 and #3 does occur in the presence of viscous sliding, but, as shown in Fig. 3a, it is much slower; HMM also picks this up.

It is well-known that the stress triaxiality in front of a crack tip is typically around 3 or so. Therefore, we have repeated the above computations but with positive Σ_2 such that $\Sigma_m / \Sigma_e = 3$. Damage development on facet #2 is now somewhat more uniform than under uniaxial stress (Fig. 2a) and again the HMM prediction agrees well with the more accurate results (Fig. 4a). Due to the higher triaxiality, there is damage development on facets #1 and #3 as well (and identical because of symmetry). According to Fig. 4b, the HMM as well as the grain element results again overestimate the damage accumulation on these inclined facets.

Close examination of the results showed that the origin for this lies in the very non-uniform distribution of effective stress that was found in the detailed analysis. As illustrated in Fig. 5a, the free sliding gives rise to localized zones of elevated effective stress in the central areas of the grains and to a shielded region around the inclined facets #1 and

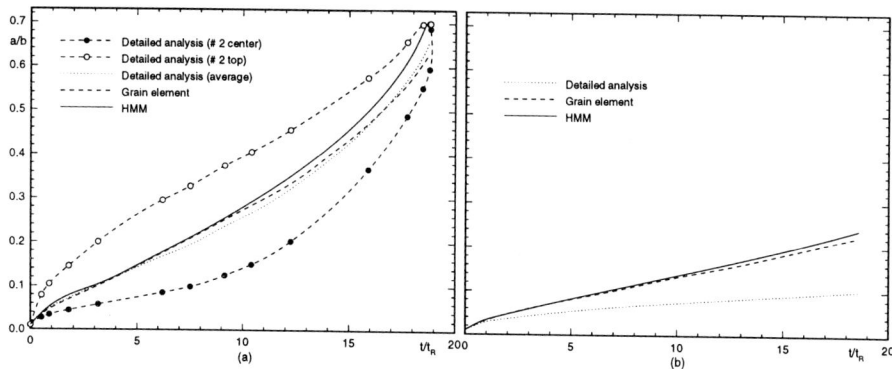

Figure 4. Evolution of damage parameter a/b according to various models under a macroscopic stress triaxiality of $\Sigma_m/\Sigma_e = 3$ for a material with free grain boundary sliding for $(a/L)_I = 0.025$ on (a) facet #2, (b) facets #1 and #3.

#3. The resolution of the grain elements (Fig. 5b) is limited and therefore does not pick up this local reduction. Evidently, the HMM model does not incorporate any of this. As a consequence, the used enhancement of the effective stress on the inclined facets based on the overall effective stress overestimates the creep contribution to cavity nucleation (and growth, but to a smaller extent) in the grain element as well as the HMM approach, Fig. 4b. Figure 5 also illustrates the non-uniformity of damage in the detailed analysis, compared to the coarser grain element description and the similar HMM representation.

6. Discussion and outlook

We have summarized the development of a novel continuum model (HMM) for creep and cavitation damage in a polycrystalline material which, contrary to previous continuum damage models, naturally includes the relevant size scales. The comparison with two microstructural models carried out here has been demonstrated that under different (uniform) stress states and different regimes of cavity growth, the HMM model succeeds in giving quite an accurate representation of the damage accumulation, including the facet-orientation dependence.

It should be noted that the HMM aims at applying to situations where cavitation damage develops on many grain facets. This is notably different from Tvergaard's (1986) continuum model which was developed for the early stages of rupture with a low density of damaged facets. The present model is expected to perform better in highly damaged regions near crack tips (cf. Onck and Van der Giessen, 1998).

Near-tip regions are also characterized by large stress gradients. There, the gradient features of the HMM model will be activated. The accuracy of the model in these situations needs to be assessed.

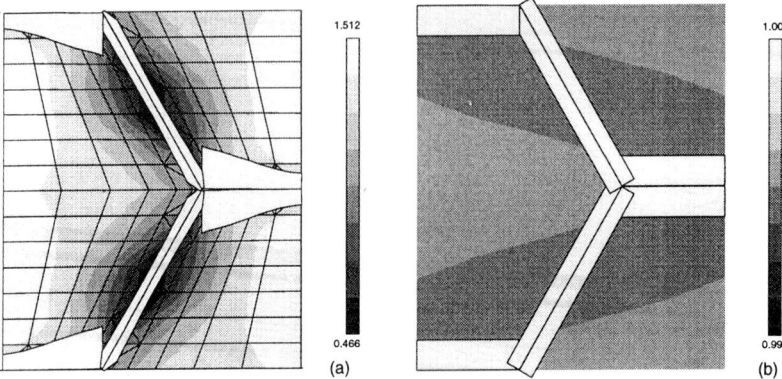

Figure 5. Snapshots of damage distribution and distribution of local effective Mises stress σ_e for the case analyzed in Fig. 4 at $t/t_R = 10.4$. Damage is shown in terms of a/b plotted on either side of the facet. The σ_e is normalized with the macroscopic Σ_e. (a) detailed analysis (mesh shown); (b) grain element model.

Acknowledgement

The research of Dr. P.R. Onck was made possible by a fellowship of the Royal Netherlands Academy of Arts and Sciences.

References

Ghahremani, F. (1980) Effect of grain boundary sliding on steady creep of polycrystals. *Int. J. Solids Struct.* **16**, 847–862.

Needleman, A. and Rice, J.R. (1980) Plastic creep flow effects in the diffusive cavitation of grain boundaries, *Acta Metall.* **28**, 1315–1332.

Nguyen, B.-N., Onck, P.R. and Van der Giessen, E. (2000) A homogenized microstructural approach to creep fracture, (submitted).

Onck, P.R. and Van der Giessen, E. (1997) Microstructurally-based Modelling of Intergranular Creep Fracture Using Grain Elements, *Mech. Mater.* **26**, 109–126.

Onck, P.R. and Van der Giessen, E. (1999) Growth of an initially sharp crack by grain boundary cavitation, *J. Mech. Phys. Solids* **47**, 99–139.

Rice, J.R. (1981) Constraints on the diffusive cavitation of isolated grain boundary facets in creeping polycrystals, *Acta Metall.* **29**, 675–681.

Tvergaard, V. (1984) On the creep constrained diffusive cavitation of grain boundary facets, *J. Mech. Phys. Solids* **32**, 373–393.

Tvergaard, V. (1984) Constitutive relations for creep in polycrystals with grain boundary cavitation, *Acta Metall.* **32**, 1977–1990.

Van der Giessen, E. and Tvergaard, V. (1994) Development of final creep failure in polycrystalline aggregates, *Acta Metall. Mat.* **42**, 959–973.

PREDICTION OF INNER CRACKING BEHAVIOR IN HEAT-RESISTANT STEEL UNDER CREEP-FATIGUE CONDITION BY MEANS OF THREE-DIMENSIONAL NUMERICAL SIMULATION

N. TADA and R. OHTANI

Department of Engineering Physics and Mechanics,
Graduate School of Engineering, Kyoto University
Yoshida-hommachi, Sakyo-ku, Kyoto, 606-8501, Japan

1. Introduction

Figure 1 shows the relationship between inelastic strain range and the number of cycles to failure for high-temperature fatigue of Type 304 austenitic stainless steel [1]. From the viewpoint of the location of small cracks, creep-fatigue fracture can be divided into two types. One is a "surface cracking type" indicated by O in Figure 1, in which small intergranular cracks are initiated only on the surface of the specimen, and their growth and coalescence bring about the final fracture. It is known that this type appears at intermediate temperatures with relatively high strain rates (or stresses). As the cracks are initiated only on the surface of the specimen in this type, creep-fatigue damage can be evaluated by surface observation of cracks. On the other hand, with increasing the temperature and/or decreasing the strain rate (or stress), the region of cracking, which was limited to the surface of the specimen, extends inward and, finally, small cracks appear almost uniformly throughout the specimen. This is called an "inner cracking

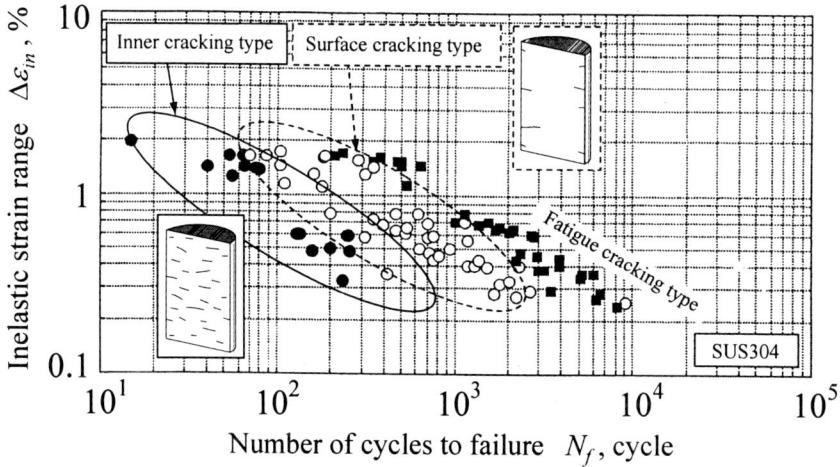

Figure 1. Relationship between inelastic strain range and the number of cycles to failure in high-temperature fatigue of Type 304 stainless steel.

S. Murakami and N. Ohno (eds.), IUTAM Symposium on Creep in Structures, 75–84.

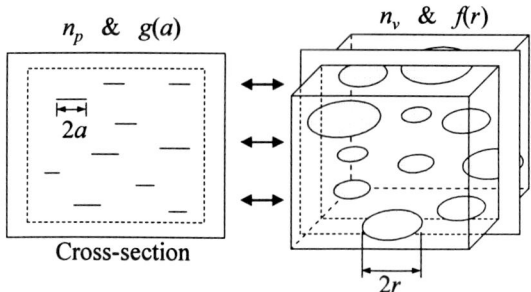

Figure 2. Circular cracks distributed inside the material and their cross-sectional image.

type" and is indicated by ● in Figure 1. As the distribution of inner cracks cannot be evaluated accurately by any non-destructive methods that are available today, it should be predicted based on the two-dimensional distribution of cracks on the cross-section.

In this paper, numerical simulation of initiation and growth of cracks was carried out for creep-fatigue fracture of inner cracking type, and the cracking behavior was predicted continuously from the beginning to the end of the life.

2. Conventional Method for Predicting the Distribution of Inner Cracks

It was clarified by a probabilistic analysis [2] that, when circular cracks are distributed uniformly at random locations as shown in Figure 2, the distribution of the truncated cracks appeared on the cross-section (*i.e.*, cracks cut by a plane perpendicular to the crack plane) is related to that of circular cracks by the following formulae,

$$g(a) = \frac{a}{[r]_m} \int_a^\infty \frac{f(r)}{\sqrt{r^2 - a^2}} dr , \qquad (1)$$

$$[r]_m = \int_0^\infty r \cdot f(r) dr , \qquad (2)$$

$$n_p = 2 \cdot [r]_m \cdot n_v , \qquad (3)$$

where n_v is volumetric density of circular cracks (*i.e.*, the number of cracks in a unit volume) and n_p is areal density of the truncated cracks (*i.e.*, the number of cracks in a unit area) on the cross-section. r and a are the radius of a circular crack and the length of the truncated crack on the cross-section, and their probabilistic density functions are assumed to be $f(r)$ and $g(a)$, respectively. As indicated by Equation (1), radius distribution of inner cracks is different from length distribution of the truncated cracks. Moreover, it is found from Equation (3) that the number of the truncated cracks on the cross-section is in proportion to both n_v and $[r]_m$. In other words, increase in the number of cracks observed on the cross-section reflects the increase in the number of inner cracks and that in the mean radius of inner cracks.

Solving Equation (1) with respect to $[r]_m$ and using Equation (3), the following equations are obtained,

$$[r]_m = \frac{\pi}{2 \left[\dfrac{1}{a}\right]_m}, \tag{4}$$

$$n_v = \frac{n_p \left[\dfrac{1}{a}\right]_m}{\pi}, \tag{5}$$

where $[1/a]_m$ is the mean of the reciprocal of a. These are the equations that relate the cross-sectional distribution of inner cracks to actual or three-dimensional one. Although the distribution of inner cracks can be predicted by these equations, predicted results obtained at different cracking stages are not related to each other, as schematically shown in Figure 3. Therefore, this method is not appropriate for discussing the behavior

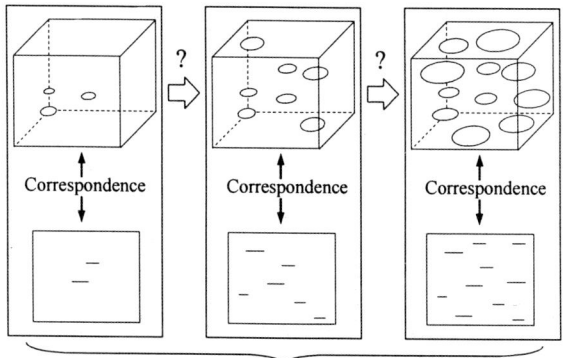

Observation Results on Cross-section

Figure 3. Conventional method for predicting the distribution of inner cracks based on probabilistic relationship.

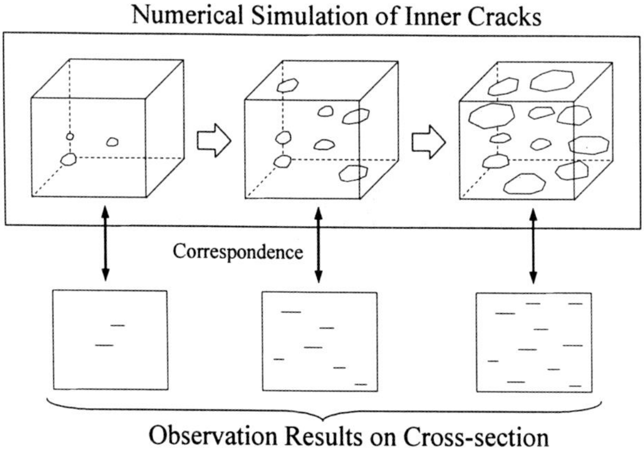

Numerical Simulation of Inner Cracks

Correspondence

Observation Results on Cross-section

Figure 4. Continuous prediction of distribution of inner cracks by means of numerical simulation.

of each crack.

3. Continuous Prediction of Distribution of Inner Cracks by Means of Numerical Simulation

Concept of a continuous prediction of the distribution of inner cracks is schematically shown in Figure 4. In this method, numerical simulation of inner cracking is carried out continuously through the life, and conditions for the cracking simulation are determined referring to the result of the cross-sectional observation of inner cracks at different cracking stages. Therefore, various information on initiation and growth behavior of inner cracks can be obtained after the simulation is completed.

3.1. EXPERIMENTAL RESULTS FOR THE BASIS OF CRACKING MODEL

Distribution of inner cracks in creep-fatigued smooth bar specimens was continuously predicted by this method. The tests were carried out under the control of the total strain at 1073K. The strain waveform was composed of very slow tension (1×10^{-4} %/s) and fast compression (1 %/s), and the total strain range was 0.7%. The failure life was 133 cycles. The tests were interrupted at different fatigue cycles, $0.1N_f$, $0.17N_f$, $0.25N_f$, $0.5N_f$, $0.75N_f$ and N_f, and cracks were observed on the longitudinal cross-section of each interrupted specimen. Details of the tests were reported previously [3]. Result of the observation of inner cracks can be summarized as follows.

(a) Small cracks of the size of a grain boundary facet are initiated at random locations. It can be said that a grain boundary facet is a unit of crack initiation.
(b) Cracks are initiated only on the grain boundary facets that are perpendicular to the stress axis.

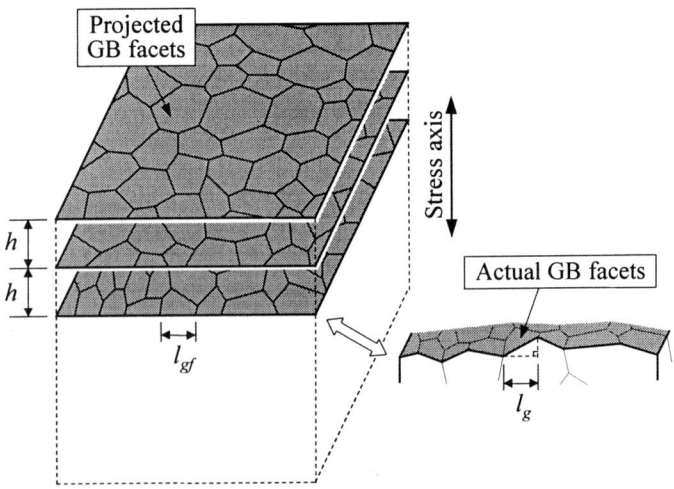

Figure 5.   Sheets of projected grain boundary facets layered perpendicular to the stress axis.

(c) The number of small cracks increases with the number of fatigue cycles increasing.
(d) Crack growth takes place by a unit of a grain boundary facet. Namely, cracks show discrete growth.
(e) Cracks grow in a direction perpendicular to the stress axis.

3.2. GENERATION OF GRAIN BOUNDARY FACETS

Before the cracking simulation, grain boundary facets are numerically generated in three dimensions as follows.

(a) Three-dimensional facets are approximated by sheets of projected grain boundary facets which are layered parallel to each other as shown in Figure 5.
(b) Grain boundary facets on each sheet are generated by Monte Carlo simulation based on an isotropic grain growth model [4,5]. The number of nuclei and their growth speed are determined so that the distribution of grain boundary length on the edge of the sheet, l_{gf}, coincides with that of projected length of actual grain boundaries, l_g.
(c) The space between each sheet is approximately calculated by

$$h = \sqrt{(\text{mean area of grains})}$$

$$= \frac{1}{\sqrt{n_g}} \tag{6}$$

where n_g is the number of actual grains in a unit area on the longitudinal cross-section. In the case of Type 304 stainless steel used in this study, n_g is equal to 300 (mm^{-2}) and yields $h \approx 0.06$ (mm).
(d) The size of the simulation body is set to be 1 by 1 by 1 (mm) and 16 facet sheets are layered.

3.3. MODEL FOR INNER CRACKING SIMULATION

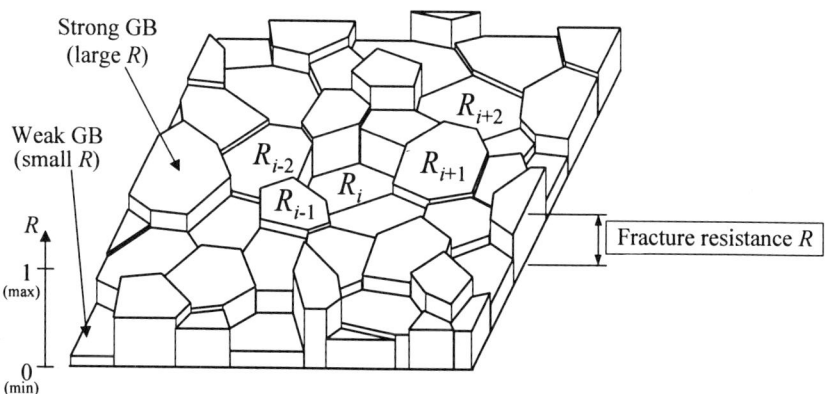

Figure 6.   Distribution of fracture resistance of grain boundary facets.

Initiation and growth behavior of inner small cracks is simply modeled as follows.

(a) As shown in Figure 6, every grain boundary facet has its own fracture resistance, R, which is assumed to be given by a random number 0 to 1. R represents all the structural inhomogeneities of a grain boundary facet such as the angle against the stress axis, the structure, mechanical interaction with neighboring grains and so on.

(b) The driving force for crack initiation, F, is given to every grain boundary facet, and the fracture resistance, R, decreases by the amount of F in every fatigue cycle. F is assumed to be constant to every facet, as shown in Figure 7(a).

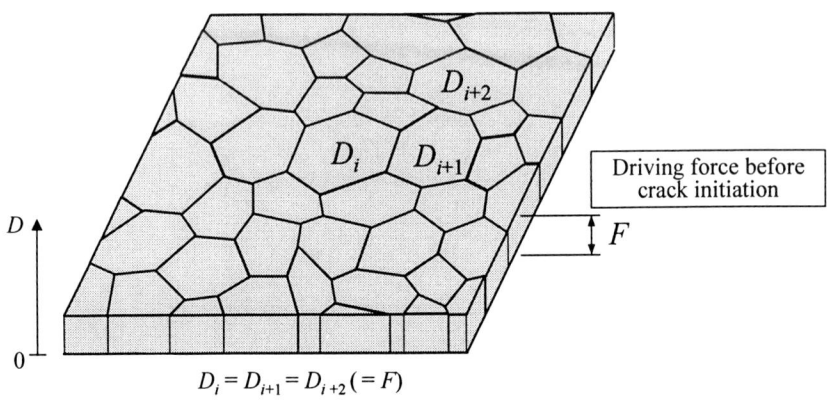

(a) Before crack initiation.

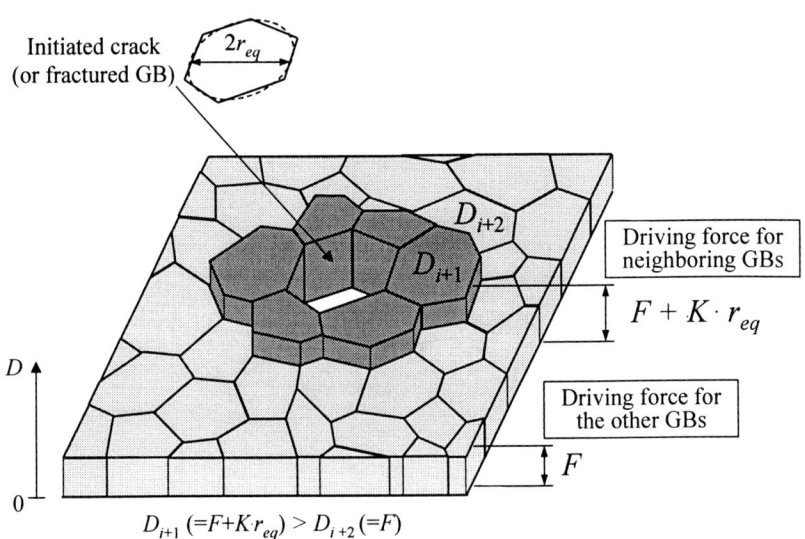

(b) After crack initiation.

Figure 7. Distribution of fracture resistance of grain boundary facets.

(c) When the remaining resistance becomes zero on a facet, the facet is assumed to break and an inner crack is initiated there.

(d) The driving force for the facets next to the initiated crack increases by $K \cdot r_{eq}$ due to stress and strain intensity where r_{eq} is an equivalent radius of the crack given by

$$r_{eq} = \sqrt{\frac{A}{\pi}} \qquad (7)$$

where A is the area of the crack. The total driving force, D, is then given by

$$D = F + K \cdot r_{eq}. \qquad (8)$$

(e) When the fracture resistance of one of the next facets becomes zero, the facet is assumed to break, which is recognized as crack growth.

As it is understood from the definition of the driving forces, the value of F determines the average increasing rate in the number of inner cracks and that of K does the average growth rate of cracks. In this study, they are determined as follows.

(a) As the mean length of cracks on the cross-section, $[a]_m$, is not affected by the value of F strongly, the value of K is firstly determined under a moderate value of F so that the value of $[a]_m$ coincides with the experimental one throughout the fatigue life.

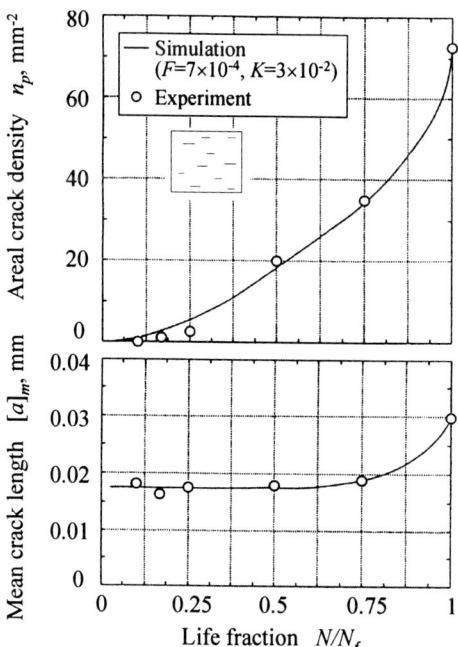

Figure 8. Change in areal crack density and that in mean crack length on the cross-section with life fraction.

(b) After the best value of K is obtained, the value of F is determined so that the areal density of cracks on the cross-section, n_p, coincides with the experimental one.

(c) Above processes are repeated until good agreement is attained for both n_p and $[a]_m$.

4. Predicted Results of Distribution of Inner Cracks

Figure 8 shows change in areal crack density and that in mean crack length on the longitudinal cross-section with the life fraction. As the minimum length of a crack was

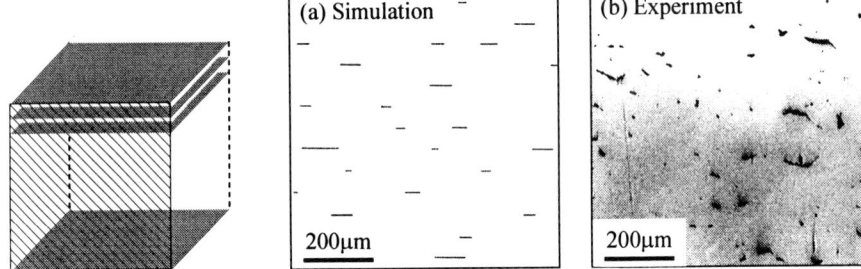

Figure 9.   Comparison of cracks on the cross-section obtained from numerical simulation with the actual ones ($N/N_f = 0.75$).

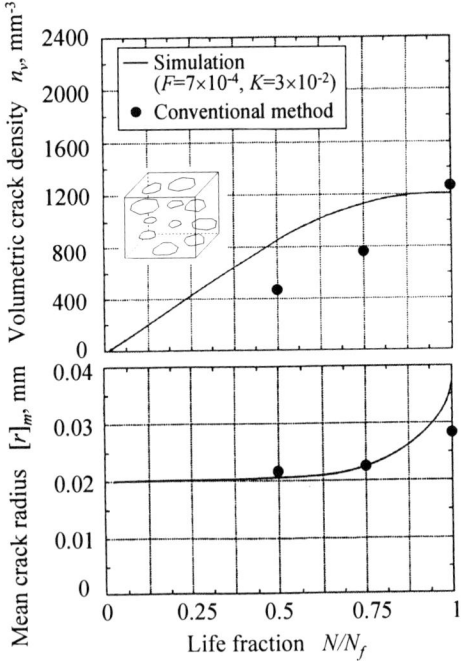

Figure 10.   Change in volumetric crack density and that in mean crack radius with life fraction.

set to be 10 (μm) in actual observation, the crack which was equal to or larger than 10 (μm) in length was recognized as a crack. When $F=7 \times 10^{-4}$ and $K=3 \times 10^{-2}$, result of the simulation agreed well with the experimental result.

Cracks on the longitudinal cross-section obtained from the simulation is compared with actual cracks in Figure 9. Both crack distributions look similar to each other. The present cracking model seems to simulate a cracking phenomenon very well.

Figure 10 shows change in volumetric crack density and that in mean crack radius with the life fraction. These are three-dimensional results and cannot be obtained directly from two-dimensional observation of cracks on the cross-section. Although areal crack density increases throughout the fatigue life, volumetric one saturates at the latter stage of the life. This shows that three-dimensional data are different from two-dimensional ones not only in the value but in the increasing tendency. Predicted results by the conventional method are also plotted as ● in Figure 10. Both results agree with each other relatively well. The difference between them might be caused by disregard of the minimum length of cracks and an approximation of crack shape by a circle in the conventional method.

Figure 11 shows the behavior of cracks on one of 16 sheets of grain boundary facets. Inner cracks are initiated at random locations at the early stage of the life and they grow with coalescence to the neighboring cracks. Although the observation of cracks on the cross-section was carried out at the life fractions of $0.1N_f$, $0.17N_f$, $0.25N_f$, $0.5N_f$, $0.75N_f$ and N_f, the distribution of cracks at different life fractions, $e.g.$, $0.2N_f$ and $0.7N_f$ in Figure 11, can be obtained easily.

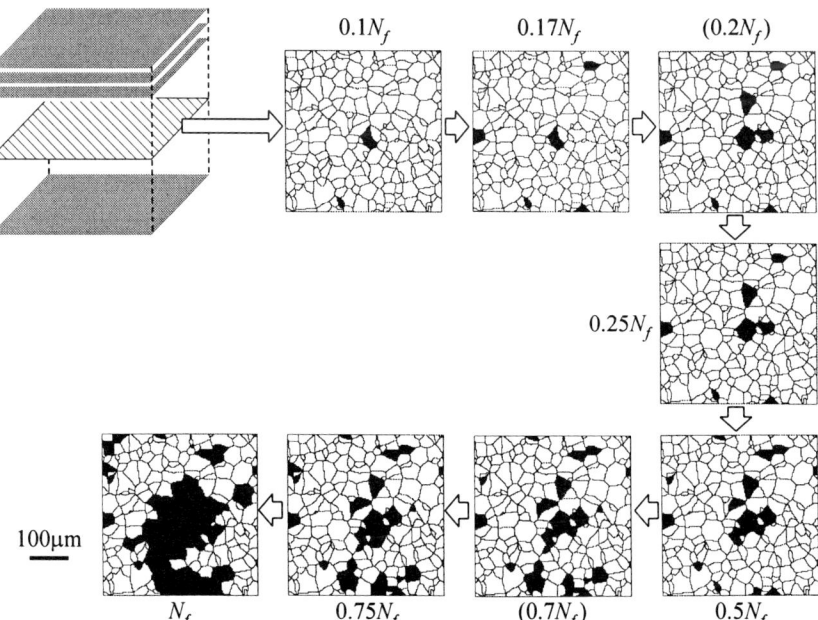

Figure 11.   Initiation and growth behavior of inner cracks on one of the layered sheets.

5. Conclusions

In this paper, cracking behavior inside the creep-fatigued specimen of Type 304 stainless steel was predicted by means of numerical simulation. The simulation was continuously carried out from the beginning to the end of the fatigue life, and the conditions were determined referring to the distribution of the truncated cracks observed on the cross-section at different fatigue cycles. Actual distribution of inner cracks were successfully predicted and cracking state inside the specimen was visualized. Although the physical meaning of fracture resistance and that of driving force are still indefinite, this method will be one of the strong tools for analysis of inner cracking.

6. References

1. Ohtani, R., Kitamura, T. and Tada, N. (1994) Experimental Mechanics on Initiation and Growth of Distributed Small Creep-Fatigue Cracks, in S. Gomes et al. (eds.), *Recent Advances in Experimental Mechanics* **2**, Balkema, Rotterdam, pp.1173-1179.
2. Tada, N., Ohtani, R. and Kitamura, T. (1994) Inverse Analysis of Distribution of Internal Small Defects, *JSME International Journal* **A-37**, 450-455.
3. Zhou, W., Ohtani, R., Kitamura, T., Tada, N. and Kosaka, A. (1995) Creep-Fatigue Intergranular Fracture of Inner Cracking Type in Type 304 Stainless Steel –Difference from Surface Cracking Type–, *Journal of Society of Materials Science, Japan* **44**, 78-83 (in Japanese).
4. Johnson, W. A. and Mehl, R. F. (1939) Reaction Kinetics in Processes of Nucleation and Growth, *Transactions of AIME* **135**, 416-458.
5. Mahin, K. W., Hanson, K. and Morris, J. W. Jr. (1980) Comparative Analysis of the Cellular and Johnson-Mehl Microstructures through Computer Simulation, *Acta Metallurgica* **28**, 443-453.

CREEP OF WELDED STRUCTURES

T. H. Hyde and W. Sun
University of Nottingham
University Park, Nottingham NG7 2RD, UK

Abstract

This paper is related to the high temperature creep analyses of welded structures, including material property generation, stress analysis, parametric studies, and failure life assessment. The methods used for generating the material constants, in the constitutive equations, for the different material zones within welds are based on experimental results from uniaxial, notched bar, cross-weld and impression creep tests. A general formulation for the creep of multi-material components is presented, from which an effective procedure for the parametric analysis of welded components is derived. Work on numerical finite element creep modelling of practical welded components, including cross-weld specimens, typical pipe welds and repaired welds, are briefly reviewed. Finally, recommendations are made for future research into the creep of welds.

1.   Introduction

At elevated temperatures, the performance of welds, in components, is often the life controlling factor and in recent years, major efforts have been made to understand the factors which affect the creep deformation behaviour of welds and to develop methods for assessing the remaining life of welded joints.

The failure at welds, rather than at positions away from welds, primarily arises because the welded zone is a region of material inhomogeneity. It is difficult to generalise the observations on creep of welds, primarily due to the variability in the material property distributions in the localised regions of welds and in the variable dimensions and shapes of welds (e.g. in weld repairs). In addition, the effects of system loading on the failure behaviour may be significant. For instance, within the low temperature HAZ region of a weld in a main steam pipeline, type IV cracking may occur [1], which is directly influenced by the local structural properties and stress state.

Analytical, numerical and experimental approaches have been widely adopted to study the effects of materials and geometric factors on weld performance. The creep deformation and creep rupture data for the parent material (PM), weld metal (WM) and

S. Murakami and N. Ohno (eds.), IUTAM Symposium on Creep in Structures, 85–94.
© *2001 Kluwer Academic Publishers. Printed in the Netherlands.*

heat-affected zone (HAZ) of welds can be obtained from laboratory creep tests. To characterise the strength reduction due to the existence of a weld, creep testing of cross-weld specimens is often adopted. Due to the complexity of the problem, closed-form analytical solutions cannot be obtained for weld situations and therefore, numerical simulations, using the finite element (FE) method, are often performed, which require accurate material properties for each material zone. Steady-state creep and continuum damage mechanics methods have been widely used for stress analysis and failure life predictions of welds.

This paper describes some recent research work, on high temperature creep of welded components. Methods used for generating the material constants in the constitutive equations, for different material zones of welds, have been developed; these are based on the experimental results obtained from uniaxial, notched bar, cross-weld and impression creep tests. A general formulation for the creep of multi-material components is used as the basis for the development of an effective procedure for performing parametric analyses of welded components. Results obtained from the finite element modelling of a series of practical welded components, including cross-weld specimens, typical pipe welds and repaired welds, are briefly described. Recommendations are made for future research on the creep of welds.

2. Experimental Techniques and Material Property Generation

2.1 Material Constitutive Equations

Two types of material behaviour models are often adopted in numerical modelling, those with a simple power law dependence, i.e.,

$$\left(\dot{\varepsilon}^c_{min}\right)_{ij} = \frac{3}{2}A'\left(\sigma_{eq}\right)^{n'-1}S_{ij} \tag{1}$$

and continuum damage constitutive equations of the form [2]:

$$\dot{\varepsilon}^c_{ij} = \frac{3}{2}A\left[\frac{\sigma_{eq}}{1-\omega}\right]^n\frac{S_{ij}}{\sigma_{eq}}t^m \quad (2a) \quad \text{and} \quad \dot{\omega} = \frac{M[\alpha\hat{\sigma} + (1-\alpha)\sigma_{eq}]^\chi}{(1+\phi)(1-\omega)^\phi}t^m \tag{2b}$$

It should be noted that more complex forms of constitutive equations have been proposed to describe the more complex metallurgical mechanisms associated with the creep of some materials [3,4]. However, equations (2) have be demonstrated to be capable of accurately representing the stress dependence of the creep strain-rate components, over a wide range of stresses and temperatures, for a number of metallic alloys [e.g. 5], and were successfully used to represent the creep deformation and failure behaviour of the parent, weld and HAZ materials in virgin and service-aged CrMoV pipe welds [6,7].

2.2 Creep Testing and Determination of Material Properties of the Weld Materials

The impression creep testing technique can be used to obtain the primary and secondary creep properties from small volumes of material, such as exist in the HAZ in welds [8]. The technique involves the application of a steady load to a flat-ended indenter placed on the surface of a material at elevated temperature. The displacement-time record from such a test is related to the creep properties of a relatively small volume of material in the immediate vicinity of the indenter, and this can be converted into the corresponding uniaxial creep strain data.

The set of the material creep constants (A, n, m, M, ϕ, χ and α, see equations (2)) for each material within a weld can be determined from the creep test data used in conjunction with FE damage modelling. A procedure has been described [7,9] and used to generate the material properties for a series of virgin, service-aged and repaired CrMoV welds [10]. For parent and weld materials, these constants are determined directly from the uniaxial creep curves obtained for different stress levels at a fixed temperature and the corresponding multi-axial creep rupture test data from notched specimens.

For the HAZ material, some of the properties (i.e. A, n and m) are obtained from the results of impression creep tests, while the other properties (i.e. M, χ, ϕ and α) are determined by creep damage modelling corresponding to data obtained from creep rupture tests of waisted and notched cross-weld specimens [7].

3. Multi-Material Creep Deformation Formulations and their Application to the Parametric Analysis of Welds under Steady-State Creep

3.1 A General Formulation for Creep of Multi-Material Components

Simplified analytical solutions have been obtained for the creep deformation and failure behaviours of welds in axisymmetric and plate models [11,12]. Also, complete analytical solutions for other multi-material component types, with simple geometries, have been used in an attempt to generalise the creep behaviour of multi-material structures [13-15]. Based on the creep solutions obtained from multi-bar structures, beams in bending, thin and thick cylinders under internal pressures etc., a general formulation, using Norton's law of the form $\dot{\varepsilon}^c / \dot{\varepsilon}_{oi} = \left(\sigma_i / \sigma_{nom} \right)^{n_i}$, was proposed for the steady-state stresses in and deformations of multi-material components [14,15]. These solutions indicate that, at a position of interest in material i of a multi-material component, the general form for the stress and deformation can be expressed by

$$\sum_{j=1}^{p} \left\{ f_j(n_1, n_2, ..., n_p, [dim]_j) \left(\frac{\dot{\varepsilon}_{oi}}{\dot{\varepsilon}_{oj}} \right)^{\frac{1}{n_j}} \left(\frac{\sigma_i}{\sigma_{nom}} \right)^{\frac{n_i}{n_j}} \right\} = 1 \qquad (3a)$$

$$\sum_{j=1}^{p} \left\{ g_j(n_1, n_2, \ldots, n_p, [dim]_j) \left(\frac{\dot{\varepsilon}_{oi}}{\dot{\varepsilon}_{oj}} \right)^{\frac{1}{n_j}} \left(\frac{\dot{u}_i}{\dot{u}_{nom}} \right)^{\frac{1}{n_j}} \right\} = 1 \qquad (3b)$$

where f_1, f_2, ..., f_p and g_1, g_2, ..., g_p are functions of the stress indices, n_i, and non-dimensional functions of dimensions, dim. It can be seen that the effects of the $\dot{\varepsilon}_{oi}/\dot{\varepsilon}_{o2}$, $\dot{\varepsilon}_{oi}/\dot{\varepsilon}_{o3}$ ratios etc, are explicitly defined in the equations.

3.2   Application to Parametric Analyses of Welds

Equations (3) can be directly applied to simplify the parametric analysis of welded components, where previously extensive numerical calculations, using the FE method, would have been required to assess the effects of material properties, geometrical sizes and loading mode on the creep stresses and deformations. The f_j and g_j values at the positions of interest, for a given n-set (n_1, n_2, ... n_p), for a component consisting of p materials, can be determined by performing a set of p FE calculations, each with different relative $\dot{\varepsilon}_{oi}$ values (usually analytical solutions are not available). Repeating this for a series of n-value sets produces f_j and g_j values which can be interpolated for any set of n-values in the range of interest. Hence a complete parametric analysis, for the stress or deformation at the positions of interest in material i, can be performed easily and quickly, for any combinations of n-values and $\varepsilon_{oi}/\varepsilon_{oj}$ ratios by solving equations (3), instead of performing further FE calculations. A detailed procedure for parametric analyses has been described and successfully applied to multi-material pipe weld situations [16,17].

4.   Creep Analysis of Welded Components Using FE Modelling

4.1   Cross-Weld Specimens and Pipe Welds

Results obtained from parametric analyses, using Norton's law for an idealized two-material axisymmetric cross-weld specimen [18,19], and continuum damage analyses for practical three-material waisted and notched cross-welds specimens [20,21], were used to identify the general influence of the material property mismatch, size and shape of the cross-weld specimens on the stress, strain rate and failure life. Observations from theses analyses showed that:

i) creep stress and strain rate distributions on the centre line of the two-material cross-weld specimen strongly depend on material property ratios [18];
ii) for practical weld situations, the influence of the stress singularity due to material mismatch is not significant [19];
iii) in general, the sizes and shapes of cross-weld specimens have significant effects on the peak stress and failure life [18,20];
iv) using practical weldment properties, realistic failure modes can be predicted from the results of waisted and notched cross-weld specimens [20,21].

Investigations of the creep behaviour of typical circumferential pipe welds in main steam thick walled pipelines, with typical V-shape and narrow gap weld configurations, based on both steady-state analyses and continuum damage modelling, showed similar behaviour [22-23]. A typical V-shape pipe weld model is shown in Fig. 1. The results obtained indicated that, in general, the effect of changing the relative creep strengths of the parent, weld and HAZ materials, on the peak rupture stresses in the pipe welds, are significant. The changes of the geometric parameters, especially the width of the weld metal and weld interface angle do not introduce significant change in the rupture stress [22,23]. In addition, the results obtained were also used to assess the accuracy of simplified methods for predicting the failure lives of pipe welds based on data obtained from tests performed on cross-weld specimens [23]. Compared with damage predictions, conservative life estimates were obtained from a simplified method, using steady-state peak rupture stresses [22,23].

4.2   Life Assessment of New, Service-Aged and Repaired Welds in Main Steam
        Pipelines

Continuum damage mechanics modelling and steady-state analysis have been used to assess the performance of repaired welds in main steam pipelines made from 1/2CrMoV steel [24-26]. Schematic diagrams of the aged and repaired welds are shown in Fig. 2. The main dimensions of pipe/weld used are the same as those shown in Fig. 1. The material properties, in the various zones of the welds, involving a number of parent and weld materials (virgin and service-aged) and several HAZ materials in different welds (virgin, service-aged and repaired), were produced from test data obtained at 640°C, see Table 1. On the basis of the results of the analyses, the creep performance of new and service-aged welds was assessed and the effects of size, system loading and various repair situations (e.g. full repairs or partial repairs) were evaluated. The resulting failure predictions are presented in Table 2.

For the particular material data used in this investigation, results obtained showed that the failure lives of repaired welds are comparable to those of the service-aged weld and are about half of those of the virgin weld. The life ratios of the new, aged, fully and partially repaired welds, obtained from damage analyses, are about 1 : 0.74 : 0.51 : 0.47. The effects of weld size and the configuration of weld repairs, in practical situations, are insignificant. However, the effect of axial (system) load on the failure life and position is extremely important. With excessive axial tensile loading, there is a high probability of type IV cracking occurring [26], which is consistent with laboratory and plant experience. Typical examples of damage variations across the HAZ, with the normalised axial stress, σ_{ax}/σ_{mh}, where σ_{mh} is the mean diameter hoop stress, are presented in Figs. 3.

5.   Discussion and Future Work

Experience indicates that the failure life of and position of failure within welded joints can be accurately predicted, using numerical modelling, provided that the material constitutive equations, which control the evolution of creep strain and damage, are accurate enough to represent the physical mechanisms [6]. Hence, there is a need for more accurate testing techniques, which will allow the variations of the material creep properties within the weld zones, particularly in the HAZ, to be determined. Creep strain data can be inferred from impression creep tests [8]. The miniature disc punch testing technique [27] may offer an alternative, particularly for obtaining the creep rupture properties of HAZ materials.

General formulations for the stresses in and deformations of simple multi-material components can be directly applied to weld situations. This is particularly useful for simplifying and presenting the results of parametric analyses of welds. The formulation allows the effects of the material properties to be easily assessed and the results of parametric analyses to be presented in compact and easily manageable forms.

FE modelling provides a useful tool for the stress analysis and failure prediction of complicated welded components, such as repaired welds. Comparison of the failure predictions obtained for various weld and repaired weld situations allows the effects of the differences in the relative properties of the constituents of the weld, the geometry and system loading, to be identified. Information of this type is essential if accurate assessments of the performance of practical service-aged/repaired welds in power plants pipework are to be obtained. They can also be useful in improving weld quality and design. Steady-state analysis, in general, provides a quicker but conservative life estimation, compared with continuum damage modelling. As a simple form of a welded joint, cross-weld specimens are commonly adopted in laboratory creep tests to characterise the effects due to the presence of a weld. However, interpretation of cross-weld rupture data still remains a difficult task, due to the difference in stress states in cross-weld specimens and in practical welded joints. Further research is necessary to establish a clearer link between them. In addition, the effects of residual stresses on the performance of a weld under creep conditions need to be considered.

Extensive development of advanced ferritic alloys has occurred over the past 10 years or so and materials such as Grade 91 are now starting to be used, with the possibility of the use of the stronger Grade 92, NF616 and E911 alloys in the future. These materials are typically high chromium content ferritic alloys and are sensitive to variations in composition, heat treatment and operating stresses and temperatures. They can be welded but their creep properties are more susceptible, than those of the lower alloyed ferritic steels, to variations which can occur during the welding process. Therefore, more care must be taken to ensure that correct metallurgical structures are present in the weld region, particularly within the HAZ, so that the required strength of the weld metal for maximum life is achieved. Also, the changes which may occur during long term service experience and the influence of these property changes on the long term performance need to be established. The techniques developed for determining the material properties for 1/2CrMoV: 2 1/4Cr1Mo low alloy welds, and for the assessment of weld and repaired weld performance, should be capable of being extended to the study of the welds or repaired welds formed in these advanced materials.

Acknowledgement

The authors wish to acknowledge Nuclear Electric, PowerGen, National Power and the Engineering & Physical Science Research Council, UK, for their financial and technical support of the work.

References

1. Gooch, D. J. and Kimmins, S. T., Type IV cracking in 1/2Cr1/2Mo1/4V/2 1/4Cr1Mo weldments. Proc. 3rd Int. Conf. on *Creep and Fracture of Engineering Materials and Structures,* eds. B. Wilshire and R. W. Evans, The Institute of materials (1987), 689-705.
2. Hayhurst, D. R., On the Role of Creep Damage in Structural Mechanics, Engineering *Approaches to High Temperature Design,* eds. B. Wilshire and D. R. Owen, Pineridge Press (1983), 85-176.
3. Hayhurst, D. R. and Perrin, I. J., CDM Analysis of Creep Rupture in Weldments, *Proc. Eng. Mechanics* **1** (1995), 393-396.
4. Perrin, I. J. and Hayhurst, D. R., Creep Constitutive Equations for 0.5Cr0.5Mo0.25V Ferritic Steel in the Temperature Range 600-675° C, *J. Strain Analysis* **31** (1996), 299-314.
5. Hayhurst, D. R., Creep Rupture under Multiaxial States of Stresses, *J. Mech. Phys. Solids* **20** (1972), 381-390.
6. Hall, F. R. and Hayhurst, D. R., Continuum Damage Mechanics Modelling of High Temperature Deformation and Failure in a Pipe Weldment, *Proc. R. Soc. London,* **A443** (1991), 383-403.
7. Hyde, T. H., Sun, W., Becker, A. A. and Williams, J. A., Creep Continuum Damage Constitutive Equations for the Parent, Weld and Heat-Affected Zone Materials of a Service-Aged 1/2Cr1/2Mo1/4V: 2 1/2Cr1Mo Multi-Pass Weld at 640° C, *J. Strain Analysis* **32** (1997), 273-285.
8. Hyde, T. H., Sun, W. and Becker, A. A., Analysis of the Impression Creep Test Method Using a Rectangular Indenter for Determining the Creep Properties in Welds, *Int. J. Mech. Sci.* **38** (1996), 1089-1102.
9. Hyde, T. H. and Sun, W., Determining high temperature properties of weld materials, Proc. Int. Conf. on *Advanced Technology in Experimental Mechanics,* Japan, July (1999), 496-502.
10. Hyde, T. H., Sun, W. and Williams, J. A., Creep behaviour of parent, weld and HAZ materials of new, service-aged and repaired 1/2Cr1/2Mo1/4V: 2 1/4Cr1Mo pipe welds at 640° C, *Materials at High Temperatures* **16** (1999), 117-129.
11. Nicol, D. A., The creep behaviour of cross-weld specimens under uniaxial loadings, *Int. J. Engng. Sci.* **23** (1985), 541-553.
12. Craine, R. E. and Hawkes, T. D., On the creep of ferritic weldments containing multiple zones in plates under uniaxial loading, *J. Strain Analysis* **28** (1993), 303-309.
13. Hyde, T H, Yehia, K and Sun, W, Observations on the creep of two-material structures, *J. Strain Analysis* **31** (1996), 441-462.

14. Hyde, T. H., Sun, W., Tang, A. and P. J. Budden, An inductive procedure for determining the stresses in multi-material components under steady-state creep, *J. Strain Analysis* (in press).

15. Hyde, T. H., Sun, W. and Tang, A., A general formulation of the steady-state creep deformation of multi-material components, 4th Int. Conf. on *Modern Practice in Stress and Vibration Analysis*, Nottingham (2000).

16. Hyde, T. H., Tang, A. and Sun, W, Parametric analyses of stresses and deformations in a pipe with a circumferential weld under creep conditions, Proc. of Int. Conf. on *Integrity of High Temperature Welds*, Nottingham (1998), 323-332.

17. Hyde, T. H., Sun, W. and Tang, A., A parametric analysis of stresses in a thick-walled pipe weld during steady-state creep, Proc. of 5th Int. Colloquium on *Ageing of Materials and Methods for the Assessment of Lifetimes of Engineering Plant*, Cape Town (1999), 231-246.

18. Hyde, T. H. and Sun, W., A method for estimating the stress distributions on the centre line of an axisymmetric, two-material, cross-weld, creep test specimens, *Int. J. Mech. Sci.* **39** (1997), 885-898.

19. Hyde, T. H. and Sun, W., Stress singularities at the free surface of an axisymmetric two material creep test specimen, *J. Strain Analysis* **32** (1997), 107-117.

20. Hyde, T. H., Sun, W., Becker, A. A., Failure prediction for multi-material creep test specimens using a steady-state creep rupture stress, *Int. J. Mech. Sci.* 42 (2000), 401-423.

21. Hyde, T. H. and Sun, W., Creep of Waisted and Notched Cross-Weld Specimens, Proc. of 7th Int. Conf. on *Creep and Fracture of Engineering Materials and Structures*, California (1997), 759-768.

22. Hyde, T. H., Williams, J. A. and Sun, W., Assessment of creep behaviour of narrow gap welds, *Int. J. Pres. Ves. & Piping* **76** (1999), 515-525.

23. Hyde, T. H. and Sun W., Creep of welded pipes, *J. Mech. Processing Eng.* **212** (1998), 171-182.

24. Hyde, T. H., Sun W. and Becker, A. A., Life assessment of weld repairs in 1/2Cr1/2Mo1/4V:2 1/4Cr1Mo main steam pipes using the finite element method, to be published in *J. Strain Analysis* (accepted).

25. Sun, W., Hyde, T. H, Becker, A. A. and J. A. Williams, Comparison of the creep and damage failure prediction of the new, service-aged and repaired thick-walled circumferential CrMoV pipe welds using material properties at 640° C, to be published in *Int. J. Pres. Ves. & Piping* (accepted).

26. Hyde, T. H., Sun, W., Becker, A. A., Effects of end loading on the creep failure behaviour of CrMoV welds in main steam pipelines, in the 6th International Conference on *Computer Aided Assessment and Control (Damage and Fracture Mechanics 2000)*, May 2000, Montreal, Canada.

27. Parker, J. D. and James, J. D., Creep Behaviour of Miniature Disc Specimens of Low Alloy Steel, *PVP* 279 (1994), Developments in a Progressing Technology, ASME, 167-172.

Table 1 Material constants for the 1/2Cr1/2Mo1/1V: 2-1/4Cr1Mo weldment materials at 640°C [10] (σ in MPa and t in hour).

	New PM	New WM	New HAZ	Aged PM	Aged WM	Aged HAZ	HAZa*
A	3.280×10^{-18}	6.459×10^{-17}	1.044×10^{-15}	6.599×10^{-16}	9.718×10^{-15}	1.708×10^{-15}	1.085×10^{-15}
n	7.269	6.430	6.108	6.108	5.208	6.108	6.108
m	0	0	0	0	0	0	0
M	4.823×10^{-12}	5.794×10^{-11}	9.66×10^{-10}	5.998×10^{-14}	8.120×10^{-13}	2.50×10^{-9}	1.90×10^{-8}
φ	4.75	4.121	4.3	4.5	4.1	4.3	4.3
χ	4.599	4.015	3.420	5.767	4.850	3.2	2.65
α	0.333	0.417	0.49	0.3	0.264	0.49	0.49

* HAZ generated in the aged parent material in repaired welds.

Table 2 Failure lives (hour) for different end loading for the new, aged and repaired main steam pipe welds (p_i = 16.55 MPa), obtained using creep properties at 640°C.

	Continuum damage prediction				
σ_{ax}/σ_{mh}	New weld	Aged weld	Full repair	Partial i	Partial ii
0.306	21,018	15,640	10,803	9,892	----
0.5	16,266	12,962	9,305	8520	----
0.75	8,274	7,197	6,206	5,632	----
1.0	4,186	3,966	3,751	3,307	----
	Steady-state prediction				
σ_{ax}/σ_{mh}	New weld	Aged weld	Full repair	Partial i	Partial ii
0.306	14,795	8,942	7,077	6,467	6,298
0.5	12,815	8,182	6,525	6,082	----
0.75	6,484	5,415	5,185	4,530	----
1.0	3,265	2,895	2,971	2,678	----

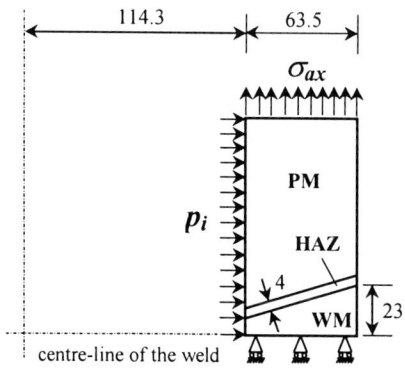

Fig. 1: Dimensions (mm) and loading of a CrMoV pipe.

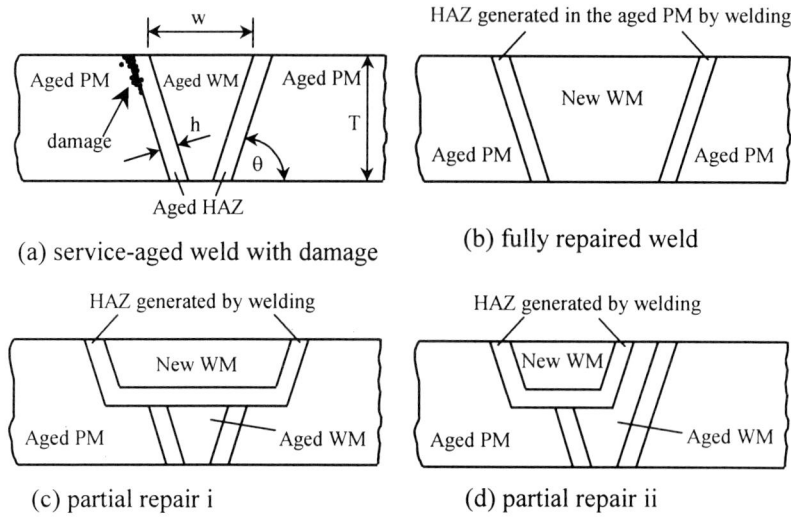

(a) service-aged weld with damage   (b) fully repaired weld

(c) partial repair i   (d) partial repair ii

Figure 2: Schematic diagrams of a service-aged weld, a fully repaired weld and two types of partial weld repairs.

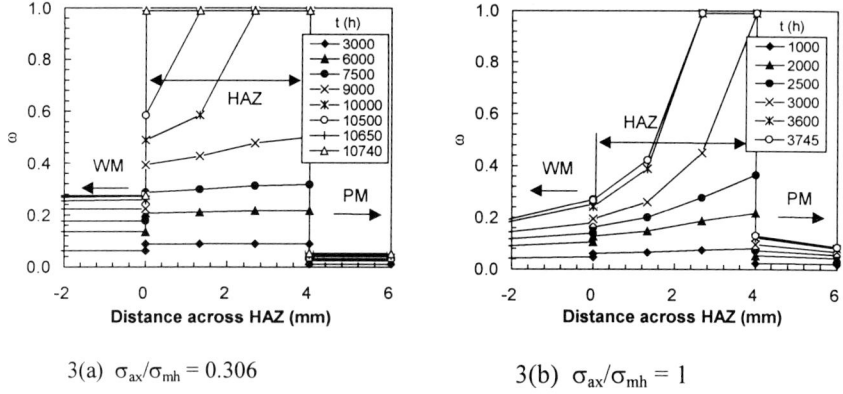

3(a) $\sigma_{ax}/\sigma_{mh} = 0.306$ 3(b) $\sigma_{ax}/\sigma_{mh} = 1$

Figure 3: Variations of damage across the HAZ, along a line parallel with and close to the outer surface of the fully repaired weld, at different times, with $\sigma_{ax}/\sigma_{mh} = 0.306$ (closed-end) and 1 ($p_i = 16.55$ MPa).

TWO PARAMETER CHARACTERISATION OF CRACK TIP FIELDS UNDER CREEP CONDITIONS

A.D.BETTINSON, N.P.O'DOWD, K.NIKBIN AND
G.A.WEBSTER
Department of Mechanical Engineering
Imperial College, London, SW7 2BX, UK.

Abstract

In this work the effect of constraint, *i.e.* specimen size and geometry, on the high temperature crack tip fields is studied. The approach extends the two parameter method developed for elastic-plastic fracture mechanics to the creep regime. Finite element distributions of crack tip stress and creep strain under transient and steady state conditions are presented and the results interpreted in terms of the Q stress. The variation of the Q value from short to long term conditions is also examined for elastic-creep materials.

1. Introduction

When analysing cracked bodies it is useful to identify parameters which can characterise the crack tip fields independent of specimen size and geometry. At low temperature either the stress intensity factor, K, (for small scale yielding) or the contour integral, J, (for elastic-plastic conditions) are used to characterise fracture. However these parameters only identify the first term in a series expansion that describes the stress at the crack tip. As the first term is the only singular term K and J can only accurately determine the magnitude of the stress fields as $r \rightarrow 0$. In elasticity the second term is usually referred to as the T stress and only contributes to a direct stress parallel to the crack plane. In plasticity, O'Dowd and Shih [1] proposed to characterise the difference between the full field solution and the J controlled HRR field with a second parameter, Q, also dependent on r and θ. Further analysis showed that this second term had a very weak dependence on the radial distance r and corresponded to an additional uniform hydrostatic stress field.

The stress fields and therefore failure are then determined by both J and Q. The Q stress quantifies the effect of constraint and geometry effects in fracture

S. Murakami and N. Ohno (eds.), IUTAM Symposium on Creep in Structures, 95–104.

toughness and its value is dependent on crack geometry, magnitude of J and the size of the specimen. An alternative method for measuring the higher order stress terms in elastic-plastic bodies has been proposed by Yang et al. [2] and identifies a second parameter, A_2, to measure the higher order terms. This approach has recently been extended by Nguyen et al. [3] for bodies in the creep regime. The present work , however, is concerned with extending the J–Q approach.

For steady state creep an analogy can be made between power law plasticity and power law creep [4, 5]. This suggests that, under creep conditions, the hydrostatic and Von Mises stress fields are given by:

$$\frac{\sigma_m}{\sigma_0} = \left(\frac{C^*}{\dot{\varepsilon}_0 \sigma_0 I_n r}\right)^{1/(n+1)} \tilde{\sigma}_m(\theta) + Q + \dots \quad , \quad \frac{\sigma_e}{\sigma_0} = \left(\frac{C^*}{\dot{\varepsilon}_0 \sigma_0 I_n r}\right)^{1/(n+1)} \tilde{\sigma}_e(\theta) + \dots \quad (1)$$

where $\dot{\varepsilon}_0$, σ_0, I_n and n are material properties and $\tilde{\sigma}_e(\theta), \tilde{\sigma}_m(\theta)$ are non-dimensional angular functions. For high constraint geometries $Q \approx 0$, at lower constraint geometries Q is negative. Steady state crack growth would now be controlled by both C^* and Q [5]. Experimental and numerical investigations are therefore required to determine the importance of Q in creep crack growth.

2. 2-D Finite Element Formulation

2.1. MODEL

An elastic power law creep material model is used where the creep strain rate is given by:

$$\frac{\dot{\varepsilon}}{\dot{\varepsilon}_0} = \left(\frac{\sigma}{\sigma_0}\right)^n \tag{2}$$

Values of n = 3 and 8 were chosen for the calculations with values of σ_0 = 1260 and 613 MPa respectively and $\dot{\varepsilon}_0$ = 0.2% h^{-1}. In the analysis the Young's Modulus (E) = 100 GPa and v = 0.3. All the finite element analysis was conducted using ABAQUS [6].

To compare the effect of specimen type on the level of constraint, centre cracked panel (CCP) and compact tension (CT) specimen types were chosen, both with a/W = 0.5. An additional CCP specimen was included, with a/W = 0.2, to study the effect of deep and shallow cracks for a given specimen type.

The finite element meshes for the CT and CCP geometries have a focused semi-circular mesh region, centred at the crack tip, with the size of the smallest element being six orders of magnitude smaller than the crack length. The meshes contain four noded, hybrid, plane strain elements [6] and a typical mesh has about 2000 elements. Only half of the CT geometry and a quarter of the CCP geometry were modelled due to symmetry. The co-ordinate system is located at the crack tip and is shown in Fig. 1.

Normalisation of the results has been carried out to make them independent of the material properties $\dot{\varepsilon}_0$ and σ_0, this is achieved by defining a normalised distance: $\bar{r} = r/(C^*/\dot{\varepsilon}_0\sigma_0)$. All stress plots will be presented in the form σ_{ij}/σ_0 vs $\bar{r}$. This normalised distance used may be considered as the distance over which the stress and strain gradients are significant, $i.e.$ within the crack tip stress dominant region. The magnitude of the applied load is given in terms of a normalised C^*, with $\overline{C}^* = C^*/a\dot{\varepsilon}_0\sigma_0$. Six different load magnitudes were applied to each of the specimens: $\overline{C}^* = 0.01, 0.1, 1, 10$ and 100.

3. Steady State Results for Q

For an elastic power law creep material, if the applied load is held constant, after sufficient time has elapsed the stress fields will no longer vary with time and will reach a steady state. In this section the results presented are for steady state conditions and as such are independent of time.

3.1. STRESS FIELDS

The variation of the hydrostatic (σ_m) and Von Mises (σ_e) stresses with distance for $n = 8$, $a/W = 0.5$ are given in Fig. 2 for the CCP geometry and Fig. 3 for the CT geometry. With increasing load the change in the magnitude of the hydrostatic stress fields is greater than for the Von Mises stress fields for both values of n. This is consistent with Eq. 1 where Q is seen only to influence the hydrostatic stress. Increasing the load has virtually no effect on either the hydrostatic or the Von Mises stress fields for the CT geometry (Fig. 3). These results are consistent with those in [1] for elastic-plastic analyses of cracked geometries. Results similar to those shown in Figs. 2 and 3 are obtained for $a/W = 0.2$.

3.2. Q VALUES

In Eq. 1, Q is defined as the difference between the hydrostatic HRR field and the finite element results. Values of Q obtained from the finite element analyses are shown in Fig. 4, evaluated at four different distances: $\bar{r} = 4\times10^{-3}$, 4×10^{-4} and $r/a = 4\times10^{-3}$, 4×10^{-4}. Values of r/a represent a fixed point in the body whereas $\bar{r}$ will depend on the load and material properties. (The distance $\bar{r} = 4\times10^{-3}$ is equivalent to the distance used in [1] for evaluating Q values for an elastic-plastic material.) It should be pointed out that Fig. 4 shows universal curves that depend only on the creep exponent n and specimen geometry (for Q only weakly dependent on distance from the crack tip).

As expected the Q values for the highly constrained CT geometry are greater (i.e. less negative) than those for the low constraint CCP geometry. It may be seen that there is a linear dependence between Q and applied load and, as in plasticity, the value of Q is only weakly dependent on the distance ahead of the crack tip. Increasing the load causes little effect on Q for the CT geometries whereas a stronger influence is seen for the CCP geometries. The loss of a linear relationship between Q and load for the $n = 3$, CCP geometries at $\bar{r} = 4\times10^{-3}$ may be because the stresses are being evaluated at the outer limits of the specimen ($r/a \approx 0.5$) away from the crack tip region, although the results for $n = 8$ are not similarly affected.

4. Transient Results

4.1. REFERENCE TIME

Steady state conditions are reached when the stress no longer varies with time. In the numerical analysis, the time at which this condition is met (designated t_{ss}) is determined by monitoring the stresses along the crack front. For the steady state results presented in the previous section, the precise value of t_{ss} was not important provided $t \gg t_{ss}$. However, to discuss results under transient loading, $0 < t < t_{ss}$, a more rigorously defined steady state or reference time is required. A transition time between the elastic response and steady state creep, defined by Reidel [7], and used elsewhere in the literature [8, 9] is defined as:

$$t^* = \frac{K^2(1-\upsilon^2)}{EC^*(n+1)} \tag{3}$$

and this quantity will be used in the subsequent section. It should be noted that the actual time taken to reach steady state within the finite element formulation is significantly greater than t^*, typically $t_{ss} > 10t^*$.

4.2. FINITE ELEMENT RESULTS

Only the centre cracked panel with $a/W = 0.5$ was analysed. Future work will include studies of other geometries and crack lengths. Two different load levels were analysed ($\overline{C}^* = 0.01$ and 10) with both $n = 3$ and 8. Q values were obtained for these loads and n values for both CCP geometries. Note that while the same definition for Q used in the steady state analysis (Eq. 1) is retained with C^* replaced by $C(t)$, the form of the field may no longer be given by Eq. 1.

Only plots of Q vs t/t^* for $n = 8$ and $a/W = 0.5$ are shown in Fig. 5. The effect of specimen geometry and creep exponent had little influence on the transient values of Q—the results for $n = 3$ and $a/W = 0.2$ are similar to those shown in Fig. 5. Values of Q are evaluated at different distances from the crack tip (it should be noted that for $\overline{C}^* = 0.01$, $\overline{r} = 4 \times 10^{-4}$ cannot always be resolved due to mesh limitations). For each curve the value of $C(t)$ at the relevant distance is used in Eq. 1 to determine Q. For both values of n, Q reaches its steady state value by $t/t^* = 4$ for the load $\overline{C}^* = 0.01$, and is within 10% of its steady state value at $t/t^* = 4$ for $\overline{C}^* = 10$. At short times, the component behaves as an elastic material leading to large positive Q values. (The large negative Q values at very short times seen for the smallest distances, $\overline{r} = 4 \times 10^{-4}$, 4×10^{-3}, may be due to numerical inaccuracies). During the early stages of creep ($0 < t/t^* < 2$) Q has a strong dependence on distance, but approaches the steady state value as t increases.

For the cases examined, the evolution of Q with time is weakly dependent on geometry and n, which may simplify the application of Q in the transient regime. However, as it is clearly dependent on distance, it will be necessary to specify a distance at which Q should be evaluated in order to make comparisons between different geometries.

5. *3-D* Finite Element Formulation

The work presented so far has been based on a 2-dimensional, plane strain analysis and is therefore independent of specimen thickness. However, as the thickness of the specimen will effect the stress fields ahead of the crack tip (plane stress vs. plane strain conditions) it is important to study this change of constraint within a 3-dimensional formulation.

5.1. MODEL

An elastic power law creep model with $n = 8$ is used in this analysis. A CT specimen with $a/W = 0.6$ and $B/W = 0.5$ was modelled (where B is the specimen thickness). The mesh has 23,000 8-noded elements in total and has 15 elements in the thickness direction; the spacing of the elements is biased through the thickness so there are more towards the free surface. The co-ordinate system is as shown in Fig. 1 with a third axis, the z-axis, measuring the out of plane dimensions. Only a quarter of the CT specimen is modelled, with the mid-plane of the specimen at $z/B = 0$ and the free surface at $z/B = 0.5$.

Only a single load, $\overline{C}^* = 10$, and steady state conditions were considered for this analysis. The *3-D* results should reflect stress conditions between the plane stress and plane strain boundaries; these boundaries were determined by two *2-D* analyses, one with plane stress and the other with plane strain element types. Both the *2-D* cases had the same applied load per unit thickness and geometry as the *3-D* specimen. To further validate the results, a second *3-D* case was modelled to generate fully plane strain conditions ahead of the crack tip. This was achieved by fixing the free surface of the specimen ($z/B = 0.5$) so that it could not move in the z-direction.

5.2. RESULTS

The stress fields and C^* vary with thickness through the mesh. The *3-D* results could therefore be considered as a series of *2-D* data sets from $z/B = 0$ to $z/B = 0.5$, each with its own value of C^* and level of constraint as measured by Q. The variation of C^* ahead of the crack tip at the mid-plane and near-free surface is shown in Fig. 6a whilst the change in C^* through the thickness of the specimen at a given distance ahead of the crack tip is shown in Fig. 6b.

As Fig. 6a shows, apart from very close to the crack tip (the first 2-3 elements) the value of C^* is constant—i.e. it is confirmed as being a path independent integral. The change in C^* through the thickness is considerable though, with mid-plane values about five times larger than free surface values.

Q values were calculated in the same manner as for the *2-D* analysis discussed previously. The variation of Q ahead of the crack tip at the mid-plane and near-free surface is shown in Fig. 6c whilst the change in Q through the thickness of the specimen at a given distance ahead of the crack tip is shown in Fig. 6d. The change in Q ahead of the crack has only a weak dependence on the distance, r, ahead of the crack tip and in this respect is similar to the *2-D* results. The Q values of the 'constrained' *3-D* mesh are identical to those from the plane strain analysis and considerably larger than the plane stress results (Fig. 6c). Although the near-free surface values of Q are between the plane stress and plane strain limits for the *3-D* mesh, the mid-plane values are above the plane strain results.

Fig. 6d shows this discrepancy more clearly. For $0 < z/B < 0.3$ the Q values from the *3-D* analysis exceed the plane strain values, an unexpected result since plane strain conditions are expected to reflect fully constrained conditions. However, as the free surface (and plane stress) conditions are approached the expected drop in constraint is seen. This issue is still being investigated.

6. Conclusions

In this paper, the stress fields ahead of a crack in a power law creeping material have been studied in order to quantify the degree of constraint imposed by different specimen types and different loads. The crack tip fields are shown to be controlled by the load, measured by C^*, and the level of constraint, measured by the Q stress.

From a *2-D*, plane strain analysis the level of constraint is shown to decrease with applied load for both CT and CCP geometries. The drop in constraint for the CCP specimens is considerably greater than that for the CT specimens showing that, for the same applied load, a CCP geometry has a lower constraint than a CT geometry.

A *3-D* analysis to examine the change in constraint through the thickness of a plain sided CT specimen shows the highest level of constaint to be along the centre line of the specimen with a rapid decrease as the free surface is approached. However, whilst this trend is as expected, the actual values of Q exceed those given by the *2-D* plane strain analysis (which should represent the upper limit of constraint)—work is currently being undertaken to resolve these issues.

7. Acknowledgements

Financial support for this work was provided by British Energy Ltd., through the Industrial Support Committee of the United Kingdom (IMC).

8. References

1. O'Dowd N.P., Shih, C.F., Two-Parameter Fracture Mechanics: Theory and Applications, *Fracture Mechanics: Twenty-Fourth Volume, ASTM STP 1207*, J.D. Landes, D.E. McCabe and J.A.M. Boulet, Eds., ASTM, Philadelphia, pp.21-47, 1994.
2. Yang S., Chao Y.J. and Sutton M.A., On the fracture of solids characterized by one or two parameters: theory and practice, *Eng. Fracture Mechanics*, **45**, pp. 1-20, 1993.
3. Nguyen B.N., Onck P.R. and Van der Giessen E., *Modelling of Microstructual Evolution in Creep Resistant Materials*, A.Strang and M.McLean, Eds., IOM Communications Ltd., London, 1999.
4. Budden P.J., The effect of constraint on creep assessments using the time dependent failure assessment diagram, Nuclear Electric Report EPD/GEN/REP/0069/96, 1996.
5. Budden P.J., The effect of Constraint on Creep Crack Initiation and Growth: A Preliminary Assessment Procedure, Nuclear Electric Report EPD/GEN/REP/0105/96, 1996.
6. ABAQUS v5.5, Hibbitt, Karlsson and Sorensen Inc., Providence, RI, 1996.
7. Riedel H, *Fracture at High Temperatures*, Springer-Verlag, Berlin, 1987.
8. Ehlers R. and Riedel H, A Finite Element Analysis of Creep Deformation in a Specimen containing a Macroscopic Crack, Advances in Fracture Research, Proceedings of the 5th International Conference on Fracture, Editor. D. Francois, Pergamon Press, Oxford and New York, Vol.2, pp691-698, 1981.
9. Linkens D., Busso E.P. and Dean D.W., Predictions of Non-steady Asymptotic Crack Tip Fields in Power Law Creeping Materials, Proc. 12th International Conference on Structual Mechanics in Reactor Technology, Stuttgart, 1993.

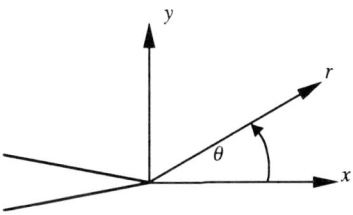

Figure 1 - Co-ordinate system for finite element meshes.

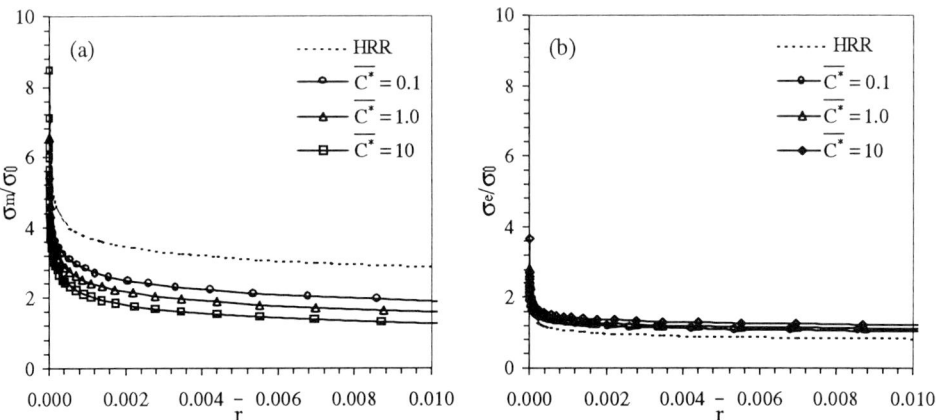

Figure 2 - CCP steady state stress distributions, $a/W = 0.5$, $n = 8$, $\theta = 0$.
(a) Hydrostatic stress, (b) Von Mises stress.

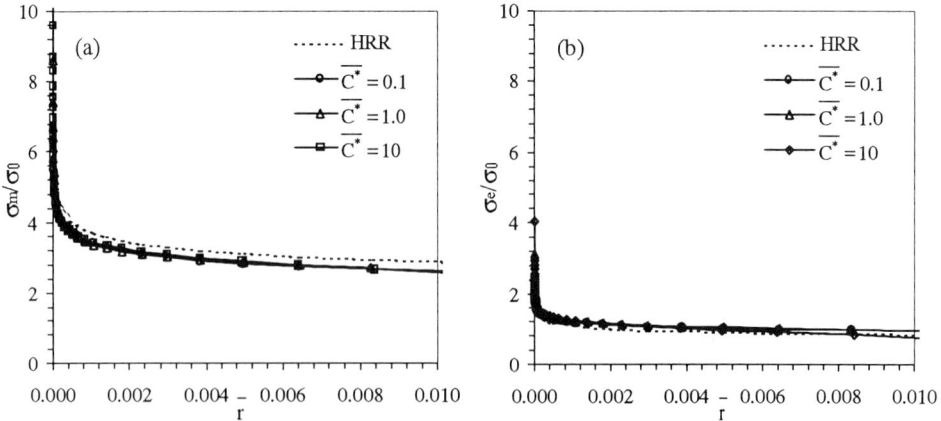

Figure 3 - CT steady state stress distributions, $a/W = 0.5$, $n = 8$, $\theta = 0$.
(a) Hydrostatic stress, (b) Von Mises stress.

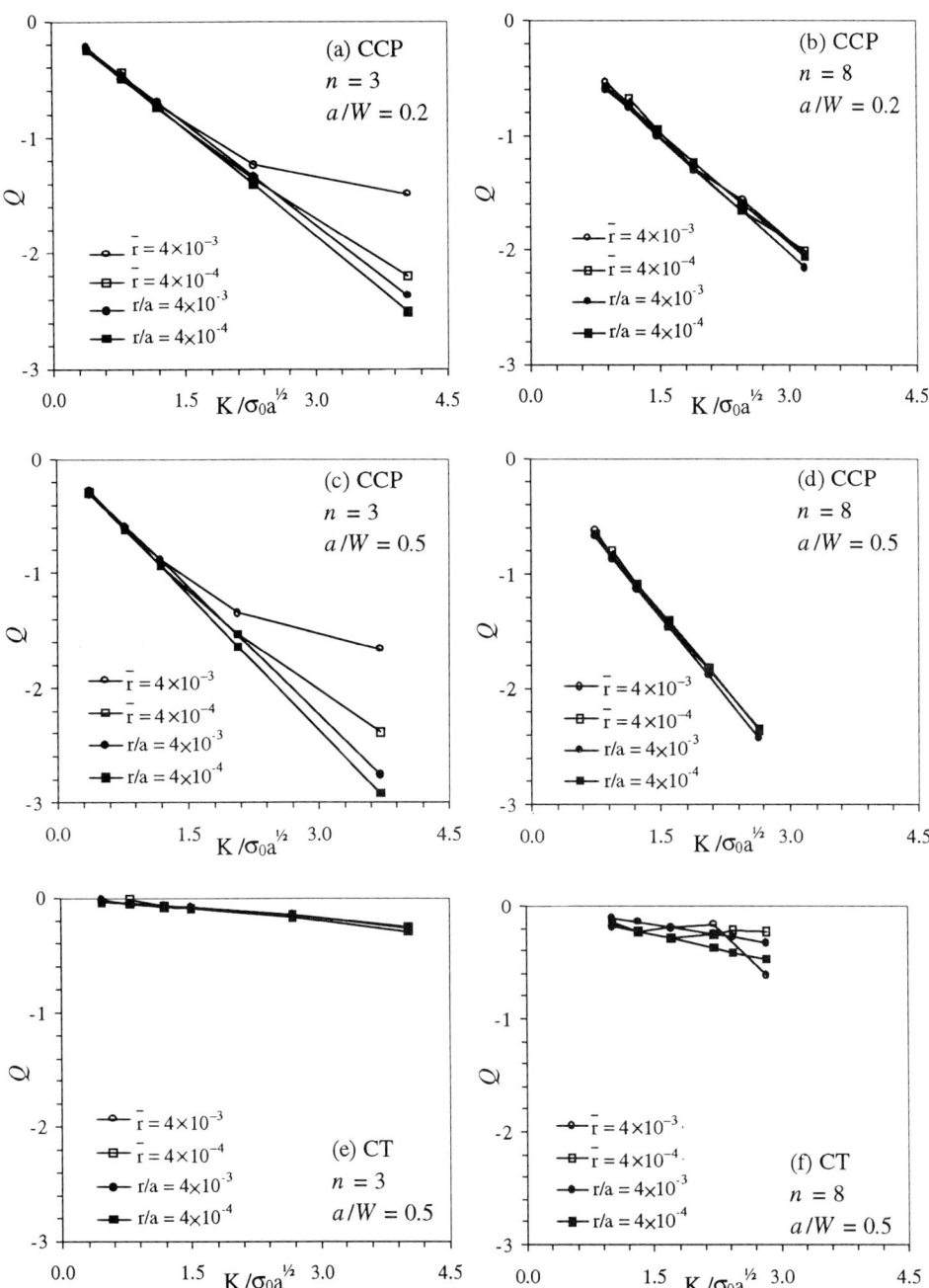

Figure 4 - Q vs normalised load for CCP and CT specimens.

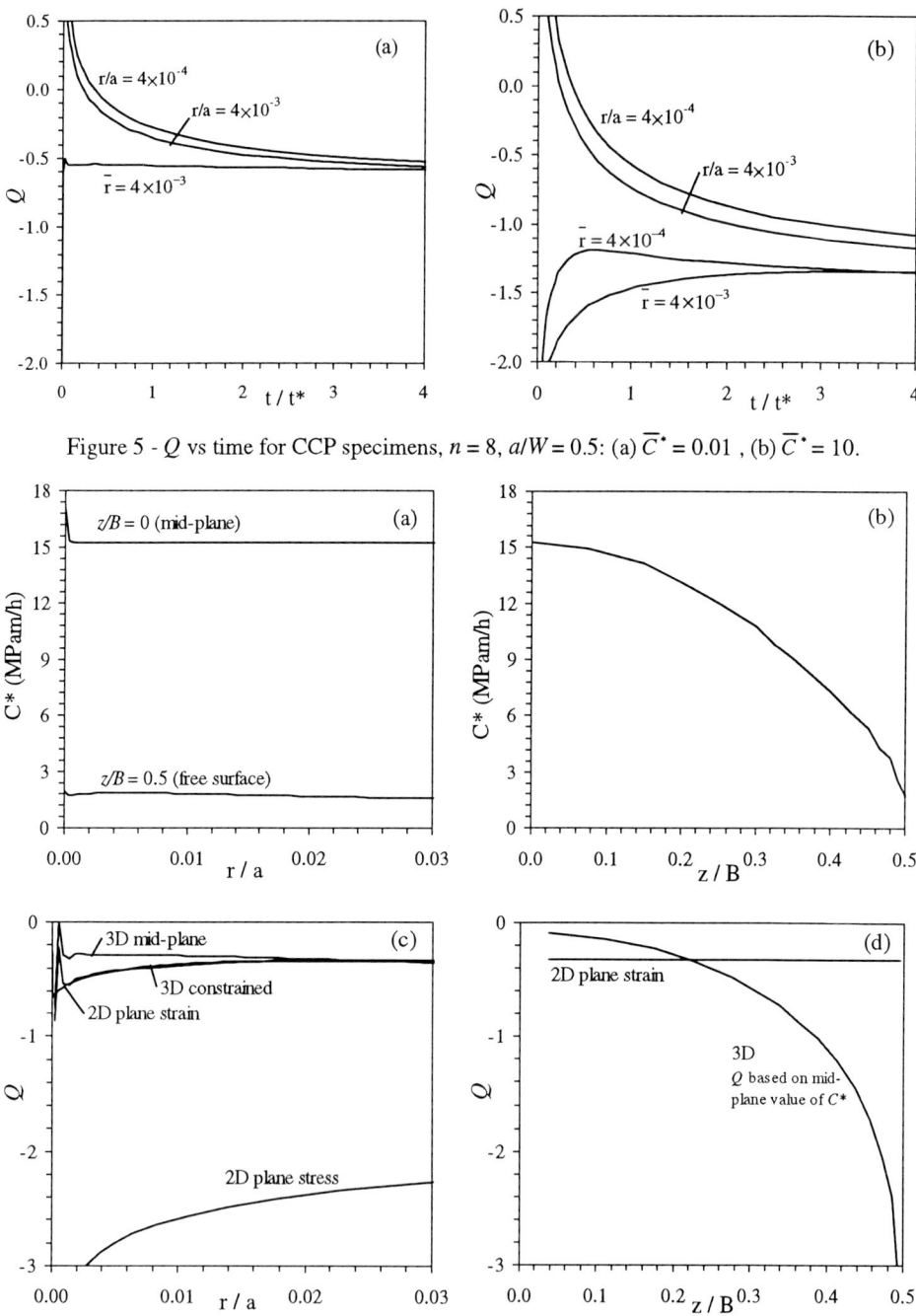

Figure 5 - Q vs time for CCP specimens, $n = 8$, $a/W = 0.5$: (a) $\overline{C}^* = 0.01$, (b) $\overline{C}^* = 10$.

Figure 6 - 3D results, $a/W = 0.6$, $n = 8$: C^* vs. (a) r/a and (b) z/B,
Q vs (c) r/a and (d) z/B (Q evaluated at $r/a = 4\times10^{-3}$).

CAVITY GROWTH INDUCED BY ELECTRIC CURRENT AND STRESS IN LSI CONDUCTOR

T. KITAMURA and T. SHIBUTANI
Kyoto University
Department of Engineering Physics & Mechanics
Graduate School of Engineering
Yoshida-honmachi Sakyo-ku Kyoto
Japan, 606-8501

1. Introduction

Reduction in dimension of microelectronic devices has been pursued in order to attain high performance, so that the conductor width in an advanced LSI is approaching 0.1 micron. Since it is exposed to high temperature conditions during the processing and in the service due to high electric current density, the atom migration along grain boundary (GB) and interface (IF) brings about serious damage such as a cavity. As only one cavity is fatal for the microcircuit, the characteristics of growth must be understood precisely.

Figure 1 shows a schematic illustration explaining the mechanism of atom migration due to the electric current in an LSI conductor. The atoms (nucleus) are transported toward the anode by the momentum of electrons, which is termed "electromigration". Because deformation of the conductor is constraint by the surrounding components such as passivation and substrate, stress is induced by the atom transport. The stress is compressive on the anode and is tensile on the cathode. Then, the stress gradient brings about another atom migration from the former to the latter, which is opposite to the electromigration. This stress-induced flow is the well-known mechanism of diffusion creep. It is, therefore, important to understand the coupling effect of migration due to the stress as well as the electric current.

In this study, the mechanics of coupling migration in a polycrystalline conductor is investigated by numerical simulation.

S. Murakami and N. Ohno (eds.), IUTAM Symposium on Creep in Structures, 105–114.

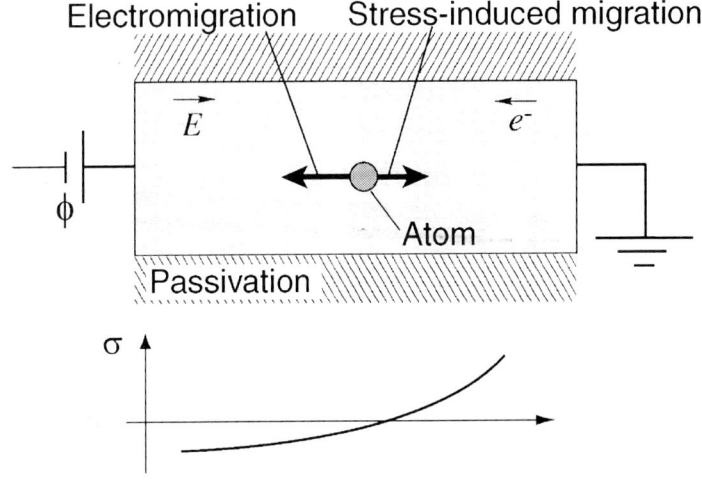

Figure 1. Coupling migrations caused by electric current in an LSI conductor.

2. Diffusion in Polycrystalline Material

2.1 DIFFUSION CREEP

The atom transport along a GB (IF) induced by the stress is formulated by

$$J_\alpha = \frac{D_\alpha \Omega}{kT}\left(\frac{\partial \sigma}{\partial x_\alpha}\right) \qquad (\alpha = b \text{ or } i) \tag{1}$$

where J_α is the atom flux along the GB (IF), D_α is the diffusion coefficient, k is the Boltzmann constant, T is the temperature, Ω is the volume of atom, σ is the normal stress acting on the GB (IF) and x_α is the local coordinate along the GB (IF). Here, $\alpha = b$ and $\alpha = i$ represent the quantities along the GB and along the IF, respectively. The conservation of matter requires

$$\delta_\alpha \frac{\partial J_\alpha}{\partial x_\alpha} + \dot{d}_\alpha = 0 \tag{2}$$

on the GB (IF) and

$$\sum_{\text{GBs at junction}} J_\alpha = 0 \tag{3}$$

at the junction of GBs (IFs) where δ_α is the GB (IF) thickness and d_α is the accumulation/erosion rate.

Combining the above equations with the principle of virtual work rate, the functional in GB (IF) diffusion creep of polycrystalline material is given by [1,2]

$$\Phi = \frac{kT}{2D_\alpha \delta_\alpha \Omega} \int_A (J_\alpha \delta_\alpha)^2 \, dA - \int_S T_i v_i \, dS + \sum_{all\ junctions} \lambda_k \sum_{GBsatjunct\ ion} J_\alpha \delta_\alpha \qquad (4)$$

where A is the all GBs (IFs), S is the boundary of the system, T_i is the traction on S, v_i is the displacement rate on S and λ_i are the Lagrange multipliers. On the basis of the variational principle, Cocks [2,3] derived the following matrix equation,

$$[G]\{U\} = \{P\} \qquad (5)$$

where $[G]$ is the stiffness and constraint matrix determined by the geometry of GB (IF) network, and $\{P\}$ is the vector of applied force. $\{U\}$ is the unknown vector which consists of the displacement/rotation rate of each grain, the Lagrange multipliers and the fluxes at the midpoint of each boundary. Since the atom flux along a GB (IF) is correlated with the relative displacement/rotation rate of the adjacent grains, the flux in Eq.(4) is converted into $\{U\}$ [3]. The matrix equation, Eq.(5), is applicable to the diffusion creep analysis of any polycrystalline with arbitrary GB network. The distribution of creep rate as well as the normal stress can be calculated by the distribution of flux, the Lagrange multipliers and Eq.(1). The derivation process of Eq.(5) and the elements in $[G]$ are shown in detail in the original paper [3].

2.2 DIFFUSION INDUCED BY ELECTRIC CURRENT AND STRESS

The atom transport due to the electric current as well as the stress is formulated by

$$J_\alpha = \frac{D_\alpha}{kT} \left(|e|Z^* \frac{\partial \phi}{\partial x_\alpha} + \Omega \frac{\partial \sigma}{\partial x_\alpha} \right) \qquad (\alpha=b\ or\ i) \qquad (6)$$

where e is the charge of an electron and Z^* is the effective charge of nucleus. The first term represents the contribution of atom transport due to the "electromigration". The following functional and the matrix equation are readily derived by the same procedure as those used in the diffusion creep [4]

$$\Phi = \frac{kT}{2D_\alpha \delta_\alpha \Omega} \int_A (J_\alpha \delta_\alpha)^2 \, dA - \int_S T_i v_i \, dS$$

$$- \frac{|e|Z^*}{\Omega} \int_A \frac{\partial \phi}{\partial x_\alpha} J_\alpha \delta_\alpha \, dA + \sum_{all\ junctions} \lambda_k \sum_{GBsatjunct\ ion} J_\alpha \delta_\alpha \qquad (7)$$

$$[G]\{U\} = \{P\} + \{C_e\} \tag{8}$$

where $\{C_e\}$ is the vector determined by the electric potential field. As $\{C_e\}$ is the only additional term to Eq.(5), Eq.(8) can be solved for an LSI conductor with arbitrary GB network if the electric potential filed is given. Because the potential obeys the well-known Laplace equation, the field can be numerically analyzed by the boundary element method (BEM). Thus, the unknown vector $\{U\}$ in the matrix equation, Eq.(6), can be numerically obtained for the LSI conductor.

3. Simulation Method

As the conductor is thin, the grains and the cavity observed are columnar. Thus, the fracture process can be simulated by the two-dimensional analysis. Figure 2 presents a model of conductor analyzed. The electric potential difference, ϕ_o, is applied at the ends of the conductor, which consists of hexagonal grains. The displacements at the sides are perfectly constraint by the surrounding glass phase and the periodic boundary condition is adopted for the longitudinal direction. No external load is applied ($\{P\}=0$ in Eq.(8)) in the simulations in order to analyze the pure effect of electric current on the fracture process. Therefore, the stress appearing in the following section is caused by the electromigration and the mass conservation as explained in Fig.1. The effect of external load was reported in elsewhere [5]. At first, the migration is analyzed for the conductor without a cavity in order to elucidate the stress concentration. Then, a circular cavity whose non-dimensional radius, a/d, is 0.1 to 0.9 is introduced at a junction of GB and IF. Here, a is the cavity radius, d is the GB length and "IF" indicates the interface between the conductor and the surrounding passivation. The simulation is carried out under the ratio of diffusivities along the IF and the GB, $D_i\delta_i/D_b\delta_b$, of 0.1, 1 and 10.

After the BEM calculation of electrical potential, the matrix equation, Eq.(8), is solved under $\{P\}=0$. The quantities in the analysis are normalized as follows;

$$\bar{a} = a/d, \quad \bar{x}_\alpha = x_\alpha/d, \quad \bar{\phi} = \phi L/\phi_o d$$

$$\bar{\sigma} = \frac{\sigma}{\sigma_o} = \frac{\sigma\Omega L}{|e|Z^*\phi_o d}, \text{ and } \bar{J}_\alpha = \frac{J_\alpha\delta_\alpha kTd}{D_b\delta_b\sigma_o} \tag{9}$$

where the variable with the upper bar indicates the quantity in the non-dimensional form.

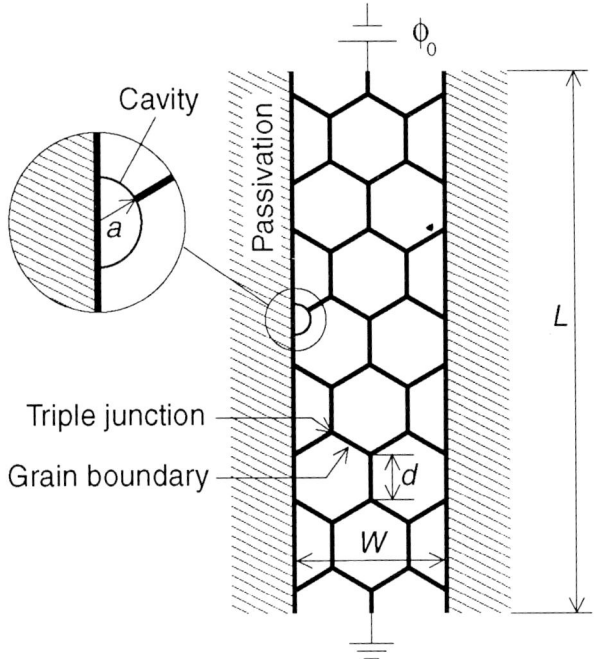

*Figure 2.*Model of straight polycrystalline conductor in an LSI.

4. Results And Discussion

Figures 3(a) and 3(b) show the electrical potential fields in the conductor without and with the cavity, respectively. The electrical current, which is perpendicular to the equi-potential lines, is disturbed near the cavity in the latter while it is uniform in the former. However, the eminent disordered region is confined in one GB length from the cavity. Figure 4(a) shows the distribution of normal stress on the GB (IF) in the conductor without the cavity under $D_i d_i / D_b d_b = 1$. The dark (light) region possesses the high tensile (compressive) stress on the GB (IF). The tensile stress concentrates near the junction along the side of conductor (IF), where an atomic flow branches off into two flows. As the diffusivity along the IF is equal to that along the GB, this indicates that the stress concentration is strongly affected by the geometry of the GB/IF network. It is also clear from the figure that the cavity is easy to initiate at the junction along the side of conductor. Figure 4(b) shows the distribution of normal stress in the conductor with a cavity at the junction. The stress concentration near the junction is relaxed by the cavity.

The cavity–affected area extends over a few GB (IF) lengths ahead of the tips, which is larger than the disordered area in the electrical potential field as shown in Fig.3(b).

Figure 5 shows the flux distribution along the GB ahead of the cavity tip. The total flow (thick line) consists of the electromigration (broken line) and the stress-induced one (thin line). Since the total flux at the tip is positive, the atoms flow out from the cavity surface into the GB. It implies that the cavity grows under the electric current without the external load. Figure 5 also points out that the electromigration is inactive

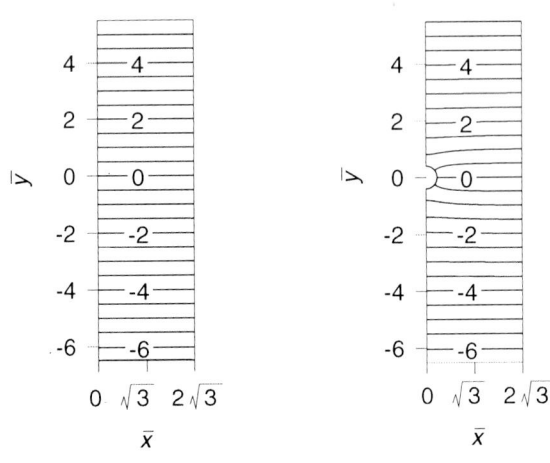

(a) Conductor without a cavity   (b) Conductor with a cavity

Figure 3.   Electrical potential.

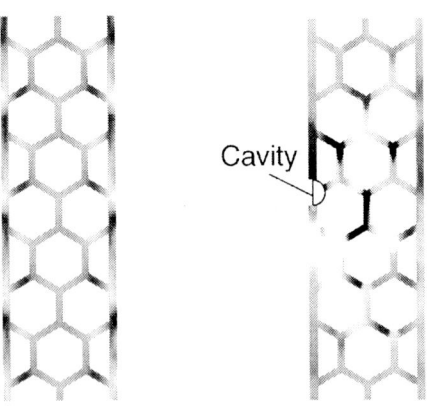

(a) Conductor without a cavity   (b) Conductor with a cavity

Figure4.   Normal stress acting on the grain boundaries and the interface

near the cavity tip so that the stress-induced migration governs the efflux from the cavity. On the other hand, the former dominates the total flux away from the cavity tip. This reveals the combined effect that the atoms are transported from the cavity into the GB by the stress and they are carried off to far GBs by the elecromigration.

In order to extract the effect of cavity on the atom flow, the "cavity flow", ΔJ_α is defined as

$$\Delta J_\alpha = J_{\alpha c} - J_{\alpha v} \tag{10}$$

where $J_{\alpha c}$ is the flux in the conductor with the cavity and $J_{\alpha v}$ is the one without it. Figure 6 shows the distribution of ΔJ_α for several cavity radii under $D_i\delta_i/D_b\delta_b=1$. The arrows indicate the magnitudes of ΔJ_α along each GB (IF) at the triple junctions. As the cavity grows, the effective zone expands ahead of the tip.

The growth rate of cavity volume, dV/dt, is given by the summation of atom efflux at the cavity tips;

$$\frac{dV}{dt} = \sum_{x_c} J_\alpha \delta_\alpha \tag{11}$$

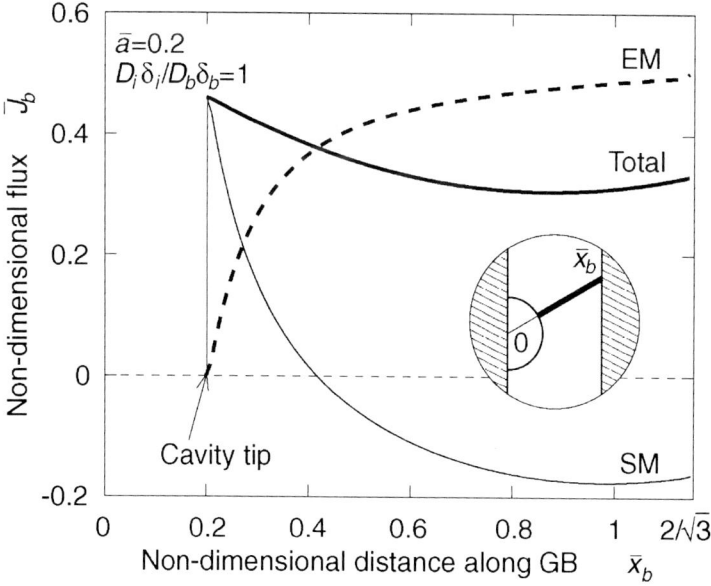

Figure 5.   Flux along the grain boundary ahead of cavity tip.

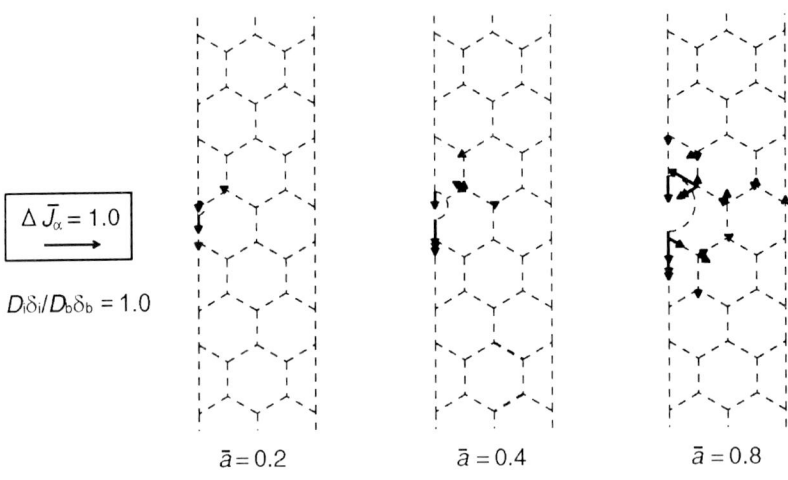

$\Delta \bar{J}_\alpha = 1.0$

$D_i\delta_i/D_b\delta_b = 1.0$

$\bar{a} = 0.2$ $\bar{a} = 0.4$ $\bar{a} = 0.8$

Figure 6. Morphology of cavity flow.

Figure 7 shows the dependence of the growth rate on the cavity radius (solid squares). It points out that the rate changes little during the cavity growth though it is decelerated at the early stage. In a previous paper [6], we analyzed the growth behavior of cavity initiated in a bulk material under the diffusion creep and clarified that the growth rate was strongly affected by the microstructure. The triple junction played as a strong barrier for the cavity growth so that the rate was slowed down as the tip approached the junction. Figure 7, however, implies that the effect of microstructure is fairly moderate in the LSI conductor under the electric current.

Figure 8 shows the effect of diffusivity along the IF on the flux inside of the conductor. For a fixed magnitude of $D_b\delta_b$, the IF diffusion at $D_i\delta_i/D_b\delta_b=10$ is the highest, while that at $D_i\delta_i/D_b\delta_b=0.1$ is the lowest. It is clear from the figure that the high IF diffusivity activates the atom transport along the GBs near the IF. This is because the diffusivity is correlated with the mobility of grains (see the section 2). Similar result is reported for the polycrystalline bulk under the diffusion creep [5]. The cavity growth rate is plotted in Fig.7, where the negative growth rate means that the cavity shrinks. Although the rate under $D_i\delta_i/D_b\delta_b=0.1$ is close to that under $D_i\delta_i/D_b\delta_b=1$, the cavity can not grow beyond a/d>0.7 in $D_i\delta_i/D_b\delta_b=0.1$. Thus, the low IF diffusivity represses the cavity growth. On the other hand, the growth is accelerated in $D_i\delta_i/D_b\delta_b>1$ because the IF governs the atom transport. The growth, however, is decelerated as the tip approaches the triple junction (a/d=1). The micro-structural effect is intensified in the case.

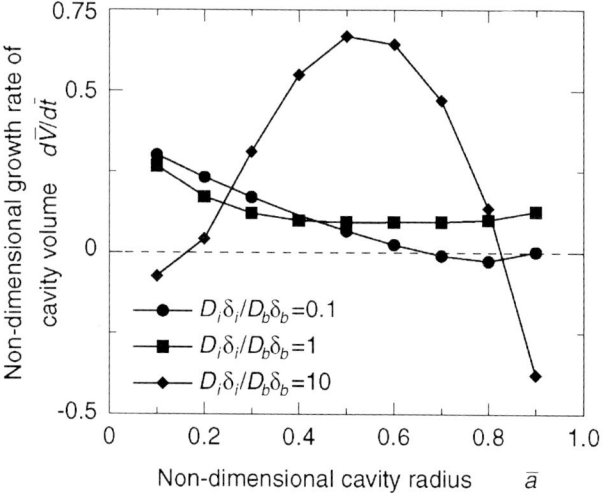

Figure 7.   Growth rate of cavity volume.

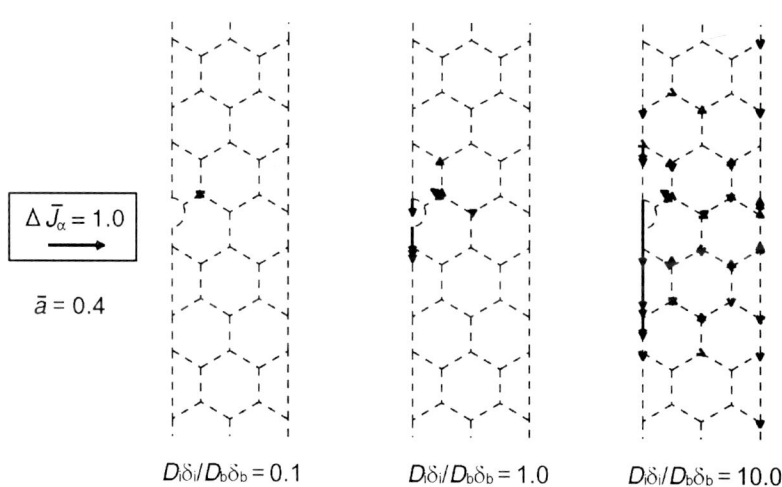

Figure 8. Dependence of the cavity flow on the diffusivity ratio.

5. Conclusion

Numerical simulations are conducted for the cavity growth in the polycrystalline LSI conductor under the electromigration and the stress induced by the electric current, and the mechanism and the mechanics are analyzed. No external load is applied in the simulation, so that the stress is caused by the electromigration and the mass conservation. The results obtained are summarized as follows.

(1) The matrix equation is derived for the atom transport caused by the electric current as well as the stress in a polycrystalline material. The atom flux along the GB (IF) near a cavity in an LSI conductor with arbitrary GB/IF network can be numerically analyzed by the equation.

(2) The high stress concentration is induced by the combined migration at the junction of GB and IF in the conductor without a cavity. When a cavity is introduced at the junction, the atom flow is activated in the vicinity of it. As the cavity grows, the activated region expands. The cavity grows under the electric current.

(3) The efflux at the cavity tip is governed by the induced stress while the flow-out atoms are carried away by the elecromigration.

(4) The cavity growth rate is dependent not only on the combined diffusivity, $D_i d_i / D_b d_b$, and cavity size but also on the GB/IF network.

This investigation was supported by the Grant-in-aids for scientific research; B(2) No.11555031 and No.00114433.

References
[1] A. Needleman J.R. Rice, Acta Metllurgica, **28**, 1315(1980).
[2] A.C.F. Cocks, Applied Solid Mechanics, Elsevier Applied Science, 30(1989).
[3] A.C.F. Cocks and A.A. Searle, Mechanics of Materials, **12**, 279(1991).
[4] T. Kitamura and T. Shibutani, Stress Induced Phenomena in Metallization, AIP Conf. Proc. **491**, American Institute of Physics, 168(1999).
[5] T. Shibutani, T. Kitamura and R. Ohtani, Trans. JSME, A, **65**, 2497(1999) (in Japanese).
[6] T. Kitamura, R. Ohtani, T. Yamanaka and Y. Hattori, JSME International Journal, A, **38**, 581(1995).

DISLOCATION DENSITY SIMULATIONS FOR BULK SINGLE CRYSTAL GROWTH PROCESS USING DISLOCATION KINETICS MODEL

N. MIYAZAKI

Department of Materials Process Engineering

Graduate School of Engineering, Kyushu University

6-10-1 Hakozaki, Higashi-ku, Fukuoka, 812-8581 JAPAN

1. Introduction

Dislocation-free or low-dislocation density bulk semiconductor crystals such as Si, GaAs and InP are required for substrates of high-performance electronic or optical devices, because dislocations existing in substrate have adverse effects on the performance of devices. They are usually manufactured by the CZ (=Czochralski) or the LEC (=Liquid Encapsulated Czochralski) method. GaAs and InP crystals have low resolved shear stresses and are easy to generate and multiply dislocations due to thermal stress during the LEC growth, so it is an important technical problem to reduce the dislocations as low as possible. One of the methods to grow low-dislocation density crystals is to increase the crystal resistance to thermal stress by doping impurity atoms. As for a Si single crystal, growing a dislocation-free bulk single crystal of large diameter such as 12- or 16-inch remains to be solved in future.

It is known that thermal stress during CZ or LEC growth plays an important role in the generation of dislocations, especially in the case of InP and GaAs which have low critical resolved shear stresses. Qualitative relation between dislocation density and thermal stress in such single crystals has been studied so far by comparing resolved shear stresses obtained from an elastic thermal stress analysis with critical resolved shear stress (Jordan *et al.*, 1980 ; Miyazaki *et al.*, 1992). In elevated temperatures, solids undergo time-dependent inelastic deformation. Hereafter such deformation is called creep in the present paper. The constitutive equations for creep of single crystals are known as the Haasen-Sumino model (Alexander and Haasen, 1968 ; Suezawa *et al.*, 1979). The dislocation density during growth process can be quantitatively predicted by using this model together with a thermal stress analysis. Although the results of dislocation density obtained from analytical methods (Volkl and Muller, 1989 ; Tsai *et al.*, 1990 ; Maroudas and Brown, 1991) and quasi-static finite element analysis (Tsai, 1991 ; Tsai *et al.*, 1992 ; Tsai *et al.*, 1993 ; Subramanyam and Tsai, 1995) have been reported so far, all these analyses assume a constant growth

S. Murakami and N. Ohno (eds.), IUTAM Symposium on Creep in Structures, 115–124.

rate and a constant radius of a bulk single crystal and give only steady state values of dislocation density.   There has been no study concerning the time variations of dislocation density and stresses during growth process.

In the present study, we develop a computer code for a transient and continuous simulation of dislocation density in a bulk single crystal during growth process using the Haasen-Sumino model.   This code can deal with the case where the pulling rate and radius of a bulk single crystal are dynamically changed and provides the time variations of dislocation density and stresses during growth process.   Dislocation density simulations are performed using this computer code for 8-inch diameter Si, InP and GaAs single crystals to show the effectiveness of this computer code. Dislocation density simulations are also performed for an InP bulk single crystal to examine the effect of doping atoms on the dislocation density.   In the simulations, the effect of doping atoms which increases crystal resistance to thermal stress on dislocation density is examined by comparing the result of a low doped InP with that of a highly doped InP.

2.   Method of Analysis

2.1.   CONSTITUTIVE EQUATION FOR CREEP

At elevated temperatures, the total strain rate $\dot{\varepsilon}_{ij}$ of a solid is assumed to be given by

$$\dot{\varepsilon}_{ij} = \dot{\varepsilon}^{e}_{ij} + \dot{\varepsilon}^{t}_{ij} + \dot{\varepsilon}^{c}_{ij} \tag{1}$$

where $\dot{\varepsilon}^{e}_{ij}$, $\dot{\varepsilon}^{t}_{ij}$ and $\dot{\varepsilon}^{c}_{ij}$ are elastic strain rate, thermal strain rate and creep strain rate, respectively.   The creep strain rate can be obtained by extending the Haasen-Sumino model to a multiaxial stress state in the same way as the isotropic flow theory of plasticity, such that (Tsai et al., 1992) :

$$\dot{\varepsilon}^{c}_{ij} = f S_{ij} \tag{2}$$

where

$$f = \frac{bk_0 N_m \exp\left(\frac{-Q}{kT}\right)\left(\sqrt{J'_2} - D\sqrt{N_m} - \tau_d\right)^p}{2\sqrt{J'_2}} \tag{3}$$

$$S_{ij} = \sigma_{ij} - \frac{\sigma_{kk}\delta_{ij}}{3} \quad , \quad J'_2 = \frac{S_{ij}S_{ij}}{2} \tag{4}$$

In the above, σ_{ij}, S_{ij} and J'_2 are stresses, deviatoric stresses and the second invariant of the deviatoric stresses, respectively. The dislocation density is denoted by N_m, and its rate is given by

$$\dot{N}_m = K k_0 N_m e^{-Q/kT} \left(\sqrt{J'_2} - D\sqrt{N_m} - \tau_d \right)^p \left(\sqrt{J'_2} - D\sqrt{N_m} \right)^\lambda \qquad (5)$$

The τ_d in the above equations is a drag-stress caused by the interaction between dislocation density and impurity atoms and is set equal to 3.61×10^7 dyn/cm^2 (1 MPa=10^7 dyn/cm^2), which is given by Tsai (1991) for Si, in all cases in the present analyses except for the analyses performed to examine the effect of doping atoms in an InP single crystal on the dislocation density. We consider the following two cases of the drag-stress to examine the effect of doping atoms (Jordan $et\ al.$, 1985) :

Case 1 : $\tau_d = 10^{(4.213+2959.0/T)}$ dyn/cm^2, for 1.5×10^{17}cm^{-3} Ge doped InP   (6)
Case 2 : $\tau_d = 6.42 \times 10^7$ dyn/cm^2, for 1.3×10^{19}cm^{-3} Ge doped InP   (7)

The Case 1 and Case 2 are a low doped case and highly doped case, respectively. The relations between the drag-stress and temperature are given in Figure 1. In the Case 1, the τ_d decreases with temperature, and takes a very small value around the melting point, compared with the Case 2.

The parameters in the Haasen-Sumino model are summarized in TABLE 1. The

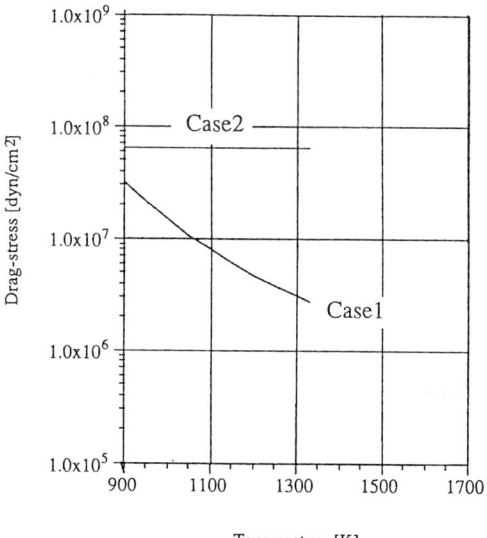

Figure 1.   Relation between drag-stress and temperature.

TABLE 1.   Parameters in the Haasen-Sumino model

N_m	density of moving dilocations	cm^{-2}
b	magnitude of Burgers vector	cm
Q	Pierls potential	eV
k	Boltamann's constant	8.617×10^{-5} ev/K
T	absolute temperature	K
K	material constant	cm/dyn
D	strain hardening factor	dyn/cm
k_0	material constant	$cm^{2p+1}/dyn^p s$
τ_d	drag-stress	dyn/cm
p	material constant	—
λ	material constant	—

TABLE 2.   Values of parameters in the Haasen-Sumino model

	Q	b	K	D	k_0	p	λ
Si	2.2	3.84×10^{-6}	3.1×10^{-7}	4.3×10^3	6.815×10^{-3}	1.1	1.0
InP	1.0	4.15×10^{-6}	1.2×10^{-5}	3.0×10^3	5.57×10^{-8}	1.4	1.0
GaAs	1.5	4.0×10^{-6}	7.0×10^{-6}	3.1×10^3	3.59×10^{-8}	1.7	1.0

values of parameters in the Haasen-Sumino model are given in TABLE 2 for Si   (Tsai *et al.*, 1990), InP (Völkl and Müller, 1989) and GaAs (Maroudas and Brown, 1991)

2.2.   ELASTIC CONSTANTS AND THERMAL EXPANSION COEFFICIENT

In addition to the material parameters for the Haasen-Sumino model, we require the elastic constants C_{11}, C_{12} and C_{44} and the thermal expansion coefficient α for the respective single crystals.   These material properties used are quoted from literatures. They are given as follows :

(a)   Si single crystal (INSPEC, 1988)

$C_{11} = 16.564 \times 10^{11} \exp(-9.4 \times 10^{-5}(T - 298.15))$

$C_{12} = 6.394 \times 10^{11} \exp(-9.8 \times 10^{-5}(T - 298.15))$

$C_{44} = 7.951 \times 10^{11} \exp(-8.3 \times 10^{-5}(T - 298.15))$

$\alpha = 3.725 \times 10^{-6}(1.0 - \exp(-5.88 \times 10^{-3}(T - 124.0))) + 5.548 \times 10^{-10} T$

(b)   InP single crystal (Jordan, 1985)

$C_{11} = 10.76 \times 10^{11} - 1.397 \times 10^8 T$

$$C_{12}=6.080\times10^{11}-8.344\times10^{7}T$$
$$C_{44}=4.233\times10^{11}-4.035\times10^{7}T$$
$$\alpha=4.896\times10^{-6}-5.164\times10^{-9}T+6.048\times10^{-12}T^{2}$$

(c) GaAs single crystal (Jordan, 1980)

$$C_{11}=12.16\times10^{11}-1.39\times10^{8}T$$
$$C_{12}=5.43\times10^{11}-5.76\times10^{7}T$$
$$C_{44}=6.18\times10^{11}-7.01\times10^{7}T$$
$$\alpha=4.68\times10^{-6}+3.82\times10^{-9}T$$

where the units of C_{ij}, α, and T are dyn/cm^2, K^{-1} and K, respectively. In the present analyses, bulk single crystals are assumed to be isotropic, so only the Young's modulus and Poisson's ratio in the {111} plane calculated from C_{11}, C_{12} and C_{44} (Brantley, 1973) are used because they are invariant in this plane.

2.3.   METHOD FOR SIMULATION OF CRYSTAL GROWTH PROCESS

Dislocation density simulations are performed for 8-inch diameter Si, InP and GaAs bulk single crystals. Figure 2 shows the temperature distributions and crystal shapes at the several stages of CZ crystal growth obtained from a transient heat conduction of a Si single crystal (Fujioka et al., 1992). The temperature distributions in InP and GaAs single crystals are so determined that the temperature difference from the solid-liquid interface, the temperature at which is the melting point T_m of each single crystal shown in Figure 2. The melting points T_m of Si, InP and GaAs are, respectively, 1685, 1331 and 1511 K. In order to show the effectiveness of the computer code, we examine how the computer code provides the difference on dislocation density among

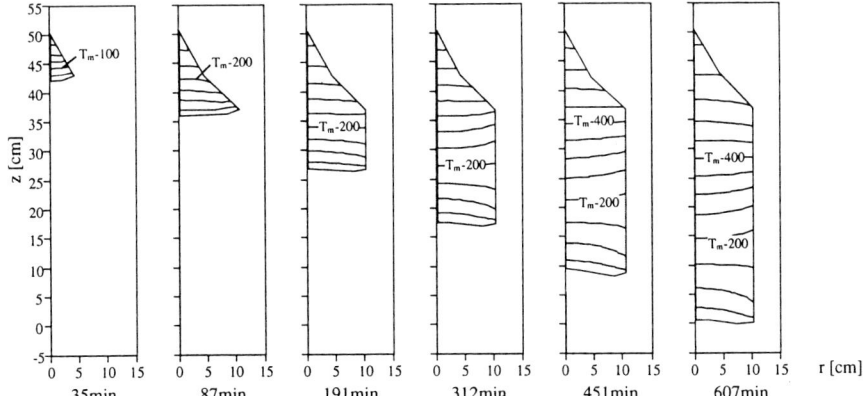

Figure 2.   Temperature distributions in bulk single crystal and crystal shapes

Si, InP and GaAs, whether it provides W-type dislocation density distributions across diameter in InP and GaAs single crystals, as can be observed in actually grown single crystals, and whether it provides reasonable values of dislocation density.

The results of a heat conduction analysis are obtained at discrete time steps. In the present case, the temperature distribution and crystal shape are obtained only at a 17 minutes interval. We divide a time interval between two adjacent time steps into several sub-steps in order to deal with the growth process as continuously as possible. The shape of crystal-melt interface and the temperature in a crystal at an arbitrary sub-step are determined by linear interpolation of the results at two adjacent time steps.

We solve a finite element equilibrium equation for a creep problem and integrate Eq.(5), then we can obtain such state variables as displacement, strains, stresses and dislocation density.

Figure 3 shows a finite element mesh at the final stage, i.e., at 607 minutes after crystal growth starts. A half of a bulk single crystal is modeled by the eight-noded isoparameteric element for an axisymmetric body. The total numbers of elements and nodes are 1539 and 4976, respectively. The finite element mesh shown in Figure 3 includes all the shapes of crystal melt interface obtained from a heat conduction analysis and the finite elements are activated downward as crystal growth process proceeds.

3. Results and Discussion

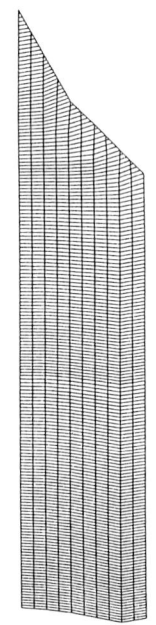

Figure 4 shows the dislocation density distributions for Si, InP and GaAs bulk semiconductor single crystals at 35, 87, 191, 312, 451 and 607 minutes after crystal growth starts. In this analysis, the drag-stress τ_d is assumed to be 3.61×10^7 dyn/cm^2 for all single crystals. High dislocation density appears at the central and peripheral regions of the InP and GaAs single crystals. On the other hand, dislocation density remains the lowest level of the gray scale in the whole region of a Si bulk single crystal. Therefore, InP and GaAs single crystals are much easier to multiply dislocations than a Si single crystal.

Figure 5 shows the dislocation density distributions across the radius in InP and GaAs single crystals at the height (z) of 25 cm and at the time of 607 minutes after crystal growth starts. The results indicate W-type dislocation distributions across the diameter. In actual

Figure 3. Finite element mesh

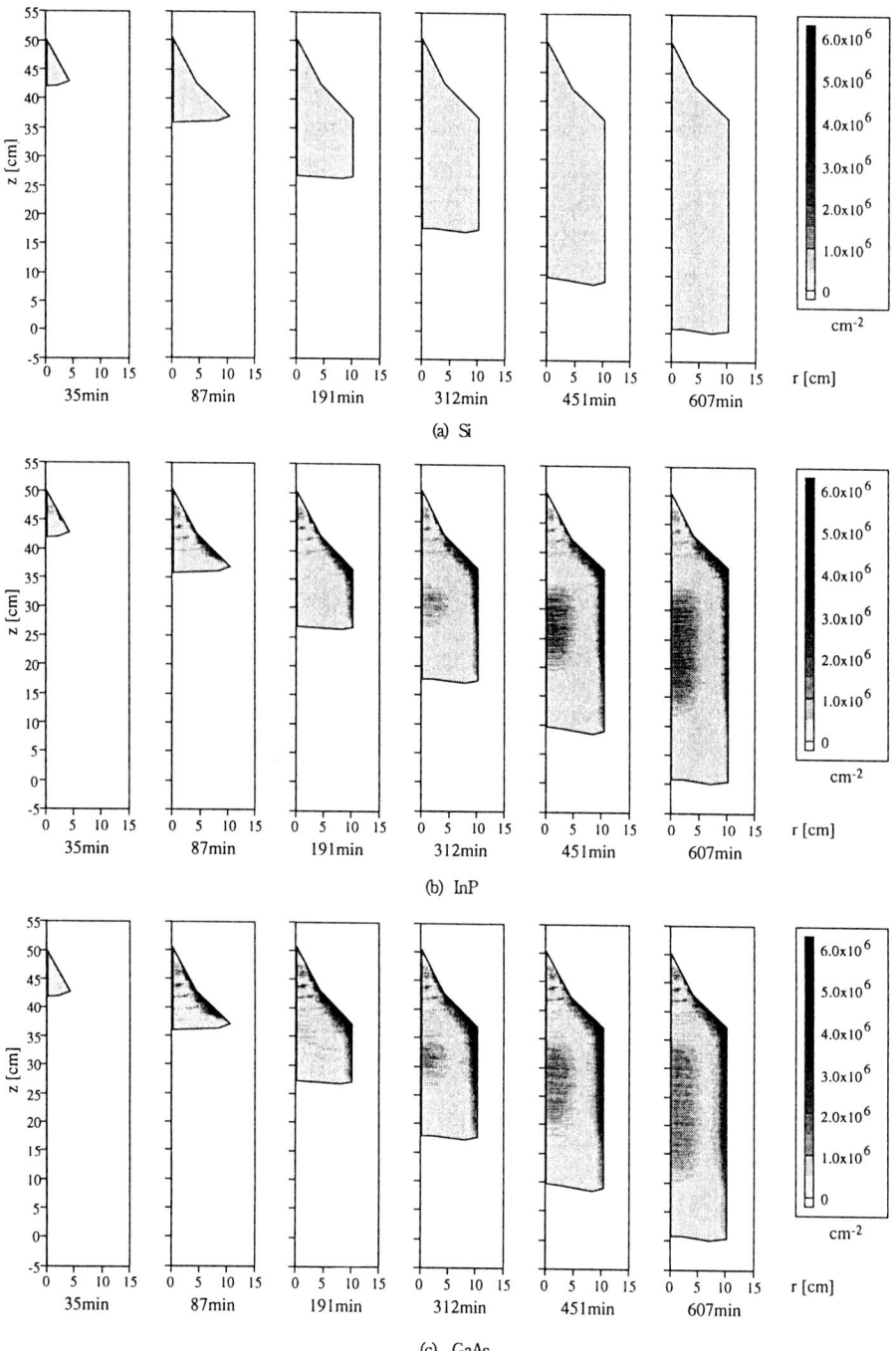

Figure 4.   Dislocation density distributions

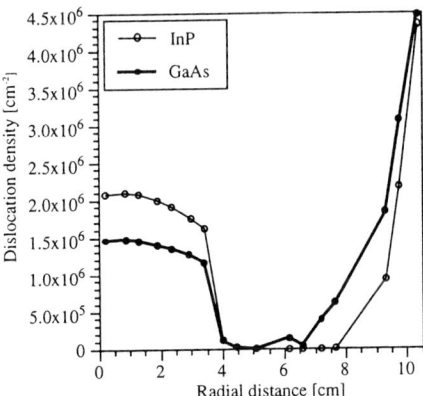

Figure 5. Dislocation density distributions across radius at z=25 cm

single crystals, dislocations are visualized as etch pit obtained from the chemical etching of single crystals. Experimental results of etch pit density (EPD) for InP and GaAs single crystals indicate W-type distribution across the diameter (Kawase *et al.*, 1990 ; Barrett, *et al.*, 1984). Therefore, the calculated dislocation density distributions for InP and GaAs single crystals agree well with those observed in actual single crystals. The values of dislocation density for InP and GaAs single crystals range, roughly speaking, from 10^4 to 10^6 cm^{-2} according to the measured EPD data (Kawase *et al.*, 1990 ; Barrett, *et al.*, 1984). The calculated dislocation density is not so bad, compared with the measured EPD data.

The results of the effect of doping atoms in an InP single crystal are shown hereafter. Figure 6 shows the dislocation density distributions across the radius at the height (z) of 25 cm and at the time of 607minutes after crystal growth starts both for the Case 1, low Ge-doped InP, and for the Case 2, highly Ge-doped InP, the drag-stresses of which are given by Eqs.(6) and (7), respectively. In the Case 1, relatively high dislocation density appears entirely across the radius. On the other hand, the regions with high dislocation density are localized at the center and periphery in the Case 2. In this case, a dislocation-free region appears between the center and periphery of the crystal. The Case 2 has the much larger τ_d than the Case 1. Therefore, the stress level causing dislocation multiplication is much larger in the Case 2 than in the Case 1. That may be the reason why the Case 2 provides the higher dislocation density at center and periphery of the crystal than the Case 1.

Figure 7 shows the dislocation density distributions depicted by using the logarithmic gray scale for the Case 1 and Case 2 at the time of 607 minutes after crystal growth starts. We can see clearly that relatively high dislocation density spreads in the whole of the crystal in the Case 1, while regions with high dislocation

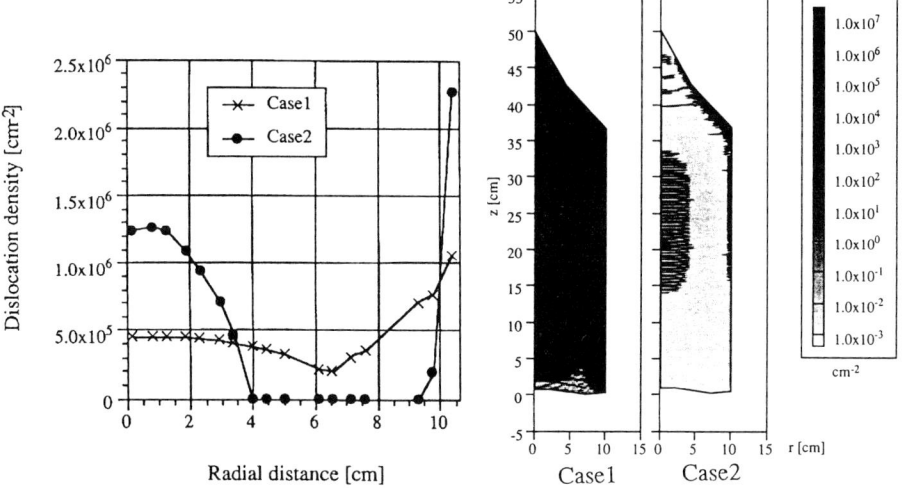

Figure 6.   Dislocation density distributions across radius for Case 1 and Case 2 at z=25 cm

Figure 7.   Dislocation density distributions for Case 1 and Case 2

density are localized in the crystal in the Case 2.   Conclusively the present dislocation density simulations indicate that low-dislocation density InP bulk single crystals can be obtained by doping impurity atoms.

4.   Concluding Remarks

A computer code developed for a transient simulation of dislocation density in a bulk single crystal during growth process gives reasonable results both qualitatively and quantitatively.   As for the effect of doping atoms on the dislocation density, dislocations are distributed in the entire region of the crystal in the case of a low doped InP single crystal, and they are localized at the central and peripheral regions of the crystal in the case of a highly doped InP single crystal.

This study was supported by a Grant-in-Aid for Science Research from the Ministry of Education and Culture.

5.   References

Alexander, H. and Haasen, P. (1968) Dislocations and plastic flow in the diamond structure, *Solid State Physics* **22**, 27-156.

Barrett, D.L., McGuigan, S., Hobgood, H.M., Eldridge, G.W. and Thomas, R.N. (1984) Low dislocation, semi-insulating In-doped GaAs crystals, *J. Crystal Growth* **70**, 179-184.

Brantley, W.A. (1973) Calculated elastic constants for stress problems associated with semiconductor devices, *J. Applied Physics* **44**, 534-535.

Fujioka, K., Nakayama, W. and Sugino, Y. (1992) Numerical simulation of thermal history for Czochralski growth of silicon single crystals, *Trans. Japan Society of Mechanical Engineers Series B* **58**, 3173-3180.

INSPEC (1988) Properties of Silicon, *EMIS Data Review Series 4*, 14-16.

Jordan, A.S. (1980) A new evaluation of the thermal and elastic constants affecting GaAs crystal growth, *J. Crystal Growth* **49**, 631-642.

Jordan, A.S., Caruso, R. and Von Neida, A.R. (1980) A thermoelastic analysis of dislocation generation in pulled GaAs crystals, *The Bell System Technical J.* **59**, 593-637.

Jordan, A.S. (1985) Some thermal and mechanical properties of InP essential to crystal growth modeling, *J. Crystal Growth* **71**, 559-.565.

Jordan, A.S., Brown, G.T., Cockayne, B. and Bonner, W.A. (1985) An analysis of dislocation reduction by impurity hardening in the liquid-encapsulated Czochralski growth of <111> InP, *J. Applied Physics* **58**, 4383-4389.

Kawase, T., Arai, T., Miura, Y., Iwasaki, T., Yamabayashi, N., Tatsumi, M., Murai, S., Tada, K. and Akai, S. (1990) Characterization of high quality InP single crystal, *Sumitomo Electric Technical Review No.29*, 228-235.

Maroudas, D and Brown, R.A. (1991) On the prediction of dislocation formation in semiconductor crystals grown from the melt : analysis of the Haasen model for plastic deformation dynamics, *J. Crystal Growth* **108** (1991) 399-415.

Miyazaki, N., Uchida, H., Munakata, T., Fujioka, Y. and Sugino, Y. (1992) Thermal stress analysis of silicon bulk single crystal during Czochralski growth, *J. Crystal Growth* **125**, 102-111.

Subramanyam, N. and Tsai, C.T. (1995) Dislocation reduction in GaAs crystals grown from the Czochralski process, *J. Materials Processing Technology* **55** (1995) 278-287.

Suezawa, M., Sumino, K. and Yonenaga, I. (1979) Dislocation dynamics in the plastic deformation of silicon crystals Ⅱ. Theoretical analysis of experimental results, *Physik Status Solidi (a)* **51**, 217-226.

Tsai, C.T., Dillon, Jr. O.W. and DeAngelis, R.J. (1990) The constitutive equation for silicon and its use in crystal growth modeling, *J. Engineering Materials and Technology* **120**, 183-187.

Tsai, C.T. (1991) On the finite element modeling of dislocation dynamics during semiconductor crystal growth, *J. Crystal Growth* **113**, 499-507.

Tsai, C.T., Yao, M.W. and Chait, A. (1992) Prediction of dislocation generation during Bridgman growth of GaAs crystals, *J. Crystal Growth* **125**, 69-80.

Tsai, C.T., Gulluoglu, A.N. and Hartley, C.S. (1993) A crystallographic methodology for modeling dislocation dynamics in GaAs crystals grown from melt, *J. Applied Physics* **73**, 1650-1656.

Völkl, J. and Müller, G. (1989) A new model for the calculation of dislocation formation in semiconductor melt growth by taking into account the dynamics of plastic deformation, *J. Crystal Growth* **97**, 136-145.

MULTIAXIAL CREEP FATIGUE
UNDER ANISOTHERMAL CONDITIONS

J. P. SERMAGE, J. LEMAITRE
LMT-Cachan
61 avenue du président Wilson
F-94235 Cachan Cedex, France

AND

R. DESMORAT
Laboratoire de Modélisation et Mécanique des Structures
Université Paris 6
8 rue du Capitaine Scott
F-75015 Paris, France

Abstract. Since creep-fatigue is mainly studied in uniaxial tension, it is shown here how to proceed to perform both experiments and calculations under multiaxial loading and when the temperature varies both in time and space. The constitutive equations used are those of elasto-visco-plasticity coupled or not, to damage, with isotropic and kinematic hardening. It is shown that the unified damage law first proposed for ductile failure and then for fatigue may also be applied to multiaxial creep-fatigue interactions with a new expression for the damage threshold. The procedure for the identification of material parameters is described. Finally, it is shown that the uncoupled calculation procedure, where damage is calculated as a post-processing of an elasto-visco-plastic computation, gives satisfactory results in comparison to the fully coupled analysis ; the latter being more acurate but very expensive in computer time.

1. Extended summary

The full paper was accepted for publication in "Fatigue and Fracture of Engineering Materials" before it was asked for a general lecture in the IUTAM Symposium. It will appear within the year 2000 (see reference).

S. Murakami and N. Ohno (eds.), IUTAM Symposium on Creep in Structures, 125–130.
© 2001 *Kluwer Academic Publishers. Printed in the Netherlands.*

In order to be able to calculate the number of cycles to reach a meso-crack initiation in a component we need elasto-visco-plastic and damage constitutive equations, and a method of structure calculation.

2. Elasto-visco-plasticity coupled to damage

The coupling between strains and damage (represented by the scalar isotropic variable D, surface density of micro-cracks) is written through the concept of effective stress

$$\tilde{\sigma}_{ij} = \frac{\sigma_{ij}}{1 - D} \tag{1}$$

Associated with the postulate of strain equivalence, the law of elasticity is

$$\epsilon_{ij}^e = \frac{1 + \nu}{E} \tilde{\sigma}_{ij} - \frac{\nu}{E} \tilde{\sigma}_{kk} \delta_{ij} \tag{2}$$

where E is the material Young's modulus and ν the Poisson ratio (both temperature dependant) and where the total stain ϵ_{ij} is split into an elastic part ϵ_{ij}^e and a plastic part ϵ_{ij}^p. The visco-plasticity constitutive equations are the classical equations where Norton's viscosity law and two non-linear strain hardenings are considered,

$$\dot{\epsilon}_{ij}^p = \frac{3}{2} \frac{\tilde{\sigma}_{ij}^D - X_{ij}}{(\tilde{\underline{\sigma}} - \underline{X})_{eq}} \dot{p} \qquad \text{if} \quad f > 0 \tag{3}$$

$$\dot{p} = \left\langle \frac{f}{K} \right\rangle^N, \qquad f = (\tilde{\underline{\sigma}} - \underline{X})_{eq} - R - \sigma_y \tag{4}$$

$$R = R_\infty \left(1 - e^{-br}\right) \qquad \text{with} \quad \dot{r} = \dot{p}\,(1 - D) \tag{5}$$

$$\frac{d}{dt}\left(\frac{X_{ij}}{\gamma X_\infty}\right) = \frac{2}{3}\dot{\epsilon}_{ij}^p(1 - D) - \frac{X_{ij}}{X_\infty}\dot{p}\,(1 - D) \tag{6}$$

R is the isotropic strain hardening stress, X_{ij} the kinematic hardening or back stress. The temperature dependent material properties are represented by the yield stress σ_y, the plasticity parameters R_∞, b, X_∞, γ, and by Norton's parameters K and N.

The damage law

$$\dot{D} = \left(\frac{Y}{S}\right)^s \dot{p} \qquad \text{if} \quad p > p_D$$
$$D = D_c \rightarrow \text{crack initiation} \tag{7}$$

has been applied with success to ductile damage, to low cycle and to high cycle fatigue in a two scale model, but never, up to now, to creep-fatigue interaction. It is one of the original results of this paper to show the ability of this unified damage law to represent also the non-linear creep-fatigue interaction. $Y = \tilde{\sigma}_{eq}^2/2E$ is the energy release rate density introducing the triaxiality function $R_\nu = 2(1+\nu)/3+3(1-2\nu)(\sigma_H/\sigma_{eq})^2$, $\tilde{\sigma}_{eq} = (\frac{3}{2}\tilde{\sigma}_{ij}^D\tilde{\sigma}_{ij}^D)^{1/2}$ is the effective von Mises stress, $\sigma_H = \sigma_{kk}/3$ the hydrostatic stress and $\dot{p} = (\frac{2}{3}\dot{\epsilon}_{ij}^p\dot{\epsilon}_{ij}^p)^{1/2}$ the accumulated plastic strain rate. S, s, D_c, ϵ_{pD} are material and temperature dependent parameters and the new expression for the damage threshold is:

$$p_D = \epsilon_{pD}\left[\exp\left(\frac{\sigma_u - \sigma_y}{\sigma_{eq}^{Max} - \sigma_y}\ln 2\right) - 1\right] \qquad (8)$$

in which $\sigma_u = \sigma_y + R_\infty + X_\infty$ is the ultimate stress and σ_{eq}^{Max} is the maximum von Mises stress for cyclic loading at zero mean stress.

The identification of the temperature dependent parameters of a steel alloy 2 1/4 Cr Mo used in headers of electrical power fossil plants has been obtained from cyclic tension-compression tests up to fracture at 20, 300, 400, 500, 600 C. The validation of the procedure has been made on uniaxial anisothermal and stress driven tests.

3. Methods of structure calculation

Several levels of structure calculation may be performed depending upon the accuracy needed and the cost accepted.

The fully coupled analysis is the procedure where the fully coupled constitutive equations are solved simultaneously to obtain strains and damage as functions of time. They have been implemented into a fully coupled scheme as the ABAQUS UMAT VISCOENDO.

A simpler analysis (uncoupled) consists in a first step to calculate the strains history by an elasto-visco-plastic computation, the same as above, but with $D = 0$. The second step is the time integration of the damage law alone in which the stresses $\sigma_{ij}(t)$ and the plastic strains $\epsilon_{ij}^p(t)$ (results of the first step calculation) are the inputs of this post-processor.

4. Application to a representative structure

Both coupled and uncoupled analyses are compared to experimental results concerning a Maltese cross shape specimen loaded by two forces in its plane drawn in Fig. 1 and having 4 notches of stress concentrations. The geometry has been chosen in order to represent loading conditions close to practical cases, where most of the time cracks initiate on the edges along which the state of stress is uniaxial, and to be able to perform non proportional loading and, last but not the least, to be a possible experimental specimen of the Triaxial machine ASTREE of the LMT-Cachan.

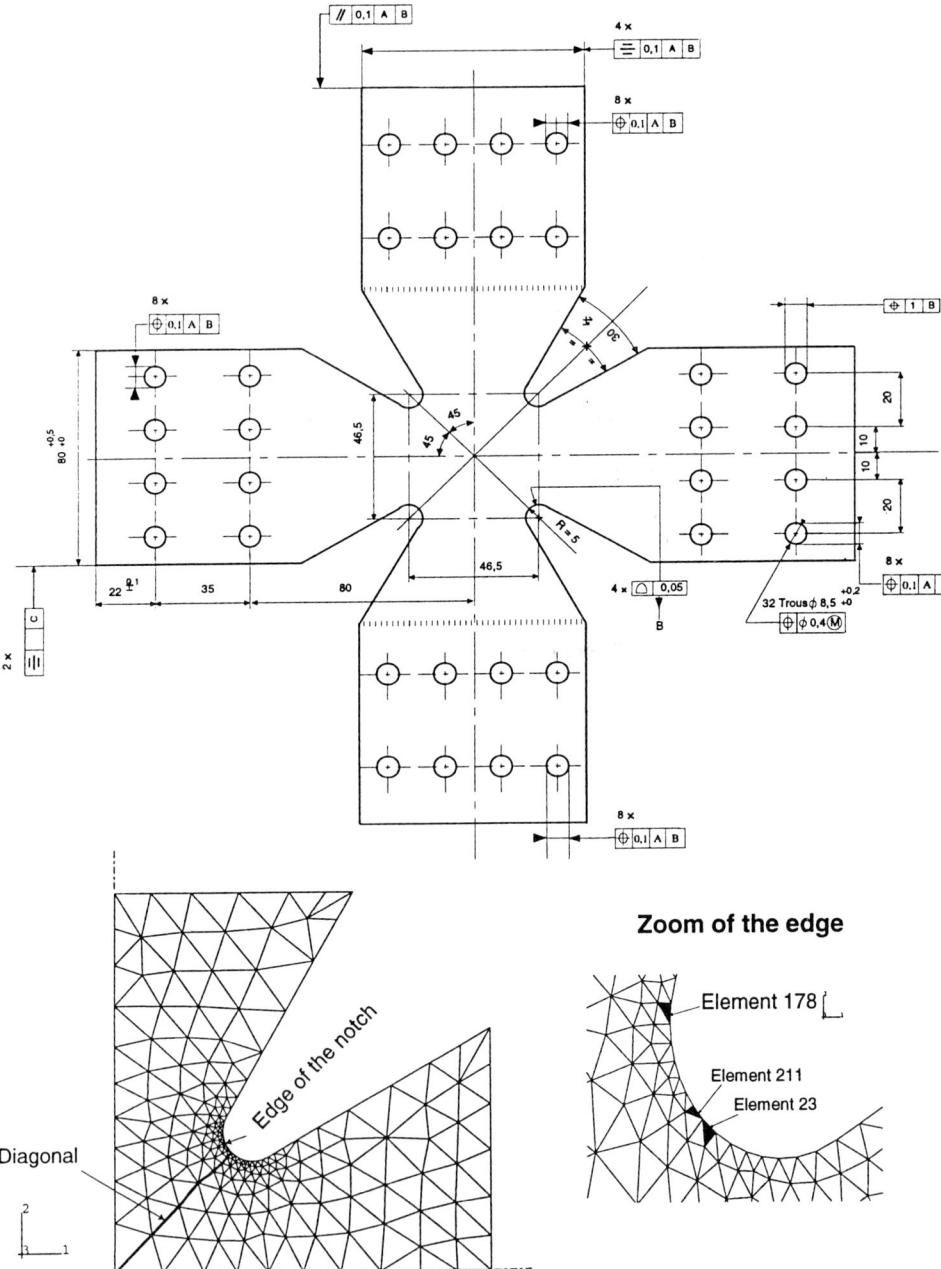

Figure 1. Maltese cross specimen (thickness 4.5 mm) and optimized mesh (320 six nodes triangles)

Five tests have been performed with different forces $F_1(t)$, $F_2(t)$ and temperature histories,

- two in-phase creep-fatigue interaction tests at constant temperature T=580 C and T=620 C (tests 2 and 3),
- two non-proportional loading tests at T=580 C (tests 4 and 6),
- one thermal fatigue test at constant loads $F_1 = F_2$ and variable temperature (test 5).

The corresponding numbers of cycles to failure are indicated in table 1.

For the structure calculation the histories of loading have been given as inputs

- in the VISCOENDO procedure of ABAQUS FEM code,
- in the VISCOPLAST procedure of ABAQUS from which the results were then introduced as inputs in DAMAGE97.

TABLE 1. Results and comparaisons

Test performed	Experiments	Fully coupled analysis	Uncoupled analysis
Tests 2 and 3 Creep fatigue	$N_R(580\ C) = 331$ $N_R(620\ C) = 100$	$N_R = 420$ $N_R = 115$	$N_R = 266$
Test 4 - Out of phase non proportional loading	$N_R = 2356$	$N_R = 2135$	$N_R = 1130$
Test 5 Thermal fatigue	$N_R = 56$	$N_R = 48$	$N_R = 27$
Test 6 - Sequential non proportional loading	$N_R = 196$	$N_R = 230$	$N_R = 273$

5. Conclusions

The results are also indicated in table 1 from which several conclusions may be pointed out :

1. First of all both biaxial creep fatigue experiments and calculations are possible.
2. Elasto-viscoplasticity coupled to damage computations relating to a structure submitted to variable loading and temperature are possible but it takes several days of computing to achieve several hundred of cycles. The results in terms of the number of cycles to crack initiation are in close concordance with the experimental results even for the

two cases of non-proportional loading. The difference never exceed 27% which is the order of magnitude of the discrepancy of the tests.

3. Fortunately, as far as computer time is concerned, the uncoupled method (which is more than 50 times faster) gives results still acceptable in fatigue evaluations since the difference between predictions and experiments never exceeds a factor of 2.

4. The extension of the unified damage law $\dot{D} = (Y/S)^s \dot{p}$ to creep-fatigue interaction may be considered as achieved. Only 4 material parameters (temperature dependent) need to be identified. A good ratio with respect to quality/computer time cost.

6.   Reference

Sermage, J.P., Lemaitre, J., Desmorat, R. (2000), Multiaxial creep-fatigue under anisothermal conditions, *Fatigue and Fracture of Engineering Materials and Structures*.

CONSTITUTIVE MODELING OF VISCOPLASTIC DAMAGE IN SOLDER MATERIAL[*]

Y. WEI and **C. L. CHOW**
Department of Mechanical Engineering
University of Michigan-Dearborn
Dearborn, Michigan 48128, USA

M.K. NEILSEN and **H.E. FANG**
Sandia National Laboratories,
Albuquerque, NM 87185, USA

ABSTRACT

This paper presents a constitutive modeling of viscoplastic damage in 63Sn-37Pb solder material taking into account the effects of micro-structural change in grain coarsening. Based on the theory of damage mechanics, a two-scalar damage model is developed by introducing the damage variables and the free energy equivalence principle. An inelastic potential function based on the concept of inelastic damage energy release rate is proposed and used to derive an inelastic damage evolution equation.

The validation of the model is carried out for the viscoplastic material by predicting monotonic tensile behavior and tensile creep curves at different temperatures. The softening behavior of the material under monotonic tension loading can be characterized with the model. The results demonstrate adequately the validity of the proposed viscoplastic constitutive modeling for the solder material.

1. Introduction

The Pb-Sn eutectic alloy is widely used as a joining material in the electronic industry. In this application the solder acts as both electrical and mechanical connection within and among different packaging levels in an electronic device. Advances in packaging technologies driven by the desire for miniaturization and increased circuit speed result in severe operating conditions for the solder joint, thus causing solder joint reliability concerns. Solder joints are usually subjected to the operating temperature as high as 0.5 to 0.8 times its melting temperature (T_m). It is well known that the behavior of Pb-Sn

---

[*] This work was supported by the United States Department of Energy under Contract DE-AC04-94AL85000. Sandia is a multiprogram laboratory operated by Sandia Corporation, a Lockheed Martin Company, for the United States Department of Energy.

S. Murakami and N. Ohno (eds.), IUTAM Symposium on Creep in Structures, 131–140.
© 2001 *Kluwer Academic Publishers. Printed in the Netherlands.*

eutectic solder material is highly rate sensitive and time dependent. For a reliability analysis, knowledge on the mechanical behavior of solder joints under thermo-mechanical loading is necessary. In past decade, many researchers have studied constitutive modeling and reliability of solder joints (Solomon, 1986, Busso, et al., 1994, McDowell, et al., 1994, Shine, et al., 1994, Frear, et al., 1997, Lau, et al., 1997 and Shi, et al. 1999). However, the previous work focuses on either the experimental results or the conventional analytical approaches based on the assumption that all materials are perfect or defect-free during a loading process. There are only a few investigations examining the effects of damage on the behavior of solder material (Basaran, et al., 1998, Wei, et al., 1999 and Qian, et al., 1999).

The primary objective of this paper is to present a damage model which takes into account the effects of damage accumulation on mechanical behavior of the solder material. The methodology is based on the theory of Damage Mechanics. The theory enables a macroscopic description of material deterioration and stiffness degradation due to the presence of micro-defects that initiate, coalescence and grow under load. The accumulation of damage is manifested as a reduction in the effective load bearing area of solder joints, causing eventually macro-crack initiation and failure in a solder material. A two-scalar damage model is for the first instance introduced to quantify the degree of damage. Then the model is applied to characterize the effects of damage on mechanical behavior of 63Sn-37Pb solder material, including its softening behavior. The validation of the damage mode is carried out by comparing the predicted and measured results.

2. Damage Variables

Based on the concept of damage mechanics, the effective stress $\bar{\sigma}$ is defined as

$$\bar{\sigma} = \mathbf{M(D)} : \sigma \tag{1}$$

where σ is the true (Cauchy) stress tensor, $\mathbf{M(D)}$ is the damage effect tensor and $\mathbf{D}$ is the damage variable. The damage variable, known as an internal state variable, is introduced to characterize the gradual deterioration of the material. The effective stress provides an essential mechanism by which the damage variable is coupled with constitutive models through the damage effect tensor. Therefore, a proper definition of $\mathbf{M}$ in Equation (1) is important. The damage effect tensor may be established with two scalar variables D and μ as proposed by Chow, et al., 1999

$$\mathbf{M} = \frac{1}{1-D} \begin{bmatrix} 1 & \mu & \mu & 0 & 0 & 0 \\ \mu & 1 & \mu & 0 & 0 & 0 \\ \mu & \mu & 1 & 0 & 0 & 0 \\ 0 & 0 & 0 & 1-\mu & 0 & 0 \\ 0 & 0 & 0 & 0 & 1-\mu & 0 \\ 0 & 0 & 0 & 0 & 0 & 1-\mu \end{bmatrix} \tag{2}$$

where D and μ are two independent components of damage tensor **D**. Obviously, this damage effect tensor reduces to the form $\mathbf{M} = \mathbf{I}/(1-D)$ when the value of μ is zero. This corresponds to a special case of isotropic damage for which the value of Poisson's ratio remains constant under load.

In order to consider the effects of damage, the free energy equivalence principle is postulated as: *the free energy for a damaged material may be expressed similar in form as a virgin/as-received material except that the true (Cauchy) stress is replaced by the effective stress.* Therefore, the free energy for a damaged solder material is postulated as

$$
\begin{aligned}
\Psi &= \frac{1}{2}\bar{\sigma}^T : \mathbf{C}_0^{-1} : \bar{\sigma} + \frac{3}{4}C_{k0}^{-1}\mathbf{X}^T : \mathbf{X} \\
&= \frac{1}{2}\sigma^T : \mathbf{M}^T : \mathbf{C}_0^{-1} : \mathbf{M} : \sigma + \frac{3}{4}C_{k0}^{-1}\mathbf{X}^T : \mathbf{X} \\
&= \frac{1}{2}\sigma^T : \mathbf{C}^{-1} : \sigma + \frac{3}{4}C_{k0}^{-1}\mathbf{X}^T : \mathbf{X}
\end{aligned}
\tag{3}
$$

where **X** is the back stress tensor, $\mathbf{C}_0^{-1}$ is the fourth order elastic tensor without damage, C_{k0}^{-1} is the material parameter for kinematic hardening, and $\mathbf{C}^{-1}$ is the effective elastic tensor for a damaged material derived as :(Chow, et al., 1999)

$$
\mathbf{C}^{-1} = \mathbf{M}^T : \mathbf{C}_0^{-1} : \mathbf{M} = \frac{1}{E}
\begin{bmatrix}
1 & -v & -v & 0 & 0 & 0 \\
-v & 1 & -v & 0 & 0 & 0 \\
-v & -v & 1 & 0 & 0 & 0 \\
0 & 0 & 0 & 2(1+v) & 0 & 0 \\
0 & 0 & 0 & 0 & 2(1+v) & 0 \\
0 & 0 & 0 & 0 & 0 & 2(1+v)
\end{bmatrix}
\tag{4}
$$

E and v in Equation (4) are the Young's modulus and the Poisson's ratio for damaged materials expressed in terms of the damage variables D and μ as

$$
E = \frac{E_0(1-D)^2}{1-4v_0\mu+2(1-v_0)\mu^2} \qquad v = \frac{v_0-2(1-v_0)\mu-(1-3v_0)\mu^2}{1-4v_0\mu+2(1-v_0)\mu^2}
\tag{5}
$$

E_0 and μ_0 are the Young's modulus and the Poisson's ratio for undamaged or as-received material.

With the theory of irreversible thermodynamics, the damage energy release rates corresponding to the damage variables D and μ are defined with the free energy in Equation (3) as

$$Y_D = -\frac{\partial \Psi}{\partial D} = -\frac{1}{1-D} \sigma^T : \mathbf{C}^{-1} : \sigma \tag{6}$$

$$Y_\mu = -\frac{\partial \Psi}{\partial \mu} = -\frac{1}{1-D} \sigma^T : \mathbf{Z} : \sigma \tag{7}$$

where

$$\mathbf{Z} = \frac{1}{E_0(1-D)} \begin{bmatrix} z_1 & z_2 & z_2 & 0 & 0 & 0 \\ z_2 & z_1 & z_2 & 0 & 0 & 0 \\ z_2 & z_2 & z_1 & 0 & 0 & 0 \\ 0 & 0 & 0 & 2(z_1 - z_2) & 0 & 0 \\ 0 & 0 & 0 & 0 & 2(z_1 - z_2) & 0 \\ 0 & 0 & 0 & 0 & 0 & 2(z_1 - z_2) \end{bmatrix} \tag{8}$$

$$z_1 = 2\mu(1-v_0) - 2v_0 \qquad\qquad z_2 = (1+\mu)(1-v_0) - 2\mu v_0$$

In order to formulate deformation and damage evolution equations, an inelastic dissipation potential function ϕ is postulated for solder material to consist of two independent processes, i.e. a deformation process and a damaging process. Accordingly, the potential may be expressed as

$$\phi = \phi^{in}(\sigma, \mathbf{X}) + \phi^d(Y_{ind}) \tag{9}$$

where Y_{ind}, the equivalent inelastic damage energy release rate, is defined as

$$Y_{ind} = \left[\frac{1}{2}(Y_D^2 + \gamma Y_\mu^2) \right]^{1/2} \tag{10}$$

Y_D and Y_μ are the thermodynamic conjugate forces of the inelastic damage variables D and μ in Equations (6) and (7). The inelastic damage part of the dissipation potential is postulated as

$$\phi^d = \frac{Y_h}{b_1 + 1} \left(\frac{Y_{ind}}{Y_h} \right)^{b_1 + 1} \tag{11}$$

where b_1 is a material constant, and Y_h is the inelastic damage hardening variable. Therefore, the inelastic damage evolution equations can be derived from Equation (9) as

$$\dot{D} = -\lambda_{in}\frac{\partial\phi}{\partial Y_D} = -\dot{w}\frac{Y_D}{2Y_{ind}} \qquad \dot{\mu} = -\lambda_{in}\frac{\partial\phi}{\partial Y_\mu} = -\dot{w}\frac{\gamma Y_\mu}{2Y_{ind}} \tag{12}$$

where $\dot{w}$, the overall inelastic damage rate, is defined as:

$$\dot{w} = \lambda_{in}\frac{d\phi^d}{dY_{ind}} = \lambda_{in}\left(\frac{Y_{ind}}{Y_h}\right)^{b_1} \tag{13}$$

λ_{in} is a multiplier, which is related to inelastic deformation to be described in the next section. The inelastic damage hardening variable Y_h may be expressed in terms of the overall inelastic damage w and the absolute temperature T as

$$Y_h(w, T) = Y_0 e^{(b_2 w + \frac{b_3}{T})} \tag{14}$$

where Y_0, b_2 and b_3 are material constants. The temperature effects on the inelastic damage evolution in equation (13) are included through the multiplier λ_{in} and the damage hardening variable Y_h.

3.   Constitutive Equations

The damage coupled elastic constitutive equation is derived with the free energy in Equation (3) as:

$$\varepsilon^e = \frac{\partial\Psi}{\partial\sigma} = \mathbf{C}^{-1}:\sigma \qquad \sigma = \mathbf{C}:\varepsilon^e \tag{15}$$

where ε^e is the elastic strain. The kinematic hardening equation is derived from the free energy in Equation (3) as:

$$\alpha_k = \frac{\partial\Psi}{\partial\mathbf{X}} = \frac{3}{2}C_{k0}^{-1}\mathbf{X} \qquad \mathbf{X} = \frac{2}{3}C_{k0}\alpha_k \tag{16}$$

where α_k is the back strain.

For inelastic deformation, the dissipation potential is written for the eutectic material as

$$\phi^{in} = J_2(\mathbf{S} - \mathbf{X}) \tag{17}$$

where J_2 is a second invariant of the stress difference and defined as

$$J_2 = \left\{ \frac{3}{2} (\mathbf{S} - \mathbf{X})^T : (\mathbf{S} - \mathbf{X}) \right\}^{\frac{1}{2}}$$

(18)

$\mathbf{S}$ is the deviatoric stress. Therefore, the inelastic strain rate can be expressed as

$$\dot{\varepsilon}^{in} = \lambda_{in} \frac{\partial \phi}{\partial \sigma} = \lambda_{in} \frac{\partial \phi^{in}}{\partial \sigma} = \lambda_{in} \frac{3}{2} \frac{\mathbf{S} - \mathbf{X}}{J_2}$$

(19)

The multiplier λ_{in} is equal to the equivalent inelastic strain rate $\dot{p}^{in}$ that may be expressed as (Frear, et al., 1997 and Wei, et al., 1999)

$$\lambda_{in} = \dot{p}^{in} = \frac{1-\mu}{1-D} f \exp\left(\frac{-Q}{RT}\right) \left(\frac{\lambda_0}{\lambda}\right)^p \sinh^m\left(\frac{1-\mu}{1-D} \frac{J_2}{\alpha(c+\hat{c})}\right)$$

(20)

where f, p, m and Q are material parameters, R is the gas constant, T is the absolute temperature, λ is the current grain/phase size, λ_0 is the initial grain/phase size, α is a scalar function of the absolute temperature, c and $\hat{c}$ are state variables.

The total strain is defined as

$$\varepsilon = \varepsilon^e + \varepsilon^{in}$$

(21)

It is worthy noting that the effects of kinematic hardening, grain/phase size coarsening and damage accumulation are taken into account in Equation (20). The back strain tensor coupled with inelastic damage can be expressed as

$$\dot{\alpha}_k = \frac{(1-D)^2}{(1-\mu)^2} \left\{ \dot{\varepsilon}^{in} - C_{k0}(A_5 \dot{p}^{in} + A_6)\alpha_k \sqrt{\frac{2}{3} \alpha_k^T : \alpha_k} \right\}$$

(22)

where A_5 and A_6 are material constants. The scalar state variable c, known as deformation hardening variable, is given by

$$\dot{c} = A_1 \dot{p}^{in} - (A_2 \dot{p}^{in} + A_3)(c - c_0)^2$$

(23)

where c_0, A_1, A_2 and A_3 are material parameters. The State variable $\hat{c}$ is related to the grain/phase size λ by

$$\hat{c} = A_7 \left(\frac{\lambda_0}{\lambda}\right)^{A_8}$$

(24)

where A_7 and A_8 are positive material parameters. The evolution of the grain/phase size for 63Sn-37Pb is given with experimental tests by

$$\lambda = \lambda_0 + \{[4.1 \times 10^{-5} e^{-11023/T} + 15.6 \times 10^{-8} e^{-3123/T} \dot{p}^{in}]t\}^{0.256} \tag{25}$$

where t is the time.

4. Applications

Several monotonic tension and tensile creep tests have been designed for the determination of material parameters in the proposed model. The monotonic tension tests were carried out at three temperatures (25^0C, 75^0C and 100^0C) under different strain rates (10^{-5}/s and 10^{-4}/s) by means of the strain control approach. The tensile creep tests were carried out at three temperatures (25^0C, 75^0C and 100^0C) under applied stress 4.14 MPa by the load control approach. From our experimental observation on tension/creep tests, the measured damage values are very small, i.e. less than 0.02 for both D and μ, when the total strain applied is less than 3%. It is therefore reasonable to assume that the material degradation or damage can be considered negligible below the total strain of 3% and all the viscoplastic parameters can be determined without the damage consideration. The nonlinear least squares fit to the experimental data (less than 3% strain) was obtained using the Levenburg-Marquardt Algorithm (More, 1978). The measurement for Young's modulus and Poisson's ratio were carried out at certain strain intervals until final rupture of specimens. Accordingly, the damage parameters in the model were determined with the measured data. The material parameters determined for the 63Sn-37Pb solder material are summarized in Table 1. The predicted results are shown in Fig.1 for monotonic tension and in Fig.2 for tensile creep.

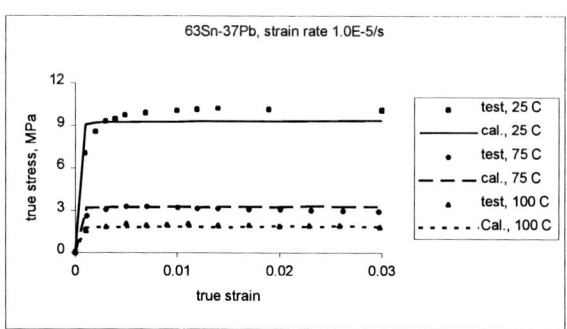

Figure 1a. Monotonic tension behavior at strain rate 10^{-5}/s

The damage model was applied to predict the monotonic tension behavior at true strain rate 10^{-4}/s under three different temperatures, namely 25^0C, 75^0C and 100^0C. The predicted results are compared with test data as shown in Fig.3. The softening behavior of 63Sn-37Pb solder material is observed from experimental tests and can be successfully characterized with the damage-coupled constitutive equations.

Y. WEI et al.

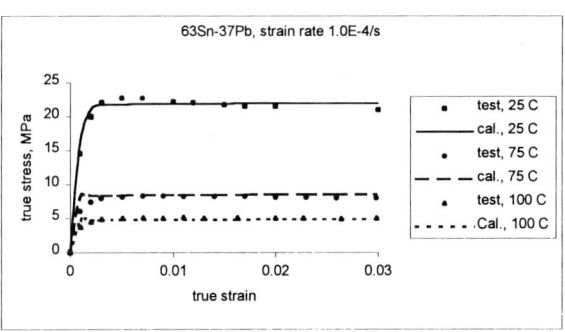

Figure 1b. Monotonic tension behavior at strain rate 10^{-4}/s

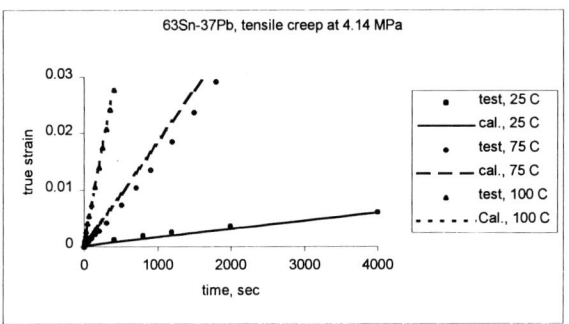

Figure 2. Tensile creep behavior at applied stress 4.14 MPa

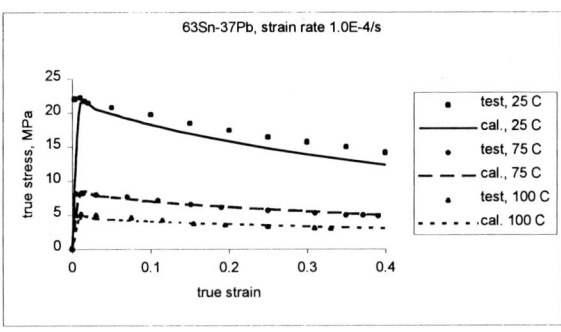

Figure 3. Softening behavior under monotonic tension at strain rate 10^{-4}/s

The uniaxial tensile creep is considered in this study to illustrate the potential applications of the proposed model. The applied stress is 4.14 MPa at three different temperatures, 25^0C, 75^0C and 100^0C respectively, under load control approach. Figure 4 shows a satisfactory comparison of the predicted creep curves with the test results.

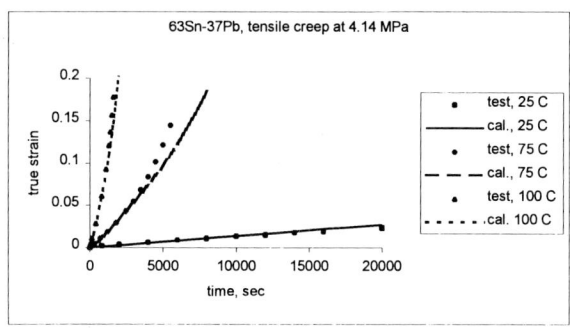

Figure 4. Prediction on tensile creep behavior at applied stress 4.14 MPa

TABLE 1 63Sn-37Pb solder material parameters

Temperature (^{0}C)	25	75	100
Temperature (^{0}K)	298	348	373
Poisson's ratio v_0	0.4	0.4	0.4
Young's modulus E_0 (GPa)	33.26	33.26	33.26
A_1 (MPa)	0.00E+00	0.00E+00	0.00E+00
A_2 (1/MPa)	0.00E+00	0.00E+00	0.00E+00
A_3 (1/MPa-sec)	0.00E+00	0.00E+00	0.00E+00
C_{k0} (GPa)	28.15	22.08	19.32
A_5 (1/MPa)	44.95	103.0	153.7
A_6 (1/MPa-sec)	8.42E-03	5.95E-02	0.2
A_7 (MPa)	5.66	5.66	5.66
A_8	0.5	0.5	0.5
Flow rate f (1/sec)	1.802E+06	1.802E+06	1.802E+06
Sinh exponent m	3.04	3.04	3.04
Growth exponent p	3.00	3.00	3.00
Flow stress c_0 (MPa)	2.83	2.83	2.83
Phase size λ_0 (mm)	2.257E-03	2.257E-03	2.257E-03
Activation energy Q (cal/mol)	1.376E+4	1.376E+4	1.376E+4
Gas constant R (cal/mol-K)	1.987	1.987	1.987
Damage constant b_1	0.4	0.4	0.4
Damage constant b_2	5.0	5.0	5.0
Damage constant b_3 (K)	3.34E+03	3.34E+03	3.34E+03
Damage constant γ	-0.2	-0.2	-0.2
Damage constant Y_0 (MPa)	3.77E-08	3.77E-08	3.77E-08

5.   Conclusions

A damage-coupled constitutive model is proposed with the theory of damage mechanics. The model was applied to predict the tensile behavior of the 63Sn-37Pb solder material under several different strain rates and the tensile creep curves under different stress levels and temperatures. The softening behavior of the material under monotonic tension load was observed experimentally and can be adequately characterized with the proposed model. An excellent agreement of the predictions with experimental data is achieved.

6. References

Basaran, C. and Yan, C.Y. (1998) A thermodynamic framework for damage mechanics of solder joints, *Journal of Electronic Packaging* 120, 379-384.

Busso, E. P., Kitano, M. and Kumazawa, T. (1994) Modeling complex inelastic deformation processes in IC packages' solder joints, *Journal of Electronic Packaging* 116, 6-15.

Chow, C.L. and Wei, Y. (1999) Constitutive modeling of material damage for fatigue failure prediction, *International Journal of Damage Mechanics* 8, 355-375.

Frear, D. R., Burchett, S. N., Neilsen, M. K. and Stephens, J. J. (1997) Microstructurally based finite element simulation of solder joint behaviour, *Soldering & Surface Mount Technology* 2, 39-42.

Lau, J. H. and Pao, Y. H. (1997) *Solder Joint Reliability of BGA, CSP, Flip Chip, and Fine Pitch SMT Assemblies*, McGraw-Hill.

More, J.J. (1978) The Levenburg-Marquardt algorithm: implementation and theory, *Lecture Notes in Mathematics* 630, edited by Watson, G.A., Springer-Verlag.

McDowell, D. L., Miller, M. P. and Brooks, D. C. (1994) A unified creep-plasticity theory for solder alloys, *Fatigue of Electronic Materials*, ASTM STP 1153, 42-59.

Qian, Z., Ren, W. and Liu, S. (1999) A damage coupling framework of unified viscoplasticity for the fatigue of solder alloys, *Journal of Electronic Packaging* 121, 162-168.

Shi, X.Q., Zhou, W., Pang, H.L.J. and Wang, Z.P. (1999) Effect of temperature and strain rate on mechanical properties of 63Sn/37Pb solder alloy, *Journal of Electronic Packaging* 121, 179-185.

Shine, M. C. and Fox, L. R. (1994) Fatigue of solder joints in surface mount devices, *Low Cycle Fatigue*, ASTM STP 942, 588-610.

Solomon, H. D. (1986) Creep, strain rate sensitivity and low cycle fatigue of 60/40 solder, *Brazing and Soldering* 11, 68-75.

Wei, Y., Chow, C.L., Fang, H.E. and Neilsen, M.K. (1999) Characteristics of creep damage for 60Sn-40Pb solder material, *ASME 99-IMECE/EEP-15*.

CONSIDERATION OF STRESS STATE INFLUENCES IN THE MATERIAL MODELLING OF CREEP AND DAMAGE

H. ALTENBACH
Lehrstuhl für Technische Mechanik
Fachbereich Ingenieurwissenschaften
Martin-Luther-Universität Halle-Wittenberg
D-06099 Halle (Saale), Germany

Abstract. Creep tests performed for many engineering materials show an independent behaviour on the kind of loading during the primary and secondary creep. In this case the creep curves obtained for tensile or compressive loads are approximately the same (only the sign is different), the equivalent strain vs. time curves for tension, compression and torsion are identical, the influence of the mean stress can be ignored, etc. From this follows that the classical von Mises concept can be used in the constitutive modelling. On the other hand, the tertiary creep is strongly influenced by the kind of stress state because, for instance, of the voids opening and closing.

Considering constant temperature, small strains and isotropic behaviour a set of constitutive and evolution equations describing the creep-damage including non-classical effects is introduced. The creep equations are based on a potential depending on three linear-independent stress tensor invariants. The damage is taken into account using the equivalent stress concept. The unknown coefficients in the equations must be identified by tests. The identification procedure is discussed, and as the result the coefficients as dependencies of the material properties are defined.

1. Introduction

At elevated temperatures, stresses imposed on structural elements produce increasing strains even if the equivalent stress level is below the yield point. This phenomenon is named creep, and the constitutive behaviour can be classified as time-dependent material behaviour. The necessary temperature level above which the creep starts is 0,3 - 0,6 T_m (melting temperature) and depends on the material considered.

There are many proposals to formulate a suitable creep theory [1]. The main problem is that the experimental tests are mostly performed as uniaxial tests (simple tests), but the constitutive equations must be multiaxially because only in this case they can be the basis of calculation the mechanical state of a structural element which is, in general, applied by complex stress states. The equivalence cannot be established by some physical principle. Therefore an engineering ap-

S. Murakami and N. Ohno (eds.), IUTAM Symposium on Creep in Structures, 141–150.

proach is commonly used, for example, based on the intuitive formulation of the constitutive equations assuming an equivalent stress. As a result of this approach we obtain the creep equations which are non-unique.

In the engineering practice creep equations based on the von Mises equivalent stress are well established. In many situations the secondary creep behaviour can be modelled with a sufficient accuracy. The extended application of light alloys, polymers, composites, ceramics, etc., [2], and the necessity to analyse specific creep problems like the creep of rock salt [3] require a reformulation of the classical phenomenological constitutive equations. The extension of the classical equations is necessary with respect to some effects which are obtained experimentally: different behaviour in tension and compression, shear stresses result in axial strains, significant influence of the mean stress, etc. All these non-classical effects can be observed more or less during primary, secondary and tertiary creep. Some of them (e.g., the different behaviour in tension and compression or the mean stress sensitivity) cannot be neglected in tertiary creep models considering certain damage evolution.

The von Mises equivalent stress is a quadratic form which leads to tensorial linear constitutive equations. For the description of the non-classical effects we have to use as a minimum a formulation which is sensitive to the sign of loading, and results in the general case in tensorial non-linear equations. For this purpose a set of equations is introduced which is based on three linear independent invariants of the stress tensor, for the dependence of the damage evolution the equivalent stress expression requires a similar formulation.

2. Non-classical creep law

Considering isotropic material behaviour a suitable creep equation can be derived from a potential as follows

$$\dot{\varepsilon}^{cr} = \dot{\eta}\frac{\partial \Phi}{\partial \sigma} \tag{1}$$

$\dot{\varepsilon}^{cr}$ denotes the creep strain rate tensor, σ is the stress tensor, and $\dot{\eta}$ is an unknown factor. The creep potential Φ itself depends on the stress tensor and some material properties [4]. In addition, one can include as arguments structural parameters which describe hardening, softening, damage, etc. In the case of isotropic material behaviour the potential should contain only isotropic arguments. For the comparison of uniaxial and multiaxial stress states we assume the equivalence of the dissipated power for both. Mathematically these suggestions result in an equivalent stress based formulation (instead of the direct use of the stress tensor). The equivalent stress itself should be a function of three linear-independent stress tensor invariants.

The set of linear independent invariants can be defined in different ways [5]. Let us introduce the linear, the quadratic and the cubic basic invariants as

$$I_1(\sigma) = \sigma \cdot\cdot I, I_2(\sigma) = \sigma \cdot\cdot \sigma, I_3(\sigma) = (\sigma \cdot \sigma) \cdot\cdot \sigma$$

I denotes the unit tensor and $\cdot$ the scalar (inner) product. The I_i $(i = 1, 2, 3)$ are the irreducible isotropic functional bases [6]. Combining these invariants we can formulate, for example, the following equivalent stress

$$\sigma_{eq} = \tilde{\alpha}\sigma_1 + \tilde{\beta}\sigma_2 + \tilde{\gamma}\sigma_3 \tag{2}$$

with

$$\sigma_1 = \mu_1 I_1, \sigma_2^2 = \mu_2 I_1^2 + \mu_3 I_2, \sigma_3^3 = \mu_4 I_1^3 + \mu_5 I_1 I_2 + \mu_6 I_3$$

Here the μ_i $(i = 1, \ldots, 6)$ are material parameters which should be identified by tests. The additional coefficients $\tilde{\alpha}, \tilde{\beta}, \tilde{\gamma}$ impose the weight of the invariants, and they can be introduced for a better curve fitting. It is easy to proof that $\beta = 1$ leads to a suitable comparison with the classical von Mises approach. Here we set $\tilde{\beta} = 1$ and, in addition, we presume the values $\tilde{\alpha}$ and $\tilde{\gamma}$ the same as that for $\tilde{\beta}$.

Neglecting the influence of hardening, softening, damage or other internal processes the creep strain rate tensor can be calculated from Eq. (1) assuming the dependence of the potential Φ on the introduced equivalent stress expression (2). After some calculations we obtain

$$\dot{\varepsilon}^{cr} = \dot{\eta}\frac{\partial\Phi}{\partial\sigma_{eq}}\left[\mu_1 I + \frac{\mu_2 I_1 I + \mu_3\sigma}{\sigma_2} + \frac{\left(\mu_4 I_1^2 + \frac{\mu_5}{3}I_2\right)I + \frac{2}{3}\mu_5 I_1\sigma + \mu_6\sigma\cdot\sigma}{\sigma_3^2}\right]$$

The unknown scalar factor $\dot{\eta}$ can be determined by the assumption of the equivalence of the dissipated power in both the uniaxial and the three-dimensional creep state

$$\sigma \cdot\cdot \dot{\varepsilon}^{cr} = \sigma_{eq}\dot{\varepsilon}_{eq}^{cr}$$

From this assumption follows [7]

$$\dot{\varepsilon}^{cr} = \dot{\varepsilon}_{eq}^{cr}\left[\mu_1 I + \frac{\mu_2 I_1 I + \mu_3\sigma}{\sigma_2} + \frac{\left(\mu_4 I_1^2 + \frac{\mu_5}{3}I_2\right)I + \frac{2}{3}\mu_5 I_1\sigma + \mu_6\sigma\cdot\sigma}{\sigma_3^2}\right] \tag{3}$$

The function $\dot{\varepsilon}_{eq}^{cr} = \dot{\varepsilon}_{eq}^{cr}(\sigma_{eq})$ is known from uniaxial creep experiments. Suitable approximations ofhe experimental data are the power, the hyperbolic sine or the exponential function. Here we prefer the power law. The discussion of other possible representations is presented, e.g., in [8]. Note that the generalised creep equation (3) has the same representation as the general dependence between two co-axial second-rank tensors [4].

The generalised non-classical creep equations can be reduced if special values of the material parameters μ_i are assumed. For example, the classical creep theory (similar to the von Mises flow theory in plasticity) follows if $\alpha = \gamma = 0$ (no influence of the linear or the cubic invariant that means equal behaviour in tension and compression and mean stress insensitive material behaviour) or $\mu_1 = \mu_4 = \mu_5 = \mu_6 = 0$ and $\mu_2 = -\frac{1}{2}, \mu_3 = \frac{3}{2}$. In this case the equivalent stress is the von Mises stress $\sigma_{eq} = \sqrt{\frac{3}{2} s \cdot\cdot s} = \sigma_{vM}$ with $s = \sigma - \frac{1}{3} I_1 I$, and the creep strain rate tensor follows as [9]

$$\dot{\varepsilon}^{cr} \equiv \dot{e}^{cr} = \frac{3}{2} \frac{\Phi(\sigma_{vM})}{\sigma_{vM}} s \quad \text{with} \quad e = \varepsilon - \frac{1}{3} \varepsilon \cdot\cdot II$$

Taking into account the usual approximation for $\Phi(\sigma_{vM})$ in the form of the Norton's creep law $\dot{\varepsilon}^{cr}_{eq} = A \sigma^n_{vM}$ (A and n denote material parameters which should be established experimentally) we conclude

$$\dot{e}^{cr} = \frac{3}{2} A \sigma^{n-1}_{vM} s,$$

which is the same equation like in the classical textbooks [10]. This creep equation is tensorial linear and does not reflect any non-classical material behaviour.

3. Identification of the creep model

The proposed generalised isotropic creep equation (3) contains 6 unknown parameters μ_i which should be identified by tests. Various possibilities of identification procedures are known. A simple procedure can be presented with the help of creep tests by constant loading and temperature. This approach is widely used in engineering applications, but should be carefully handled because the material dependent parameters are determined for a given constant load level. The use of the state equations with these parameters is allowed only in a small range of stresses in the neighbourhood of the given load level [11]. Other possibilities are discussed, for example, in [12]. The choice of the tests depends on the experimental facilities and on the possibility to achieve analytical solutions for comparison with the experimental results.

Assuming a Norton-type creep law with a creep exponent n which is constant with respect to the kind of loading (this was established experimentally [13]) let us suggest the following tests

- uniaxial tension ($\sigma_{11} > 0$)

$$\dot{\varepsilon}^{cr}_{11} = A_+ \sigma^n_{11}, \dot{\varepsilon}^{cr}_{22} = \dot{\varepsilon}^{cr}_{33} = -A_T \sigma^n_{11}$$

- uniaxial compression ($\sigma_{11} < 0$)

$$\dot{\varepsilon}^{cr}_{11} = -A_- |\sigma_{11}|^n$$

- simple shear ($\sigma_{12} = \sigma_{21} \neq 0$)

$$2\dot{\varepsilon}_{12}^{cr} = 2\dot{\varepsilon}_{21}^{cr} = A_S \sigma_{12}^n, \dot{\varepsilon}_{11}^{cr} = A_{PS} \sigma_{12}^n$$

- hydrostatic pressure ($\sigma_{11} = \sigma_{22} = \sigma_{33} < 0$)

$$\dot{\varepsilon}_{11}^{cr} = \dot{\varepsilon}_{22}^{cr} = \dot{\varepsilon}_{33}^{cr} = -3A_P |\sigma_{11}|^n$$

Providing analytical solutions for these simple stress states the comparison with the tests results in the following coefficients μ_i [7]

$$
\begin{aligned}
\mu_3 &= \tfrac{1}{2} A_S^{2r}, \mu_1 = \frac{A_{PS}}{(\sqrt{2\mu_3})^n}, \mu_2 = X^2 - \mu_3, \\
6\mu_4 &= (\sqrt{9\mu_2 + 3\mu_3} - 3\mu_1 - A_P^r)^3 - 3(T - \mu_1)^3 \\
&\quad + 18 \left(\frac{\mu_2}{\sqrt{\mu_2 + \mu_3}} + \mu_1 + A_T A_+^{-nr} \right)(T - \mu_1)^2, \\
2\mu_5 &= 3(T - \mu_1)^3 - (\sqrt{9\mu_2 + 3\mu_3} - 3\mu_1 - A_P^r)^3 \\
&\quad - 24 \left(\frac{\mu_2}{\sqrt{\mu_2 + \mu_3}} + \mu_1 + A_T A_+^{-nr} \right)(T - \mu_1)^2, \\
\mu_6 &= (T - \mu_1)^3 - \mu_4 - \mu_5
\end{aligned}
\tag{4}
$$

with $T = \tfrac{1}{2}(A_+^r - A_-^r), X = \tfrac{1}{2}(A_+^r + A_-^r), r = \frac{1}{n+1}$.

The proposed model takes into account various nonclassical effects. For instance, if $A_{PS} \neq 0$ the Poynting-Swift effect [14] can be modelled and volumetric creep strain rates take place as a result of shear loads (similar to the Kelvin effect in elasticity [15]), if $A_T \neq 0$ - the independent transverse contraction can be described. On the other hand, these effects are second order effects. Their experimental observation is limited (experimental difficulties). Another non-classical effect is the neglecting the creep incompressibility. The classical model is based on the assumption that $\dot{\varepsilon}_{ii}^{cr} = 0$. At the same time A_P tends to 0.

Let us discuss special cases of the generalised creep equation. The above discussed von Mises-type theory we obtain from the following considerations. At first we assume that the behaviour in tension and compression is the same ($A_+ = A_- \equiv A$) and we get $T = 0$ and $X = A^r$. The classical theory cannot reflect the Poynting-Swift effect ($A_{PS} = 0$) and is based on the second axiom of rheology [16] - the volumetric strains are purly elastic ($A_P = 0$). Finally we get

$$\mu_3 = \frac{1}{2} A_S^{2r}, \mu_1 = 0, \mu_2 = A^{2r} - \mu_3$$

Taking into account the values of μ_2 and μ_3 in section 2 which are based on the assumption of the Norton creep law with the material characteristics A and n we estimate that the classical creep theory is restricted by the relation $A_S^{2r} = 3A^{2r}$. For the last three values μ_4, μ_5, μ_6 we conclude: $\mu_6 = 0$ (the von Mises theory is tensorial linear), at the same time we find $\mu_4 = \mu_5 = 0$. In this case $3\mu_2 = \mu_3$ and we get again the restriction $A_S^{2r} = 3A^{2r}$.

Another simplification can be introduced by $\mu_1 = \mu_4 = \mu_5 = \mu_6 = 0$. This extended von Mises-type theory is restricted by $A_{PS} = 0$ (no Poynting-Swift effect), $T = 0$ (identical behaviour in tension and compression with $A_+ = A_- \equiv A$) and $9A^{2r} - 3A_S^{2r} = A_P^{2r}$. This simplified theory is based on two tests: tension and torsion or tension and transverse contraction. The first possibility was prefered in [7].

Finally let us discuss variants of the generalised creep law based on three material parameters. The following possibilities can be introduced. A tensorial linear creep law can be formulated with the help of the assumptions $\mu_4 = \mu_5 = \mu_6 = 0$. From these assumptions follow the restrictions for the material characteristics

$$T = \frac{1}{2}(A_+^r - A_-^r) = A_{PS}A_S^{-rn}, 9(A_+^r + A_-^r)^2 - 12A_S^{2r} = 4(A_P^r + 3A_{PS}A_S^{-nr})^2$$

Another tensorial linear creep law we obtain if $\mu_1 = \mu_4 = \mu_6 = 0$. On the other hand, a tensorial nonlinear creep law can be established by $\mu_1 = \mu_4 = \mu_5 = 0$. Providing similar calculations as in the first case the restrictions for the material characteristics for the two last cases can be deduced. A detailed description of the possibilities to formulate simplified equations with three parameters starting from the generalised creep law is presented in [2].

4. Creep-damage coupling

The presence of damage processes can be observed during the whole exploitation time of a structural element, but only the tertiary creep state the damage process is dominant. Following [17, 18] the damage process is connected with stress increase. Introducing the inner damage variable ω this stress increase can be expressed by substitution the stress σ by $\frac{\sigma}{1-\omega}$. Considering isotropic creep-damage this approach can be used in the multiaxial case too. In this case instead of σ we have to take into account $\frac{\sigma}{1-\omega}$.

The main problem now is the introduction of a suitable damage evolution equation. The damage evolution is a function of the stress state and the damage itself (the temperature is fixed). With respect to the assumed isotropy the stress state is characterised by an equivalent stress which depends on the stress tensor invariants. The equivalent stress function is, in general, different from the equivalent stress for the creep process.

Let us suppose the following damage evolution equation

$$\dot{\omega} = R(\omega)H[< \sigma_{eq}^{\omega}(\sigma) >]$$

Here σ_{eq}^{ω} is a equivalent stress expression which can differ from the equivalent stress for the creep equation. The brackets $< (\ldots) >$ denote

$$< (\ldots) > = \begin{cases} (\ldots) & \text{if } (\ldots) > 0 \\ 0 & \text{if } (\ldots) \leq 0 \end{cases}$$

The specification of the equivalent stress can be realised in a similar form proposed in [19] for a generalised strength and yield criterion. Introducing the first invariant I_1, the von Mises equivalent stress σ_{vM} and as a cubic invariant

$$\sin 3\xi = -\frac{27}{2}\frac{(s \cdot s) \cdots s}{\sigma_{vM}^3}, \quad -\frac{\pi}{6} \leq \xi \leq \frac{\pi}{6}$$

the following equivalent stress can be formulated

$$\sigma_{eq}^{\omega} = \lambda_1 \sigma_{vM} \sin \xi + \lambda_2 \sigma_{vM} \cos \xi + \lambda_3 \sigma_{vM} + \lambda_4 I_1 + \lambda_5 I_1 \sin \xi + \lambda_6 I_1 \cos \xi$$

The unknown coefficients λ_i should be identified experimentally. It is easy to show that the generalised damage criterion contains as a special case criteria for which the damage controlled by the von Mises stress is dominant ($\lambda_1 = \lambda_2 = \lambda_4 = \lambda_5 = \lambda_6 = 0$). If the damage process is controlled by the von Mises stress, the mean stress and the first (maximum) principal stress σ_1 the Hayhurst criterion [20] can be recommended

$$\sigma_{eq}^{\omega} = \alpha \sigma_I + \beta \sigma_{vM} + (1 - \alpha - \beta) I_1$$

Taking into account [21]

$$\sigma_I = \frac{1}{3}\left[2\sigma_{vM}\sin\left(\xi + \frac{2\pi}{3}\right) + I_1\right] = -\frac{1}{3}\sigma_{vM}\sin\xi + \frac{\sqrt{3}}{3}\sigma_{vM}\cos\xi + \frac{1}{3}I_1$$

we obtain

$$\lambda_1 = -\frac{1}{3}\alpha, \lambda_2 = \frac{\sqrt{3}}{3}\alpha, \lambda_3 = \beta, \lambda_4 = 1 - \frac{2}{3}\alpha - \beta, \lambda_5 = \lambda_6 = 0$$

The constitutive model for the creep-damage coupled behaviour can be formulated as a set of first order differential equations for the creep rate tensor presented in section 2 and the damage evolution equation discussed in this section. The function $R(\omega)$ must be specified in this case. Convenient approximations are $(1 - \omega)^{-l}$ [4] with one additional material parameter l or $(1 - \omega^k)^{-l}$ as it was introduced, e.g., in [22]. It is easy to show that the proposed constitutive model contains as a special case the classical creep-damage model proposed in [4] or [17]. This model does not reflect nonclassical creep effects (it is adequate to the von Mises flow theory discussed in section 2) and the damage evolution is controlled by the von Mises stress.

5. Identification of the creep-damage model

Let us briefly discuss the identification procedure in the case of creep-damage behaviour. A simplified procedure assuming the generalised equivalent stress expression in the creep constitutive equation only was presented, e.g., in [23]. The damage evolution has been included as a measure related to the dissipation power. In this case it was necessary to identify 6 material parameters. The values are the same as discussed in section 3. This model provides a good agreement with experimental results for the steady state creep and some differences for the tertiary creep. The reason for such differences is that the presented in [23]damage measure reflects only the creep-damage coupling.

The model established in section 4 is more sophisticated because it contains 12 unknown parameters. A suitable identification procedure will be discussed in a forthcoming paper. The conventional creep-damage model are based on the von Mises stress controlled creep rate. The damage rate depends on the Hayhurst equivalent stress [20]. This model reflects the stress state dependence of the tertiary creep only. Some applications to the structure analysis of thin shell and plates are discussed in [24] and [25].

6. Conclusions and Outlook

For various materials a dependence of the creep-damage behaviour upon the kind of the stress state is experimentally established. The classical approaches of modelling based on the von Mises flow theory (creep) do not reflect such a dependence. The Rabotnov-Kachanov theory with a scalar damage parameter allows to consider the stress state dependence of tertiary creep only. In this situation an improved model for the constitutive behaviour must be formulated.

The stress state dependend behaviour can be introduced, e.g., by modifying the equivalent stress expression considering fixed temperature and isotropic behaviour. Extended formulaes based on all three stress tensor invariants are introduced. In contrast to previous works different expression for the creep behaviour and for the damage are in use.

The main problem is the identification of 12 unknown parameters in the general case. Particular cases (for example pure creep behaviour or only stress state dependence for the damage part) are well investigated and verified. It can be shown that many proposals known from the literature can be deduced from the generalised creep-damage equations.

Further investigations should be directed to the following items:

- The creep-damage behaviour can be (in some cases) anisotropic. This experimentally established result should be reflected by the constitutive equations.
- The identification procedure should be developed for the general case.

- The application of the special cases of the generalised creep-damage equations in the analysis of creeping thin-walled structures results needs a modification of the conventional structural mechanics models. The consideration of stress state effects in the material behaviour requires the use of the refined cross section assumptions in the models of beams, plates or shells [26].

References

1. Altenbach, H. and Skrzypek, J.J. (Eds.) (1999) *Creep and Damage in Materials and Structures*. Springer, Wien, New York (CISM Courses and Lectures No. 399)
2. Altenbach, H., Altenbach, J., and Zolochevsky, A. (1995) *Erweiterte Deformationsmodelle und Versagenskriterien in der Werkstoffmechanik*. Dt. Verlag für Grundstoffindustrie, Stuttgart
3. Jin, J. and Cristescu, N.D. (1998) An elastic/viscoplastic model for transient creep of rock salt, *Int. J. Plasticity*, **14 no. 1-3**, pp. 85–107
4. Rabotnov, Yu.N. (1969) *Creep Problems in Structural Members*. North Holland, Amsterdam
5. Życzkowski, M. (1981) *Combined Loadings in the Theory of Plasticity*. PWN, Warszawa
6. Boehler, J.P. (1987) Representations for isotropic and anisotropic non-polynomial tensor functions, in: *Applications of Tensor Functions in Solid Mechanics* (ed. by J.P. Boehler), Springer, Wien, New York, pp. 13–53 (CISM Courses and Lectures No. 292)
7. Altenbach, H., Schieße, P., and Zolochevsky, A. (1991) Zum Kriechen isotroper Werkstoffe mit komplizierten Eigenschaften, *Rheologica Acta*, **30**, pp. 388–399
8. Voyiadjis, G.Z. and Zolochevsky, A. (1998) Modeling of secondary creep behaviour for anisotropic materials with different properties in tension and compression, *Int. J. Plasticity*, **14 no. 10-11**, pp. 388–399
9. Altenbach, H. (1999) Creep-damage behaviour of plates and shells, *Mechanics of Time-Dependent Materials*, **3**, pp. 103–123
10. Odqvist, F.K.G. and Hult, J. (1962) *Kriechfestigkeit metallischer Werkstoffe*, Springer, Berlin u.a.
11. Gummert, P. (1998a) *Contributions and discussions during the CISM Course No. 399*
12. Gummert, P. (1998b) General Constitutive Equations for Simple and Non-simple Materials, in: *Creep and Damage in Materials and Structures* (ed. by H. Altenbach and J.J. Skrzypek), Springer, Wien, New York
13. Malinin, N.N. (1981) *Raschet na polzuchest' konstruk cionnykh elementov (Creep Calculations in Structural Elements)*, Mashinostroenie, Moskva
14. Billington, E.W. (1985) The Poynting-Swift effect in relation to initial and post-yield deformation, *Int. J. Solids & Structures*, **21**, pp. 355-372
15. Backhaus, G. (1983) *Deformationsgesetze*, Akademie-Verlag, Berlin
16. Reiner, M. (1969) *Deformation and Flow. An Elementary Introduction to Rheology*, H.K. Lewis & Co., London
17. Kachanov, L.M. (1986) *Introduction to Continuum Damage Mechanics*, Martinus Nijhoff, Dordrecht et al.
18. Lemaitre, J. (1996) *A Course on Continuum Damage Mechanics*, Springer, Berlin et al.
19. Altenbach, H. and Zolochevsky, A. (1996) A generalized failure criterion for three-dimensional behaviour of isotropic materials, *Engng Fract. Mech.* **54**, pp. 75–90
20. Hayhurst, D.R. (1972) Creep rupture under multiaxial states of stress, *J. Mech. Phys. Solids*, **20**, pp. 381–390
21. Novozhilov, V.V. (1951) On the stress-strain relations in non-linear elastic continua, *Prikl. Mat. i Mekh.*, **XV no. 2**, pp. 183–194

22. Konkin, V.N. and Morachkovskij, O.K. (1987) Long-term strength of light alloys with anisotropic properties (in Russian), *Probl. Prochn.*, **no. 6**, pp. 38–42

23. Altenbach, H. and Zolochevsky, A.A. (1994) Eine energetische Variante der Theorie des Kriechens und der Langzeitfestigkeit für isotrope Werkstoffe mit komplizierten Eigenschaften, *ZAMM*, **74 no. 3**, pp. 189–199

24. Altenbach, H., Morachkovsky, O., Naumenko, K., and Sychov, A. (1997) Geometrically nonlinear bending of thin-walled shells and plates under creep-damage conditions, *Arch. Appl. Mech.*, **67**, pp. 339–352

25. Altenbach, H., Kolarov, G., Morachkovsky, O.K., and Naumenko, K. (2000) On the accuracy of creep-damage predictions in thin-walled structures using the finite element method, *Comp. Mech.*, **25**, pp. 87–98

26. Altenbach, H., Kushnevsky, V., and Naumenko, K. (2000) On the use of solid and shell type finite elements in creep-damage predictions of thinwalled structures, *Arch. Appl. Mech.*, (in press)

STRAIN, STRESS AND DAMAGE FIELDS IN DAMAGED AND CRACKED SOLIDS

A. BENALLAL
Laboratoire de Mécanique et Technologie
ENS Cachan / Université Paris 6 / CNRS
61 avenue du président Wilson
F-94235 Cachan Cedex, France

AND

L. SIAD
Laboratoire de Physique et Mécanique des Matériaux,
Université de Metz
BP 80794
F-57045 Metz Cedex, France

1. Introduction

One of the goals of Continuum Damage Mechanics is the development of an integrated and consistent theory of fracture. Ingredients of damage mechanics are the proper definition of a continuous damage variable able to represent the physical mechanisms of degradation, a kinetic law for its evolution during any loading process, its couplings with the deformation behaviour, a rupture criterion and finally a failure law for the material. Once these ingredients are defined, initiation, propagation and bifurcation of a crack (damaged zone) are in principle handled in a natural and consistent way.

Continuous degradation of materials generally leads to a softening behaviour. The presence of this softening is known to lead to physical as well as mathematical difficulties and to promote the phenomenon of localization of strain and damage. In the framework of local continua considered herein, this localization phenomenon takes place in a surface which is supposed to represent ultimately the crack.

The main purpose of this paper is to investigate the states of strain, stress and damage around a sharp crack in a longitudinally sheared solid.

S. Murakami and N. Ohno (eds.), IUTAM Symposium on Creep in Structures, 151–163.

This is done for a degrading elastic material:the material behaves linearly to the yield strain (sound material), hardens with damage to a peak load, softens until an ultimate shear strain is reached. Beyond, the material has a constant residual load carrying capacity in shear. The case of a completely degrading material is obtained by the limit when this residual load tends to zero. For simplicity, we limit the analysis to small scale damage conditions and also to the situation where no unloading is permitted. This last assumption should surely be avoided for practical purposes.

The solution to the mode III problem is constructed using the classical hodograph approach (see e.g. Rice (1967), Knowles and sternberg (1980). We underline through this construction the difficulties that arise due to the presence of softening but also those linked to the limit of a vanishing residual load capacity.

Asymptotic solutions near the crack tip for damaged materials were considered by Hao, Zhang and Hwang (1992), Zhang, Hwang and Hao (1993,1994), Gao and Bui (1995). Damaged zones in elastic-brittle materials were studied by Bui and Ehrlacher (1980).

2. Constitutive equations

Continuum damage mechanics (CDM) deals with the load carrying capacity of solids without major cracks but where the material itself is damaged due to the presence of microscopic defects such as microcracks or voids. Herein we restrict ourselves to isotropic damage formulation (see Lemaitre (1992)). The material under consideration is elastic and degrades with loading. The behavior of such a material may be discribed by a set of constitutive relations modelling elasticity coupled with damage. The Helmoltz free energy Ψ for a small deformation and isothermal conditions provides the law of elasticity coupled with damage and the definition of the associated variable related to the scalar-valued internal damage variable D. Ψ depends on the strain tensor ϵ_{ij} and the internal variable D.

$$\rho\Psi = \rho\Psi(\epsilon_{ij}, D) = \frac{1}{2}(1 - D)E_{ijkl}\,\epsilon_{ij}\,\epsilon_{kl} \qquad (1)$$

where ρ is the mass density and E_{ijkl} is the forth-order tensor of elastic sound material. The parameter D reflects the amount of damage which the material has experienced. It starts at zero (sound material) and grows to one (fully damaged material corresponding to complete loss of coherence). From the hypothesis of strain equivalence the stress tensor is given by

$$\sigma_{ij} = \rho\frac{\partial\Psi}{\partial\epsilon_{ij}} = (1 - D)E_{ijkl}\,\epsilon_{kl} \qquad (2)$$

and the force Y conjugated to the damage variable D is

$$Y = -\rho \frac{\partial \Psi}{\partial D} = \frac{1}{2} E_{ijkl} \, \epsilon_{ij} \, \epsilon_{kl} \tag{3}$$

The damage criterion (see Marigo (1981))(or the damage loading function) f depends on the force Y and the damage variable D

$$f = f(Y; D) \tag{4}$$

f is such that during progressive damage evolution, the identity $f = 0$ holds, otherwise we have $f < 0$. The kinetic law of damage evolution is derived from the potential of dissipation $\Psi^*(Y; D)$

$$\dot{D} = \dot{\lambda} \frac{\partial \Psi^*}{\partial Y}(Y; D) \tag{5}$$

where $\dot{\lambda}$ is the damage multiplier calculated from the consistency condition $\dot{f} = 0$. We will use in this paper

$$f(Y; D) \equiv \sqrt{Y} - \sqrt{Y_o} - M\,D \tag{6}$$

where M is a material constant. From the damage criterion $f(Y; D) = 0$ we have then

$$D = \frac{1}{M} \left(\sqrt{Y} - \sqrt{Y_o} \right) \tag{7}$$

3.   The antiplane crack problem

3.1. STATEMENT OF THE PROBLEM

For convenience, herein we treat the anti-plane deformation problem. We consider a mode III crack in a homogeneous isotropic elastic-damaged material. The semi infinite crack occupies the negative portion of the x axis. Referring to the rectangular coordinates system (x, y) shown in Fig. 1, all stress and strain components are zero except $\bar{\tau}_{xz} \equiv \bar{\tau}_x$, $\bar{\tau}_{yz} \equiv \bar{\tau}_y$, $\bar{\gamma}_{xz} \equiv \bar{\gamma}_x$ and $\bar{\gamma}_{yz} \equiv \bar{\gamma}_y$, where z is the coordinate axis perpendicular to the plane of Fig. 1. For such a deformation, we obtain $Y = \frac{1}{2}G\,\bar{\gamma}^2$ and hence provided the damage criterion is satisfied, we have $D = \frac{1}{M}\sqrt{\frac{G}{2}}\,(\bar{\gamma} - \bar{\gamma}_o)$ where $\bar{\gamma}_o$ is the yield shear strain.

The damage law is then given in terms of the shear strain $\bar{\gamma}$ by the following relation

$$D = \begin{cases} 0 & \text{if} \quad \bar{\gamma} < \bar{\gamma}_o \\ \dfrac{\bar{\gamma} - \bar{\gamma}_o}{\bar{\gamma}_{cr} - \bar{\gamma}_o} & \text{if} \quad \bar{\gamma}_o \leq \bar{\gamma} < \bar{\gamma}_{cr} \\ 1 & \text{if} \quad \bar{\gamma}_{cr} \leq \bar{\gamma} \end{cases} \tag{8}$$

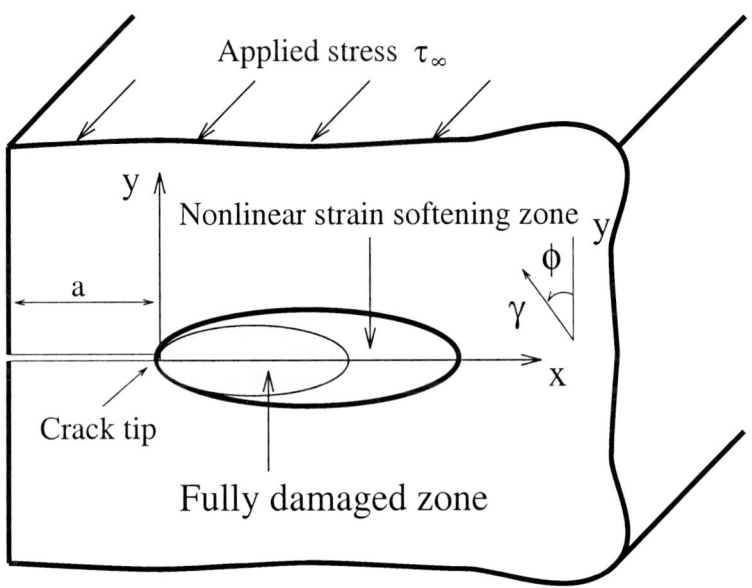

Figure 1.   The mode III crack problem

It is worth noting that for $\bar{\gamma}_{cr} \leq \bar{\gamma}$ the material is fully-damaged. The fully-damaged zone is treated as a void inside which the stresses are zero and subsequently cannot sustain any load. Furthermore displacements and strains cannot be described. In other words

$$\bar{\tau} = 0 \quad , \quad D = 1 \quad \text{and} \quad \bar{\gamma} \text{ is indeterminate}$$

Alternatively, we also consider the situation where beyond a given ultimate shear strain $\bar{\gamma}_u$, the shear stress is constant and equal to $\bar{\tau}_u$. Obviously, the former case cprresponds to the limit $\bar{\gamma}_u \to 0$

It is more convenient to introduce the following non-dimensional quantities

$$\tau = 4\gamma_o(1 - \gamma_o)\frac{\bar{\tau}}{\bar{\tau}_o} \quad , \quad \gamma = \frac{\bar{\gamma}}{\bar{\gamma}_{cr}} \quad , \quad \gamma_o = \frac{\bar{\gamma}_o}{\bar{\gamma}_{cr}}$$

The constitutive equations adopted here are then the following. The shear stress τ is given in terms of the shear strain γ by

$$\tau = \begin{cases} 4(1 - \gamma_o)\gamma & \text{if} \quad \gamma < \gamma_o \\ 4(1 - \gamma)\gamma & \text{if} \quad \gamma_o \leq \gamma < \gamma_u \\ \tau_u & \text{if} \quad \gamma_u \leq \gamma \end{cases} \tag{9}$$

The completely degrading material is obtained by the limit $\gamma_u \to \frac{1}{2}$. The previous constitutive relations are only valid for loading situations of the material point. We assume in this paper that unloading does not take place and then will not be invoked hereafter.

3.2. FIELD EQUATIONS

The equilibrium equation reduces to (no body forces)

$$\frac{\partial \bar{\tau}_x}{\partial x} + \frac{\partial \bar{\tau}_y}{\partial y} = 0 \tag{10}$$

and the strain-displacement relations are written as

$$\bar{\gamma}_x = \frac{\partial w}{\partial x} \quad , \quad \bar{\gamma}_y = \frac{\partial w}{\partial y} \tag{11}$$

Then we have a single compatibility equation which is

$$\frac{\partial \bar{\gamma}_x}{\partial y} = \frac{\partial \bar{\gamma}_y}{\partial x} \tag{12}$$

The effective stress and strain are related to the stress and strain components by

$$\bar{\tau}^2 = \bar{\tau}_x^2 + \bar{\tau}_y^2 \quad , \quad \bar{\gamma}^2 = \bar{\gamma}_x^2 + \bar{\gamma}_y^2$$

The components γ_x and γ_x are related to the effective strain γ by (See inset of Fig. 1)

$$\bar{\gamma}_x = -\bar{\gamma} \sin \phi \quad \text{and} \quad \bar{\gamma}_y = \bar{\gamma} \cos \phi \tag{13}$$

The components of stress and strain are related by

$$\bar{\tau}_x = \frac{\bar{\tau}(\bar{\gamma})}{\bar{\gamma}} \bar{\gamma}_x \quad \text{and} \quad \bar{\tau}_y = \frac{\bar{\tau}(\bar{\gamma})}{\bar{\gamma}} \bar{\gamma}_y \tag{14}$$

3.3. BOUNDARY CONDITIONS

Referring to a polar cylindrical coordinates system (r, θ) with origin at the crack tip, the boundary conditions for the mode III problem are

- The crack faces are stress free : $\tau_{\theta z}(r, \pm \pi) = 0$
- At infinity for $\bar{\tau}_y = \bar{\tau}_\infty < \bar{\tau}_o$, where $\bar{\tau}_o$ is the yield shear stress, the stress and strain distributions remain elastic.

4. Solution Construction

4.1. THE HODOGRAPH TRANSFORMATION

In order to obtain the solution to this problem we use the hodograph transformation used by Rice (1967). It can be readily shown that the anti-plane displacement w is given in terms of a potential function $\psi(\gamma, \phi)$ by

$$w(\gamma, \phi) = \gamma \frac{\partial \psi}{\partial \gamma}(\gamma, \phi) - \psi(\gamma, \phi) \tag{15}$$

The potential function ψ satifies, see for instance Rice (1967)

$$\frac{\tau(\gamma)}{\tau'(\gamma)} \frac{\partial^2 \psi}{\partial \gamma^2}(\gamma, \phi) + \frac{\partial \psi}{\partial \gamma}(\gamma, \phi) + \frac{1}{\gamma} \frac{\partial^2 \psi}{\partial \phi^2}(\gamma, \phi) = 0 \tag{16}$$

the boundary conditions along the crack, i.e.

$$\frac{\partial \psi}{\partial \phi}(\gamma, \pm \frac{\pi}{2}) = 0 \tag{17}$$

and eventually the conditions at infinity. In equation (16) $\tau'(\gamma)$ is the derivative of τ with respect to γ

At the crack tip $x = y = 0$, the strain is singular when $\tau_u \neq 0$. Thus $r \to 0$ as $\gamma \to \infty$.

5. The small-scale damage solution

We consider here the situation when $\tau_u \neq 0$. In the small-scale damage situation, the solution should match the classical elastic far-field , i.e

$$w(r, \theta) \to \frac{2K}{\mu} \sqrt{\frac{r}{2\pi}} \sin \frac{\theta}{2} \tag{18}$$

A particular solution ψ of the governing equation (16) which meets the boundary conditions $\frac{\partial \psi}{\partial \phi}(\gamma, \pm \frac{\pi}{2}) = 0$ can be constructed through a separation of variables of the form

$$\psi(\gamma, \phi) = f(\gamma) \sin \phi \tag{19}$$

By means of (16) the unknown function f satisfies the ordinary differential equation

$$\gamma \frac{\tau(\gamma)}{\tau'(\gamma)} f''(\gamma) + \gamma f'(\gamma) - f(\gamma) = 0 \tag{20}$$

where $(')$ and $('')$ stand for the first and second derivative of f with respect to γ respectively. As mentioned before, a strain singularity holds at the crack tip so that $\gamma \to \infty$ as $r \to 0$. It follows from

$$r = \left[f'(\gamma) \sin^2 \phi + \frac{1}{\gamma^2} f^2(\gamma) \cos^2 \phi \right]^{\frac{1}{2}} \tag{21}$$

that $f(\gamma)$ is such that

$$f'(\gamma) \to 0 \text{ as } \gamma \to \infty \tag{22}$$

to which we may add another condition to make f definite. The physical coordinates $x(\gamma, \phi)$ and $y(\gamma, \phi)$ are given in terms of the function f by

$$
\begin{aligned}
x &= -\frac{1}{2}\left[f'(\gamma) + \frac{1}{\gamma}f(\gamma)\right] + \frac{1}{2}\left[f'(\gamma) - \frac{1}{\gamma}f(\gamma)\right]\cos 2\phi \\
y &= \frac{1}{2}\left[f'(\gamma) - \frac{1}{\gamma}f(\gamma)\right]\sin 2\phi \\
0 &< \gamma < \infty, \quad -\frac{\pi}{2} \le \phi \le \frac{\pi}{2}
\end{aligned}
\tag{23}
$$

If one sets

$$
\begin{aligned}
R(\gamma) &= \frac{1}{2}\left[f'(\gamma) - \frac{1}{\gamma}f(\gamma)\right] \\
X(\gamma) &= -\frac{1}{2}\left[f'(\gamma) + \frac{1}{\gamma}f(\gamma)\right]
\end{aligned}
\tag{24}
$$

equations (23) may be put in the more compact form

$$
\begin{aligned}
x &= X(\gamma) + R(\gamma)\cos 2\phi \\
y &= R(\gamma)\sin 2\phi
\end{aligned}
\tag{25}
$$

which geometrical interpretation is immediate : the lines along which γ and $\tau(\gamma)$ have constant and positive values are circles

$$(x - X(\gamma))^2 + y^2 = R^2(\gamma) \tag{26}$$

with radius $R(\gamma)$ and centered on the x-axis at the abscissa $X(\gamma)$. By means of (15) the anti-plane displacement $w(\gamma, \phi)$ has the form

$$w(\gamma, \phi) = 2\gamma R(\gamma)\sin \phi \tag{27}$$

5.1. SOLUTIONS IN ELASTIC, DAMAGED AND FULLY DAMAGED ZONES

Solution of equation (20) imposes to treat separatly the types of behavior that a material point could meet.Here and in the sequel, the subscripts e, d and f for the function f stand for elastic, damaged and fully damaged respectively.

5.1.1. *Elastic zone*
Far away from the crack tip the material is elastic and therefore the potential ψ is harmonic, that is $\Delta_\gamma\psi(\gamma, \phi) = 0$. The corresponding solution is

that given by Rice (1967):

$$f_e(\gamma) = -A \left(\frac{1}{\gamma} + B\gamma \right) \tag{28}$$

with $A = \frac{1}{2\pi} \left(\frac{K\gamma_o}{\tau_o} \right)^2$, K is the loading parameter (stress intensity factor) and B will be determined later on.

Thus, inside the elastic region characterized by $\gamma < \gamma_o$ the iso-strain circles are

$$(x - X_e(\gamma))^2 + y^2 = R_e^2(\gamma) \tag{29}$$

with

$$R_e(\gamma) = \frac{A}{\gamma^2} \quad \text{and} \quad X_e(\gamma) = AB \tag{30}$$

5.1.2. Damaged zone

We do not discern between the hardening ($\gamma_o \leq \gamma \leq \frac{1}{2}$) and softening ($\frac{1}{2} \leq \gamma \leq \gamma_u$) parts of the damaged zone and the damaged-softening zone. In the circumstances where $\gamma \geq \gamma_o$, a solution of equation (20) is

$$f_d(\gamma) = C\gamma \int_\gamma^\gamma \frac{du}{u^2 \tau(u)} + D\gamma \tag{31}$$

as can be checked by substitution. The constants C and D are, as will be seen later, related to the constants A and B associated to the elastic solution through the continuity requirement of the physical coordinates $x(\gamma, \phi)$ and $y(\gamma, \phi)$ with respect to γ. Inside the damaged zone, where $\gamma \geq \gamma_o$ the iso-strain lines are, once again, circles

$$(x - X_d(\gamma))^2 + y^2 = R_d^2(\gamma) \tag{32}$$

with

$$R_d(\gamma) = \frac{C}{8} \frac{1}{\gamma^2(1 - \gamma)}$$

$$X_d(\gamma) = \frac{C}{8} \left\{ \frac{(1 - 2\gamma)}{\gamma(1 - \gamma)} + 2\ln \frac{\gamma_o(1 - \gamma)}{\gamma(1 - \gamma_o)} - \frac{1 + 2\gamma_o}{\gamma_o^2} \right\} - D \tag{33}$$

The following geometrical comments may be made. Both the radius and the center position on the x-axis depend upon the values of γ, and also upon the material constant γ_o. The circles shrink (ie R_d decreases) as γ increases from γ_o until the value $\gamma = \frac{2}{3}$. Beyond this later value the radius increases and for $\gamma = \gamma_u$ the corresponding circle is tangent at the origin O to the y-axis. As γ increases from γ_o to γ_u, the left-intercept of the iso-strain-circle corresponding to γ and the x-axis moves first to the right ($\gamma_o \leq \gamma \leq \frac{1}{2}$), and then to the left ($\gamma > \frac{1}{2}$).

5.1.3. *Fully damaged zone*
Inside the yet undetermined fully damaged zone for which $\gamma_u \leq \gamma$ and $\tau = \tau_u$, a solution of equation (20) is readily obtained

$$f_f(\gamma) = E\gamma + F \tag{34}$$

where E and F are constants to be determined. The requirement

$$f_f'(\gamma) \to 0 \quad \text{as} \quad \gamma \to \infty$$

imposes $E = 0$. The constant F will be determined later on from the contiguity conditions. Consequently the iso-strain circles in the (x, y) plane corresponding to the fully-dammaged zone are

$$[x - X_f(\gamma)]^2 + y^2 = R_f^2(\gamma) \tag{35}$$

with

$$R_f(\gamma) = -\frac{F}{2\gamma} \quad \text{and} \quad X_f(\gamma) = R_f(\gamma) \tag{36}$$

5.2. CONTIGUITY CONDITIONS

So far we have formally solved the governing equation (16). It remains to precise the constants of integration A, B, C, D, E. and the domain of the physical plane where the constructed solutions are valid. First of all, the physical coordinates $x(\gamma, \phi)$ and $y(\gamma, \phi)$ must be continuous functions with respect to their arguments. Furthermore, at the boundaries separating different zones, the displacement and the tractions should be continous.

5.2.1. *Continuity of the physical coordinates*
The continuity of the physical coordinates $x(\gamma, \phi)$ and $y(\gamma, \phi)$ with respect to γ holds if we have, on one hand

$$\begin{aligned} f_e(\gamma_o) = f_d(\gamma_o) \quad &\text{and} \quad f_e'(\gamma_o) = f_d'(\gamma_o) \\ f_d(\gamma_u) = f_f(\gamma_u) \quad &\text{and} \quad f_d'(\gamma_u) = f_f'(\gamma_u) \end{aligned} \tag{37}$$

The requirements (37), together with the solutions (28), (31) and (34) yield after some algebra to the relations

$$
\begin{aligned}
\frac{C}{A} &= 2\frac{T_o}{\gamma_o} = 8(1-\gamma_o) \\
\frac{D}{C} &= -\frac{1}{\gamma_u \tau_u} - \frac{1}{8}\left\{ \frac{(1+2\gamma_o)}{\gamma_o^2} - \frac{(1+2\gamma_u)}{\gamma_u^2} + 2\ln\frac{\gamma_u(1-\gamma_o)}{\gamma_o(1-\gamma_u)} \right\} \quad (38) \\
B &= -8(1-\gamma_o)\frac{D}{C} - \frac{1}{\gamma_o^2} \\
\frac{F}{C} &= -\frac{1}{\tau_u}
\end{aligned}
$$

Notice from the above relations that C, D, F are all proportional to A and that B is a constant independent of the loading parameter. Note also that B is such that

$$
\lim_{\tau \to 0} B = +\infty \qquad (39)
$$

5.2.2. Determination of the boundaries between the different zones

We turn now to the question of the determination of the interfaces between the different zones as defined above. These zones meet at a common boundary denoted by $\mathcal{S}$ which can be determined from the continuity of the displacement $w(\gamma, \phi)$ and the traction. It is obvious that $\mathcal{S}$ is symmetric with respect to the x-axis. The expression of the displacement $w(\gamma, \phi)$ is readily obtained from the combination of equations (27) and (25)

$$
w(\gamma, \phi) = 2\gamma R(\gamma) \left[\frac{X(\gamma) + R(\gamma) - x}{2R(\gamma)}\right]^{\frac{1}{2}} = \gamma \left[2R(\gamma)\right]^{\frac{1}{2}} \left[X(\gamma) + R(\gamma) - x\right]^{\frac{1}{2}}
$$

$$(40)$$

Our purpose is to obtain a parametric representation of the curve $\mathcal{S}$. In this goal, it is expedient to introduce the function g_k, as suggested by Knowles and Sternberg (1980) defined by:

$$
g(t) = t^2 R(t) \left[X(t) + R(t)\right]
$$

where R and X stands for the radius and the center of the iso-strain circle respectively. Let (x, y) be a point on the sought curve $\mathcal{S}$. Since the displacement w is to be continuous at (x, y), it follows that on the curve $\mathcal{S}$, we have

$$
\gamma_i \left[2R(\gamma_i)\right]^{\frac{1}{2}} \left[X(\gamma_i) + R(\gamma_i) - x\right]^{\frac{1}{2}} = \gamma_j \left[2R(\gamma_j)\right]^{\frac{1}{2}} \left[X(\gamma_j) + R(\gamma_j) - x\right]^{\frac{1}{2}}
$$

$$(41)$$

and this leads to

$$x(\gamma_i, \gamma_j) = \frac{g(\gamma_j) - g(\gamma_i)}{\gamma_j^2 R(\gamma_j) - \gamma_i^2 R(\gamma_i)} \tag{42}$$

To obtain the expression for y we recall that the point (x, y) of the curve $\mathcal{S}$ is the intersect of two iso-strain-circles the equations of which equations are

$$\begin{aligned}
(x - X(\gamma_i))^2 + y^2 &= R^2(\gamma_i) \\
(x - X(\gamma_j))^2 + y^2 &= R^2(\gamma_j)
\end{aligned} \tag{43}$$

The first equation yields the expression of y in terms of γ_i and γ_j on the the the curve $\mathcal{S}$.

$$y(\gamma_i, \gamma_j) = \pm\sqrt{R^2(\gamma_i) - \left[\frac{g(\gamma_j) - g(\gamma_i)}{\gamma_j^2 R(\gamma_j) - \gamma_i^2 R(\gamma_i)} - X(\gamma_i)\right]^2} \tag{44}$$

Consequently, the parametric representation of the sought curve $\mathcal{S}$ is

$$\begin{aligned}
x(\gamma_i, \gamma_j) &= \frac{g(\gamma_j) - g(\gamma_i)}{\gamma_j^2 R(\gamma_j) - \gamma_i^2 R(\gamma_i)} \\
y(\gamma_i, \gamma_j) &= \pm\sqrt{R^2(\gamma_i) - \left[\frac{g(\gamma_j) - g(\gamma_i)}{\gamma_j^2 R(\gamma_j) - \gamma_i^2 R(\gamma_i)} - X(\gamma_i)\right]^2}
\end{aligned} \tag{45}$$

The expressions for the radius $R(\gamma)$ and the center $X(\gamma)$ in the foregoing expression depend on the point, and hence on γ, where these quantities are evaluated. Thus, the both expressions $R(\gamma)$ and $X(\gamma)$ corresponding to the elastic, damaged and fully damaged zones are given by (30), (33) and (36) respectively. To determine the curves $\mathcal{S}$ we follow the procedure suggested by Knowles and Sternberg (1980).

$$\Phi(\gamma_i, \gamma_j) = 0 \tag{46}$$

with Φ defined by

$$\begin{aligned}
\Phi(\gamma_i, \gamma_j) = \gamma_j^2 R(\gamma_j)[X(\gamma_j) + R(\gamma_j) - X(\gamma_i) + R(\gamma_i)] \\
- \gamma_i^2 R(\gamma_i)[X(\gamma_i) + R(\gamma_i) - X(\gamma_j) + R(\gamma_j)]
\end{aligned} \tag{47}$$

First, for a given γ_i, we solve the nonlinear equation (46) with the unknown γ_j. The root γ_j, if such a solution exists, is function of γ_i. The substitution of the expression $\gamma_j(\gamma_i)$ into (45) yields the parametric representation of the sought curve $\mathcal{S}$ (the parameter is precisely γ_i).

$$\begin{aligned}
x &= x(\gamma_i, \gamma_j) = x(\gamma_i, \gamma_j(\gamma_i)) = x(\gamma_i) \\
y &= y(\gamma_i, \gamma_j) = y(\gamma_i, \gamma_j(\gamma_i)) = y(\gamma_i)
\end{aligned} \tag{48}$$

This procedure is followed and applied between the different zones. The full derivation may be found in Benallal and Siad (1997).The location of the different zones are displayed graphically in Fig. 2. Note in passing that the different branches of the curve $\mathcal{S}$, obtained separately, fit continuously one to each other. It remains to check that the tractions are continuous accross the curve $\mathcal{S}$. The proof is exactly the same as the one given by Knowles and Sternberg (1980) and will not be repeated here.

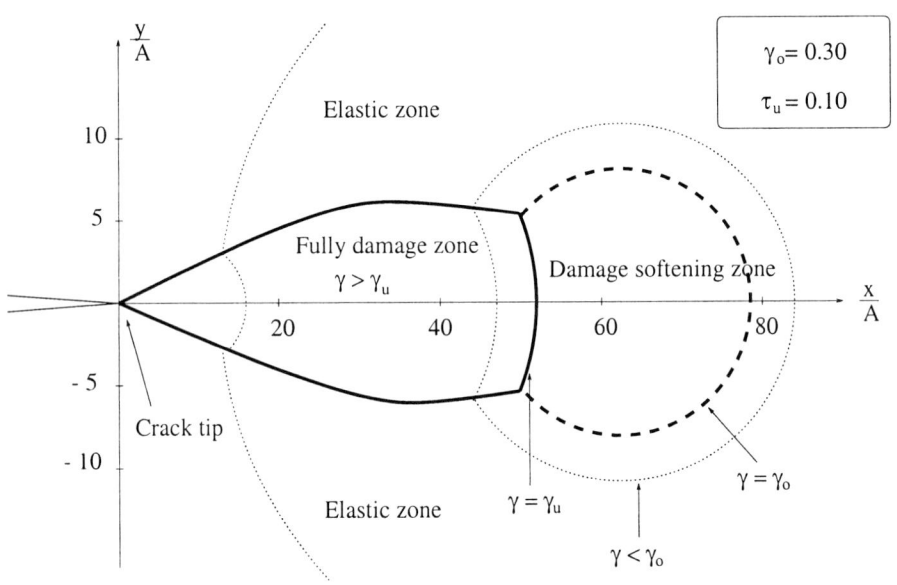

Figure 2. Shape and properties of the damaged zone with different shear strain iso-contours

The location of the different zones are displayed graphically in Fig. 2. Note in passing that the different branches of the curve $\mathcal{S}$, obtained separately, fit continuously one to each other. It remains to check that the tractions are continuous accross the curve $\mathcal{S}$. The proof is exactly the same as the one given by Knowles and Sternberg (1980) and will not be repeated here.

6. Conclusion

Antiplane shear fields around a stationnary crack have been obtained for an elastic degrading material in the case of small scale yielding and without unloading. This leads with softening to discontinuous strains. The size of the damaged zone ahead of the crack tip is found to be proportional to the loading parameter (elastic stress intensity factor) and to the inverse of the residual shear stress. The damaged zone extends then unboundedly in the limit of a completely degrading material, leading to a highly unstable behaviour. The results obtained here may explain (to some extent) the shielding effect due the damage existing ahead of the crack. The same analysis, taking into account unloading is under investigation.

7. Reference

Bui, H. D., Ehrlacher, A. (1980), Propagation dynamique d'une zone endommagée dans un solide élastique en mode III et en régime permanent, *Comptes Rendus de l'Académie des Sciences* **290**, 273-276.

Gao, Y. C., Bui, H. D. (1995), "Damaged Field Near a Stationary Crack Tip, *International Journal of Solids and Structures* **34**, 1979-1984.

Hao, T. H., Zhang, X. T. Hwang, K. C. (1992), Inhomogeneous behavior of anti-plane shear crack in softening material with local unloading, *Theoretical and Applied Fracture Mechanics* **17**, 163-175.

Knowles, J. K., Sternberg, Eli (1980), Discontinuous deformation gradients near the tip of a crack in finite anti-plane shear fields : an example, *Journal of Elasticity* **10**, 81-110.

Knowles, J. K., Sternberg, Eli (1981), Anti-plane shear fields with discontinuous deformation gradients near the tip of a crack in finite elastostatics, *Journal of Elasticity* **11**, 129-164.

Lemaitre, J. (1992), A Course on Damage Mechanics, *Springer-Verlag.*

Marigo, J. J. (1982),Modelling of brittle and fatigue damage for elastic material by growth of microvoids, *Engineeringrg Fracture Mechanics* **21**, 257-267.

Rice, J. R. (1967), Stresses Due to a Sharp Notch in a Work-Hardening Elastic-Plastic Material Loaded by Longitudinal Shear, *Journal of Applied Mechanics* **32**, 287-298.

Zhang, X. T., Hwang K. C., Hao T. H. (1993), Asymptotic solution of mode III crack in damaged softening materials, *Internatioanl Journal of Fracture* **62**, 269-281.

Zhang, X. T., Hao, T. H., Hwang, K. C. (1994), Unloading characteristics of anti-plane shear crack in softening material, *Theoretical and Applied Fracture Mechanics* **21**, 233-240.

Benallal, A., Siad,L. (1997), Stress and strain fields in cracked damaged solids, *Technische Mechanic* **17**, 295-304.

EFFECTS OF DAMAGE ON THE ASYMPTOTIC FIELDS OF A MODE I CREEP CRACK IN STEADY-STATE GROWTH

Sumio MURAKAMI and Toshiyuki HIRANO
Department of Mechanical Engineering, Nagoya University
Furo-cho, Chikusa-ku, Nagoya, 464-8603, Japan

Yan LIU
Centre for Surface Transportation Technology
National Research Council Canada
U-89, Alert Road, Uplands, Ottawa, Ontario K1A0R6, Canada

1. Introduction

Creep fracture process in usual polycrystalline materials is brought about by nucleation, growth and coalescence of distributed grain-boundary cavities in front of a crack-tip, and this damage field gives significant influence on the stress field near the crack-tip. The analyses of the effects of material damage on the stress and strain fields near a crack-tip in non-linear creep materials, therefore, have been discussed in a number of papers (Bassani and Hawk [1]; Astafjev, Grigorova and Pastukhov [2]; Astafjev and Grigorova [3]; Lu, Lee, Kim and Mai [4]). However, because of difficulties of the stress-damage coupled analysis, they could not derive complete and consistent results [2-4], nor could obtain systematic information about the effects of material damage on the asymptotic crack-tip fields [1].

The present authors (Liu and Murakami [5][6], Murakami, Hirano and Liu [7]; Hirano, Murakami and Liu [8], Liu, Hirano and Murakami [9]), on the other hand, developed a continuum damage mechanics analysis of mode I and mode III crack by means of semi-inverse method. They discussed the effects of material damage on the asymptotic fields of stress, strain (rate) and damage by employing power law strain-hardening and creep damage theory, and showed the relation among the exponent p of the asymptotic stress field, the exponent n of the power law constitutive law, and the exponent m of damage law.

The present paper is concerned with a brief review and further elaboration of Reference [6], [7] and [9] to take account of a more general damage evolution equation in the case of a mode I creep crack.

2. Governing Equations and Asymptotic Stress Field

2.1 GOVERNING EQUATIONS

Let us take a mode I creep crack extending at a constant rate v in a stationary Cartesian

S. Murakami and N. Ohno (eds.), IUTAM Symposium on Creep in Structures, 165–174.

coordinates $O - X_1 X_2 X_3$ as shown in Fig.1, and assume that the material in the vicinity of the crack is in the state of plane strain or of plane stress. Then, we further take moving Cartesian coordinates $o - x_1 x_2 x_3$ and moving polar coordinates $o - r\theta z$ with the origins o at the tip of the moving cracks, where the direction x_1 and that of $\theta = 0$ are in the direction of crack extension. By denoting the components of stress and strain with respect to the moving coordinates by σ_{ij} and ε_{ij}, the governing equations for mode I creep crack in steady-state extension are given as follows:

Components of stress

$$\sigma_{rr} = \frac{1}{r^2}\frac{\partial^2 \Phi}{\partial \theta^2} + \frac{1}{r}\frac{\partial \Phi}{\partial r}, \quad \sigma_{\theta\theta} = \frac{\partial^2 \Phi}{\partial r^2}, \quad \sigma_{r\theta} = -\frac{\partial}{\partial r}\left(\frac{1}{r}\frac{\partial \Phi}{\partial \theta}\right) \tag{1}$$

where $\Phi = \Phi(r, \theta, z)$ is the Airy stress function.

Condition of compatibility

$$r\frac{\partial}{\partial r}\left(\frac{\partial(r\dot{\varepsilon}_{\theta\theta})}{\partial r}\right) + \frac{\partial^2 \dot{\varepsilon}_{rr}}{\partial \theta^2} - r\frac{\partial \dot{\varepsilon}_{rr}}{\partial r} - 2\frac{\partial}{\partial r}\left(r\frac{\partial \dot{\varepsilon}_{r\theta}}{\partial \theta}\right) = 0 \tag{2a}$$

$$(\dot{\ }) = \frac{\partial}{\partial t} - \left(\cos\theta\frac{\partial}{\partial r} - \frac{\sin\theta}{r}\frac{\partial}{\partial \theta}\right)v \tag{2b}$$

where $(\dot{\ })$ denotes the material time derivative with respect to the time t.

Constitutive equation of creep
If we represent the damage state of the material by an isotropic damage variable $D (0 \le D \le 1)$, and employ the hypothesis of strain equivalence in damage mechanics, a creep constitutive equation of a damaged material can be given as follows (Kachanov [10]; Lemaitre and Chaboche [11]; Lemaitre [12]):

$$\dot{\varepsilon}_{ij} = (3/2)A\sigma_{EQ}^{n-1}s_{ij} / (1-D)^n \quad (i, j = 1, 2, 3) \tag{3}$$

where $s_{ij} = \sigma_{ij} - (1/3)\sigma_{kk}\delta_{ij}$ and $\sigma_{EQ} = (3s_{ij}s_{ij}/2)^{1/2}$ are the deviatoric stress and the equivalent stress. The symbols n and A are a creep exponent and a material constant.

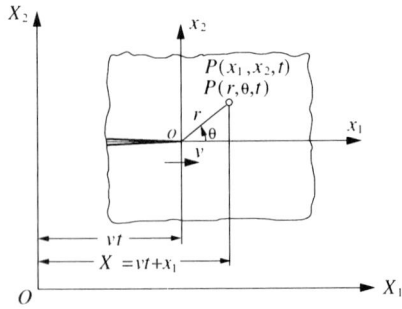

Figure 1. Stationary and moving coordinate systems

Evolution equation of creep damage

By assuming that the creep damage is governed by the equivalent stress σ_{EQ}, maximum principal stress σ_I and hydrostatic stress $(1/3)\sigma_{kk}$, the damage evolution equation in multi-axial state of stress may be given as follows [10-12].

$$\dot{D} = B \frac{[(1-\alpha-\beta)\sigma_{EQ} + \alpha\sigma_I + \beta\sigma_{kk}]^m}{(1-D)^q} \tag{4}$$

where B, $m\,(m>0)$, $q\,(q>0)$, $\alpha\,(0\le\alpha\le1)$ and $\beta\,(0\le\beta\le1)$ denote material constants. In the particular case of steady-state crack growth, we have

$$\frac{\partial D}{\partial t} \equiv 0 \tag{5}$$

and hence equation (4) leads to

$$-\cos\theta \frac{\partial D}{\partial r} + \frac{\sin\theta}{r}\frac{\partial D}{\partial\theta} = \frac{B}{v} \frac{[(1-\alpha-\beta)\sigma_{EQ} + \alpha\sigma_I + \beta\sigma_{kk}]^m}{(1-D)^q} \tag{6}$$

2.2 ASYMPTOTIC STRESS FIELD

In order to elucidate the effect of material damage on the asymptotic fields near a creep-crack tip, we will assume the following asymptotic solution for the crack-tip stress:

$$\Phi(r,\theta) = Kr^s f(\theta) \tag{7}$$

where K and s are undetermined constants, while $f(\theta)$ is an unknown function of θ. By substituting equation (7) into (1), we have the components of the asymptotic stress field as follows:

$$\sigma_{rr}(r,\theta) = Kr^p[s \cdot f(\theta) + f''(\theta)] = Kr^p \tilde{\sigma}_{rr}(\theta)$$

$$\sigma_{\theta\theta}(r,\theta) = Kr^p[s(s-1)f(\theta)] = Kr^p \tilde{\sigma}_{\theta\theta}(\theta)$$

$$\sigma_{r\theta}(r,\theta) = Kr^p[(1-s)f'(\theta)] = Kr^p \tilde{\sigma}_{r\theta}(\theta) \tag{8}$$

$$\sigma_{EQ}(r,\theta) = Kr^p \tilde{\sigma}_{EQ}(\theta)$$

where $p = s - 2$ represents the exponent of stress field, and $\tilde{\sigma}_{rr}(\theta)\ \cdots\ \tilde{\sigma}_{EQ}(\theta)$ are given by the expressions in the corresponding brackets [ ] in equation (8). In view of equation (8), the undetermined constant K corresponds to the stress intensity factor of a non-linear material and depends on the exponent n of the creep constitutive equation (3).

3. Semi-Inverse Solution of Non-Linear Differential Equations

3.1 SEMI-ELLIPTIC DAMAGE DISTRIBUTION AT CRACK-TIP

The numerical simulation and experimental observation on the damage distribution around a mode I creep crack in OFHC copper at $250\,^\circ$C in steady-state growth (Liu, Murakami,

Kojima and Matsushima [13]) shows that the contour of the damage field can be represented by a semi-ellipse in front of the crack and by a wake parallel to the crack plane behind the crack.

Then, we will assume the damage distribution of Fig.2 represented as follows:

$$1 - D = h(\theta)(r/r_0)^{\ell} \qquad (0 \le \ell < 1) \tag{9a}$$

$$h(\theta) = \begin{cases} [(\cos\theta/k)^2 + (\sin\theta)^2]^{\ell/2} & 0 \le \theta \le \pi/2 \\ (\sin\theta)^{\ell} & \pi/2 < \theta \le \pi \end{cases} \tag{9b}$$

where ℓ and r_0 are parameters characterizing the damage distribution, while $h(\theta)$ and k denote the θ-distribution of the damage field and its aspect ratio. The locus $r = r_0 h(\theta)^{-1/\ell}$, in particular, represents the boundary of the damage field where $D = 0$, and hence is the boundary between the damaged and the undamaged region.

3.2 DIFFERENTIAL EQUATIONS OF THE ASYMPTOTIC STRESS FIELD FOR SEMI-ELLIPTIC DAMAGE DISTRIBUTION

By substituting the stress and damage fields of equations (8) and (9) into the creep constitutive equation (3), and then by substituting the resulting creep rate $\dot{\varepsilon}_{ij}$ into the equation of compatibility (2), we can readily obtain the differential equations governing unknown function $f(\theta)$. The differential equations can be written for the states of plane strain and plane stress as follows:

State of plane strain:

$$[\frac{\partial}{\partial\theta^2} - n(s - \ell - 2)\{n(s - \ell - 2) + 2\}][h(\theta)^{-n}\tilde{\sigma}_{EQ}^{n-1}\{s(2 - s)f(\theta) + f''(\theta)\}]$$

$$+ 4(s - 1)[n(s - \ell - 2) + 1]\frac{\partial}{\partial\theta}[h(\theta)^{-n}\tilde{\sigma}_{EQ}^{n-1}f'(\theta)] = 0 \tag{10a}$$

$$\tilde{\sigma}_{EQ} = \{3[s(2 - s)f(\theta) + f''(\theta)]^2/4 + 3(1 - s)^2 f'(\theta)^2\}^{1/2} \tag{10b}$$

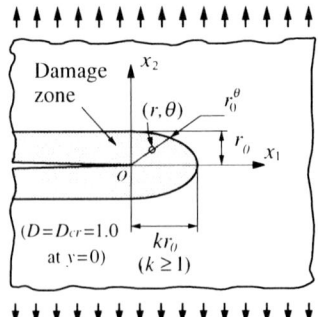

Figure 2. Mode I crack and damage fields

State of plane stress:

$$[n(s-\ell-2)-\frac{\partial}{\partial\theta^2}][h(\theta)^{-n}\tilde{\sigma}_{EQ}{}^{n-1}\{s(s-3)f(\theta)-2f\,'(\theta)\}]$$

$$+[n(s-\ell-2)+1]n(s-\ell-2)h(\theta)^{-n}\tilde{\sigma}_{EQ}{}^{n-1}[s(2s-3)f(\theta)-f\,'(\theta)]$$

$$+6[n(s-\ell-2)+1](s-1)\frac{\partial}{\partial\theta}[h(\theta)^{-n}\tilde{\sigma}_{EQ}{}^{n-1}f\,(\theta)]=0 \tag{11a}$$

$$\tilde{\sigma}_{EQ}(\theta)=[s^2(3-3s+s^2)f(\theta)^2+(3-6s+3s^2)f\,'(\theta)^2$$

$$+s(3-s)f(\theta)f''(\theta)+f''(\theta)^2]^{1/2} \tag{11b}$$

3.3 BOUNDARY CONDITIONS OF THE DIFFERENTIAL EQUATIONS AND THEIR NUMERICAL CALCULATION

Because of the symmetry of the stress components σ_{rr} and $\sigma_{\theta\theta}$ at $\theta=0$ and of the vanishing condition of the stress components $\sigma_{r\theta}$ and $\sigma_{\theta\theta}$ at the crack plane $\theta=\pi$, we have the following two boundary conditions for the differential equations (10) and (11):

$$f'(0)=0,\quad f'''(0)=0 \tag{12}$$

$$f(\pi)=0,\quad f'(\pi)=0 \tag{13}$$

The two-point boundary value problem of equations (10)-(13) for the asymptotic stress field $f(\theta)$ can be solved by a shooting method as discussed in detail in Reference [5-7], and gives a $p-\ell-n$ relation among the exponent p of stress distribution, the exponent ℓ of damage distribution and the creep exponent n.

3.4 COUPLED SOLUTION OF THE ASYMPTOTIC STRESS AND DAMAGE FIELDS

The damage field (9) employed in the above analysis is not always consistent with the damage evolution equation (6) and the resulting asymptotic stress field. Moreover the determination of the damage field which satisfies the damage evolution equation for the entire region of the problem is not easy. Thus we will satisfy this condition approximately by specifying the damage field of equation (9) so that it may be consistent with the evolution equation (6) in the region in front of the crack, because this region has the largest influence on the asymptotic fields of stress and damage.

On the crack plane $\theta=0$ in front of the crack, the evolution equation (6) is reduced to

$$\frac{B}{v}\frac{[(1-\alpha-\beta)\sigma_{EQ}+\alpha\sigma_I+\beta\sigma_{kk}]^m}{(1-D)^q}=-\frac{\partial D}{\partial r} \tag{14}$$

Substitution of equations (8) and (9) into this equation gives

$$\frac{B}{v}\frac{1}{\ell}(kr_0)^{\ell(q+1)}K^m[(1-\alpha-\beta)\tilde{\sigma}_{EQ}(0)+\alpha\tilde{\sigma}_I(0)+\beta\tilde{\sigma}_{kk}(0)]^m=r^{(q+1)\ell-pm-1} \tag{15}$$

In order that this condition could be satisfied always, we have the following relations:

$$\ell = \frac{pm+1}{q+1} \tag{16a}$$

$$v = \frac{B}{\ell}(kr_0)^{\ell(q+1)} K^m \left[(1-\alpha-\beta)\tilde{\sigma}_{EQ}(0) + \alpha\tilde{\sigma}_I(0) + \beta\tilde{\sigma}_{kk}(0)\right]^m \tag{16b}$$

which specifies the exponent ℓ of the damage distribution (9) and the steady-state crack rate v. Equation (16b) implies that the crack rate v is proportional to the coefficient B of the damage evolution equation (6) and to the stress $(kr_0)^{\ell(q+1)} K^m \left[(1-\alpha-\beta)\tilde{\sigma}_{EQ}(0) + \alpha\tilde{\sigma}_I(0) + \beta\tilde{\sigma}_{kk}(0)\right]^m$. As observed from equation (9a), the exponent of damage distribution has the value $0 \le \ell < 1$, and hence equation (16b) predicts the infinite crack rate $v = \infty$ in the case of uniform damage field $\ell = 0$.

As observed in equation (16), the stress criterion for creep damage evolution , (i.e., the values of the material constants of α and β) has no influence on the exponent p of stress singularity, but gives direct effects on the crack growth rate v.

4. Results of Analysis and Discussions (in the case of $m = q$)

4.1 EFFECTS OF MATERIAL DAMAGE ON THE ASYMPTOTIC FIELD AT CRACK-TIP

According to the numerical calculation of Section 3.3, we have the $p - \ell - n$ relation among the exponents of p, ℓ and n. Equation (16a) of Section 3.4, on the other hand, specifies the $p - \ell - m$ relation in the particular case of $m = q$. By the use of these two relations of $p - \ell - n$ and $p - \ell - m$, we can readily calculate the value of the exponent p as a function of the creep exponent n and the damage exponent m [7].

Since it has been ascertained that the value of k in equation (9) has not significant effects on the $p - \ell - n$ relation [6], we will assume $k = 1$ throughout the following analysis.

The small circles in Fig.3 show the numerical results for the exponent p of the asymptotic stress field (8) of mode I creep crack in steady-state growth, for the case of $k = 1$. The dotted lines, on the other hand, represent the exponent p of the HRR (Hutchinson [14]; Rice and Rosengren [15]) stress field for undamaged non-linear materials:

$$p = -\frac{1}{n+1} \tag{17}$$

Since the exponent $p < 0$ implies the stress singularity at the crack-tip $r = 0$, the HRR field has always stress singularity for any finite value of the creep exponent n. However, as observed in Fig.3, the presence of material damage increases significantly the value of the exponent p, and may give non-singular stress field even for finite values of n. Namely, Fig.3 shows that the stress singularity of the asymptotic field is determined by a relative relation between n and m, and this is a very important result as the effect of the material damage on the asymptotic fields at a crack-tip.

By comparing Fig.3(a) and (b), it will be observed that, for a given set of n and m, the case of plane strain has larger stress singularity than that of the plane stress, which is different from the case of undamaged non-linear materials [14, 15].

Finally, the solid lines in Fig.3 represent an approximate expression to the corresponding

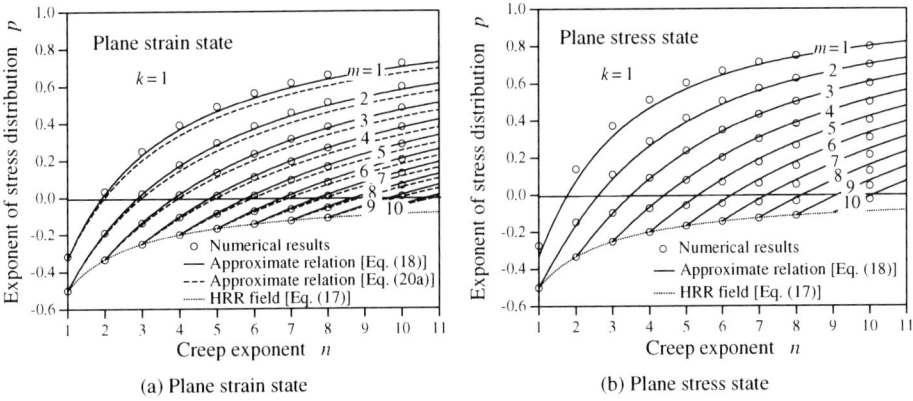

Figure 3. Exponent of asymptotic stress field for a mode I creep crack in stedy-state growth $(m = q)$

numerical results, and are given by the following relations [7]:

$$p = \frac{n^{(1+c/n)} - (m+1)}{n + m[1 + n - n^{(1+c/n)}] + 1} \tag{18a}$$

$$c = 0.100 \quad \text{(plane strain state)}$$
$$c = 0.370 \quad \text{(plane stress state)} \tag{18b}$$

4.2 STRESS SINGULARITY AT CRACK-TIP IN THE STATE OF PLANE STRAIN

In the case of plane strain state, the $p - \ell - n$ relation of equation (18) for $k = 1$ can be expressed by the following approximate relation [6]:

$$p = \frac{\ell n - 1}{n + 1} \tag{19}$$

Equation (19) together with equation (16a) gives the following relations:

$$p = \frac{n - (m+1)}{n + m + 1}, \quad \ell = \frac{n - (m-1)}{n + m + 1} \tag{20a), (20b}$$

The relations (20a) and (20b) give the approximate relations among the creep exponent n, the damage exponent m, the exponent of stress distribution p and the exponent ℓ of the damage distribution consistent to the damage evolution equation (14).

Since the damage variable has the range $0 \leq D \leq 1$, equation (9) specifies the exponent ℓ to be $\ell \geq 0$. Then, equation (20b) imposes the condition

$$n \geq m - 1 \tag{21}$$

which is usually satisfied because the relation

$$n \geq m \tag{22}$$

is ascertained for most metallic materials (Kachanov [10]).

According to equations (20a) and (21), the exponent of stress distribution p is

$$p < 0, \quad m - 1 \leq n < m + 1 \tag{23a}$$

$$p \geq 0, \quad m + 1 \leq n \tag{23b}$$

Namely, this result implies that the asymptotic stress field at a crack tip may be singular or non-singular when the set of the exponents n and m is in the range of equation (23a) or of equation (23b).

In the case of the plane stress, on the other hand, it is difficult to make a similar argument as above, because we could not obtain a simple expression for p as equation (19).

5. Results of Analysis and Discussions (in the case of $m \neq q$)

5.1 RESULTS OF ANALYSIS FOR THE STATES OF PLANE STRAIN AND PLANE STRESS

According to the numerical calculation of Sections 3.3 and 3.4, we can analyse the coupled asymptotic fields of stress and damage for a more general damage law of equation (6) with $m \neq q$.

Figs 4(a) and (b) show the relations between the exponent p of stress singularity and the creep exponent n for the states of plane strain and plane stress, respectively. These figures show the example of $k = 1$ and a typical value of $m = 5$.

The solid and the dashed lines in Fig.4 represent the numerical results of equations (1)-(16) and the corresponding approximate relation (24a) to be discussed below while the dotted line shows the HRR field of equation (17). The solid lines of $q = 5$ in Figs 4(a) and (b) coincide with the results of Fig.3. It will be observed in these figures that the parameter q has significant influence on the $p - n$ relations. Though the approximate $p - m - n - q$ relations entered by dashed lines in Fig.4(a) show significant deviation from the corresponding numerical results (solid lines) in the range of $p \gg 0$, they coincide well with the solid lines in the range of $p \sim 0$ or $p < 0$.

5.2 STRESS SINGULARITY AT CRACK-TIP IN THE STATE OF PLANE STRAIN [9]

Since the numerical analysis of Section 3.3 is not related to a damage evolution equation,

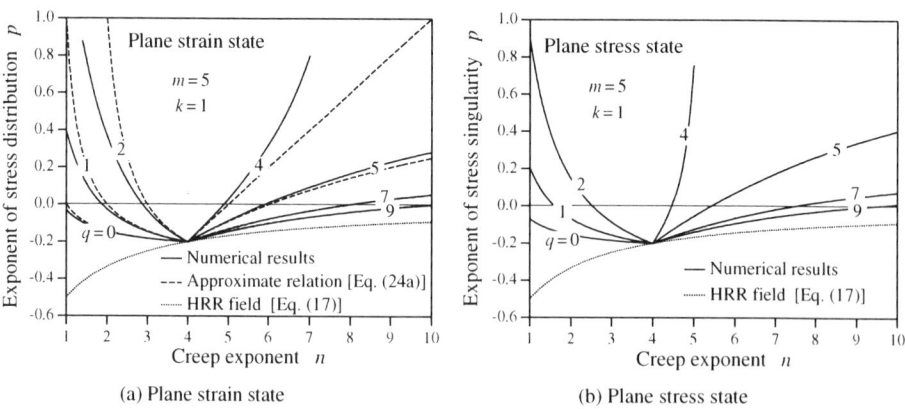

(a) Plane strain state   (b) Plane stress state

Figure 4. Exponent of asymptotic stress field for a mode I creep crack in steady-state growth $(m \neq q)$

the approximate relation (19) for the case of plane strain and $k = 1$ holds also in the case of $m \neq q$. Then, by the use of equations (16a) and (19), we have

$$p = \frac{n - (q+1)}{(n+1)(q+1) - mn}, \qquad \ell = \frac{(n+1) - m}{(n+1)(q+1) - mn} \qquad \text{(24a), (24b)}$$

Since $\ell \geq 0$ as mentioned in Section 4.2, equation (24b) gives the conditions

$$q + 1 > mn / (n+1) \quad \text{if} \quad n + 1 > m \qquad \text{(25a)}$$

$$q + 1 < mn / (n+1) \quad \text{if} \quad n + 1 < m \qquad \text{(25b)}$$

According to equation (25), we have the following three cases:

(1) $(n + 1) > m$
We have $n > mn / (n+1)$ in this case. Thus in view of equation (25a), there may exist the values of q such that

$$n > (q+1) > mn / (n+1) \qquad \text{(26)}$$

Substitution of this relation into equation (24a) gives the condition of vanishing stress singularity, $p > 0$.

On the other hand, if equation (26) does not hold, i.e.

$$q + 1 > n \qquad \text{(27)}$$

we have $p < 0$, and the asymptotic stress field is singular. For larger values of $q + 1$, the exponent p tends to that of HRR field of equation (17).

(2) $(n + 1) < m$
According to a similar argument as above, we have the condition of vanishing stress singularity $p > 0$;

$$n < (q+1) < mn / (n+1) \qquad \text{(28)}$$

On the other hand, in the case of

$$(q+1) < n \qquad \text{(29)}$$

we have $p < 0$, and the stress field is singular. In the particular case of very small $q + 1$, the exponent p again tends to that of HRR field of equation (17).

(3) $n + 1 = m$
In the particular case of $n + 1 = m$, if $q + 1 \neq n$ equation (24b) implies $\ell = 0$, and thus we have uniform damage field. In this case the asymptotic stress field at a crack tip is reduced to HRR field of equation (17).

Fig.5 shows the relations of equation (24) for the cases of $n + 1 > m$ and $n + 1 < m$.

The above results show the effects of damage field on the asymptotic stress field of a mode I creep crack in steady-state growth, and give very important information for the discussion and elucidation of creep crack behavior in structural elements.

6. Conclusions

The effects of material damage on the asymptotic stress field of a mode I creep crack in steady-state growth were analysed on the basis of continuum damage mechanics by postulating a power law creep damage theory. The resulting governing differential equations were solved by semi-inverse method. The relations among the exponent p of

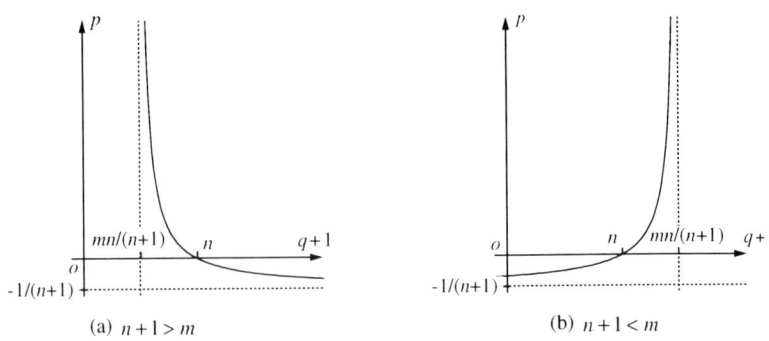

Figure 5. Approximate relation between p and $(q+1)$ for a given set of n and m

the asymptotic stress field, the exponent n of the power law creep law and the exponents m, q of the power creep damage law were elucidated in detail.

References

1. Bassani, J. L. and Hawk, D. E. (1990) Influence of damage on crack-tip fields under small-scale-creep conditions, *Int. J. Fracture*, **42**, 157-172.
2. Astafjev, V. I., Grigorova, T. V. and Pastukhov, V. A. (1991) Influence of continuum damage on stress distribution near a tip of a growing crack under creep conditions, In *Mechanics of Creep and Brittle Materials 2* (ed. by. Cocks, A. C. F. and Ponter, A. R. S.), Elsevier, London, 49-61.
3. Astafjev, V. I., Grigorova, T. V (1995) Stress and damage distribution near the tip of a crack growing under creep, *Mechanics of Solids*, **30**, 144-150.
4. Lu, M., Lee, S. B., Kim, J. Y. and Mai, H. C. (1997) An asymptotic analysis to a tensile crack in creeping solids coupled with damage–Part II. Large damage region very near the crack tip, *Int. J. Solids Structures*, **34**, 1183-1197.
5. Liu, Y., Murakami, S. (1997) Effects of damage field on elastic stress singularity at a mode III crack, *Trans. Japan Soc. Mech. Eng.*, **63**, A, 2084-2091 (in Japanese).
6. Liu, Y., Murakami, S. (1998) Asymptotic stress field of a mode I crack in a nonlinear -hardening damage material, *Trans. Japan Soc. Mech. Eng.*, **64**, A, 1183-1191 (in Japanese).
7. Murakami, S., Hirano, T. and Liu, Y. (2000) Asymptotic fields of stress and damage of a mode I creep crack in steady-state growth, *Int. J. Solids Structures*, **37** (in press).
8. Hirano, T., Murakami, S. and Liu, Y. (1999) Damage field and stress singularity of a mode III creep crack in steady-state growth, *Trans. Japan Soc. Mech. Eng.*, **65**, A, 1587-1592 (in Japanese).
9. Liu, Y., Hirano, T. and Murakami, S. (2000) Analysis of damage and stress fields of a mode I creep crack in steady-state growth, *Trans. Japan Soc. Mech. Eng.*, **66**, A, 604-611 (in Japanese).
10. Kachanov, L. M. (1986) *Introduction to Continuum Damage Mechanics*, Martinus-Nijhoff, Dordrecht.
11. Lemaitre, J. and Chaboche, J. L. (1990) *Mechanics of Solid Materials*, Cambridge University Press, Cambridge.
12. Lemaitre, J. (1996) *A Course on Damage Mechanics*, Springer-Verlag, Berlin.
13. Liu, Y., Murakami, S., Kojima, Y. and Matsushima, H. (2000) Observation and quantification of damage field for mode I creep cracks, *Trans. Japan Soc. Mech. Eng.*, **66**, A, 595-603 (in Japanese).
14. Hutchinson, J. W. (1968) Singular behaviour at the end of a tensile crack in a hardening materials, *J. Mech. Phys. Solids*, **16**, 13-31.
15. Rice, J. R. and Rosengren, G. (1968) Plane strain deformation near a crack tip in a power -law hardening material, *J. Mech. Phys. Solids*, **16**, 1-12.

COMPUTATIONAL CONTINUUM DAMAGED MECHANICS: ITS USE IN THE PREDICTION OF CREEP IN STRUCTURES - PAST, PRESENT AND FUTURE

D. R. HAYHURST
Department of Mechanical Engineering,
UMIST, Manchester, M60 1QD, U.K.

Abstract

The development is reviewed of the field of computational Creep Continuum Damage Mechanics. Emphasis is placed on four principal drivers: firstly, high-quality laboratory data; secondly, mechanisms-based constitutive equations and their calibration against laboratory data; thirdly, recent applications of computational CDM in structural analysis; and, fourthly, the rapidly increasing power of supercomputers and associated techniques. The need for high-quality laboratory data is expressed through an examination of the sources of errors in uni-axial creep testing; in particular, testpiece design and variations in test temperature. The importance is stressed of using mechanisms-based constitutive equations which reflect the underlying physics of the processes of deformation, damage, and failure. Strategies are also discussed for the identification of the dominant mechanisms for inclusion in the constitutive equation set and calibration against laboratory data. A recent Continuum Damage Mechanics solution is presented to the problem of failure in a welded pressure vessel obtained using two and three-dimensional finite element techniques. Finally, advances are addressed which are likely to take place over the next decade in both computer processor speed and in large-scale fast access computer memory. The impact of these likely developments is addressed on the ability to numerically solve problems in the creep of structures. Barriers to progress are identified and their solution discussed.

1. Introduction

The foundations of Continuum Damage Mechanics, CDM, were laid by Kachanov [1958] and by Rabotnov [1969] using the simple constitutive and damage equations:

$$d\varepsilon / dt = K\sigma^n /(1-\omega)^n$$
$$d\omega / dt = M\sigma^x /(1-\phi)(1-\omega)^\phi \tag{1}$$

where ω is the continuum damage variable which takes the values $\omega = 0$, when $t = 0$ and $\omega = 1$, when failure takes place. Towards to end of the 1960's, when digital computers were becoming more widely available, CDM Finite Element–based solutions were obtained using more complex equations calibrated for real materials, for plates containing stress raisers [Hayhurst, (1970) & (1973)]. The predictions were shown to be in close accord with

175

S. Murakami and N. Ohno (eds.), IUTAM Symposium on Creep in Structures, 175–188.
© 2001 *Kluwer Academic Publishers. Printed in the Netherlands.*

the results of experiments carried out on model structures. Multi-axial stress rupture criteria were introduced [Hayhurst, 1972] and used to predict the behaviour of axisymmetrically notched bars [Hayhurst, Dimmer and Morrison, 1984].

The research showed the effect of two phenomena: namely, stress redistribution due to damage or tertiary creep; and, the multi-axial stress rupture criterion which could either have the effect of strengthening or weakening structures. The former depends on whether a material is creep ductile [Hayhurst, 1999]; and, the latter on the multi-axial stress rupture criterion. A maximum effective stress rupture criterion can yield strengthening and a maximum principal tension stress criterion can produce weakening [Hayhurst, 1972].

These same principles and techniques have been used to predict creep crack growth in a range of materials which includes copper, aluminium alloys and stainless steels [Hayhurst, Brown and Morrison, 1984] and [Hall, Hayhurst and Brown, 1996]. The investigations showed a close tie-up between the results of experiments and Finite Element CDM predictions based on uni-axial creep data and appropriate multi-axial stress rupture criteria.

The same techniques have been used to study the behaviour of Ferritic Steel welded pipes and pressure vessels, from a knowledge of the constitutive equations of the parent, weld, Type IV and Heat Affected Zone materials [Hall and Hayhurst, 1991; Wang and Hayhurst, 1994; and Perrin and Hayhurst, 1999].

The approach has led to close correlations between the results of experiments and theory. Hence the use of creep computational CDM has been shown to be capable of predicting the behaviour of a wide range of components and structures which fail by creep. The approach can only be used satisfactorily provided that one has:

(i)   accurate materials data

(ii)   calibrated mechanisms-based constitutive equations capable of representing materials behaviour over a wide range of stress, stress-states and temperatures.

Given this information, it can be used to: (i) life structures, (ii) perform remanent lifetime assessments, and (iii) assess deformation histories. More generally it may be used to carry out supercomputer simulation of in-service component/structural behaviour.

In the next section, the general philosophy of the approach is examined.

2.   Use of CDM in Component Analysis and Simulation for Design

The diagram given in Figure 1 is similar to that presented by Hayhurst [1998] and the same philosophy was expounded almost one decade ago [Dyson and Hayhurst, 1989] to provide an integrating forward view of developments which might take place. Although significant change has occurred in the ensuring period, the diagram is still relevant, since developments are still taking place within the same methodology. It is therefore worth revisiting the structure of the diagram.

The left-hand column of the diagram, labelled Wealth Creation, contains two important elements which related to high-temperature creep design, they are concept design and detailed design. The former requires approximate or bounding type calculations which can be done quickly at little cost, and may require the input of raw materials data, or occasionally more complete constitutive equations. The latter, or detailed design process requires more accurate descriptions of materials behaviour provided by Mechanisms or

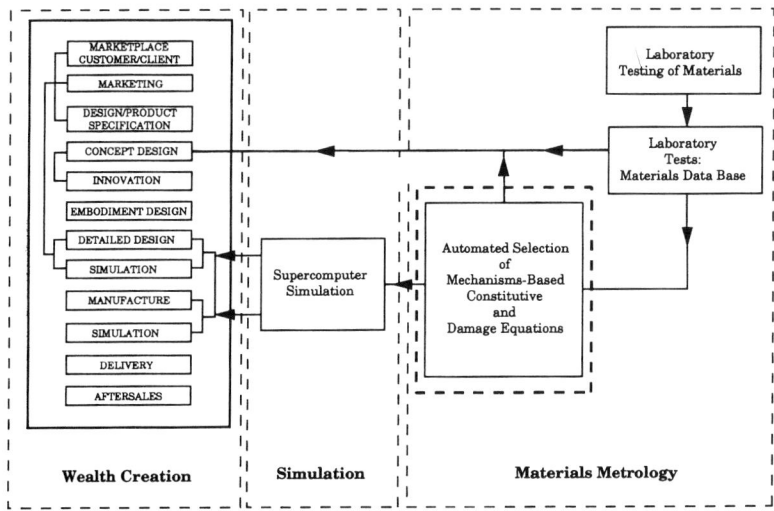

Figure 1. Diagrammatic representation of the route leading from Materials Testing, through Materials Metrology and Supercomputer Simulation, to the Wealth Creation Process.

Physics-Based Constitutive and Damage equations [Hayhurst, 1998]; the figure defines the domain of the Materials Metrology discipline, and shows how it relates to supercomputer simulation and to simulation in both detailed design and manufacturing processes. This paper will concern the former as related to computational creep CDM. The paper addresses the following aspects each in turn:

(i)   Laboratory material testing and high-quality data.
(ii)   Selection and calibration of physics/mechanisms-based constitutive equations.
(iii)   Recent applications of computational CDM in component/structural analysis.
(iv)   Advances in supercomputing and in computational/numerical techniques.

3.   Experimental Methods for the Determination of Accurate Creep Data

3.1   TEMPERATURE DEPENDENCE

Both creep strain rates and rupture lifetimes are exponentially dependent on temperature. It is therefore important that the control of test temperature is sufficiently good to achieve the desired accuracy on lifetime, then the required accuracy on strain rate is usually automatically achieved. The effects of errors, expressed as a percentage temperature change, upon change in lifetime is shown in Figure 2 for a range of materials. For example, a 4% change in the test Temperature of 150°C for aluminium results in a 35% change in lifetime. Even with the use of modern temperature sensors and temperature controllers over the long times involved in creep testing it is difficult to achieve near perfect test control. The unwanted feature associated with inadequate temperature control is that each creep curve, which might be used to calibrate a set of constitutive equations, will have significant errors present. This will make it difficult for the numerical routine or computer-based optimisation procedure, to work satisfactorily.

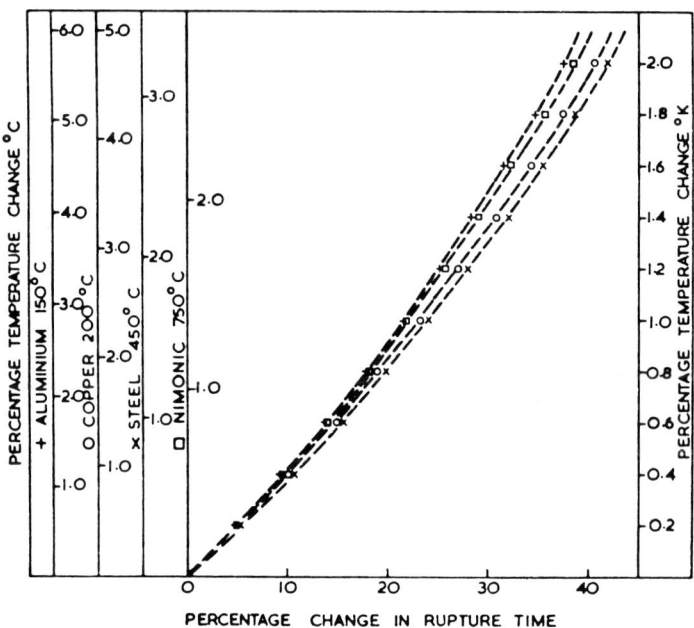

Figure 2. Change in creep rupture lifetime resulting from percentage change in test temperature.

An alternative procedure recently proposed is to log temperatures along with strains and to compute a mean temperature for the test. The constitutive equations are then expressed in temperature dependent form and the equations fitted for both mean temperatures and creep strain histories [Hayhurst, Lin, Kowalewski and Dyson, 2000].

3.2   TESTPIECE BENDING

Testpiece bending is quantified by the axial components of strain, ε_1 and ε_2, measured at opposite ends of a diagonal on the mid-plane of the testpiece perpendicular to its axis. It is defined as:

$$Bending\ (\%) = \{(\varepsilon_1 - \varepsilon_2)/(\varepsilon_1 + \varepsilon_2)\}\ 100 \qquad (2)$$

Computational studies carried out by Hayhurst [1974] have shown that provided the maximum initial bending is less than six percent then the creep curve is relatively insensitive to bending.

3.3   EXTENSOMETER RIDGE DESIGN

The measurement of uni-axial creep strain in laboratory tests is frequently carried out using cylindrical bar testpieces on which ridges have been machined to identify the gauge length over which strain is to be measured. Mechanical extensometers are fitted to these ridges which are used to transfer the displacements occurring during creep to a location outside of the high-temperature furnace where transducers can accurately measure displacements at ambient temperatures. The measured displacements are then used to compute the variation of strain with time.

A previous theoretical study carried out by Lin, Hayhurst and Dyson [1993a] has shown that the uni-axial strain measured in tensile testpieces using ridged specimens and extensometers do not agree with the true strains in the parallel section of the testpiece. The error levels have been shown in Lin, Hayhurst and Dyson [1993b] to be dependent upon the applied stress level, but principally upon the size of the gauge length. For Nickel superalloy specimens, of diameter 7.65 mm, with gauge lengths of 51 mm errors have been predicted in excess of 10%. For gauge lengths of 10 mm the error increases to typically 30%.

Figure 3.   Slitted extensometer ridged testpiece.

The reason for these high errors has been shown by Lin, Hayhurst and Dyson [1993a] to be due to the circumferential reinforcement of the testpiece provided by the extensometer ridges. The ridge perturbs the stress, strain, and damage fields above and below the ridge, with the perturbations extending a distance of typically 1.5 times the diameter of the parallel sided region of the testpiece. The circumferential stress generated in the extensometer ridge is predominantly compressive and Lin, Hayhurst and Dyson [1993b] have shown how these stresses may be relieved by the introduction of slits into the ridges, as shown in Figure 3; and how the errors in measured creep strains can, as a consequence, be reduced.

3.4   RECOMMENDATIONS

(i)   To establish a database of creep test results which can be used to calibrate constitutive equations, it is recommended that test temperature not only be accurately controlled; but, that temperatures be continuously measured. From these measurements a mean temperature can be determined and used in the calibration process.

(ii)   Initial testpiece percentage bending levels be controlled to below 6%.

(iii)   A testpiece design be used with a long gauge length and with slitted extensometer ridges which do not constrain the deformation of the testpiece.

It is in this way that high-quality creep test data can be obtained which will enable accurate calibration of constitutive equations.

4.   Selection and Calibration on Physics-Based Constitutive Equations

The single damage-state variable Equation (1) introduced by Kachanov [1958] and Robotnov [1969] may be generalised [Hayhurst, 1973] to the multi-dimensional form:

$$\frac{d\varepsilon_{ij}}{dt} = K \frac{3\sigma_e^{n-1} s_{ij} t^m}{2(1-\omega)^n}$$

$$\frac{d\omega}{dt} = \frac{M[\alpha\sigma_1 + (1-\alpha)\sigma_e]^x t^m}{(1+\phi)(1-\omega)^\phi}$$

(3)

where $\sigma_e\ (=(3s_{ij}s_{ij}/2)^{1/2})$ is the stress deviator, σ_1 is the maximum principal tension stress, and α is the multi-axial stress rupture criteria. The term in t^m has been introduced to model primary creep. Equations (3) will accurately predict both creep strains and lifetimes provided that there is either only one physical softening mechanism, or that there are several but only one of them is dominant. In the latter case, creep strains are usually not well predicted, but lifetimes are.

To overcome the deficencies of this approach, multi-state variable theories have been introduced [Dyson, Hayhurst and Lin, 1996] in which the kinetics of each softening process is described by a simple equation. In this paper one such hardening process and three softening processes will be considered where synergy is developed between the mechanisms through the deformation rate process. The set of equations used to describe this is given in Equations (4)-(8) and each will now be discussed in turn.

$$\frac{d\varepsilon_{ij}}{dt} = \frac{3}{2} \frac{A}{(1-\omega_1)} \left\{ \frac{S_{ij}}{\sigma_e} \right\} Sinh \left\{ \frac{B\sigma_e(1-H)}{(1-\Phi)(1-\omega_2)} \right\}$$

(4)

$$\frac{dH}{dt} = \Theta_H \frac{h}{\sigma_e} \frac{A}{(1-\omega_1)} Sinh \left\{ \frac{B\sigma_e(1-H)}{(1-\Phi)(1-\omega_2)} \right\} \left\{ 1 - \frac{H}{H^*} \right\}$$

(5)

$$\frac{d\Phi}{dt} = \Theta_K \frac{K_c}{3}(1-\Phi)^4 \qquad (6)$$

$$\frac{d\omega_1}{dt} = \Theta_{\omega_1} CA(1-\omega_1) Sinh\left\{\frac{B\sigma_e(1-H)}{(1-\Phi)(1-\omega_2)}\right\}\left\{1-\frac{H}{H*}\right\} \qquad (7)$$

$$\frac{d\omega_2}{dt} = \Theta_{\omega_2}\frac{DA}{(1-\omega_1)}\left\{\frac{\sigma_1}{\sigma_2}\right\}^v Sinh\left\{\frac{B\sigma_e(1-H)}{(1-\Phi)(1-\omega_2)}\right\} \qquad (8)$$

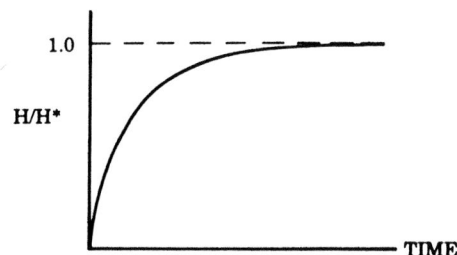

Figure 4. Change of Primary Creep State variable with time

Equation (4) describes the multi-axial strain rate behaviour, and in Equation (5) primary creep is represented by the parameter H; for constant stress, and with all other variables negligible, its evolution with time is shown in Figure 4. At zero time its value is zero and it then increases to its saturation value $H*$ at the end of primary creep. The parameter H models the change in dislocation structures during primary creep. Equation (6) models stress independent ageing due to particle coarsening which is shown schematically at the grain level, in Figure 5 the paramater Φ monotonically increases from zero to unity.

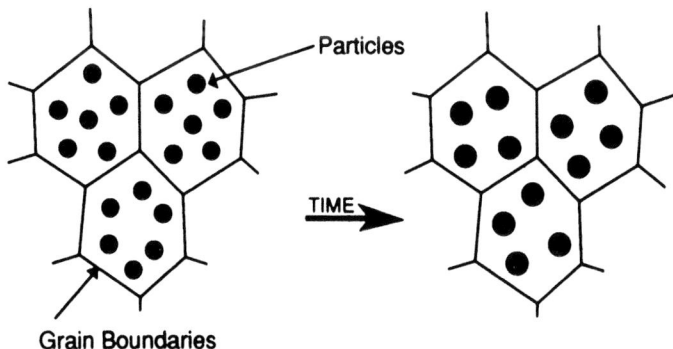

Figure 5. Stress independent ageing and particle coarsening

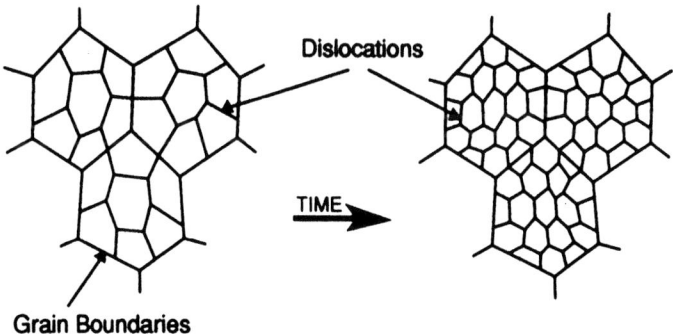

Figure 6. Schematic representation of creep softening by multiplication of dislocation sub-structures

The first damage state variable ω_l, described by Equation (7), models softening due to multiplication of dislocation sub-structures, as shown schematically in Figure 6. Equation (8) describes grain boundary creep constrained cavitation which is shown schematically in Figure 7. Creep constrained cavitation can be either nucleation or growth controlled. Equations (4)-(6) and (8) have been used to model Aluminium alloys, [Kowalewski, Hayhurst and Dyson, 1994] with $\Theta_{\omega l} = 0$, and shown to accurately describe the creep size behaviours; and, the creep behaviour of the Nickle-based superalloy Nimonic 80A, [Dyson, Hayhurst and Lin, 1996], with $\Theta_H = \Theta_\phi = 0$, and also the behaviour of Ferritic steels, with $\Theta_{\omega l} = 0$, [Perrin and Hayhurst, 1999].

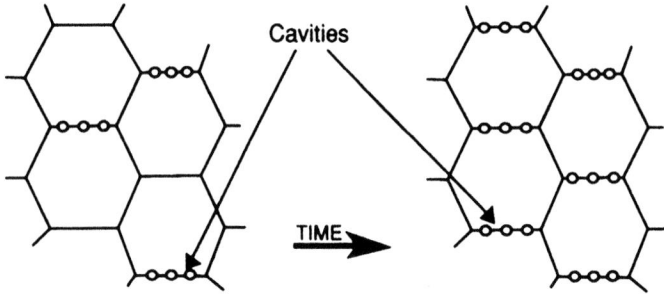

Figure 7. Schematic representation of creep softening by grain boundary cavity nucleation and growth

4.   Calibration of Constitutive Equations

The single state damage and constitutive equations may be calibrated against uniaxial data using most numerical optimisation schemes [Dunne, Othman, Hall and Hayhurst, 1990], and when the appropriate optimisation functional has been established they converge well and achieve good accuracy. However equations (4)–(8) contain nine independent constants, each of which has to be determined to provide an adequate fit to the experimental data. Attempts to use available numerical, schemes such as those employed by Dunne et al [1990] for simpler equation sets, have not proved successful. Entry to the computer optimisation routines with best estimates of initial values always resulted in a failure to identify a set of equation constants which provided a global minimum solution. To overcome this difficulty Kowalewski at al [1994] developed a new approach. It involves taking each mechanism or its allied equation and isolating it from the others in such a way that a good estimate of the constants associated with the mechanism can be determined. When good estimates have been determined for as many parameters as possible, the values are used as starting values in the numerical optimisation of the Kernal function:

$$\text{Minimum} \quad = \quad \sum_{i=1}^{a}\left[\left\{\sum_{j=1}^{b}\left(\varepsilon_j^p - \varepsilon_j^{exp}\right)^2\right\}_i\right] + Z_i\left(t_i^p - t_i^{exp}\right)/t_i^{exp} \tag{9}$$

where a is the number of creep curves, b is the number of points per curve, ε^p and ε^{exp} are the predicted and experimental values of strain respectively, Z_i is an amplification constant, t^p and t^{exp} denote predicted and experimental lifetimes, respectively. The second term in Equation (9) is invoked only when $t^p > t^{exp}$.

Proceeding in this way, and by exclusion or inclusion of each of the equations the quality of fit to the experimental data can be judged by the magnitude of the Kernal function (9). Hence it is possible to identify those profound or dominant physical mechanisms. It is in this way that the appropriate mechanisms-based constitutive equations can be identified from the available materials data base.

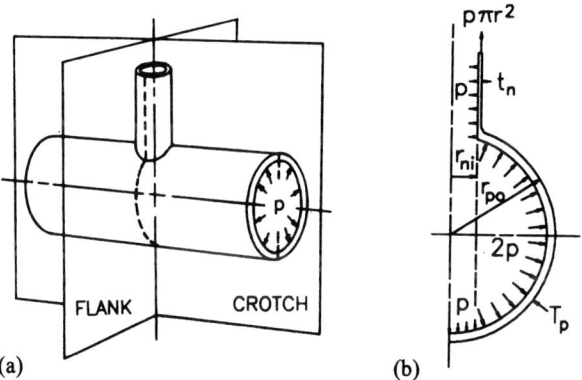

Figure 8. Definition of branch Crotch and Flank Sections (a) and idealised sphere-cylinder equivalent applied pressure boundary conditions for Flank section (b) with $r_{ni}=55.5$, $r_{po}=288$, $t_n=8$, $T_p=18$, all dimensions are in mm.

5. Recent Application of Computational Creep CDM in Structural Analyses

The constitutive Equations (4)–(6) & (8) introduced in the previous section are used here for ferritic steels at 590°C; the materials are 0.5Cr 0.5Mo 0.25V pipe material and 2.25Cr 1.0Mo weld material. The component investigated is a branch pipe connection which has been idealised as a cylinder-sphere intersection subjected to a constant internal pressure of 4MPa, as shown in Figure 8. The constitutive Equations (4)–(6) & (8), Equation (7) being

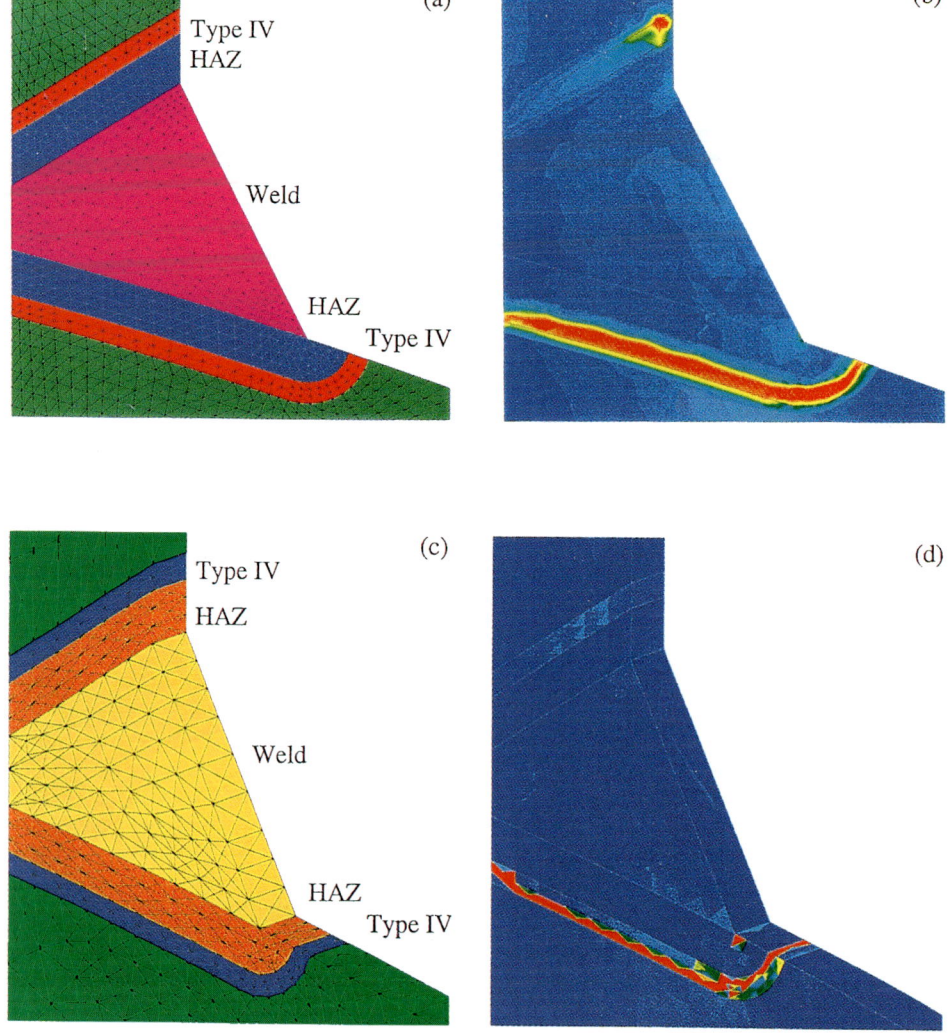

Figure 9. Finite Element meshes for the Flank section of the sphere-cylinder idealisation of the Branch connection. Two dimensional model (a) and (b); and a three-dimension 6° slice (c) and (d)

eliminated with $\Theta_{\omega l}$=0 since this mechanism is inoperative, have been used by Hayhurst and Miller [1998] to carry out two dimensional axi-symmetric vessel analyses using the two dimensional CDM solver DAMAGE XX. Materials data are given in Table 1. The weld geometry and finite element mesh is shown in Figure 9(a). The widespread propagation of damage in the Type IV material region, shown in red in Figure 9(b), where the damage has reached its critical value at the time of 14,759 hr.

The same vessel has more recently been analysed by Wong [1999] using the three-dimensional CDM solver DAMAGE XXX. This has been carried out by expressing the axi-symmetric boundary conditions on a 6° slice cut from the vessel. The finite element model has 10,995 degrees of freedom compared with 5,584 for the two dimensional case. The finite element mesh for the slice in the region of the weld is shown in Figure 9 (c).

Tetrahedral elements have been selected for the analysis, and the damage field solution close to failure is presented in Figure 9(d). The red regions denote failed material, ω=0·99. The predicted failure zones of the two and three dimensional solutions are in close agreement, although in the three dimensional solution the damage has not spread through the thickness of the Type IV zone. A failure time of 18,254 hr was predicted which is 19% higher than that for the two dimensional solution. This result is commensurate with the different mesh sizes, and the manner in which the circumferential symmetry is modelled in the three dimensional case.

What is evident from the finite element solutions is that the computer data storage requirements and the central processor run times are both significantly increased for the three dimensional case. In addition, the amount of data which requires interpretation is considerably increased, with a persistent need to be able to examine and interrogate data at points within the body.

TABLE 1: Material parameters for constitutive equations sets (4)-(8) for each of the weldment materials at the temperature of 590°C

Parameter	Parent	Weld Metal	HAZ	Type IV
A (MPa h^{-1})	$2·1618 \times 10^{-9}$	$2·589 \times 10^{-10}$	$2·1618 \times 10^{-9}$	$6·8568 \times 10^{-8}$
B (MPa^{-1})	0·20524	0·19106	0·20524	0·10414
C (-)	1·8537	1·8537	1·8537	4·5550
H (MPa)	$2·4326 \times 10^{5}$	$3·9070 \times 10^{4}$	$2·4326 \times 10^{5}$	$1·7441 \times 10^{4}$
H^* (-)	0·5929	0·4625	0·5929	0·6500
K_c (MPa^{-3}h^{-1})	$9·2273 \times 10^{-5}$	$2·7289 \times 10^{-4}$	$9·2273 \times 10^{-5}$	$7·2572 \times 10^{-4}$
v (-)	2·8	2·8	2·8	2·8
ω_f (-)	1/3	1/3	1/3	1/2

6. Advances in Computational and Numerical Techniques

There are two areas where the application of computational Continuum Damage Mechanics is limited by computational, numerical analysis capability, they are:

a) Two-dimensional structures subjected to combined cyclic plasticity and creep, where many cycles of loading are required to study lifetime scenarios.

b)   Three-dimensional structures where the total number of degrees of freedom used in the non-linear Finite Element analysis is large, often in excess of 200,000.

To obtain solutions in both of these areas the principal requirement is high computational speed, and in the latter case large amounts of Random Access Memory. These aspects and other problem areas are now discussed.

6.1   Increased central processor unit, cpu, speed

Over the next five years or so, an increase in cpu speed of the order of ten is required to permit the solution of the problems outlined above. This translates into an increase in high-performance supercomputer speed, over the same period, from the current 10 to 100 Tera Flops; and, an increase in memory from 5 to approximately 30 Tera Bytes [Mesina, 1999].

6.2   Increased sources of computational speed

Such increases are not expected to come from the increased speed of an individual computer processor; but, from a range of individual sources, which are now discussed. Firstly, it is expected that advances in computer chip design, and manufacturing techniques, will continue to yield faster individual processors. However, it is likely that the most significant increases will come from other sources. Secondly, the use of massively parallel computer architectures which can be coupled in a manner which enhances processing speed. And, thirdly, coupled with the latter, are numerical techniques and software for: (i) faster solution of large sets of equations on both single and multi-processor machines; and, (ii) distribution, in a scaleable way, of the computational task over thousands of processors.

It will be a combination of several of these factors, tailored to suit particular problems, which will achieve the greatest enhancements of computational speed.

6.3   Visualisation and interpretation of computer modelling output

Three dimensional creep Continuum Damage Mechanics computer simulations, which model structural behaviour over large periods of time, generate enormous quantities of data. The task of examining and interrogating the data requires the use of new visualisation, animation and virtual reality techniques. Translation from two to three dimensions results in a massive increase in the size of the data sets; which, in turn, requires both new visualisation hardware, increased processing speed, and new software.

6.4   Drivers for change

The Department of Energy of the United States of America currently has a programme of research in progress which is aimed at the development of their nuclear capability by the peaceful, environmentally friendly, means of supercomputer simulation alone. It involves the Los Alamos, Lawrence Livermore and Sandia Laboratories, and is known as the Accelerated Strategic Computing Initiative, ASCI, [Mesina, 1999]. It will provide a driver for change across the entire physical sciences. The developments on this programme, particularly with computer hardware manufacturers, will initiate changes in computing, hardware and numerical techniques which will map over directly into the creep computational CDM field.

7.   Developments Over the Next Decade and Conclusions

The provision of and access to enhanced supercomputing power will mean that over the next ten years significant advances will take place in the analysis, modelling and simulation of structure and components manufactured in:

(i)   Metallic alloys;
(ii)   Fibre, Particulate, and Reinforced composites; and
(iii)   Laminates.

And, in addition, in the design of components and structures subjected to:

(i)   Combined Creep and Fatigue
(ii)   Thermo-mechanical loadings in three-dimensions.

In all cases the barriers to progress will be:

(i)   High quality long-term materials test data
(ii)   Physics-Based constitutive equations which can be used to extrapolate reliably from short-term data to the long-term
(iii)   High performance computing and associated applications software.

An ability to overcome such barriers to progress will generate a competitive edge and open up potential for wealth creation.

The way ahead, therefore, is to set up forward looking materials testing programmes, particularly where good long-term data is needed; and, at the same time, establish techniques for automated selection and calibration of physics-based constitutive equations. It will then be possible over the next decade to take advantage of the use of computational CDM techniques through high-performance supercomputing power.

7. References

Dyson, B.F. and Hayhurst, D.R. (1989) *Materials and engineering design: the next decade*, Ed. B.F. Dyson and D.R. Hayhurst, The Institute of Metals, London.

Dyson, B.F., Hayhurst, D.R. and Lin, J. (1996) The ridged uniaxial testpiece: creep and fracture predictions using large-displacement finite-element analyses, *Proc. R. Soc. Lond.* **A452**, 655–676.

Dunne, F.P.E., Othman, R., Hall, F.R. and Hayhurst, D.R. (1990) Representation of uni-axial creep curves using continuum damage mechanics, *I. J. Mech. Sci.* **32, 11,** 445–957.

Hall, F.R. and Hayhurst, D.R. (1991) Continuum damage mechanics modelling of high temperature deformation and failure in a pipe weldment, *Proc. Roy. Soc. Land.* **A433**, 383–403.

Hall, F.R., Hayhurst, D.R. and Brown, P.R. (1996) Prediction of plane-strain creep-crack growth using continuum damage mechanics, *Int. Jnl. Damage Mech.* **5**, 353–393.

Hayhurst, D.R. (1970) *Isothermal creep deformation and rupture of structures*, PhD Thesis, University of Cambridge.

Hayhurst, D.R. (1972) Creep rupture under multi-axial states of stress, *J.Mech. Phys. Solids* **20**, 381–390.

Hayhurst, D.R. (1973) Stress redistribution and rupture due to creep in a uniformly stretched thin plate containing a circular hole, *J. Appl. Mech.* 1, **40**, 244–256.

Hayhurst, D.R. (1974) The effect of test variables on scatter in high-temperature tensile creep-rupture data, *Int. J. Mech. Sci.* **16**, 829–841.

Hayhurst, D.R. (1998) *Thermodynamic modelling and materials data engineering*, Springer-Verlag, Berlin Heidelberg, Chapter 4, 189–224.

Hayhurst, D.R. (1999) *Materials data bases and mechanisms-based constitutive equations, in creep and damage in materials and structures*, Edited by H. Altenbach and J.J. Skrzypek, Springer Wien New York.

Hayhurst, D.R., Brown, P.R. and Morrison, C.J. (1984) The role of continuum damage in creep crack growth, *Phil. Trans. R. Soc.* **A 311**, 131–158.

Hayhurst, D.R., Dimmer, P.R. and Morrison, C.J. (1984) Development of continuum damage in the creep rupture of notched bars, *Phil. Trans. R. Soc. Lond.* **A311**, 103–129.

Hayhurst, D.R. and Miller, D.A. (1998) *The use of creep continuum damage mechanics to predict damage evolution and failure in welded vessels*, *I.Mech.E. Seminar Pub., Remanent Life Prediction, 1998-4, 117, 132.*

Hayhurst, D.R., Lin, J., Kowalewski, Z.L. and Dyson, B.F. (2000) *Creep strain uncertainties associated with testpiece extensometer ridges: their identification and reduction*, Proc. Of the 5th IUTAM Symposium on *Creep in Structures*, Nagoya, Japan, April 2000.

Kachanov, L.M. (1958) Time of the fracture process under creep conditions, *Azv. Akad. Nauk. SSSR, Otd. Teck. Nauk.* **8**, 26.

Kowalewski, Z.L., Hayhurst, D.R. and Dyson, B.F. (1994) Mechanisms-based creep constitutive equations for an aluminium alloy, *J. Strain Analysis* **29**, 309–316.

Lin, J. Hayhurst, D.R. and Dyson, B.F. (1993a) The standard ridged uniaxial testpiece: computed accuracy of creep strain, *J. of Strain Analysis* **28**, 2, 101–115.

Lin, J., Hayhurst, D.R. and Dyson, B.F. (1993b) A new design of uniaxial creep testpiece with slit extensometer ridges for improved accuracy of strain measurement, *I. J. Mech. Sci.* **35**, 1, 63–78.

Mesina, P. (1999) *The accelerated strategic computing initiative (ASCI) and its impact on computational science*, EPSRC High Performance Computer User meeting, Westminster, London.

Perrin, I.J. and Hayhurst, D.R. (1999) Continuum damage mechanics analyses of the Type IV creep failure in ferritic steel, *Int. J. Press Vess. & Piping* **76**, 599–617.

Rabotnov, Yu. N. (1969) *Creep problems in structural members* (English translation, Ed. F.A. Leckie) North-Holland, Amsterdam.

Wang, Z.P. and Hayhurst, D.R. (1994) The use of super-computer modelling of high temperature failure in pipe weldments to optimise weld and heat affect zone, Material Property Selection, *Proc. Roy. Soc. Land.* **A446**, 127–148.

Wong, M.T. (1999) *Three-dimensional finite element analysis of creep continuum damage growth and failure in weldments*, PhD Thesis, UMIST.

CRACKING OF CREEPING STRUCTURES
DESCRIBED BY MEANS OF CDM

A. BODNAR, M.CHRZANOWSKI
Cracow University of Technology
Warszawska 24, 31-155 Kraków, Poland

1. Introduction

Engineering design of a structure should sustain all loading history. Consequently, two aspects of structure's usability are usually studied i.e. structure's deformability and its integrity. The latter one is of prime importance for structure's reliability.

A branch of mechanics named Fracture Mechanics has proved to be an extremely useful and convenient tool for evaluation of load carrying capacity of structures which contain (one or several) crack(s). Over several last decades this was demonstrated for both elastic and elasto-plastic materials, as well as for materials which exhibit time-dependent properties. In terms of formulae the best representation is given here by J and C^* integrals introduced by Rice [1] and Kubo et al. [2], respectively. The main advantage of the Fracture Mechanics approach is that it gives evaluation of material toughness against crack propagation. However, Fracture Mechanics can be applied only to the analysis of structures with pre-existing crack(s).

For creeping structures one has to deal with a material without any pre-existing cracks. These will develop due to material deterioration induced by loading history. Three stages of this process can be distinguished:

- The growth and coalescence of micro-defects (voids, cavities, micro-cracks). This first stage is terminated by the formation of macro-defect in a given point (or points) of structure's body at the time denoted here as t_1. This instant of time when failure condition is fulfilled in a body point will be referred to as a time of First Crack Appearance (FCA for short).
- The development of macro-defects to form a network of cracks penetrating structure's body until a crack spanning characteristic dimension of a structure is formed at the time t_2. This instant of time when failure occurs over characteristic dimension of a body will be referred to as a time of Through-body Crack Appearance (TCA).
- Spreading of through-body cracks formed in the previous stage, to make structure kinematically unstable at time t_3. This instant of time when failure causes the structure to collapse by formation of specific networks of cracks will be called Time of Structure Collapse (TSC).

S. Murakami and N. Ohno (eds.), IUTAM Symposium on Creep in Structures, 189–196.

The description of the above process is available since the Continuum Damage Mechanics emerged in late 50-ties [3, 4] and finally took its full recognition in 1977[5]. In a series of works [6, 7, 8, 9] going back as far as to early 70-ties [10] authors have pursued the idea of safety margin defined as ratio of TCA/FCA (i.e. t_2 / t_1). Here, this factor will be analysed for creeping rectangular plates with different boundary conditions and different material properties.

The numerical formalism used by authors also allows the description of the geometry of progressive cracking (crack evolving in time) which takes place within the second stage of the whole process. In particular, the information on the position FCA, subsequent cracking, and position of TCA can be gained. The further step in this analysis could incorporate a kind of optimisation with respect to: safety margin, and relative positions of FCA and TCA.

The present analysis does not cover the third stage, i.e. when a network of macrocracks is being formed to cause ultimate collapse of a structure. Formally this analysis is possible, though numerical difficulties may make it highly cumbersome. Therefore, the safety margin defined as ratio t_2 / t_1 can be viewed as a lower limit of structure's reliability.

2. Cracking of creeping plates

2.1. Constitutive equations

The following assumptions with respect to the constitutive equations are made:
- the total strain can be decomposed into two components: elastic and creep ones,
- steady state creep theory is modified by coupling with a damage variable,
- damage is represented by a scalar parameter, with the appropriate evolution law given below.

The above assumptions yield the set of equations which describe material properties:

$$\underline{\varepsilon} = \underline{\varepsilon}^e + \underline{\varepsilon}^c , \tag{1}$$

$$\underline{\sigma} = \underline{D}\underline{\varepsilon}^e \tag{2}$$

$$\underline{\dot{\varepsilon}}^c = \underline{\Gamma}(\underline{\sigma},\omega)\underline{\sigma} \tag{3}$$

where: $\underline{\varepsilon}^e, \underline{\varepsilon}^c, \underline{\sigma},$ ω denote elastic and creep strains, stress and damage parameter ($0 \le \omega \le 1$), respectively, $\underline{D}$, $\underline{\Gamma}(\underline{\sigma},\omega)$ are elastic constants and creep-rupture material functions matrices that will be specified later, and dots stand for time derivative.

The evolution law for damage parameter ω used here is [11, 12]:

$$\dot{\omega} = A \left[\alpha \, max \left\{ \frac{\langle \sigma_{max} \rangle}{1-\omega}, \frac{\langle -\sigma_{min} \rangle}{1-h\omega} \right\} + (1-\alpha) \frac{\sigma_{eff}}{1-\omega} \right]^m \tag{4}$$

where: A and m are damage evolution law constants, α is parameter which characterises type of local failure mechanism ($0 \leq \alpha \leq 1$), σ_{max} and σ_{min} - maximal positive and minimal negative principal stresses, h - parameter responsible for direct influence of negative principal stress upon deterioration process ($0 \leq h \leq 1$), σ_{eff} - Huber-Mises effective stress, and $< >$ denote McAuley brackets.

The case of $\alpha \neq 1$ and $h \neq 1$ in Eq. (4) defines so called *non-unilateral* behaviour of materials, which respond differently for tensile and compressive stresses. For $h = 1$ material is equally sensitive to tension and compression independent of value of α, and is called *bilateral* one.

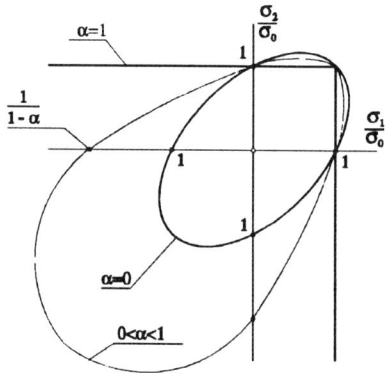

Fig. 1. Isochronous failure curves for *unilateral* material

A *unilateral* material is described by the simplified version of Eq (4), used by many authors [13]:

$$\frac{\partial \omega}{\partial t} = A\left[\alpha\frac{\sigma_{max}}{1-\omega} + (1-\alpha)\frac{\sigma_{eff}}{1-\omega}\right]^m , \tag{5}$$

in which immediate effect of compressive stress is neglected.

The different behaviour of above materials can be illustrated by the shape of isochronous curves, which are loci of different principal stresses combination to yield failure at the same given time. In Fig. 1 the influence of α is shown for a *unilateral* material. Two limiting cases $\alpha = 1$ and $\alpha = 0$ correspond to two different local failure mechanisms : inter- and trans-crystalline ones, respectively. The case of $\alpha = 1$ reduces Eq. (5) to that of classical the Kachanov-Rabotnov theory of brittle creep failure [3, 4], whereas the case of $\alpha = 0$ yields the Huber-Mises ellipse.

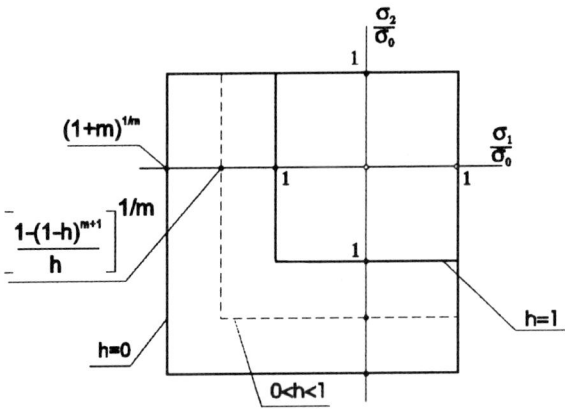

Fig. 2. Isochronous failure curves for *non-unilateral* material and $\alpha = 1$

The effect of the *h* parameter is demonstrated by Fig. 2 which shows isochronous curves for $\alpha = 1$. Finally, combined effect of both α and *h* is shown in Fig. 3: for $\alpha = 0,5$ and $m = 2$ (chosen as an example). Different values of *h* yield different curves, within limit curves - shown by dashed lines - for *unilateral* and *bilateral* materials.

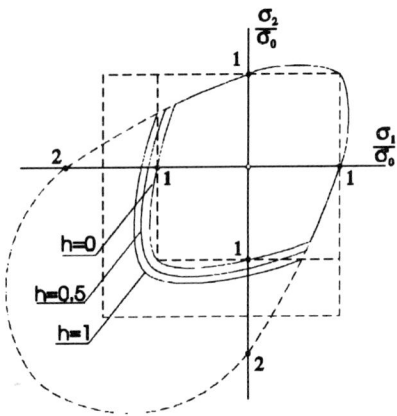

Fig. 3. Isochronous failure curves for *non-unilateral* material

2.2. Problem formulation and numerical procedures

A rectangular plate under constant pressure uniformly distributed over its upper surface is analysed. Two types of plate support is considered, namely simply supported along all edges, and with all edges clamped.

The searched quantities are: times t_1, t_2 and the ratio of t_2/t_1, as it can be seen as safety margin for a given structure. The analysis for times greater then t_2 was not undertaken; as it required a combined Fracture/Damage Mechanics approach to be used; such an analysis was out of the scope of the present paper.

Out of many material constants which influence results the following two were chosen as variable parameters, α - which defines type of local failure mechanism, and h - which reflects material sensitivity to compressive stress. In numerical analysis these parameters were set to 0, 0.5, and 1, whereas the remaining material constants were: $E = 0.102 \cdot 10^6$ [MPa], $v = 0.33$, $n = 6.8$, $m = 5.79$, $\gamma = 1.38 \cdot 10^{-24}$[(MPa)$^{-n}$ h^{-1}], $A = 1.08 \cdot 10^{-20}$[(MPa)$^{-m}$ h^{-1}] (Ti-6Al-2Cr-2Mo alloy at temperature 675 K [14]).

The analysis was carried out by means of the Finite Element Method for structure discretisation and the Euler procedure for time integration. In the computer code developed by authors, the layered shell elements derived from the three dimensional equations of continuum mechanics by use of degeneration method were employed. According to this approach the stress component normal to the midsurface of a plate was neglected and five degrees of freedom were specified at each nodal point corresponding to its three displacements and two rotations of the „normal" [15,16]. As the degrees of freedom related to the displacements and rotations are independent, thus this approach enables the analysis of thin plates as well as moderate-thick ones. Plates made of materials which consists of layers with different mechanical properties (e.g. composites) can be analysed, too.

For the theory employed the strain and stress vectors take the forms :

$$\underline{\varepsilon} = [\varepsilon_x, \varepsilon_y, \varepsilon_{xy}, \varepsilon_{xz}, \varepsilon_{yz}]^T \tag{6}$$

$$\underline{\sigma} = [\sigma_x, \sigma_y, \sigma_{xy}, \sigma_{xz}, \sigma_{yz}]^T \tag{7}$$

and elastic constants $\underline{D}$ and creep-rupture functions $\underline{\Gamma}(\underline{\sigma}, \omega)$ matrices are assumed to be :

$$\underline{D} = \frac{E}{1-v^2} \begin{bmatrix} 1 & v & 0 & 0 & 0 \\ v & 1 & 0 & 0 & 0 \\ 0 & 0 & (1-v)/2 & 0 & 0 \\ 0 & 0 & 0 & (1-v)/2 & 0 \\ 0 & 0 & 0 & 0 & (1-v)/2 \end{bmatrix} \tag{8}$$

$$\underline{\Gamma}(\underline{\sigma}, \omega) = \gamma \frac{\sigma_{eff}^{n-1}}{(1-\omega)^n} \begin{bmatrix} 1 & -1/2 & 0 & 0 & 0 \\ -1/2 & 1 & 0 & 0 & 0 \\ 0 & 0 & 3 & 0 & 0 \\ 0 & 0 & 0 & 3 & 0 \\ 0 & 0 & 0 & 0 & 3 \end{bmatrix} \tag{9}$$

where: E, v, γ and n are material constants.

Each quadrant of the plate with sides length set to 1.0 and 2.0 m, and thickness -

0.10 m, was divided into 64 elements, and ten layers and two-point Gaussian quadrature for volume integration were used. The time t_1 (FCA) was identified with $\omega = 1$ condition fulfilled at any Gaussian point. When this condition was satisfied in all ten layers of the same Gaussian point, the time was referred to as t_2 (TCA). For time $t > t_1$ calculation was performed for modified structure geometry i.e. excluding integration points in which $\omega = 1$.

The results were then transformed by a computer program which drew the lines connecting consecutive points in which condition $\omega = 1$ was fulfilled to provide visualisation of plates cracking process for time $t > t_1$.

3. Results of numerical analysis

The results of calculations are given in Table 1 for simply supported plates, and in Table 2 for clamped ones.

TABLE 1. Safety margins in simply supported plates made of *non-unilateral* and *unilateral* materials

α	h	*Non-unilateral* and *bilateral* material (immediate compression effect)			*Unilateral* material (no compression effect [17])		
		t_1 [10^5 hrs]	t_2 [10^5 hrs]	t_2/t_1	t_1 [10^5 hrs]	t_2 [10^5 hrs]	t_2/t_1
0	-	-	-	-	0.4212	0.4578	1.0870
0.5	0	0.4264	0.5217	1.2235	0.5334	0.6645	1.2458
	0.5	0.4087	0.4932	1.2068			
	1.0	0.3626	0.4056	1.1186			
1.0	0	0.3647	0.4731	1.2972	0.4306	0.5603	1.3012
	0.5	0.3489	0.4496	1.2886			
	1.0	0.2949	0.3380	1.1462			

TABLE 2. Safety margins in clamped plates made of *non-unilateral* and *unilateral* materials

α	h	*Non-unilateral* and *bilateral* material (immediate compression effect)			*Unilateral* material (no compression effect [17])		
		t_1 [10^5 hrs]	t_2 [10^5 hrs]	t_2/t_1	t_1 [10^5 hrs]	t_2 [10^5 hrs]	t_2/t_1
0	-	-	-	-	1.0566	1.2644	1.1967
0.5	0	0.7259	1.5399	2.1214	0.8709	1.9720	2.2644
	0.5	0.6995	1.4370	2.0543			
	1.0	0.6177	1.0344	1.6746			
1.0	0	0.4509	1.1725	2.6004	0.5220	1.3593	2.6041
	0.5	0.4334	1.1139	2.5702			
	1.0	0.3682	0.7176	1.9489			

TABLE 3. Clamped against simply supported plates comparison

α	h	Non-unilateral and bilateral material (immediate compression effect)			Unilateral material (no compression effect)		
		FCA	TCA	Safety margin	FCA	TCA	Safety margin
0	-	-	-	-	2.5085	2.7619	1.1009
0.5	0	1.7024	2.9517	1.7339	1.6327	2.9676	1.8176
	0.5	1.7115	2.9136	1.7023			
	1.0	1.7035	2.5503	1.4970			
1.0	0	1.2364	2.4783	2.0046	1.2123	2.4260	2.0013
	0.5	1.2422	2.4775	1.9946			
	1.0	1.2486	2.1231	1.7003			

The cracking patterns and location of FCA and TCA for the plates analysed above have been discussed elsewhere (cf. [18]); such a discussion is out of main goal of the present paper, which concerns safety margin evaluation.

4. Concluding remarks

The comparison of quantitative results given in Tables 1 and 2 suggests following conclusions:

1. Boundary conditions influence both: values of t_1 and t_2 (which are smaller for simply supported plates), and the ratio of t_2/t_1 , as well. This observation applies to all analysed values of α and h and is obvious taking into account lower redundancy of simply supported plate. Table 3 gives the ratios of searched values for clamped plate divided by corresponding values for simply supported plate (to enable comparison the applied load has been chosen to cause the same maximum equivalent stress at time $t = 0$ for all plates analysed).

2. The type of local failure mechanism, represented in equivalent stress formulation (Eqs.4 and 5) by parameter α has also essential, but less pronounced influence than that of boundary conditions. This influence is affected also by the value of h parameter (immediate compression effect) but the following general tendency is preserved: lower the value of α higher the values of t_1, t_2 and ratio of t_2/t_1. In the other words, with prevailing effective stress the life-time of structure is prolonged, as well as greater is safety factor t_2/t_1.

3. The immediate compression effect is visible from Tables 1 and 2: it reduces the structure's lifetime for all analysed cases. The relative position of FCA and TCA are also highly influenced by compression effect; in most cases these points converge for high value of h, whereas for lower values of h - as well as for a unilateral material - the positions of FCA and TCA are different.

5. References

[1] Rice, J.R. (1968) A path independent integral and the approximate analysis of strain concentration by notches and cracks, *J. of Applied Mechanics* **35**, 379-386.

[2] Kubo, S., Ohji, K. and Ogura, K (1979) An analysis of creep crack propagation on the plastic singular stress field, *Engineering Fracture Mechanics* **11**, 315-329.

[3] Kachanov, L.M. (1958) Time to failure under creep conditions (in Russian), *Izv. AN SSSR OTN* **8**, 26-31.

[4] Rabotnov, Y.N. (1959) On a long-time failure mechanism (in Russian), In *Problems of Strength of Materials and Structures, Izv AN SSSR*, 5-9.

[5] Janson, J., Hult, J. (1977) Fracture mechanics and damage mechanics, a combined approach, *J. de Mecanique appliqué* **1**, 69-84.

[6] Bodnar, A., Chrzanowski, M. and Latus, P. (1992) Lifetime evaluation of creeping structures, *Proc. 5th Int. Conf. on Creep of Materials,* Grosvenor Resort, Florida, 461-469.

[7] Bodnar, A., Chrzanowski, M. and Latus, P. (1994) Safety of materials and structures in creep conditions, *Int. J. Pres. & Piping* **59**, 161-174.

[8] Bodnar, A., Chrzanowski, M. (1994) Cracking of creeping plates in terms of Continuum Damage Mechanics, *J. of Theoretical and Applied Mechanics* **32**, 32-42.

[9] Bodnar, A., Chrzanowski, M. and Nowak, K. (1996) Brittle failure lines in creeping plates, *Int. J. Pres. & Piping* **66**, 253-261.

[10] Piechnik, S., Chrzanowski, M. (1970) Time of total creep rupture of a beam under combined load, *Int. J. Solids and Structures* **6**, 453-477.

[11] Bodnar, A., Chrzanowski, M. (1991) A non-unilateral damage in creeping plates, *Trans. Fourth IUTAM Symp. on Creep in Structures, Cracow 1990,* Springer Verlag, Berlin - Heidelberg-New York , 287-293.

[12] Lemaitre, J.(1986) Damage constitutive equations, *CISM Course on Damage Continuum Mechanics.*

[13] Hayhurst, D.R. (1972) Creep rupture under multi-axial states of stress, *J. Mech. Phys. Solids* **20**, 381-390.

[14] Walczak, J. (1986) On an energy creep rupture criterion, *Int. J. Mech. Sci.* **2**, 71-81.

[15] Ahmad, S.,Irons, B.M. and Zienkiewicz, O.C. (1970) Analysis of thick and thin shell structures by curved finite elements, *Int.J.Num.Meth.Eng.* **2**, 419-451.

[16] Figureiras, J.A.,Owen, D.R.J. (1984) Analysis of elasto-plastic and geometrically nonlinear anisotropic plates and shells, in E. Hinton and D.R.J. Owen (eds.), *Finite element software for plates and shells* , Peneridge Press Ltd, 235-326.

[17] Bodnar, A., Chrzanowski, M. (1995) The analysis of life extension for creeping plates, *Proc. Inter. Symp. Materials Ageing and Component Life Extension*, Milan, 1221-1229.

[18] Bodnar, A., Chrzanowski, M. (1999) Development of non-unilateral damage field in creeping plates, *Proc. IUTAM Symp. on Rheology of Bodies with Defects, Beijing 1997,* Kluwer Academic Publishers, Dordrecht-Boston-London, 267-276.

A COUPLED FORMULATION FOR THERMO-VISCOPLASTICITY AT FINITE STRAINS : APPLICATION TO HOT METAL FORMING.

L. ADAM AND J.P. PONTHOT
FRIA Ph.D. Student and Senior assistant
LTAS-Continuum Mechanics and Thermomechanics
University of Liège, 1 Chemin des Chevreuils, B52/3
B4000 Liège-1, Belgium. Phone : +32-4-366 93 10
Email : L.Adam@ulg.ac.be & JP.Ponthot@ulg.ac.be

1. Introduction

Our goal in this paper is to present a complete thermo-viscoplastic formulation at finite strains and its implementation in a finite element code. The formulation is derived from the classical J2-plasticity with a flow criterion expressed in the current configuration in terms of the Cauchy stresses and the temperature distribution.

A stagerred scheme will be used for the resolution of the thermomechanical problem as well as an extension of the radial return algorithm for the integration of the constitutive law.

In the next section we will briefly formulate the fundamentals of the kinematics for a large deformation approach, and the basic equations governing our model.

After this, we will explain, in more details, the foundation of the staggered scheme and of the radial return algorithm introduced before.

Finally, in the last section, we will show the good behavior of the formulation in a numerical simulation of the forming of a sheet at different temperatures.

S. Murakami and N. Ohno (eds.), IUTAM Symposium on Creep in Structures, 197–206.

2. Mathematical foundation of the thermo-viscoplastic model at finite strains

2.1. KINEMATICS FOR LARGE DEFORMATION CONTINUUM

Let us consider two configurations of a body : first, the reference configuration (not necessarily the initial configuration) at a certain time t_0 where the position of a material particle at this time is denoted by its *position vector* X and second, the current configuration, at time t, where the position of the same material particle is x. Then there exists a mapping between x and X of the form

$$x = x(X, t) \tag{1}$$

The *velocity* of the reference point X is the material time derivative of the position vector and is defined by

$$v = \dot{x} = \frac{\partial x(X, t)}{\partial t} \tag{2}$$

The *deformation gradient* of the motion at X is the second-rank two-point tensor F such that

$$F = \frac{\partial x}{\partial X} \qquad \text{with} \qquad J = \det F > 0 \tag{3}$$

By the polar decomposition, we can uniquely decompose F as

$$F = RU \qquad \text{with} \qquad R^{\mathrm{T}} R = I \qquad \text{and} \qquad U = U^{\mathrm{T}} \tag{4}$$

The corresponding *spatial gradient of velocity* is given by

$$L = \frac{\partial v}{\partial x} = \dot{F} F^{-1} \tag{5}$$

It can be decomposed into the symmetric and the antisymmetric parts, $L = D + W$ with

$$D = \tfrac{1}{2}(L + L^{\mathrm{T}}) \qquad \text{the rate of deformation} \tag{6}$$
$$W = \tfrac{1}{2}(L - L^{\mathrm{T}}) \qquad \text{the spin tensor} \tag{7}$$

2.2. CONSERVATION EQUATIONS

In this section, we will briefly formulate the fundamental set of conservation equations of a thermomechanical formulation.

2.2.1. *Mechanical part*

The equations used in this part of the formulation are the classical, and well known, conservation equations of the mass and of the momentum. Thus we won't write these in this section (see e.g. Malvern [2]).

2.2.2. *Thermal part*

The equation of heat is derived from the first principle of thermodynamics. We have decided, given a certain choice of state variables and a model for the description of the kinematics of the body, to express that equation as (see [8, 6] for more details) :

$$\rho c_v \overset{.}{T} - \overset{. \ \text{thel}}{W} + \rho \frac{\partial}{\partial \boldsymbol{\alpha}}[\psi - T\frac{\partial \psi}{\partial T}] \cdot \overset{.}{\boldsymbol{\alpha}} = \overset{. \ \text{irr}}{W} + \rho r - \text{div } \boldsymbol{q} \qquad (8)$$

where ρ is the density, c_v is the specific heat at constant volume, T is the temperature, $\overset{. \ \text{thel}}{W}$ is the thermoelastic dissipation term, ψ is the Helmholtz's free energy, $\boldsymbol{\alpha}$ is the vector of internal state variables, $\overset{. \ \text{irr}}{W}$ is the viscoplastic dissipation term, r is the heat source and $\boldsymbol{q}$ is the thermal flux referring to the current configuration and linked to the temperature gradient by the well known Fourier's equation. Our choice of observable and internal state variables is $[\boldsymbol{\epsilon}^{\text{rev}}, T, \boldsymbol{\alpha}]$ where $\boldsymbol{\epsilon}^{\text{rev}}$ is a representative tensor of the reversible part of the deformation, thus $\psi(\boldsymbol{\epsilon}^{\text{rev}}, T, \boldsymbol{\alpha})$. In a further section we will describe in more details the two dissipation terms of this equation.

2.3. CONSTITUTIVE EQUATIONS

2.3.1. *General formulation*

It is generally assumed, see e.g. Whertheimer [9] and Wriggers et al. [10] for details, that the rate of deformation can be additively decomposed into an elastic (reversible), an inelastic (irreversible) and a thermal parts, i.e. $\boldsymbol{D} = \boldsymbol{D}^{\text{e}} + \boldsymbol{D}^{\text{vp}} + \boldsymbol{D}^{\text{th}}$ and that the hypoelastic stress-strain relation is given, for elasto-viscoplastic materials, by a relation of the type

$$\overset{\triangledown}{\sigma}_{ij} = H(T)_{ijkl}(D_{kl} - D_{kl}^{\text{vp}} - D_{kl}^{\text{th}}) \qquad (9)$$

where $\boldsymbol{H}(T)$ is the Hooke stress-strain tensor at the temperature T given by

$$H(T)_{ijkl} = K(T) \, \delta_{ij}\delta_{kl} + 2G(T) \, \delta_{ik}\delta_{jl} - \frac{1}{3}\delta_{ij}\delta_{kl}) \qquad (10)$$

in which

$\overset{\triangledown}{\boldsymbol{\sigma}}$ is an objective rate of Cauchy stress tensor,

$\boldsymbol{D}$ is the rate of deformation,

$\boldsymbol{D}^{\mathrm{vp}}$ is the viscoplastic part of $\boldsymbol{D}$,

$\boldsymbol{D}^{\mathrm{e}}$ is the elastic part of $\boldsymbol{D}$,

$\boldsymbol{D}^{\mathrm{th}}$ is the thermal part of $\boldsymbol{D}$,

$\boldsymbol{\delta}$ is the Kronecker delta symbol

$K(T)$ is the bulk modulus of the material at T

$G(T)$ is the shear modulus of the material at T

Classicaly, for a $J2$ von Mises material with isotropic hardening, we assume the existence of a *yield function* f given by

$$f(\boldsymbol{\sigma}, \sigma^v, T) = \bar{\sigma} - \sigma^v(T) = 0 \tag{11}$$

where

$\bar{\sigma}$ is the effective stress, i.e. $\bar{\sigma} = \sqrt{\frac{3}{2} \boldsymbol{s} : \boldsymbol{s}}$;

$\boldsymbol{s}$ is the deviator of the stress tensor;

$\sigma^v(T)$ is the current yield stress.

With this approach, the admissible stress states are constrained to remain on or within the elastic domain ($f \leq 0$).

In this elasto-viscoplastic formulation, the effective stress $\bar{\sigma}$ is no longer constrained to remain less or equal to the yield stress but one can have $\bar{\sigma} \geq \sigma^v(T)$. Therefore we define the *overstress* as

$$d = \langle \bar{\sigma} - \sigma^v(T) \rangle \tag{12}$$

where $\langle x \rangle$ denotes the Mac Auley brackets defined by $\langle x \rangle = 1/2(x + |x|)$. Clearly, an inelastic process can only take place if, and only if, the overstress d is positive. In that case, $f \geq 0$.

The notion of irreversibility, linked to the viscoplastic part, is built into the formulation by introducing nonsmooth equations of evolution for $\boldsymbol{D}^{\mathrm{vp}}$ (flow rule) and $\sigma^v(T)$ (hardening laws) as follows.

2.3.2. *Flow rule*

When viscoplastic deformation occurs, $f \geq 0$ and one can write, in the case of associative viscoplasticity:

$$\boldsymbol{D}^{\mathrm{vp}} = \Lambda \boldsymbol{N} \qquad \text{where} \qquad \boldsymbol{N} = \frac{\partial_\sigma f}{\|\partial_\sigma f\|} \tag{13}$$

is the unit outward normal ($\boldsymbol{N} : \boldsymbol{N} = 1$) to the yield surface and Λ is a positive parameter called the *consistency parameter*. In a viscoplastic formulation, Λ cannot be determined, as in J2-plasticity, by expressing the so-called consistency condition.

So one more equation is needed to be able to express that consistency parameter, and so on the viscoplastic part of the rate of deformation. Various model were proposed to express Λ in terms of the current stress level. For example, classical viscoplastic models of the Perzyna type [3, 4] use a consistency parameter of the form :

$$\Lambda = \sqrt{\frac{3}{2}} \left\langle \frac{\bar{\sigma} - \sigma^v(T)}{\eta(\bar{\epsilon}^{\mathrm{VP}}, T)^{1/n(T)}} \right\rangle^{m(T)} \tag{14}$$

where
$$\begin{array}{ll} n(T) & \text{is a hardening exponent} \\ m(T) & \text{is a rate sensitivity parameter} \\ \eta(T) & \text{is a viscosity parameter.} \end{array}$$

This consideration can be viewed as a penalty regularization of rate-independent plasticity ($f \leq 0$) where the consistency parameter has been replaced by an increasing function of the overstress.

2.3.3. *Isotropic hardening law*
The evolution equation of the internal variable σ^v (which, in this case, is the only internal variable) is given by:

$$\dot{\sigma^v} = \sqrt{\frac{2}{3}} h(T) \, \Lambda \tag{15}$$

with $h(T)$ is called the *plastic modulus* and corresponds to the slope of the effective stress vs. effective plastic strain curve under uniaxial loading conditions. Generaly, h is a function of the effective viscoplastic strain, leading to a non-linear evolution equation for σ^v. Equation (15) can also be rewritten, in this more general case, as

$$\dot{\sigma^v} = h(T, \bar{\epsilon}^{\mathrm{VP}}) \, \dot{\bar{\epsilon}}^{\mathrm{VP}} \tag{16}$$

where $\dot{\bar{\epsilon}}^{\mathrm{VP}}$ is the rate of effective viscoplastic strain defined as

$$\dot{\bar{\epsilon}}^{\mathrm{VP}} = \sqrt{\frac{2}{3} \boldsymbol{D}^{\mathrm{VP}} : \boldsymbol{D}^{\mathrm{VP}}} = \sqrt{\frac{2}{3}} \Lambda = \left\langle \frac{\bar{\sigma} - \sigma^v(T)}{\eta(\bar{\epsilon}^{\mathrm{VP}}, T)^{1/n(T)}} \right\rangle^{m(T)} \tag{17}$$

so that, in the viscoplastic range, one can define a new constraint

$$\bar{f} = \bar{\sigma} - \sigma^v(T) - \eta(\bar{\epsilon}^{\mathrm{VP}}, T)^{1/n(T)} (\dot{\bar{\epsilon}}^{\mathrm{VP}})^{1/m(T)} = 0 \tag{18}$$

This criterion is a generalization of the classical von-Mises criterion $f = 0$ for rate-dependent materials. The latter can simply be recovered by imposing $\eta = 0$ (no viscosity effect).

2.3.4. Thermal Part

The equation governing the evolution of the thermal part of the tensor of the rate of deformation is a generalization of the equation used in infinitesimal strain theory, which is given by :

$$D_{ij}^{\text{th}} = \beta \, \dot{T} \, \delta_{ij} \tag{19}$$

where β is the linear thermal expansion coefficient.

2.4. DISCUSION OF THE DISSIPATION TERMS OF THE HEAT EQUATION

2.4.1. Viscoplastic heating

The viscoplastic heating $\dot{W}^{\text{irr}}$, as introduced above, is the most important heating source due to a mechanical deformation. As explained in [8] a part of the viscoplastic deformation does not create heat but induces the stocking of energy in the materials through the creation of micro-stress fields linked to the developement of dislocations and another microscopic defects. That part of non recoverable energy is expressed by the term :

$$\rho \frac{\partial}{\partial \alpha}[\psi - T \frac{\partial \psi}{\partial T}] \, \dot{\alpha} \tag{20}$$

in equation (8). This term is usually managed in considering that it represent, in the area of metal forming, between 5% and 15% of $\dot{W}^{\text{irr}}$. So, these two terms merge in a unique expression written as $\chi \, \dot{W}^{\text{irr}}$ where χ is a multiplicative factor which classicaly takes its value between 0.85 and 0.95.

Considering the flow rule introduced before we can express the viscoplastic heating as :

$$\dot{W}^{\text{irr}} = \bar{\sigma} \, \dot{\bar{\epsilon}}^{\text{vp}} \tag{21}$$

2.4.2. Thermoelastic heating

The thermoelastic heating, represented by the term $\dot{W}^{\text{thel}}$ in equation (8), has a minor contribution to the thermal equation. Many authors have neglected this term but, as a precise analyse can show, it has a stabilizing effect on the solution schemes used in the field of thermo-elasto-viscoplastic problem.

It can be expressed, using some properties of the viscoplastic flow, as :

$$\dot{W}^{\text{thel}} = 3K(T)\beta T \frac{\dot{J}}{J} \tag{22}$$

3. Integration procedure

In this section we will briefly describe the main steps of numerical schemes able to solve these equations in time :

3.1. FORMULATION OF THE PROBLEM TO SOLVE

Let us consider $[t_n, t_{n+1}]$ a sub-interval of the global time interval of interest, and ϕ the vector of primary variables defined as :

$$\phi = (\boldsymbol{x}, \boldsymbol{v}, T) \tag{23}$$

To determine ϕ at the time t_{n+1} we need to integrate the consitutive law, using the value of the internal variables $\boldsymbol{\alpha}$ at t_n, to evaluate $\boldsymbol{\alpha}$ at t_{n+1}. This part of the integration procedure will be describe in section 3.3 .

The problem under consideration can be formulated as a first order problem of evolution written as :

$$\dot{\boldsymbol{x}} = \boldsymbol{v} \tag{24}$$

$$\dot{\boldsymbol{v}} = \frac{\text{div}\boldsymbol{\sigma}}{\rho} + \boldsymbol{b} \tag{25}$$

$$\dot{T} = \frac{1}{\rho c_v}[\rho r - \text{div}\boldsymbol{q} + \overset{.\ \text{irr}}{W} + \overset{.\ \text{thel}}{W}] \tag{26}$$

in the interval $[t_n, t_{n+1}]$ and on the whole body.

3.2. STAGERRED SCHEMES

In our model we have decided to use the backward-Euler formula for the approximation of the time derivative, and the finite element technique for the spatial discretization. Under above assumptions it's easy to show that a monolithic scheme, if applied to this problem, will induce the resolution of a linear unsymmetric system of order $(n_{\text{dim}} + 1).n_{\text{node}}$, where n_{dim} is the dimension of the space and n_{node} is the number of node of the finite element mesh. Staggered schemes, based on a split of the operator defined by the equations (24, 25, 26), have been developped to reduce the computational cost of such resolution.

3.2.1. *Isothermal split :*
The simplest split consists in the decomposition of the global thermome-chanical operator in a mechanical operator at fixed temperature and a

thermal operator at fixed geometry. This split leads to the resolution of two symmetric linear systems, the first of order $n_{\mathrm{dim}} \cdot n_{\mathrm{node}}$ and the second of order n_{node} involving respectively the two equations (24, 25) and the equation (26). This scheme of resolution results only in a conditional stability which can be improved by the modification introduced by Armero & Simo [1] and described below.

3.2.2. *Adiabatic split* :

This split consists in the decomposition of the global thermomechanical operator in a mechanical operator at fixed entropy and a thermal operator at fixed geometry. This modification consists in the resolution of one part of the equation (26) during the first phase to transform that phase into an adiabatic one. That part of the equation can be solved analytically, a fact that enables us to keep the same order for the linear systems of both phases.

3.3. THE RADIAL RETURN MAPPING ALGORITHM

For a given configuration defined by its known set of positions $x(t_n)$ at time t_n the problem is now to update all state variables to a new configuration defined by its set of position $x(t_{n+1})$ (which are supposed to be known) at time t_{n+1}. These incremental motions are, in turn, used to calculate the incremental strain history by means of the kinematic relations.

The problem dealt with in this paragraph is, for the given incremental strain history, to find the new values of the variables $(\sigma_{n+1}, \bar{\epsilon}_{n+1}^{\mathrm{vp}}, \sigma_{n+1}^{v})$ at t_{n+1}. These are obtained by integration of the local constitutive equations with initial conditions given by $(\sigma_n, \bar{\epsilon}_n^{\mathrm{vp}}, \sigma_n^v)$ at t_n. To integrate these equations in time, we will rely on the general methodology of elastic-predictor /plastic-corrector (return mapping algorithm), as synthesized by Simo & Hughes [7] but here, we will extend this methodology to the time-dependent case.

In a first step, the elastic predictor problem is solved with initial conditions that are the converged values of the previous time step while keeping irreversible variables frozen. This produce a trial elastic stress state σ_{tr} which, if outside the yield surface f is taken as the initial conditions for the solution of the viscoplastic corrector problem. The objective of this second step is to restore consistency by returning back the trial stress to the generalized criterion $\bar{f}$ (and not on the yield function f as is done in the rate-independent case!). For more details about this algorithm see Ponthot [5].

4. Application : Hot metal forming

4.1. VISCOUS BEHAVIOUR OF METALS AT HIGH TEMPERATURE.

Most metals exhibit a partial viscous behavior during hot forming. At high temperatures some metals can develop important viscous deformations (superplastic behavior) that permits their forming as if it were a polymer or a glass. Another important observation is that the flow stress, during and after forming, is substantially reduced when compared to cold metal forming.

4.2. NUMERICAL RESULTS : FORMING OF A QUARTER CYLINDER.

This numerical test has, especially, the aim of showing the importance of the thermal spring-back, and so of the unloading phase, at different forming temperatures. The material parameters are typicals of an aluminium alloy which has a superplastic behaviour at high temperature.

The test conditions are represented in the Figure 1. The right end of the sheet is clamped and the die has a fixed position. The pressure and the temperature are, in a first phase, increased to their forming values, then they are maintained during the forming of the sheet, and finaly lowered to their room values. In the results exposed below, the thickness to die radius ratio is $\frac{1}{50\pi}$. The forming time is 3100s (300s for the loading, 2700s for the stress relaxation and 100s for the unloading).

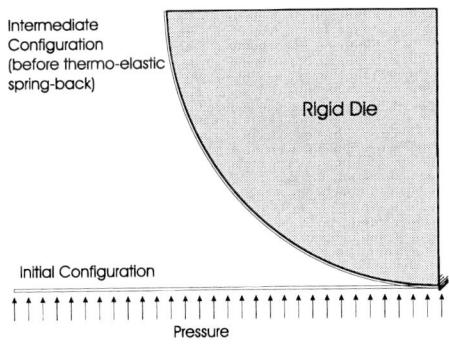

Figure 1. Description of the numerical test

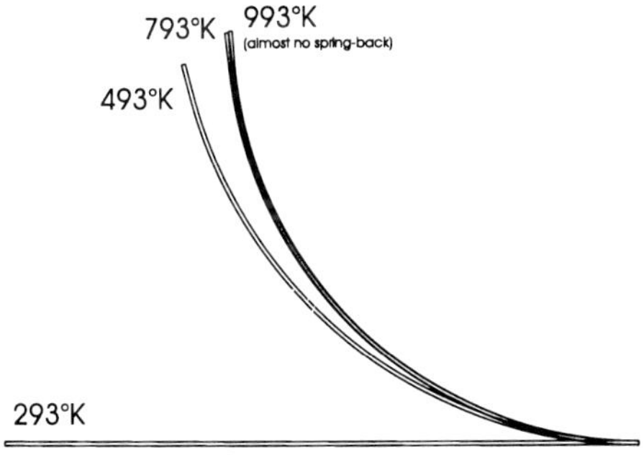

Figure 2. Final configurations for different forming temperatures

Figure 2 shows final configurations for the different forming temperatures. We observe that the cold forming of the sheet induce few irreversible deformations and so the spring-back is almost total. On the other hand, for the forming at high temperature, there is almost no spring back. The final stress level is a decreasing function of the forming temperature, as we can

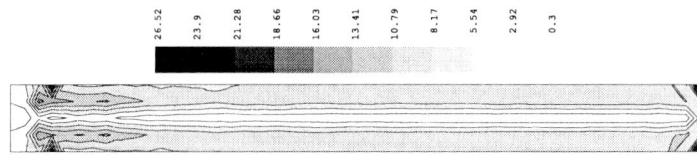

Figure 3. Final distribution of the J2 stress in MPa (forming at 493°K)

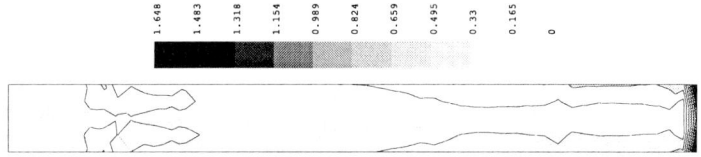

Figure 4. Final distribution of the J2 stress in MPa (forming at 993°K)

see in the next two Figures (Figures 3 and 4). These stress distributions are plotted in the initial configuration with a dilatation factor of 15 along the vertical axis.

References

1. F. Armero and J.C. Simo. A new unconditionnaly stable fractionnal step method for non-linear coupled thermomechanical problem. *International Journal of Numerical Methods in Engineering*, 35:737–766, 1992.
2. L.E. Malvern. *Introduction to the Mechanics of Continuous Medium*. Prentice-Hall, 1969.
3. P. Perzyna. Fundamental problems in visco-plasticity. In G. Kuerti, editor, *Advances in Applied Mechanics*, volume 9, pages 243–377. Academic Press, 1966.
4. P. Perzyna. Thermodynamic theory of plasticity. In Chia-Shun Yih, editor, *Advances in Applied Mechanics*, volume 11, pages 313–355. Academic Press, 1971.
5. J.P. Ponthot. Unified stress update algorithms for the numerical simulation of large deformation elasto-plastic and elasto-viscoplastic processes. *International Journal of Plasticity*, Accepted for publication.
6. D. Rozenwald. *Modélisation thermomécanique des grandes déformations. Application aux problèmes de mise à forme des métaux, des élastomères et des structures mixtes métal-elastomère*. PhD thesis, University of Liège, Liège, Belgium, 1996.
7. J.C. Simo and T.J.R Hughes. General return mapping algorithms for rate-independent plasticity. In *Constitutive Laws for Engineering materials : Theory and Applications*. Elsevier Science Publishing Co, 1987.
8. J.C. Simo and C. Miehe. Associative coupled thermoplasticity at finite strains formulation, numerical analysis and implementation. *Computer Methods in Applied Mechanics and Engineering*, 98:41–104, 1992.
9. T.B. Whertheimer. Thermal mechanically analysis in metal forming processes. In Pittman Wood Alexander Zienkiewicz, editor, *Numerical Methods in Industrial Forming Processes*, pages 425–434, 1982.
10. P. Wriggers, C. Miehe, M. Kleiber, and J.C. Simo. On the coupled thermo-mechanical treatment of necking problems via FEM. In D.R.J. Owen, E. Hinton, and E. Oñate, editors, *International Conference on Computational Plasticity (COMPLAS2)*, pages 527–542, Barcelona, Spain, 1989. Pineridge Press.

THICK AXISYMMETRIC PLATE SUBJECTED TO THERMO-MECHANICAL DAMAGE

A. GANCZARSKI
Institute of Mechanics and Machine Design
Cracow University of Technology
Jana Pawła II 37, PL 31-864 Kraków, Poland

1. Introduction

A creep process and the associated material deterioration are temperature sensitive phenomena. A classical approach consists in accounting for the effect of temperature on the material functions in the constitutive and evolution equations of a damaged solid, cf. Ganczarski and Skrzypek Ref. [1], whereas the temperature field is considered as the steady state. In the general case, when thermo-mechanical loadings are applied to the structure, in addition to the constitutive and evolution equations with the appropriate mechanical boundary conditions, the damage coupled heat transfer equation must simultaneously be solved to yield a transient temperature field which satisfies the thermal boundary conditions. Thick plate is considered here as the 3D rotationally symmetric problem, without the additional constraints imposed when the simplified mid-thick plate theory is used Refs [2], [3].

2. Basic equations

2.1. ASSUMPTIONS

The general assumptions of rotational symmetry of both temperature and displacement fields hold

$$\partial_\varphi T = 0, \quad \partial_\varphi \mathbf{u} = 0 \tag{1}$$

Additionally, the linearized total strain tensor is decomposed into the elastic, creep and thermal parts

$$\boldsymbol{\varepsilon} = \boldsymbol{\varepsilon}^e + \boldsymbol{\varepsilon}^{th} + \boldsymbol{\varepsilon}^c, \quad \boldsymbol{\varepsilon}^{th} = \mathbf{1}\alpha\theta \tag{2}$$

S. Murakami and N. Ohno (eds.), IUTAM Symposium on Creep in Structures, 207–216.

where its inelastic part is incompressible

$$\text{Tr}\varepsilon^{c} = 0 \tag{3}$$

and $\theta = T - T_{\text{ref}}$ denotes the difference between actual and reference temperatures.

2.2. HEAT TRANSFER IN DAMAGED MATERIAL

In the present paper, an extension of the Tanigawa concept Ref. [4], originally derived for the time-independent nonhomogeneous but isotropic material, to the time-dependent partly damaged materials is presented. Namely, the coefficient of thermal conductivity λ of heat flux through partly damaged material in the homogeneous quasi-stationary Fourier equation (slow temperature field change following damage) is assumed to be the linear function of the scalar damage parameter D, Refs [5], [6], [7]:

$$\text{div}\left(\lambda \mathbf{grad}T\right) = 0 \qquad \lambda = \lambda_0 \left(1 - D\right) \tag{4}$$

A more general approach, when second-rank tensor of thermal conductivity $\mathbf{\Lambda}$ with damage tensor $\mathbf{D}$ as an argument is due to Skrzypek and Ganczarski Ref. [8]. However, the effect of damage on the thermal expansion coefficient α is not taken into account. This problem was discussed in the paper by Ganczarski Ref. [10]. Additionally heat transfer between a fluid (gas) and solid (plate) is assumed to be affected by damage in the Newton law of convection.

2.3. EQUATIONS OF MECHANICAL STATE

The general equations of mechanical state take the form of displacement equations without body forces but enriched by the thermal and creep terms

$$G\nabla^2\mathbf{u} + \frac{G}{1 - 2\nu}\mathbf{grad}\ \text{div}\mathbf{u} = \frac{E\alpha}{1 - 2\nu}\mathbf{grad}\theta + 2G\ \text{Div}\varepsilon^{c} \tag{5}$$

where strain and stress are defined as follows

$$\varepsilon = \frac{1}{2}\left(\nabla\mathbf{u} + \mathbf{u}\nabla\right)$$
$$\sigma = 2G\left[\varepsilon - \varepsilon^{c} - \mathbf{1}\alpha\theta + \frac{\nu}{1 - 2\nu}\left(\text{Tr}\varepsilon - 3\alpha\theta\right)\mathbf{1}\right] \tag{6}$$

This formulation leads to certain inconsistency. Namely, the completely damaged zone is free from mechanical stress and heat flux but, simultaneously, it must be able to carry the thermal stresses that result from a residual temperature field. A consistent formulation, where damage affects not only the coefficient of thermal conductivity Eq. (4), but also Kirchhoff's modulus and Poisson's ratio requires extension of the mechanical state equations (5) in a similar fashion as previously proposed by authors in Ref. [8] for the Love-Kirchhoff plate of a variable thickness.

2.4. CONSTITUTIVE EQUATIONS

The essence of thermo-damage coupling results from the reciprocal coupling between the processes of creep, microcrack growth and change of thermal field Eq. (4), which are described by the similarity of deviators based on the flow theory

$$\dot{\varepsilon}^c = \frac{3}{2} \frac{\dot{\varepsilon}^c_{\text{eff}}}{\sigma_{\text{eff}}} \mathbf{s} \tag{7}$$

the time hardening hypothesis

$$\dot{\varepsilon}^c_{\text{eff}} = \left(\frac{\sigma_{\text{eff}}}{1 - D} \right)^m \dot{f}(t) \tag{8}$$

and Kachanov's type isotropic brittle rupture law Ref. [8]

$$\frac{dD}{dt} = C \left[\frac{\chi(\boldsymbol{\sigma})}{1 - D} \right]^n, \tag{9}$$

respectively. In Eq. (9) the state of damage is measured by the scalar internal variable D, whereas damage growth is controlled by the damage equivalent stress $\chi(\boldsymbol{\sigma})$ according to Hayhurst. For the aluminium alloy $\chi(\boldsymbol{\sigma}) = \sigma_{\text{eff}}$ is applicable Ref. [11]. In the above formulae the following definitions hold:

$$\sigma_{\text{eff}} = \sqrt{\frac{3}{2} \mathbf{s} : \mathbf{s}} \qquad \dot{\varepsilon}^c_{\text{eff}} = \sqrt{\frac{2}{3} \dot{\mathbf{e}}^c : \dot{\mathbf{e}}^c} \tag{10}$$

3. Formulation of boundary problems

3.1. THERMAL BOUNDARY PROBLEM

The temperature distribution of a medium over the upper plate surface is not uniform. Assuming spherical symmetry of the temperature front propagation (Fig. 1a) we find that the temperature distribution of the gas neighbouring the upper plate surface exhibits cylindrical symmetry and is given by the parabola $T(x) = T_s - \frac{q_v}{6\lambda_g} x^2$, where T_s stands for the gas temperature at the heat source, λ_g denotes thermal conductivity of the gas, whereas x is a distance between the center of the temperature front (heat source) and the particle of a gas neighbouring current point of the plate. A typical

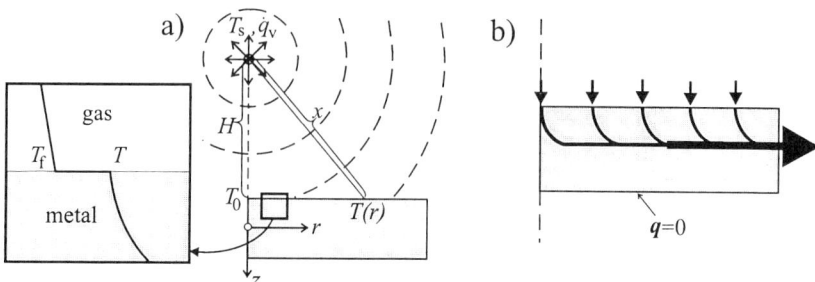

Figure 1.   Temperature front propagation a) and schematic heat flow through plate b)

heat flux through the plate is presented in Fig. 1b. It is assumed that the whole heat flux which enters the upper plate surface goes towards the side wall, whereas the lower surface subjects to the adiabatic conditions. Above assumptions allow to formulate the mixed and the Neumann integral type boundary conditions for Eq. (4) in a following form, Ref. [12]:

$$
\begin{aligned}
&\beta_0\,(1 - D)\,(T_{\mathrm{f}} - T) = -\lambda_0\,(1 - D)\,\partial_n T & &\text{upper surface} \\
&\partial_n T = 0 & &\text{lower surface} \\
&\partial_n T = 0 & &\text{symmetry axis} \quad\quad (11) \\
&\int_A [\lambda_0\,(1 - D)\partial_n T]\,\mathrm{d}A = C & &\text{sidewall}
\end{aligned}
$$

where $\beta_0\,(1 - D)$ denotes a damage coupled coefficient of the Newtonian convection between the gas and the upper surface of plate. The heat flux, that enters the top surface of the plate and is sent through the side wall is hold constant throughout the process ($C = \text{const}$).

3.2. MECHANICAL BOUNDARY PROBLEM

System of equations (5) requires appropriate boundary equations. Simple support (Fig. 2) is assumed for the plate under mechanical load p. This condition, however, constraints essentially thermal transverse expansion at the sidewall so, for the thermal problem a separate boundary conditions are proposed with $u_z^{\mathrm{th}} = 0$ at the lower corner (Fig. 3). Finally, two sets of the boundary conditions are used to solve mechanical problem separately for the purely mechanical and the purely thermal loading (Table 1), which requires a decomposition of the radial displacement into purely mechanical and purely thermal components $u_z = u_z^{\mathrm{m}} + u_z^{\mathrm{th}}$.

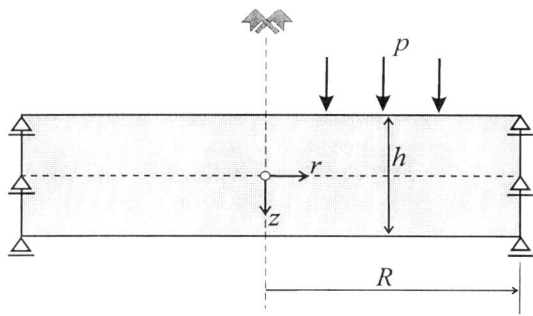

Figure 2. Boundary conditions for purely mechanical load p

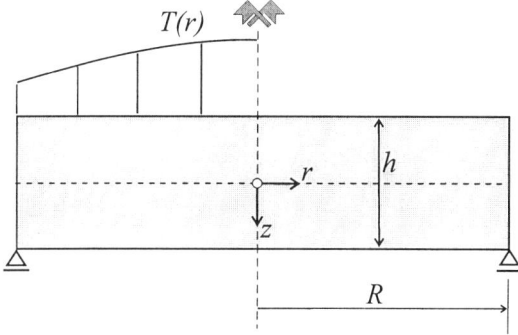

Figure 3. Boundary conditions for purely thermal load $T(r)$

TABLE 1. Boundary conditions for purely mechnical and purely thermal loads

mechanical state		thermal state		boundary
$\sigma_z = -p$	$\sigma_{rz} = 0$	$\sigma_z = 0$	$\sigma_{rz} = 0$	upper surface
$\sigma_z = 0$	$\sigma_{rz} = 0$	$u_r = 0$	$\partial_r u_z^{\mathrm{th}} = 0$	lower surface
$u_r = 0$	$\partial_r u_z^{\mathrm{m}} = 0$	$u_r = 0$	$\partial_r u_z^{\mathrm{th}} = 0$	symmetry axis
$\sigma_r = 0$	$u_z^{\mathrm{m}} = 0$	$\sigma_r = 0$	$\sigma_{rz} = 0$	sidewall
			$u_z^{\mathrm{th}} = 0$	lower corner

4. Numerical algorithm for creep-damage problem

To solve the initial-boundary problem by FDM, we discretize time by inserting N time intervals Δt_k, where $t_0 = 0$, $\Delta t_k = t_k - t_{k-1}$ and $t_N = t_I$

(macrocrack initiation) Ref. [8]. Hence, the initial-boundary problem is reduced to a sequence of the quasistatic boundary-value problems, the solution of which determines unknown functions at a given time t_k, e.g., $T(\mathbf{x}, t_k) = T^k(\mathbf{x})$, $\mathbf{u}(\mathbf{x}, t_k) = \mathbf{u}^k(\mathbf{x})$. At each time step the Runge-Kutta II method is applied to yield updated functions T^k, $\mathbf{u}^k$. To account for primary and tertiary creep regimes, dynamically controlled time step Δt_k is required, the length of which is defined by the bounded maximum damage increment:

$$\Delta D^{\text{lower}} \leq \max_{(\mathbf{x})} \left\{ \left[\dot{D}^k(\mathbf{x}) - \dot{D}^{k-1}(\mathbf{x}) \right] \Delta t_k \right\} \leq \Delta D^{\text{upper}}. \qquad (12)$$

Discretizing also spatial coordinates $\mathbf{x} = [r, z]_{i,j}$, by inserting a mesh $\Delta r = r_i - r_{i-1}$, $\Delta z = z_j - z_{j-1}$, we rewrite the equations of thermal state Eq. (4) and mechanical state Eq. (5) for a time step t_k in terms of FDM with respect of r_i and z_j, respectively. Applying stage algorithm, first for the damage $[D]_{i,j} \equiv 0$, the equation of thermal problem Eq. (4) with the boundary conditions Eqs (11) is solved by using the combination of procedures `bandec.for` and `banbks.for` Ref. [13], and the initial "elastic" temperature $[T^e]_{i,j}$ is found. Then, the system of equations of mechanical state Eqs (5) with the known temperature field $[T^e]_{i,j}$ as the right hand side and the boundary conditions (cf. Table 1) is numerically solved by using the combination of procedures `ludcmp.for` and `lubksb.for`, and the elastic displacements are determined $[\mathbf{u}^e]_{i,j}$. Next, the program enters the creep damage loop which requires the effective stress intensity and components of the damage and strain rates $[\sigma_{\text{eff}}, \dot{D}, \dot{\varepsilon}^c]_{i,j}$, Eqs (7, 9). Repeating again the stage algorithm a solution of discretized thermo-creep-damage problem, rates of temperature $[\dot{T}]_{i,j}$ and displacements $[\dot{\mathbf{u}}]_{i,j}$ are computed. In the next time step the "new" temperature $[T]_{i,j}$ and displacements $[\mathbf{u}]_{i,j}$ are found, and the process is continued until the maximum value of damage reaches the critical level $\max[D]_{i,j} \cong 1$.

5. Results

5.1. MATERIAL DATA

Numerical results presented in this paper deal with plate made of Al-Mg-Si alloy of the following thermo-mechanical properties at temperature 483 K Ref. [14]: $T_0 = 300°C$, $T_{\text{ref}} = 0°C$, $E = 60.06$ GPa, $\sigma_0 = 149.5$ MPa, $\nu = 0.3$, $\alpha = 2.5 \times 10^{-5}$, $\lambda_0 = 203$ W/m°C, $\beta_0 = 75$ W/m²°C, $m = 3.7$, $n = 7.0$, $C = 8.44 \times 10^{-17}$ MPa^{-n}h^{-1}. Characteristic parameters of the plate (thickness ratio to midsurface diameter ratio) and the magnitude of pressure are: $h/R = 0.2$, $p = 6.7 \times 10^{-3}\sigma_0$, respectively.

5.2. EXAMPLE

The effective stress type sensitivity to damage of the aluminium alloy and combined thermo-mechanical loads cause that the first macrocrack appears at the central, upper plate surface which is a region of the compressive stress concentration (Fig. 4). The zone of the most advanced damage is more dispersed over the shallow-wide area, when compared to the Reissner's mid-thick plate Ref. [3].   Damage accumulation causes essential decrease

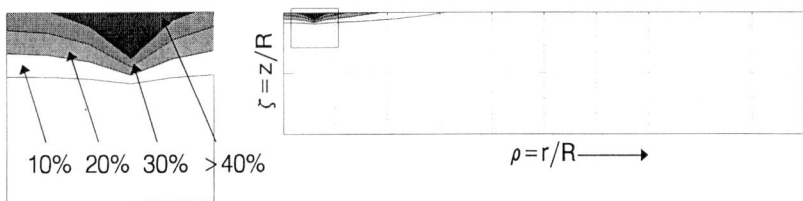

Figure 4.   Damage distribution at the instant of macrocrack initiation

of the coefficient of thermal conductivity $\lambda = \lambda_0 (1 - D)$ and consequently decrease of the temperature (Fig. 5) and the local heat flux (Fig. 6). Region of the most advanced damage becomes unable to conduct any heat (Fig. 6).
However, in the model under consideration, where mechanical moduli

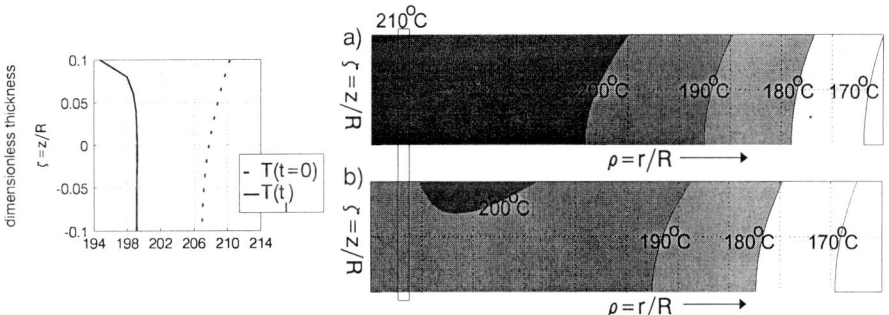

Figure 5.   Evolution of temperature field: a) initial, b) at the instant of macrocrack initiation

are not affected by damage, the total effective stress in the most damaged zone (top) that results from both the bending compression and the thermal compression is not released to zero, but remains almost unchanged, whereas in the less damaged tensile zone (bottom) the effective stress at t_I is about three times as high as at the beginning ($t = 0$), Fig. 7.   Nevertheless, on the contrary to the Reissner's approach Refs [2], [3], in the present 3D

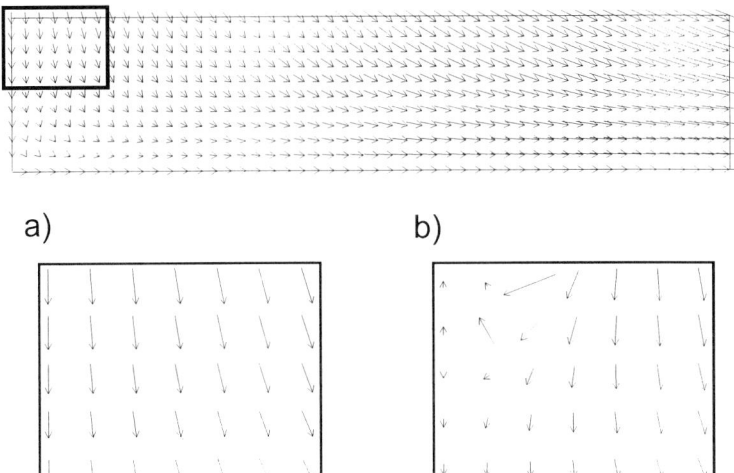

Figure 6. Heat flux $\mathbf{q} = -\lambda_0 (1 - D) \mathbf{grad} T$: a) in case of $t = 0$, b) at the moment of rupture

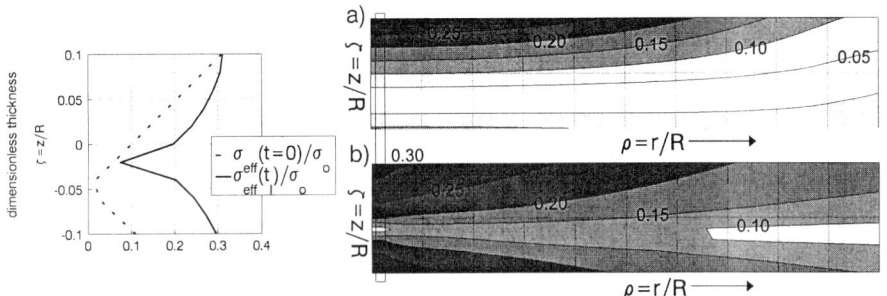

Figure 7. Redistribution of von Mises effective stress a) initial "elastic", b) at the moment of rupture

rotationally symmetric formulation the shear stress σ_{rz} exhibits qualitative changes: it is no longer symmetric Ref. [15] and, what is the most essential, it follows the actual damage state (Fig. 8). For higher values of depth to radius ratio, the 3D approach will show not only qualitative but also essential quantitative effect of both the shear stress σ_{rz} and the axial stress σ_{zz} on the damage process, stress and temperature redistribution. A comparison of the time to macrocrack initiation, when either the exact (3D) or the Reissner approach is used, shows negligible discrepancies $t_I^{3D}/t_I^R \cong 1.08$

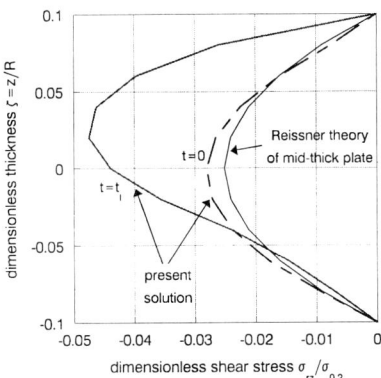

Figure 8. Comparison of shear stress distribution for 3D and Reissner's mid-thick plates

6. Conclusions

- Heat flux through the damaged material is sensitive to current damage state. To describe this effect in the most general case both the tensor of thermal conductivity $\widetilde{\lambda}$ and the tensor of thermal expansion $\widetilde{\alpha}$ in the Fourier heat flux and the mechanical state equations, respectively, should be functions of the damage tensor $\mathbf{D}$ Refs [10], [9]. However, in the special case of the aluminium alloy, damage growth of which is controlled by the effective stress σ_{eff}, the scalar damage parameter D is applicable and, consequently, material remains isotropic during the creep-damage process.
- In the model under consideration, a completely damaged particle is free from the heat flux, but it has to carry non-zeroth stress resulting from the residual temperature field. To avoid this inconsistency it is necessary to incorporate the effect of damage on the mechanical constitutive moduli $\widetilde{\mathbf{E}}$ Ref. [7]. In this more consistent formulation some additional terms associated with derivatives of stiffnesses, as functions of damage parameter D, with respect to both spatial co-ordinates r, z and time t have to be taken into account.
- In the model proposed, the temperature jump at the fluid/solid interface is used to approximate the real heat exchange conditions. To precisely describe fluid-solid interaction it is suggested to extend the model by taking into account boundary layer effect as well as an environmental corrosion.

- Comparison of the Reissner and the exact (3D) approach exhibits local changes in the stress field, especially when transverse shear and normal stresses σ_{rz} and σ_{zz} considered as "live" stress components change following damage process. However, the effect of the exact formulation on a global lifetime prediction as compared to the Reissner's formulation, is not so essential.

References

1. Ganczarski, A and Skrzypek, J. (1991), On optimal design of disks with respect to creep rupture, Proc. of *IUTAM Symp. Creep in Structures IV*, Ed. M. Życzkowski, Springer, Berlin, pp. 571–577.
2. Ganczarski, A., Freindl, L. and Skrzypek, J. (1997), Orthotropic brittle rupture of Reissner's prestressed plates, *COMPLAS 5*, CIMNE, Barcelona, pp. 1904-1909.
3. Ganczarski A., Skrzypek J. (2000), Damage effect on thermo-mechanical fields in a mid-thick plate, *J. Theor. Appl. Mech.*, 2, **38**, pp. 1-13.
4. Tanigawa, Y. (1995), Some basic thermoelastic problems for nonhomogeneous structural materials, *ASME*, **48**, 6, pp. 287–300.
5. Ganczarski, A and Skrzypek, J. (1995), Concept of Thermo-Damage Coupling in Continuum Damage Mechanics, *First Int. Symp. Thermal Stresses'95*, Hamamatsu, pp. 83–86.
6. Ganczarski, A and Skrzypek, J. (1997), Modeling of Damage Effect on Heat Transfer in Solids, *Second Int. Symp. Thermal Stresses'97*, Rochester, pp. 213–216.
7. Skrzypek, J. and Ganczarski, A. (1998), Modeling of damage effect on heat transfer in time-dependent nonhomogeneous solids, *J. Thermal Stresses*, **21**, 3-4, pp. 205-231.
8. Skrzypek, J.J. and Ganczarski, A. (1999), *Modeling of Material Damage and Creep Failure of Structures, Theory and Applications*, Springer-Verlag, Berlin Heidelberg.
9. Ganczarski A. (1999): Thermal anisotropy inducing brittle damage, *Technische Mechanik*, **19**, 4, 321-330.
10. Ganczarski, A. (1999), Effect of brittle damage on thermal expansion tensor, *Thermal Stresses'99*, Cracow, pp. 267–270.
11. Hayhurst, D. (1998), Materials data and mechanisms-based constitutive equations for use in design, in: *Creep and Damage in Materials and Structures*, Eds. H. Altenbach and J. Skrzypek, Springer Wien New York, pp. 167-208.
12. Moon, P. and Spencer, D.E. (1961), *Field theory for engineers*, Van Nostrand.
13. Press, W.H., Teukolsky, S.A. (1993), Vetterling, W.T. and Flannery, B.P., *Numerical Recipes in Fortran*, Cambridge Press.
14. Litewka, A. (1989), Creep rupture of metals under multi-axial state of stress, *Arch. Mech.*, **41**, 1, pp. 3-23.
15. Okumura, I.A., Oguma, Y. (1998), Series solutions for a transversely loaded and completely clamped thick rectangular plate based on the three-dimensional theory of elasticity, *Arch. Appl. Mach.*, 68, pp. 103-121.

CREEP IN SHOTCRETE TUNNEL SHELLS

CH. HELLMICH, M. LECHNER, R. LACKNER, J. MACHT and
H.A. MANG
Institute for Strength of Materials
Vienna University of Technology
Karlsplatz 13/202
A-1040 Vienna, Austria

Abstract.
    Creep of shotcrete is modelled within the framework of thermodynamics of chemically reactive
porous media. The process of creep is divided into a short-term and a long-term part. Short-term
creep stems from stress-induced water movement within the capillary pores of shotcrete, located
between the already formed hydrates which are the reaction products between cement and water.
Thus, short-term creep is related to the accumulation of initially (micro)stress-free hydrates. Hence,
it depends on increments of (macro)stress. Long-term creep results from dislocation-like processes
within the (micro)stressed hydrates. Therefore, this process depends on the total (macro)stress.
Microcracking of shotcrete is modelled by means of multisurface chemoplasticity. Moreover, the
model accounts for chemical shrinkage and hydration heat. Finally, the significance of creep of
shotcrete for real-life structures is shown by means of 3D hybrid analyses of a railway tunnel. The
term 'hybrid' reflects the combination of advanced material modelling with *in-situ* displacement
measurements in the framework of nonlinear Finite Element analyses.

1. Introduction

Creep of concrete is a material property of great importance for the design of
concrete structures like bridges or dams. A large number of macroscopic material models for creep of concrete has been developed since the advent of the
computer era. They constituted the basis for structural analyses of concrete structures subjected to creep. The basis of these models is often purely empirical.
Frequently, however, they try to bridge the gap between fundamental research on
the microstructural level and the observable macroscopic material behavior [28].
In [24], creep of concrete was modelled in the framework of thermodynamics of
chemically reactive porous media.

   This strategy was revisited, extended and applied, with special emphasis on
shotcrete, by Sercombe et al. [23]. According to [21], shotcrete is 'the name given
to mortar or concrete conveyed through a hose and pneumatically projected onto
a backup surface'. This material plays a central role in the New Austrian Tunnel-

217

S. Murakami and N. Ohno (eds.), IUTAM Symposium on Creep in Structures, 217–229.
© 2001 *Kluwer Academic Publishers. Printed in the Netherlands.*

ing Method (NATM). Of particular great importance are the creep properties of shotcrete. When driving tunnels according to the NATM, after the excavation of a cross-section of a tunnel, shotcrete is applied onto the tunnel walls, constituting a thin and flexible shell. This flexibility, together with a remarkable compliance resulting from creep, prevents the tunnel shell from being destroyed by the ground pressure which may be very large. In fact, the pressure is limited to an admissible value and the ground itself becomes a prominent part of the load-carrying system. Therefore, NATM-tunnels can be regarded as soil-shotcrete compound structures.

According to [28] [25] creep of concrete (and shotcrete) can be divided into two distinct phenomena, short-term creep and long-term creep. Short-term creep is of great importance during the first days after the installation of the tunnel shell, which usually are most critical. This paper presents a new formulation for short-term creep, extending the one of Sercombe et al. [23]. The new formulation is based on a central statement of Carol and Bažant [6] for concrete: Aging of concrete results from the accumulation (solidification) of quasi-instantaneously formed (non-aging) constituents. Hence, analogous to the incremental stress-strain relationship for aging elasticity (see, e.g. [27] [19], [11], [23]), an incremental formulation for short-term creep is introduced herein.

The influence of the new formulation on the structural response of a real-life tunnel is the other main item of this paper. Structural analyses are performed by means of a hybrid method, combining 3D thermochemomechanical Finite Element analyses with *in-situ* displacement measurements, see [9].

2. Material model for creep of shotcrete

The presented material model is cast within the framework of thermodynamics of chemically reactive porous media [7]; as applied to concrete by Ulm and Coussy [27] and to shotcrete by Hellmich et al. [13].

Microstructural phenomena are described on the macroscopic level (i.e., for an observation scale of about 10 cm for shotcrete) by means of state variables, energetically correlated thermodynamic forces, and respective evolution laws, as described in the following. The state variables are divided into external (macroscopically measurable) ones and internal ones. The latter are correlated to dissipative phenomena on the microlevel of material description.

2.1. STATE VARIABLES - MICROSTRUCTURAL PHENOMENA

Two external state variables are used in the material model. They are the absolute temperature T and the tensor of macroscopic strains, ε. A change of temperature results in (volumetric) thermal strains ε^T (thermomechanical coupling), $d\varepsilon^T = \mathbf{1}\alpha dT$, with $\mathbf{1}$ standing for the second-order unity tensor and α being the thermal dilatation coefficient.

The meaning of the internal state variables used in this work is described in the following:

—   The *hydration process* encompasses the complex chemical reactions between cement and water. The respective reaction products are called hydrates. The state variable associated with the hydration process is the *degree of hydration* ξ. The kinetics of the hydration process is determined by the diffusion of the free water through the pores between the hydrates in order to reach yet unhydrated cement grains. Once such a grain is reached, the chemical reaction between water and cement is a quasi-instantaneous event. Therefore, ξ is defined as the mass of water bound in hydrates per unit volume of concrete divided by the respective mass at the end of the hydration process.

    The hydration process leads to (volumetric) chemical shrinkage strains ε^s (chemomechanical coupling), $d\varepsilon^s = 1\beta d\xi$, with the chemical dilatation coefficient β, see Figure 1(a) for the shotcrete of Huber [14].

    Moreover, hydration is an exothermal process. The latent heat production is given by $f_\xi \dot{\xi}$ (chemothermal coupling), with the (constant) heat of hydration f_ξ.

—   Macroscopic stresses σ may lead to *microcracking*, represented by macroscopic (plastic or permanent) strains ε^p. Microstructural changes resulting from microcracking are modelled by means of the internal state variable χ, the vector of hardening variables.

—   Macroscopic stresses σ lead to a *microdiffusion process* of water within capillary pores between the hydrates [22]. These pores measure several micrometer. The internal variable w is the mass of water per unit volume engaged in this process. This water movement results in short-term (viscous) creep strains ε^v, $\varepsilon^v = \mathbf{g}w \rightarrow d\varepsilon^v = \mathbf{g}dw$. As for the (constant) second-order coupling tensor $\mathbf{g}$, see Subsection 2.2. Proper description of the viscous creep process is the central issue of this paper.

— σ also provokes *dislocation-like processes* within the hydrates, where groups of water molecules ('creep sites' according to [28]) are re-arranged in the nanopores of the hydrates, measuring several nanometer. Resulting macroscopic strains are called long-term (flow) creep strains ε^f. The microstructural changes stemming from the dislocation-like sliding processes are represented by the internal state variable γ, the viscous slip [24].

2.2. THERMODYNAMIC FORCES - STATE EQUATIONS

The change of a state variable results from the action of the energetically conjugate thermodynamic force. This is expressed by means of evolution equations, see Subsection 2.3. The thermodynamic forces may depend on all state variables. These dependencies are described by state equations. However, as a consequence of several decoupling hypothesis (see, e.g., [26]), some of these dependencies are regarded as negligible, as will be detailed in the following:

— The chemical affinity A_ξ is the driving (thermodynamic) force of the hydration reaction, i.e., it is responsible for changes of ξ. For shotcrete, the chemical affinity depends mainly on ξ, expressed by the state equation $A_\xi = A_\xi(\xi)$. The function $A_\xi(\xi)$ is depicted in Figure 1(b) for a shotcrete mixture of Lafarge, see [8].

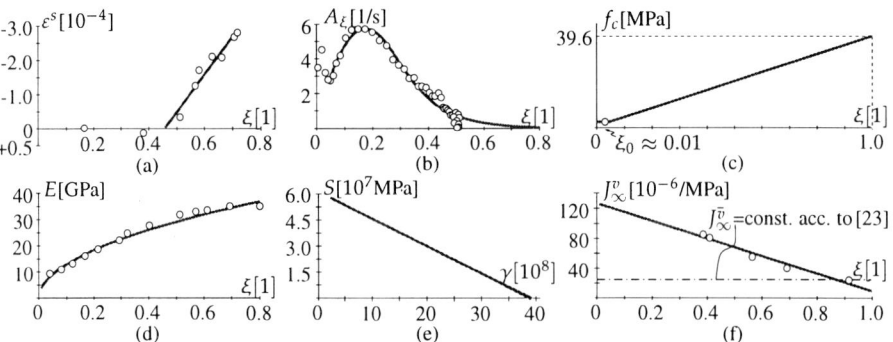

Figure 1. Intrinsic material functions for shotcrete: (a) chemical shrinkage, (b) chemical affinity, (c) strength growth, (d) aging elasticity, (e) relaxation of microprestress force, and (f) asymptotic compliance for short-term creep

— The hardening force vector ζ (representing strengths) is the thermodynamic force associated with the hardening variables χ. According to the theory of chemoplasticity [27] [13], $\zeta = \zeta(\chi, \xi)$. Figure 1(c) depicts the (quasi)linear strength growth of concrete/shotcrete [20]. For the Lafarge mixture, $f_{c,\infty} = 39.6$ MPa.

— The macroscopic stresses σ are the thermodynamic forces associated with the plastic strains ε^p and the flow strains ε^f. The state equation for the stresses in hydrating shotcrete reads

$$d\sigma = \mathbf{C}(\xi) : (d\varepsilon - d\varepsilon^p - d\varepsilon^f - d\varepsilon^v - d\varepsilon^s - d\varepsilon^T),\qquad(1)$$

with $\mathbf{C}(\xi)$ denoting the isotropic (aging) elasticity tensor, see Figure 1(d) for Young's modulus $E(\xi)$ of the Lafarge shotcrete. In contrast to the state equations considered so far, (1) constitutes an (infinitesimally) *incremental* relation. The term in parentheses on the right-hand side of (1) represents the increment of elastic (instantaneously recoverable) strains, $d\varepsilon^e$. Formulation (1) accounts for the fact that hydrates are formed (quasi-instantaneously) in a state *free of (micro)stress*, as was pointed out by Bažant [1]. Hence, each hydrate contributes to the macroscopic elastic stiffness only with respect to increments of macrostress occurring *after* the (quasi-instantaneous) formation of the hydrate.

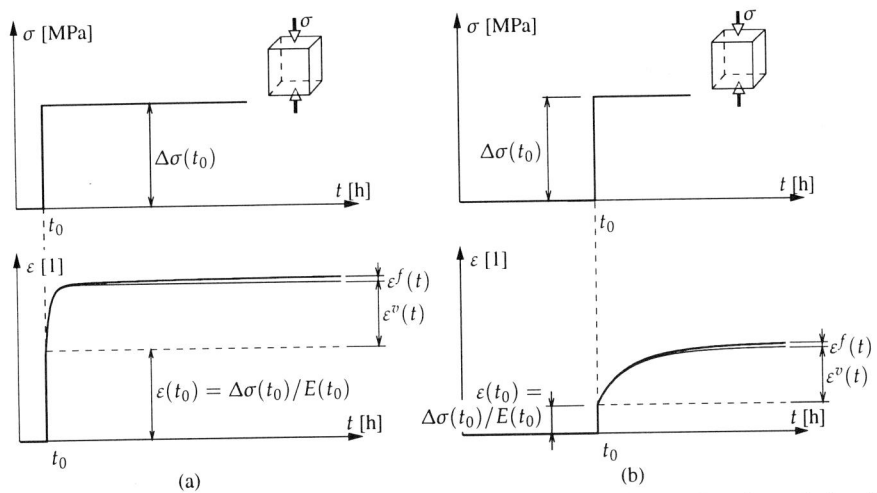

Figure 2. Classical (compressive) creep test for concrete: Stress history and strain evolution for (a) early and (b) late loading

— A_w is the thermodynamic force associated with w, provoking short-term creep. According to [24] [23], A_w practically depends only on σ, w, and ξ. In these references, a total formulation for the respective state equation was chosen,

$$A_w(t) = \frac{1}{f[\xi(t)]}\mathbf{g} : \boldsymbol{\sigma}(t) - \kappa w(t),\qquad(2)$$

with the constant of equilibrium κ and the scalar solidification function $f(\xi)$. Experimental evidence (see, e.g., [17] and the evaluation in [23]) has shown that in a classical 1D creep test (see Figure 2) the asymptotic creep strain $\varepsilon_\infty^v = gw_\infty$, $\varepsilon_\infty^v = w_\infty$ depends on the time instant of loading, t_0, and, hence, on the degree of hydration $\xi(t_0)$, reached then. As far as the total formulation (2) is concerned, for the end of short-term creep, when $A_w(t = \infty) = A_{w,\infty} = 0$, it follows that

$$w_\infty = \frac{1}{f(\xi_\infty)\kappa}\mathbf{g} : \boldsymbol{\sigma}_\infty \leftrightarrow \varepsilon_\infty^v = \frac{1}{f(\xi_\infty)\kappa}\mathbf{g}\otimes\mathbf{g} : \boldsymbol{\sigma}_\infty = \frac{1}{f(\xi_\infty)\kappa}\mathbf{G} : \boldsymbol{\sigma}_\infty,\qquad(3)$$

where $\mathbf{G} = \mathbf{g}\otimes\mathbf{g} = \mathbf{C}^{-1}E$ is a (normalized) fourth-order compliance tensor. According to (3), the asymptotic creep strain would only depend on the asymptotic value of ξ, which is in contradiction to the aforementioned experimental evidence. This difficulty can be overcome by formulating an incremental state equation, reading

$$dA_w(t',t) = \frac{1}{f[\xi(t')]}\mathbf{g} : d\boldsymbol{\sigma}(t') - \kappa dw(t',t),\qquad(4)$$

describing the evolution of the driving force dA_w for $t > t'$ resulting from a change in stress $d\sigma(t')$, which occurred at time instant t'. As usual in the context of so-called Stieltjes integrals, $d\sigma(t')$ may be an infinitesimal or even a finite value. The latter is true for the classical creep test sketched in Figure 2. $dw(t, t')$ describes the movement of water resulting from the stress change at t', in the course of time $t > t'$. The current value of the affinity driving the microdiffusion process is given by the Stieltjes integral

$$A_w(t) = \int_{t'=0}^{t} dA_w(t', t) = \int_{t'=0}^{t} \frac{1}{f[\xi(t')]} \mathbf{g} : d\sigma(t') - \kappa w(t). \quad (5)$$

Evaluation of (5) for the end of the short-term creep process, $A_{w,\infty} = 0$, gives

$$w_\infty = \frac{1}{\kappa} \int_{t'=0}^{\infty} \frac{1}{f[\xi(t')]} \mathbf{g} : d\sigma(t') \rightarrow \varepsilon_\infty^v = \frac{1}{\kappa} \mathbf{G} : \int_{t'=0}^{\infty} \frac{1}{f[\xi(t')]} d\sigma(t'). \quad (6)$$

For a classical 1D creep test, see Figure 2, the integral in (6) is simplified to

$$\int_{t'=0}^{\infty} \frac{1}{f[\xi(t')]} \mathbf{g} : d\sigma(t') = \frac{1}{f[\xi(t_0)]} \Delta\sigma(t_0). \quad (7)$$

Thus, for the 1D creep test, (6) is simplified to

$$w_\infty = \varepsilon_\infty^v = \frac{1}{\kappa f[\xi(t_0)]} \Delta\sigma(t_0). \quad (8)$$

In fact, use of (4) yields the final creep strains ε_∞^v dependent on the degree of hydration at the time instant of loading, as is evident from experiments. For the evaluation of a classical 1D creep test, Figure 2, the introduction of the compliance $J = \varepsilon/\Delta\sigma$ is useful. Because of (8), the final viscous compliance $J_\infty^v = \varepsilon_\infty^v/\Delta\sigma$ is given as

$$J_\infty^v[\xi(t_0)] = \frac{1}{\kappa f[\xi(t_0)]}. \quad (9)$$

— The microprestress force S is the thermodynamic force associated with the viscous slip γ [4] [24]. According to [23] it practically depends only on the viscous slip, $S = S(\gamma)$. Figure 1(e) depicts this state equation for the aforementioned shotcrete mixture, determined in [23] [8].

2.3. DISSIPATIVE PROCESSES - EVOLUTION LAWS

The evolution of the dissipative processes (hydration, microcracking, dislocation-like sliding, microdiffusion of water) is described by means of evolution laws, relating the rate of the respective state variable to the energetically conjugate thermodynamic force.

— Hydration is a thermo-activated process. Hence, an evolution law of the Arrhenius type is chosen [26]

$$\dot{\xi} = A_\xi(\xi) \exp\left(-\frac{E_a}{RT}\right),$$ (10)

with E_a denoting the activation energy of cement and R standing for the universal constant of ideal gas; $E_a/R \approx 4000$ K, see, e.g., [2].

— Microcracking is modelled within the framework of multisurface chemoplasticity [13]. In the stress space, all admissible stress states lie within the elastic domain,

$$\sigma \in C_E \leftrightarrow f_\alpha = f_\alpha(\sigma, \zeta(\chi, \xi)) \leq 0 \ \forall \alpha \in [1, 2, ..., N].$$ (11)

N is the number of loading functions f_α. For the sake of simplicity, the evolution laws for the (instantaneously occurring) plastic strains ε^p (the flow rule) as well as for χ (the hardening rule) follow associated multisurface plasticity [15],

$$d\varepsilon^p = \sum_{\alpha \in J_{act}} d\lambda_\alpha \frac{\partial f_\alpha}{\partial \sigma}, \ d\chi = \sum_{\alpha \in J_{act}} d\lambda_\alpha \frac{\partial f_\alpha}{\partial \zeta},$$ (12)

where $d\lambda_\alpha$ are the plastic multipliers; J_{act} denotes the set of n active yield surfaces, $n \leq N$. It is defined by $J_{act} := \{\alpha \in 1, 2, ..., N | f_\alpha(\sigma, \zeta) = 0\}$. The occurrence of microcracking is controlled by the so-called Kuhn-Tucker conditions:

$$f_\alpha \leq 0, \ d\lambda_\alpha \geq 0, \ f_\alpha d\lambda_\alpha = 0,$$ (13)

together with the consistency requirement

$$d\lambda_\alpha df_\alpha = 0.$$ (14)

Figure 3 depicts two multisurface models used for the simulations described in Section 3. In both models, a Drucker-Prager surface is used for shotcrete under uniaxial and biaxial compression. A tension-cut-off and three Rankine surfaces, respectively, are modelling the tensile behavior.

— For the description of the microdiffusion process leading to short-term creep, a linear evolution law is adopted,

$$\dot{w} = \frac{1}{\eta_w} A_w,$$ (15)

with η_w as the viscosity related to microdiffusion. Inserting (5) into (15) and accounting for $\varepsilon^v = gw$ and for (9), yields a viscoelastic law accounting for

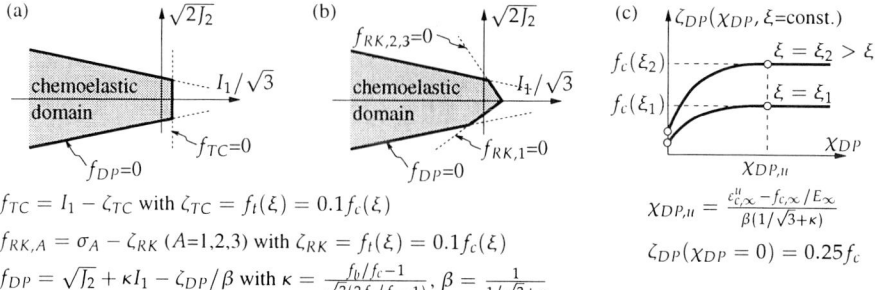

$f_{TC} = I_1 - \zeta_{TC}$ with $\zeta_{TC} = f_t(\xi) = 0.1 f_c(\xi)$

$f_{RK,A} = \sigma_A - \zeta_{RK}$ (A=1,2,3) with $\zeta_{RK} = f_t(\xi) = 0.1 f_c(\xi)$

$f_{DP} = \sqrt{J_2} + \kappa I_1 - \zeta_{DP}/\beta$ with $\kappa = \frac{f_b/f_c - 1}{\sqrt{3}(2f_b/f_c - 1)}$, $\beta = \frac{1}{1/\sqrt{3}+\kappa}$

$\chi_{DP,u} = \frac{\varepsilon_{c,\infty}^u - f_{c,\infty}/E_\infty}{\beta(1/\sqrt{3}+\kappa)}$

$\zeta_{DP}(\chi_{DP} = 0) = 0.25 f_c$

Figure 3. Multisurface plasticity for representation of microcracking: Drucker-Prager (DP) surface with (a) tension-cut-off (TC) and (b) three Rankine (RK) surfaces (I_1, J_2: invariants of σ); (c) chemoplastic hardening (f_c: uniaxial compressive strength; f_b: biaxial compressive strength; f_t: uniaxial tensile strength; $\varepsilon_{c,\infty}^u = 0.0022$)

aging,

$$\dot{\varepsilon}^v = \frac{1}{\eta_w} \left[\int_{t'=0}^t \frac{1}{f[\xi(t')]} \mathbf{G} : d\sigma(t') - \kappa \varepsilon^v \right] =$$
$$= \frac{1}{\tau_w} \left[\int_{t'=0}^t J_\infty^v[\xi(t')] \mathbf{G} : d\sigma(t') - \varepsilon^v \right], \qquad (16)$$

with $\tau_w = \eta_w/\kappa$ as the characteristic time of the short-term creep process. According to [23], $\tau_w = \tau_{w,\infty}\xi$, for the shotcrete of Lafarge, $\tau_{w,\infty} \approx 24$ h, [8]. For the same shotcrete, $J_\infty^v(\xi)$ is depicted in Figure 1(f), [18]. The evolution law for ε^v given in [23] can be interpreted as an approximation of (16) with a constant $J_\infty^{\bar{v}}$, see Figure 1(f),

$$\dot{\varepsilon}^v = \frac{1}{\tau_w} [J_\infty^{\bar{v}} \mathbf{G} : \sigma - \varepsilon^v]. \qquad (17)$$

— According to [4], the evolution of the long-term (flow) creep strains is proportional to the *total stresses* σ,

$$\dot{\varepsilon}^f = \frac{1}{\eta_f} \mathbf{G} : \sigma. \qquad (18)$$

Long-term creep is a thermo-activated process [28], which affects the flow viscosity,

$$\frac{1}{\eta_f} = cS(\gamma) \exp\left(-\frac{2U}{RT}\right), \qquad (19)$$

with $U/R = 2700$ K [3], and the evolution law for the viscous slip γ,

$$\dot{\gamma} = cS^2(\gamma) \exp\left(-\frac{U}{RT}\right), \qquad (20)$$

where $c = 1$ MPa^{-2}s^{-1}.

Generally, attention should be paid to the following: Aging elasticity and short-term creep, which both are related to the accumulation of *initially (micro-)stress-free* hydrates (solidification process), are described by relations containing *increments of (macro-)stress*, namely (1) and (16). In contrast, microcracking and long-term creep, which are related to microstructural changes within the *(micro-)stressed* hydrates, are described by relations containing *total (macro-)stress* values, namely (11), (12), and (18).

2.4. PHYSICAL SPACE - FIELD EQUATIONS

The physical space is described by means of two field equations. The first one is the equilibrium equation,

$$\text{div}\,\sigma + \mathbf{k} = \mathbf{0}, \tag{21}$$

with $\mathbf{k}$ denoting the vector of the volume-force density. The second field equation is the first law of thermodynamics. Accounting only for non-negligible terms, this law reads as [26]

$$C\dot{T} - f_\xi \dot{\xi} = -\text{div}\mathbf{q}, \tag{22}$$

with C as the specific heat capacity, f_ξ as the latent heat of hydration per unit volume of concrete, and $\mathbf{q}$ as the heat flow vector (positive for an efflux). Eqn. (22) states that the change of internal energy, $C\dot{T} - f_\xi\dot{\xi}$, equals the external heat supply, $-\text{div}\mathbf{q}$.

The algorithmic treatment of the governing equations (1) to (22) are dealt with in [13] [12] [23] [16].

3. Hybrid structural analyses of tunnels

In [9] [8] a hybrid method was presented, combining 3D *in-situ* displacement measurements with advanced material modelling of shotcrete in the framework of 3D thermochemomechanical Finite Element analyses:

A ring with a width of 1 m is fictitiously cut out of the tunnel shell and modelled by 3D Finite Elements, Figure 4(a). Displacement fields are prescribed as

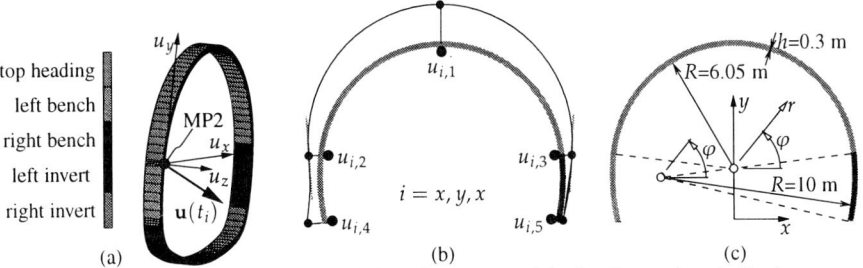

Figure 4.   Hybrid method: (a) 3D Finite Element model, (b) interpolated displacements as boundary conditions, (c) cross-section of Sieberg tunnel, Lower Austria

boundary conditions at all surfaces at which forces are introduced into the investigated part of the tunnel. These displacement fields are approximated by means of temporal and spatial interpolation of measurements, performed at discrete points (MP) on the interior surface of the tunnel shell, Figure 4(a) and (b).

This method delivers spatial fields of stresses as the main result. From these stresses, levels of loading $\mathcal{L}$, amounting to 0% for the structure without loads and to 100% when the (compressive) strength is reached, can be evaluated. For the Drucker-Prager surface depicted in Figure 3,

$$\mathcal{L} = \frac{\sqrt{J_2} + \kappa I_1}{\zeta_{DP}/\beta}. \tag{23}$$

Herein, results for the Sieberg tunnel in Lower Austria are presented, see Figure 4(c). For further details on material parameters and structural modelling, see [9] [8]. At the measurement cross-section 1452 (km 156.990), this tunnel practically does not undergo any bending [9].

The deformational state in this tunnel is illustrated in Figure 5 by means of the circumferential stretches $\epsilon_\varphi = 1/h \int_h \varepsilon_\varphi dr$ and the longitudinal stretches $\epsilon_z = 1/h \int_h \varepsilon_z dr$ at the top of the cross-section ($\varphi = 90°$) and at the medium part of the left bench ($\varphi = 182.6°$). At the top, compressive circumferential stretches

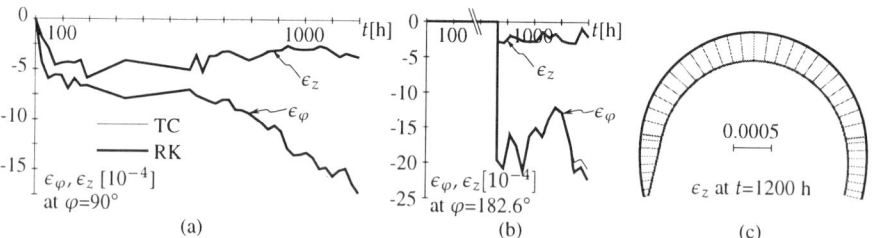

Figure 5.   Stretches at km 156.990 of Sieberg tunnel

prevail. Remarkably, they are *not* monotonically increasing. In the left bench, the compressive circumferential stretches are reduced by approximately 50%, see Figure 5(b). The longitudinal stretches lie within the compressive regime. They are significantly smaller than ϵ_φ.

Herein, the focus is on the significance of the incremental formulation for short-term creep, (16), in comparison to the total one, (17), as well as that of the Rankine plasticity model, Figure 3(b), in comparison to the tension-cut-off, Figure 3(a), for structural analyses. At early stages of the hydration, the larger short-term creep compliance resulting from (16) as compared to (17), see also Figure 1(f), leads to smaller compressive axial forces $n_\varphi = \int_h \sigma_\varphi dr$, see Figure 6(a). In the longitudinal direction, chemical shrinkage, Figure 1(a), leads to tensile longitudinal axial forces $n_z = \int_h \sigma_z dr$ [9], see Figure 7. These forces are overestimated by the tension-cut-off, whereas the Rankine model delivers the maximal bearable

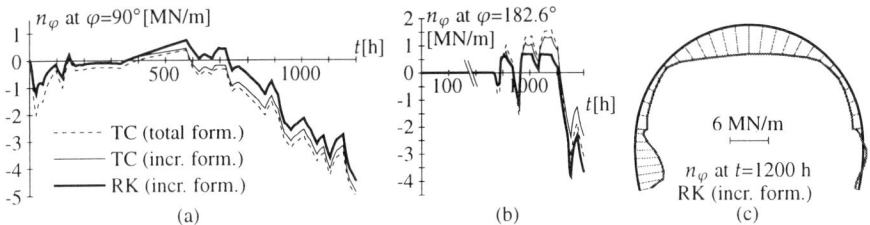

Figure 6.   Circumferential axial forces at km 156.990 of Sieberg tunnel: (a)(b) Influence of tension-cut-off (TC) versus Rankine (RK) plasticity model, and of total formulation for short-term creep (17) versus incremental one (16); (c) spatial distribution

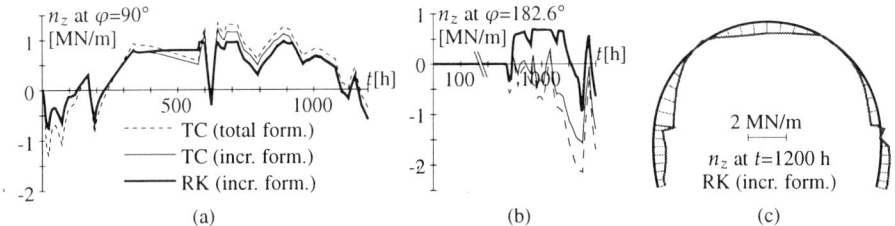

Figure 7.   Longitudinal axial forces at km 156.990 of Sieberg tunnel: (a)(b) Influence of tension-cut-off (TC) versus Rankine (RK) plasticity model, and of total formulation for short-term creep (17) versus incremental one (16); (c) spatial distribution

load $n_z \approx f_{t,\infty} h = 3 \cdot 0.3 = 0.9$ MN/m, see Figure 7(a). In consequence of the dramatic reduction of the circumferential compressive stretches, tensile forces n_φ are occurring at the left bench (Figure 6(b)), resulting in circumferential plastic tensile stretches ϵ_φ^p. As for the associated tension-cut-off plasticity model, see (12) and Figure 3(a), *volumetric* tensile plastic strains result in tensile axial forces n_φ which are by far too large (Figure 6(b)) and in *compressive* axial forces n_z (Figure 7(b)), which are physically meaningless. In contrast, the Rankine model delivers plausible results. Regarding the maximal value of $\bar{\mathcal{L}} = 1/h \int_h \mathcal{L} dr$ in the tunnel cross section, Figure 8, both the incremental formulation (16) and the Rankine model lead to a reduction of the level of loading, allowing a more economic design of the tunnel shell.

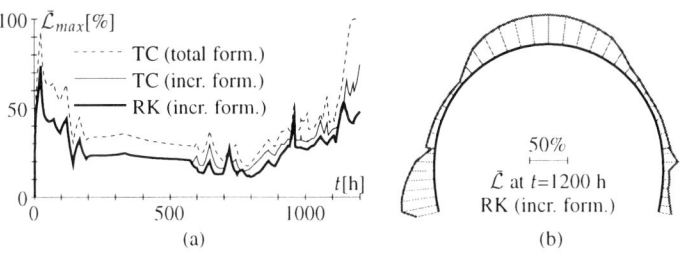

Figure 8.   Level of loading of Sieberg tunnel shell at km 156.990 (a) Influence of tension-cut-off (TC) versus Rankine (RK) plasticity model, and of total formulation for short-term creep (17) versus incremental one (16); (b) spatial distribution

References

1. Bažant, Z.: 1979, 'Thermodynamics of solidifying or melting viscoelastic material'. *Journal of the Engineering Mechanics Division, ASCE* **105**(6), 933–952.
2. Bažant, Z. (ed.): 1988, *Mathematical modelling of creep and shrinkage in concrete*. Chichester, England: Wiley.
3. Bažant, Z.: 1995, 'Creep and damage in concrete'. In: J. Skalnet and S. Mindess (eds.): *Materials Science of Concrete*. pp. 335–389.
4. Bažant, Z., A. Hauggard, S. Baweja, and F.-J. Ulm: 1997, 'Microprestress solidification theory for concrete creep, Part I: Aging and drying effects'. *Journal of Engineering Mechanics, ASCE* **123**(11), 1188–1194.
5. Bažant, Z. and F. Wittmann (eds.): 1982, *Creep and shrinkage of concrete structures*. Chichester, England: Wiley.
6. Carol, I. and Z. Bažant: 1993, 'Viscoelasticity with aging caused by solidification of nonaging constituent'. *Journal of Engineering Mechanics, ASCE* **119**(11), 2252–2269.
7. Coussy, O.: 1995, *Mechanics of porous continua*. Chichester, England: Wiley.
8. Hellmich, C.: 1999, 'Shotcrete as part of the New Austrian Tunneling Method: from thermochemomechanical material modeling to structural analysis and safety assessment of tunnels'. Ph.D. thesis, Vienna University of Technology, Vienna, Austria.
9. Hellmich, C., H. Mang, and F.-J. Ulm: 1999a, 'Hybrid Method for Quantification of Stress States in Shotcrete Tunnel Shells: combination of 3D *in-situ* displacement measurements and thermochemoplastic material law'. In: W. Wunderlich (ed.): *CD-ROM Proceedings of the European Conference of Computational Mechanics*. Munich, Germany.
10. Hellmich, C. and H. A. Mang: 1999, 'Influence of the dilatation of soil and shotcrete on the load bearing behavior of NATM-tunnels'. *Felsbau, Rock and Soil Engineering* **17**(1), 35–43.
11. Hellmich, C., F.-J. Ulm, and H. A. Mang: 1997, 'Chemoplasticity for shotcrete at early ages'. In: D. Owen, E. Oñate, and E. Hinton (eds.): *Computational Plasticity, Proceedings of the 5th International Conference*. Barcelona, Spain, pp. 1499–1507.
12. Hellmich, C., F.-J. Ulm, and H. A. Mang: 1999b, 'Consistent linearization in Finite Element analysis of coupled chemo-thermal problems with exo- or endothermal reactions'. *Computational Mechanics* **24**(4), 238 – 244.
13. Hellmich, C., F.-J. Ulm, and H. A. Mang: 1999c, 'Multisurface Chemoplasticity I: Material model for shotcrete'. *Journal of Engineering Mechanics (ASCE)* **125**(6), 692–701.
14. Huber, H.: 1991, 'Untersuchungen zum Verformungsverhalten von jungem Spritzbeton im Tunnelbau [Investigations concerning the deformation behavior of young shotcrete in tunneling]'. Master's thesis, University of Innsbruck, Innsbruck, Austria. In German.
15. Koiter, W.: 1960, *General theorems for elastic-plastic solids*, Vol. I, Chapt. IV, pp. 167–218. Amsterdam, The Netherlands: North-Holland Publishing Company.
16. Lackner, R., C. Hellmich, and H. Mang: 2000, 'Numerical treatment of multisurface chemoplasticity with special emphasis on brittle material failure and creep'. *International Journal for Numerical Methods in Engineering*. Submitted for publication.
17. Laplante, P.: 1993, 'Propriétés mécaniques des bétons durcissants: analyse comparée des bétons classiques et à très hautes performances [Mechanical properties of hardening concrete: a comparative analysis of ordinary and high performance concretes]'. Ph.D. thesis, Ecole Nationale des Ponts et Chaussées, Paris, France. In French.
18. Lechner, M.: 2000, 'Kurzzeitkriechen von Spritzbeton - thermochemoplastische Materialmodellierung und nichtlineare Analysen eines Laborversuchs sowie eines NÖT-Tunnelvortriebs mittels Finiten Elementen [Short-term creep of shotcrete - thermochemoplastic material modelling and nonlinear Finite Element analyses of a laboratory test and of a tunnel driven according to the NATM]'. Master's thesis, Vienna University of Technology, Vienna, Austria. In German.

19. Meschke, G.: 1996, 'Consideration of aging of shotcrete in the context of a 3D viscoplastic material model'. *International Journal for Numerical Methods in Engineering* **39**, 3123–3143.

20. Mindess, S., J. Young, and F.-J. Lawrence: 1978, 'Creep and drying shrinkage of calcium silicate pastes. I: specimen preparation and mechanical properties'. *Cement and Concrete Research* **8**, 591–600.

21. Neville, A.: 1981, *Properties of Concrete*. London, England: Pitman Publishing, Third edition.

22. Ruetz, W.: 1966, 'Das Kriechen des Zementsteins im Beton und seine Beeinflussung durch gleichzeitiges Schwinden [Creep of cement in concrete and the influence on it by simultaneous shrinkage]'. *Deutscher Ausschuss Stahlbeton* **Heft 183**. In German.

23. Sercombe, J., C. Hellmich, F.-J. Ulm, and H. A. Mang: 2000, 'Modeling of early-age creep of shotcrete. I: model and model parameters'. *Journal of Engineering Mechanics (ASCE)* **126**(3), 284–291.

24. Ulm, F.-J.: 1998, 'Couplages thermochémomécaniques dans les bétons : un premier bilan. [Thermochemomechanical couplings in concretes : a first review]'. Technical report, Laboratoires des Ponts et Chaussées, Paris, France. In French.

25. Ulm, F.-J. and P. Acker: 1998, 'Le point sur le fluage et la recouvrance des bétons [Concrete creep and recovery: a review]'. *Special issue of the "Bulletin des Laboratoires des Ponts et Chaussées"* **XX**, 73–82. In French.

26. Ulm, F.-J. and O. Coussy: 1995, 'Modeling of Thermochemomechanical Couplings of Concrete at early ages'. *Journal of Engineering Mechanics (ASCE)* **121**(7), 785–794.

27. Ulm, F.-J. and O. Coussy: 1996, 'Strength Growth as Chemo-Plastic Hardening in Early Age Concrete'. *Journal of Engineering Mechanics (ASCE)* **122**(12), 1123–1132.

28. Wittmann, F.: 1982, *Creep and shrinkage mechanisms*, Chapt. 6, pp. 129–161. in [5].

RUPTURE LIFE TIME PREDICTION AND DEFORMATION MECHANISMS DURING CREEP OF SINGLE-CRYSTAL NICKEL-BASE SUPERALLOYS

A. EPISHIN[1], E. KABLOV[1], E. GOLUBOVSKIY[1], I. SVETLOV[1],
T. LINK[2], U. BRÜCKNER[3], P. PORTELLA[3]

[1]*Russian Institute of Aviation Materials, 107005 Moscow, Russia*
[2]*Technische Universität Berlin, 10623 Berlin, Germany*
[3]*Bundesanstalt für Materialforschung und –prüfung, 12205 Berlin, Germany*

Abstract

Mechanisms of creep deformation of single-crystal superalloys at low and high temperature are considered. Rupture life time of a single-crystal superalloy during creep is described in a wide temperature-stress range as a function of temperature and stress taking into account a change of creep mechanism with rising temperature. This phenomenological description is confirmed by creep tests. Anisotropy of creep behavior of single-crystal superalloys is discussed with respect to a choice of the optimal crystallographic direction for solidification of the turbine blades.

1. Introduction

Blades of modern aviation engines and stationary gas turbines are usually directionally solidified as the [001] orientated single-crystal (SC) castings of nickel-base superalloys [1] possessing excellent mechanical properties at elevated temperatures. During operation the blades are subjected to different thermo-mechanical stresses including a stress due to a centrifugal force which causes creep deformation of the material along the blade axis. Therefore in designing of the blade geometry one needs an approach to estimate the blade creep strength. That makes the task of prediction of rupture life time (RLT) of SC superalloys during creep deformation to be of immediate interest. RLT has to be described as a function of temperature T and stress σ in a wide temperature-stress range because the parts of the turbine blade work at different temperatures and stress levels. In the present work authors consider mechanisms of creep of SC superalloys at different temperatures and propose a mathematical formalism to describe RLT. Creep tests confirm the phenomenological model. The problem of the optimal crystallographic orientation of turbine blades is discussed using results of creep tests of SCs grown in different orientations.

S. Murakami and N. Ohno (eds.), IUTAM Symposium on Creep in Structures, 231–240.

1. Superalloy Microstructure

The microstructure of SC nickel-base superalloys consists of a γ-matrix (solid solution of nickel), hardened by small γ'-precipitates (ordered phase on base of Ni_3Al). After standard heat treatment the precipitates have the shape of cuboids with an average size H of about 0.3-0.5 μm, faced by {001} habit planes and separated by the thin γ-layers with a thickness of about $H/10$ (Fig. 1a).

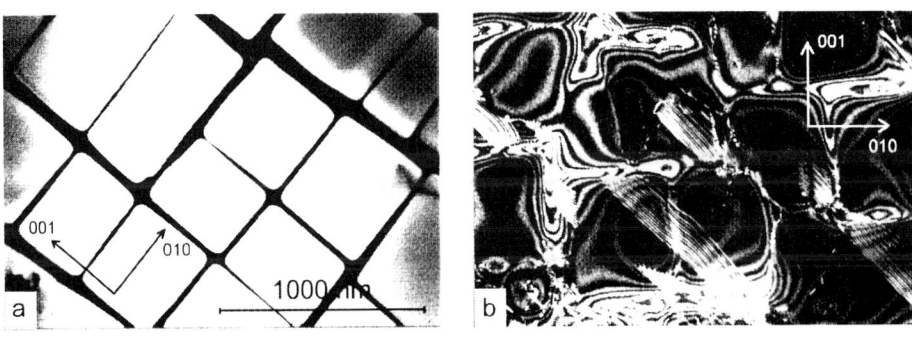

Fig. 1 Microstructure of superalloy SRR99, standard heat treated undeformed (a) and crept at LT (b).

The volume fraction of the γ'-phase is high ≈70%. The unconstrained misfit $δ=2 (a_{γ'} - a_γ)/(a_{γ'} + a_γ)$, where $a_γ$ and $a_{γ'}$ are the lattice spacings of γ- and γ'-phases respectively, is usually about –0.1- 0.2% at RT. It decreases up to –0.2 - –0.3% at temperatures ~1000°C due to the higher coefficient of thermal expansion of the γ-phase. In undeformed condition the γ/γ'-interface is coherent and the difference of the lattice spacings is compensated by the elastic strains. These coherency strains play an important role in the structural and mechanical behaviour of superalloys. For example, in undeformed material they determine the cuboidal shape of the γ'-precipitates and the precipitate alignment along the crystallographic directions <001>. The cuboidal γ/γ'-microstructure is thermally stable up to a critical temperature level of about 900-950°C. Above it, the microstructure coarsens and evolves. During mechanical testing it can undergo morphological transformations, for example, the so-called "rafting" [2]. Mechanisms of creep deformation at temperatures below and above of this critical level significantly differ. Low temperatures (LT) 700-900°C at which the microstructure is stable are typical for working conditions of the blade root and colder parts of the blade tip. High temperatures (HT) 950-1150°C corresponding to the hottest parts of the blade tip activate transformations of the microstructure.

3. Creep Mechanisms

Mechanisms of LT creep are quite well investigated [e.g. 3]. TEM analysis and X-ray diffraction of specimens crept at temperatures 700-900°C show, that creep deformation of superalloy SCs can be realised by dislocation gliding in different slip systems. This gliding can be multiple or single slip, depending on the specimen orientation. LT creep of SCs with orientation near [001] starts by a short incubation period, when $a/2<011>$ matrix dislocations multiply and move on $\{111\}$ planes in the layers between the cuboids. During primary creep they cut the γ'-phase as super-Shockley dislocations by single slip of the $a/3<112>\{111\}$ glide system with the highest Schmidt factor (Fig. 1b). This rotates the specimen axis in multiple slip orientation and the creep rate slows down to the secondary creep. Cutting of the cuboids has the character of viscous shearing therefore creep deformation at LT is a time-dependent process.

At HT the superalloy microstructure gets thermally unstable and undergoes morphological transformations. During primary creep of the [001] orientated SCs the whole γ-phase concentrates in the (001) matrix layers normal (N) to the load direction [001], while the (100) and (010) matrix layers parallel (P) to this direction become thinner and finally disappear. This coalescence of the γ'-cuboids leads to a plate-like γ/γ'-microstructure, called „raft-structure". Further deformation causes a degeneration of the structure: it coarsens and distorts. The HT creep curve of the SC superalloy SRR99 and steps of the evolution of its γ/γ'-microstructure are shown in Fig. 2.

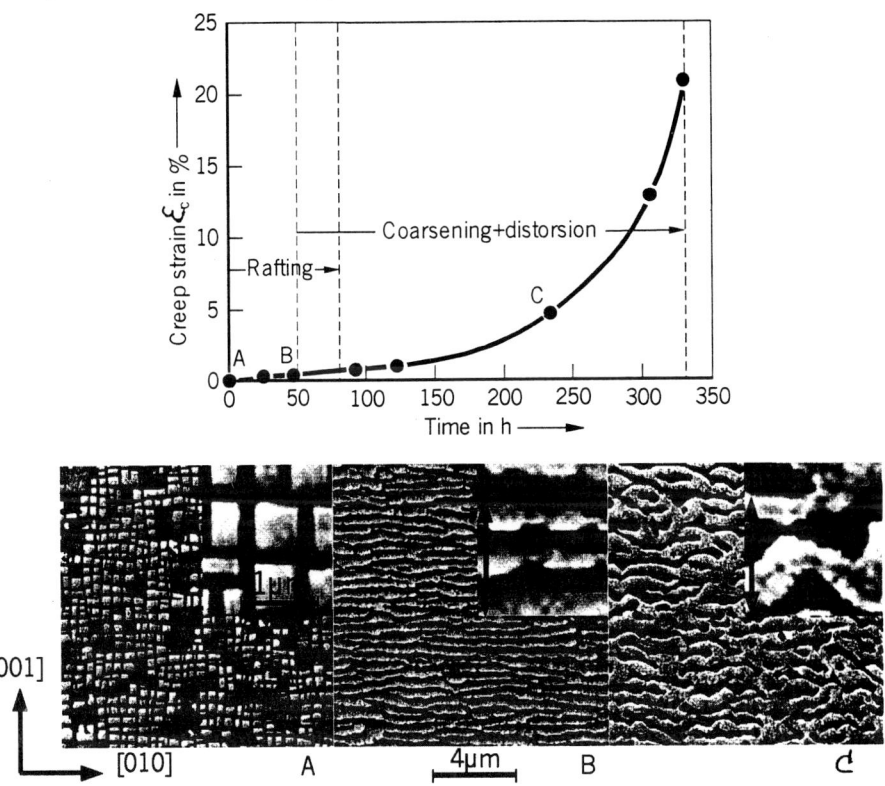

Fig. 2. Creep curve of SRR99, tested until rupture at 980°C and 200 MPa (above) and steps of evolution of its microstructure (below) at the initial state (A), after 45.4 h (B) and 243 h (C).

In this curve one can distinguish a short transient creep with an accumulated creep strain $\Delta\varepsilon_c < 0.3\%$, a quasi-stationary creep with a duration of about 150 h and an accumulated strain $\Delta\varepsilon_c \sim 2\%$ and the final accelerated creep with a duration of about 200 h and an accumulated strain $\Delta\varepsilon_c \sim 20\%$. It is known from TEM analysis that during primary creep $a/2<011>$ dislocation loops multiply and glide along $\{111\}$ planes in the N–matrix layers. In these layers the resolved shear stress τ_{rss} is the maximal because the applied σ and coherency σ_c stresses contribute to τ_{rss} with the same sign. Stresses in the γ- and γ' -phases before and after loading are shown in Fig. 3.

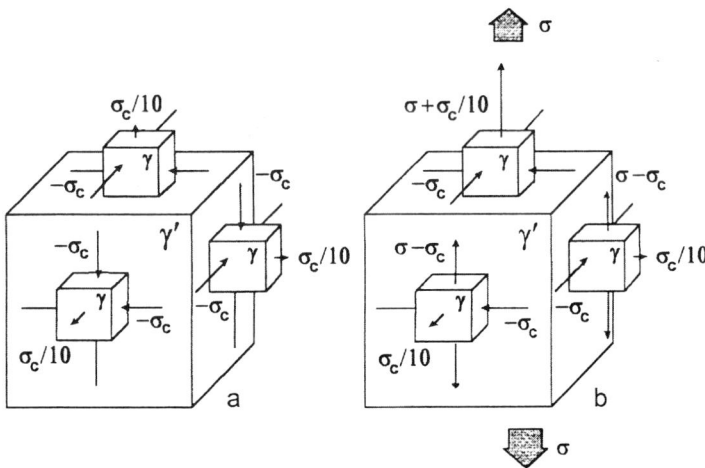

Fig. 3. Stresses in the γ- and γ'-phases before (a) and after loading (b) [4].

Dislocation loops propagating in the matrix layers leave dislocation dipoles in the γ/γ'-interfaces which form the interfacial networks (Fig. 4).

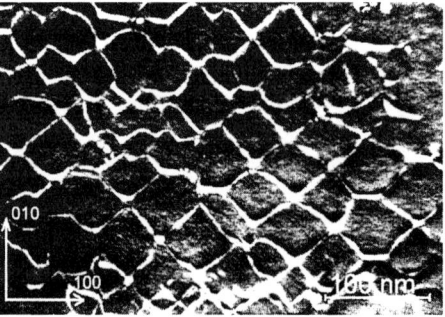

Fig. 4. Dislocation network in superalloy SRR99 crept at 980°C and 200 MPa.

These dislocation networks induce a back stress in the matrix layers which increases with rising network density. At a certain saturation density this back stress together with the Orowan stress compensate the superimposed applied and misfit stress and therefore dislocation gliding stops with the end of primary creep. Quantitative analysis of the misfit of dislocation networks by TEM and XRD showed that their saturation density does not significantly change until rupture [5]. This limited dislocation gliding has two consequences. The first one is a transient creep strain of about ε^{glide} =0.2-0.3% [5]. The second one is relaxation of coherency stresses in the N-layers [6]. The last effect results in a pressure gradient between the N- and P-matrix layers causing cross-diffusion of alloying elements between these layers. This process was investigated by EDX in STEM [7]. An concentration increase of γ'-forming elements and a decrease γ-forming elements in P-matrix layers was found (Fig. 5).

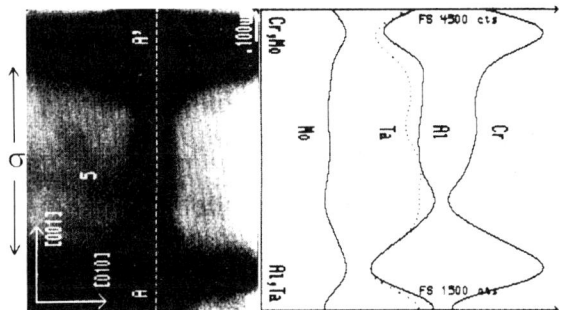

Fig. 5. A change of concentrations of alloying elements in the P-layer during rafting measured along the line A-A' by EDX in STEM.

These concentration changes mean that γ'-forming elements diffuse from N-layers to P-ones while γ-forming elements diffuse in opposite direction. According [8] in the first approximation kinetics of this cross-diffusion can be described by the equation:

$$\left| \vec{j_i} \right| = D_i \frac{C_i}{kT} \frac{p^P - p^N}{H} a_\gamma^3 3\delta^2 \qquad (1)$$

where $\vec{j_i}$ is the flux of atoms of the ith element, D_i and C_i are the diffusion coefficient and equilibrium concentration respectively of the ith element in the matrix, $(p^P - p^N)/H$ is the pressure gradient between the P- and N-layers. Due to this cross-diffusional process the N-faces of the precipitate and the P-matrix layers have a mutual phase inversion γ↔γ' and finally all γ-phase concentrates in N-layers while the γ'-phase forms the (001) orientated rafts (Fig. 6).

Fig. 6. Coalescence of the γ'-precipitates controlled by cross-diffusion (SEM).

Rafting causes some strain of the specimen $\varepsilon^{rafting} = c|\delta|$ but it is very small, because the phase volume fractions do not change during this transformation, i.e. $c \ll 1$ and $\varepsilon^{rafting} \ll |\delta| = 0.2$-$0.3\%$. The mechanism of HT creep of SC superalloys at the quasi-stationary stage is not quite clear. The extended γ'-plates block up conservative movement of dislocations. TEM investigations of the plate-like microstructure show no cutting of the γ'-plates until rupture. One could suppose that further strain accumulated during quasi-stationary creep is mostly attained by climbing of $a/2<011>$-dislocations along the γ/γ'-interfaces perpendicular to the load direction. Such dislocation movement can be caused by a component of the Peach-Koehler force along the interface equal to $\sigma a/2$. It is facilitated by pipe-diffusion through the cores of interfacial dislocations and does require the penetration γ'-plates. During such a climbing dislocations introduce additional atomic planes perpendicular to the load direction which results in both elongation of the specimen and thickening of the γ- and γ'-plates. Precise quantitative analysis of changes of the γ/γ'-microstructure during HT creep by Fourier analysis of SEM images indirectly confirms this assumption [9]. It shows that there is a linear correlation between the accumulated HT creep strain and the increase of the microstructure period measured $\lambda_{[001]}$ (total thickness of γ- and γ'-plates) in the load direction (Fig. 7).

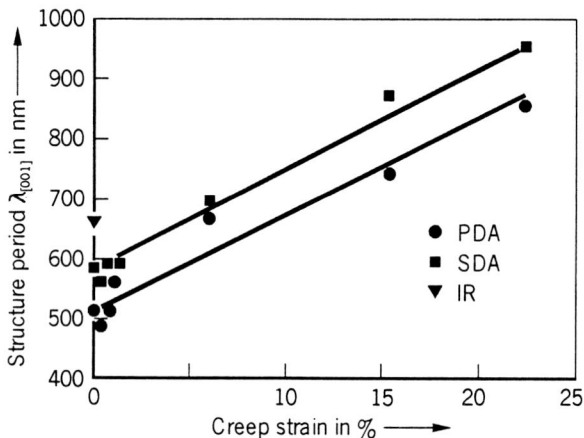

Fig. 7. A change of the structural period $\lambda_{[001]}$ vs. creep strain measured in SRR99 crept at 980°C and 200 MPa.

However a direct confirmation of the above model could be only done by a detail analysis of processes occurring in the γ/γ'-interfaces during HT creep.

4. Prediction of Rupture Life Time

Prediction of RLT means finding of a mathematical expression which describes RLT as a function of temperature T, stress σ and microstructure, i.e. $\tau_R = f(T, \sigma, q_1, q_2, ..., q_n)$, where τ_R is RLT and q_i are structure parameters. Different types of functions $f(T, \sigma, q_1, q_2..., q_n)$ were proposed to describe RLT. Widely used ones are the Larson-Miller's [10], Manson-Hafer's [11] and Dorn-Scherby-Orr's [12] dependencies. It was shown in [13] by analysis of the equations of the physical models describing deformation and rupture during creep, that all these equations can be expressed by one common expression:

$$\tau_R = \xi T^m \sigma^{-n} \exp\left(\frac{U_0 - \gamma\sigma}{RT}\right) \tag{2}$$

where R is the gas constant and ξ, m, n, U_0, γ are temperature dependant experimental parameters.

In [14] this equation was introduced for the characterisation of RLT of a wide spectrum of alloys for high temperature application. It was shown that using equation (2) RLT can be described with high accuracy and predicted for long-term tests with high reliability. Because of the large number of fitting parameters a solution of equation (2) can be unstable. To exclude that, the parameters m and n are chosen by a step method. The parameters ξ, m, n, U_0 and γ are determined by processing of RLT data obtained in different temperature ranges. A change of these parameters with rising temperature is connected with a changing deformation mechanisms.

The phenomenological approach was checked by creep tests. [001] orientated SC castings of superalloy GS40 (a superalloy with a high Mo content) were directionally solidified by the VIAM using a seed technology. Quality of the SC structure was checked by a visual control of the etched specimen surface and by X-ray diffraction. The specimen axis do not deviate from the [001] direction more than at 10°. The casting were standard heat treated and cylindrical specimens with the gage diameter 5 mm and length 25 mm were machined. The specimens were creep tested at different temperatures and stress levels: 800°C, 800-680 MPa; 900°C, 610-350 MPa; 1000°C, 310-190 MPa and 1100°C, 200-90 MPa. For these test parameters the obtained RLT are not longer than 1000 h. The arrays of experimental data were approximated by equation (2) and fitted parameters ξ, m, n, U_0 and γ were determined for different temperature ranges (see Table 1).

TABLE 1. Fitted parameters of equation (2) for SC superalloy GS40

Temperature Range, °C	Number of Processed Points	lnξ	m	n	U_0, KJ/mol	γ, J/(mol·MPa)	$S^2(\ln\tau_R)$
800-900	20	567.2	0	3	567.2	75.4	0.0337
900-1000	63	540.1	0	6	540.1	32.2	0.1080
1000-1100	58	548.4	1	2	548.1	244.3	0.1606

Using the obtained parameters RLT was predicted for low stress levels at temperatures 800 and 1000°C. Control creep tests showed a good agreement of predicted and experimental data, confirming used phenomenological model (Fig. 8).

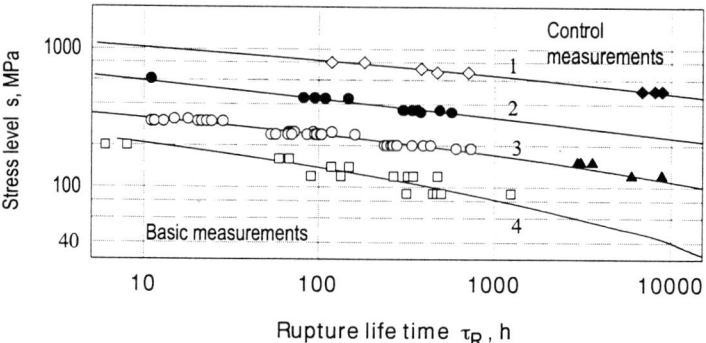

Fig. 8 Prediction of RLT for low stress levels by use of equation (2);
(curves: 1-800°C; 2-900°C; 3-1000°C; 4-1100°C).

5. Creep Anisotropy

SC superalloys show significant anisotropy of their creep behaviour. For uniaxial creep tests their RLT is a function of the axial crystallographic orientation of the specimen. Fig. 9 shows the orientation dependence of RLT for SC specimens of superalloy GS6F at T=760°C and σ=686 MPa.

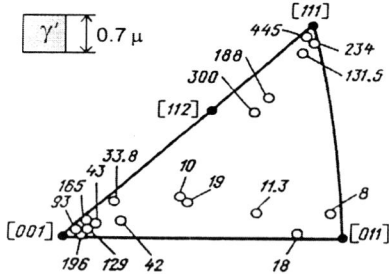

Fig. 9. Orientation dependence of RLT for SC specimens of as cast GS6F at T=760°C and σ=686 MPa.

It is seen in Fig. 9 that RLT strongly changes within the stenographic triangle. The shortest RLTs cumulate in the central part of the triangle. The strongest orientations are [001] and [111]. Anisotropy of creep strength of SC superalloys at LT is determined by the orientation dependence of the Schmidt factor of the slip systems <011>{111} and <112>{111} and by the number of activated slip systems, which defines the intensivity of deformation hardening. The longest RLT in the [001] and [111] corners of the stereographic triangle are caused by the high number of activated slip systems due to the high symmetry of these orientations. While low values of RLT in the central part of the crystallographic triangle are explained by single system gliding without deformation hardening and by high Schmidt factors near the maximum possible value 0.5. At HT the orientation dependence of RLT is determined by the restriction of dislocation mobility to N-layers caused by rafting. For [001] specimens the rafted microstructure consists of the (001) parallel plates while for [111] ones it is a three dimensional arrangement of the mutually perpendicular γ'-rafts and rods. However in spite of the change of the creep mechanism from LT to HT the character of RLT anisotropy in general remains. But some changes of creep anisotropy are also possible depending on the condition of the microstructure and alloy composition. For example, at T=1000°C and σ=140 MPa the [011] orientated SCs of as cast superalloy GS6F with a large γ'-size of about 0.7 μm show RLT≈370 h. That is essentially longer than RLT≈230 h of the [001] orientated SCs. In practice the turbine blades are usually solidified as [001] SC castings using a simple grain selection technique which allows to grow the [001] orientated crystals. An advantage of such crystals is the minimal thermal stress because of the lowest Young's modulus in the [001] direction. However it is under discussion to manufacture blades with [111] axial orientation by an alternative seed technique and therefore the relative comparison of creep strength of the [001] and [111] orientated SCs is of practical interest. This comparison can be done by use of the ratio $r = \sigma_\tau^{[111]} / \sigma_\tau^{[001]}$, where $\sigma_\tau^{[111]}$ and $\sigma_\tau^{[001]}$ are stresses rupturing the [111] and [001] orientated specimens after a time t. Fig. 10 shows changes of the parameter r for SCs of two superalloys: as cast GS6F and heat treated GS32 (a superalloy with 1 wt. % of Re [1]) as a function of temperature T and RLT.

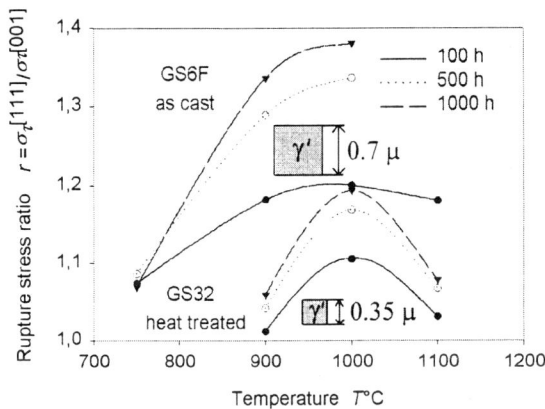

Fig. 10. Changes of the parameter r for SCs of as cast GS6F and heat treated GS32
as a function of temperature T and RLT.

It is seen that for these superalloys r always greater then 1, i.e. the [111] orientated crystals are generally stronger than the [001] ones. The ratio r is higher for as cast GS6F with the large γ'-size of about 0.7 μm and lower for heat treated GS32 with the smaller γ'-size of about 0.35 μm. At LT r increases with rising of temperature, then it has a maximum at about 1000°C and decreases at HT. For the both superalloys at all temperatures r is higher for long-term creep tests. It can be concluded that the [111] orientated SCs are preferable for long-term applications where thermal gradients in the blade are not high, otherwise they will cause strong thermal stresses along the stiff blade axis [111] superimposing with a centrifugal stress. Such working conditions are typical for the energetic gas turbines.

6. Conclusions

1. Mechanisms of creep deformation of SC superalloys differ below and above a critical temperature level of 900-950°C because above this level the superalloy microstructure becomes unstable and evolves.
2. A phenomenological model was proposed to describe the temperature-stress dependence of rupture life time of SC superalloys. Performed creep tests showed high reliability of the predicted values of rupture life time.
3. Analysis of orientation dependence of creep strength of SC superalloys GS6F and GS32 allowed to conclude that in general the [111] orientated SCs are stronger than the [001] ones, especially during long-term creep tests. Therefore [111] orientation can be recommended for long-term applications where thermal gradients in the blade are not high, for example, in the stationary gas turbines.

7. References

1. Kablov, E. N., Svetlov, I. L. and Petrushin, N. V. Nickel-base superalloys for casting of blades with directional and single-crystal structure. Part 1 (1997) *Materials Science Transactions*, **4**, 32-39.
2. Tien, J. K. and Copley, S. M. The effect of uniaxial stress on the periodic morphology of coherent gamma precipitates in nickel-base superalloy crystals (1971) *Met. Trans.*, **2**, 215-219.
3. Link, T. and Feller-Kniepmeier M. Shear mechanismus of the γ′ phase in single crystal superalloys and their relation to creep (1992) *Met. Trans. A*, **23A**, 99-105.
4. Socrate, S. and Parks, D. M. Numerical determination of the elastic force for directional coarsening in Ni-superalloys (1993) *Acta Metall. Mater.*, **41**, 2185-2209.
5. Link, T., Epishin, A., Portella, P. D. and Brückner, U. (2000) Increase of misfit during creep of superalloys and its correlation with deformation, *Acta Mater.*
6. Louchet, F. and Hazotte, A. A model for low stress cross-diffusional creep and directional coarsening of superalloys (1997) *Scripta Mater.*, **37**, 589-597.
7. Svetlov, I. L., Golovko, B. A., Epishin, A. I. and Abalakin, N. P., Diffusinal mechanism of γ′-phase particles coalescence in single crystals of nickel-base superalloys (1992) *Scripta Metall. Mater.*, **26**, 1353-1358.
8. Epishin, A.I., Svetlov, I.L., Brückner, U., Link, T., Portella, P. and Golubovskii, E.R. High temperature creep of single crystals of nickel superalloys with [001] orientation (1999) *Materials Science Transactions*, **5**, 32-42.
9. Brückner, U., Epishin, A. I., Link, T. and Portella, P. Creep induced evolution of internal stresses and structure in a single crystal nickel-base superalloy (in press), *The 5th European conference on residual stresses,* Delft-Noordwijkerhout, the Netherlands, 28-30 September, 1999.
10. Larson, F. R. and Miller, J. A. (1952) Time-temperature relationship or rupture and creep tests, *Trans. ASME,* **74**, 765-775.
11. Manson, S. S. and Haferd, A. M. (1953) A linear time-temperature relation for extrapolation of creep and stress rupture data, *NACA, TN-2890.*
12. Orr, R. L., Scherby, O. D. and Dorn, J. E. (1954) Correlation of rupture data for metals at elevated temperatures, *Trans. ASME,* **46**, 113-128.
13. Trunin, I. I. (1969) Estimation of deformation characteristics of steel during long-term rupture, *Problems of Strength*, N 6, 3-8.
14. Kablov, E. N. and Golubovskiy, E. R. (1998) *High temperature strength of nickel-base alloys,* Mashinostroeniye, Moscow.

CREEP DAMAGE ASSESSMENT AND VOID FORMATION IN ENGINEERING MATERIALS

H.C. FURTADO
CEPEL Centro de Pesquisas de Energia Elétrica
C.P. 2754, 20001-970 Rio de Janeiro, Brazil

I. LE MAY
Metallurgical Consulting Services Ltd.
P.O. Box 5006, Saskatoon,S7K 4E3, Canada

Abstract

The formation of voids at grain boundaries during creep deformation of engineering alloys is discussed. Although methods of assessment of the extent of damage and of remaining life have been developed based on the extent of void formation, it is pointed out that recent work has shown that voids, which are indicated in alloy steels by means of metallographic preparation, may be primarily artifacts produced in the preparation process. The implications of these observations are important for the reliable assessment of remaining life of plants. Alternative approaches to damage assessment are discussed, including the Kachanov-Rabotnov continuum damage accumulation method. Directions for future research to resolve remaining questions concerning the nature of creep damage and its assessment are discussed.

1. Introduction

The evolution of creep damage in metals and alloys has been modelled extensively in terms of the formation and growth of voids at grain boundaries. Early work by Hull and Rimmer (1959) indicated that cavities nucleated and grew by the production and coalescence of vacancies on grain boundaries where rapid surface diffusion, as compared to the much slower matrix diffusion, maintained spherical cavity surfaces. The grains were assumed to act as rigid blocks of material and there was no external restraint preventing grains moving apart on adjacent sides of the boundary. Since that time, models for grain boundary cavitation have been greatly refined, with Cocks and Ashby (1982) developing an expression for the growth of a cavity by power law creep. Constrained cavity growth has been modelled by Dyson (1976), Beere (1981) and others, and is considered to be the probable mode of damage on grain boundaries on CrMo steels; this is of particular relevance for power boilers. Figure 1 shows a schematic of constrained cavity growth.

241

S. Murakami and N. Ohno (eds.), IUTAM Symposium on Creep in Structures, 241-256.

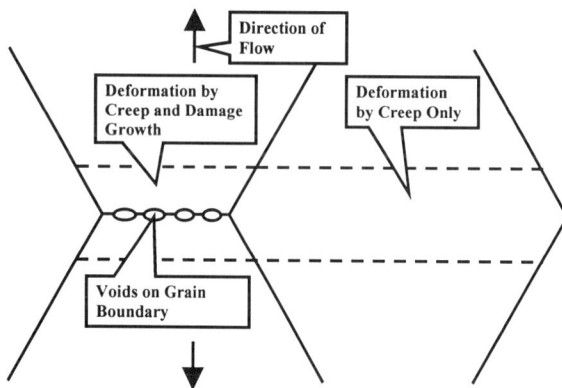

Figure 1. Schematic illustration of constrained cavity growth.

It is also well established that grain boundary voids nucleate at carbides and other particles on grain boundaries. Where there are noncoherent boundaries, these provide ready sites for nucleation, and the cavities can be considered to be present at time zero during creep deformation. Coherent grain boundary carbides can provide the conditions for continuous nucleation of voids during creep life between matrix and precipitate.

Direct observation of creep cavities has been made in specimens and components that have suffered creep damage. Observations on intergranular fracture surfaces have revealed that they are covered with ductile dimples, resulting from cavity growth. In addition, metallographic sections through failed or heavily creep-damaged specimens have shown the presence of voids. Thus, it has seemed reasonable to attempt to assess the extent of creep damage in terms of the degree of grain boundary cavitation present. Consequently, methods for the assessment of creep damage and the remaining life of engineering components have been developed in which metallographic procedures are used to examine the outer surfaces of the components (Neubauer and Wedel, 1983; Auerkari, 1983).

The assessment of creep damage in power plants by metallographic means has, in general, been accomplished through the use of replicas taken from polished and etched areas on the outer surfaces of tubes and headers. These can then be examined in the laboratory, supplementing direct *in situ* observation through a field microscope, and allowing the detection of fine voids that are not readily discernable in the field. The first practical guide to assessment of power plant components was that prepared by Neubauer and Wedel (1983), and this was considered by its authors as an essentially qualitative guide. The assessment model of Neubauer and Wedel is illustrated in Figure 2. Subsequently, other codes have been developed based on metallographic assessment of voids and microcracks, still of a qualitative nature (NORDTEST, 1991; VGB, 1992).

Attempts were made to make the process of damage assessment more quantitative, and from tests by Shammas (1987) and Cane and Shammas (1984) the A-parameter approach was developed to characterize the degree of grain boundary cavitation in terms of the number fraction of grain boundaries that contain voids (A). The values obtained for the A-parameter for Cr-Mo steels in a condition corresponding to that of a heat affected zone were correlated with the degree of damage or the remaining life fraction.

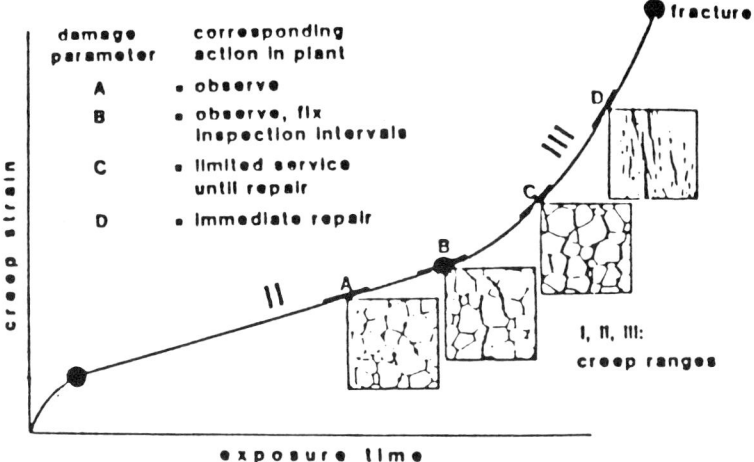

Figure 2. The assessment model of Neubauer and Wedel (1983).

The concept of "damage" was considered by Kachanov (1958). He postulated that materials suffer a loss in strength as a result of exposure to stress, temperature and resulting deformation and the parameter ψ decreases from 1 at the outset to 0 at the time of failure (t_r). Although damage does not need a direct physical model, it can be represented by a loss in cross-sectional area from the formation of voids and cracks on grain boundaries. This causes an increase in the "true" stress σ_t under constant load. The rate of damage accumulation can be considered as a function of the initial stress σ_0 and the damage. For power law creep,

$$d\psi / dt = -A'\left(\sigma_0 / \psi\right)^n \tag{1}$$

where A' and n are constants depending on temperature, although A' may also depend on ageing of the material. Integrating,

$$\psi^{n+1} = 1 - A'(n+1)\sigma_0^n t \tag{2}$$

At failure, $\psi = 0$ and $t = t_r$; hence t_r is given by

$$t_r = 1 / A'(n+1)\sigma_0^n \tag{3}$$

Rabotnov (1969) modified the approach using the parameter $\omega = 1 - \psi$, so that $\omega = 0$ at the start and $\omega = 1$ at the end of life. In many ways this is more satisfactory in that damage ω can be considered as a concept without a strict physical meaning.

The Kachanov-Rabotnov concept of continuum damage accumulation has been correlated with the damage as represented by the A-parameter by Murakami *et al.* (1992). The relations for remaining life estimation obtained were as follows.

For $\omega = A$,

$$t / t_r = 1 - \left(1 - A / A_{cr}\right)^{\lambda m /(\lambda - 1)} \tag{4}$$

For cavity nucleation complete at $t = 0$,

$$t / t_r = 1 - \left(1 - A / A_{cr}\right)^{2 - \lambda m /(\lambda - 1)} \tag{5}$$

For continuous nucleation during creep,

$$\left[1 - \left(1 - t / t_r\right)^{(\lambda - 1)\lambda m}\right]\left[1 - \left(1 - t / t_r\right)^{1/\lambda}\right] = \left(A / A_{cr}\right)^2 \tag{6}$$

The symbols m and λ are, respectively, the exponent in the Norton equation and $\varepsilon_r/\varepsilon_s$, the ratio of the strain at fracture, ε_r, to the cumulative steady state creep strain, $\varepsilon_s = \dot{\varepsilon}_m t_r$, $\dot{\varepsilon}_m$ being the minimum creep rate. A_{cr} is the critical value of A at failure.

Figure 3 shows a plot of t/t_r versus the A-parameter, together with the test data of Shammas (1998). It may be seen that the value of A for the Cr-Mo steel is small and much less than unity when the life is largely used up.

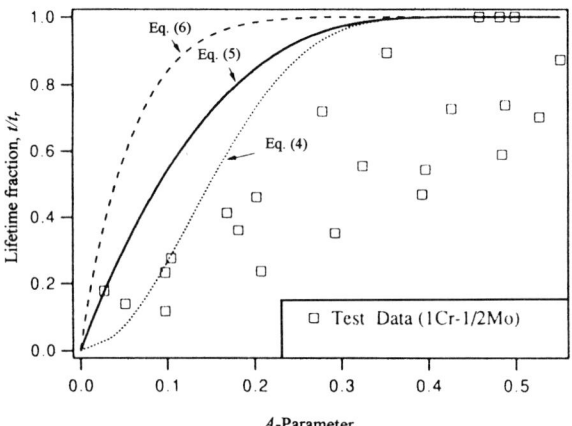

Figure 3. Lifetime fraction t/t_r versus the A-parameter. The plots are of Eqns. (4), (5) and (6) and are compared with the data of Shammas (1988). After Liu *et al.* (1994).

2. Problems in Using Metallographic Assessment

There have been many evaluations made using metallographic methods to determine the condition of boiler and other high temperature plant. These have included evaluations based on the *A*-parameter, as well as the other procedures developed for the estimation of creep voids. Additional methods that do not depend on voids include carbide distribution and size determination and changes to the morphology of the microstructure. For example, an assessment procedure based on the microstructure was proposed by Toft and Marsden (1961) in which the degree of spheroidization was classified in six stages (A to F). All such evaluations are usefully supplemented by hardness measurements, with hardness decreasing as a function of time and temperature.

However, work by da Silveira *et al* (1989), da Silveira and Le May (1992) and Samuels *et al.* (1992) has shown that predictive methods based on apparent void formation and detection may give erroneous results, in that, in many of the commonly used alloys such as the Cr-Mo steels, the voids that are used as the basis for assessment are primarily artifacts produced during the polishing and etching procedures. The implications of these observations are that void counting and assessment may be an unreliable method for the prediction of damage and remaining life. If apparent voids are produced during the metallographic preparation procedure, material and components may be condemned as being unfit for further service although they would still be safe to use, while other components in which voids are not indicated after the metallographic preparation may, in fact, have limited life.

3.   Observations in High Temperature Plant

If metallographic methods are to be used to evaluate the extent of high temperature damage in plant, they can be utilized in several ways. First, the extent of microstructural changes can be determined, usually on the basis of the changes in carbide morphology. Second, if the extent of formation of apparent voids is to be determined, the metallographic procedure used should be a standardized one involving a series of polish-etch steps. However, even such a procedure may not indicate the presence of apparent voids in material that has been significantly damaged by creep. As an example of this, an outlet steam pipe from a boiler ruptured after approximately 100,000 h of operation. The outlet steam temperature was nominally 480°C and the material was a DIN 15Mo3 alloy steel. Fracture was local with cracks having opened up causing a window to form in the tube wall.

The microstructure was of ferrite with spheroidized carbides, primarily on grain boundaries, and a distribution of graphite nodules: this is shown in Figure 4. Cracking had taken place between graphite nodules, and Figure 5 shows the tip of a crack running into the wall from the outer surface, graphite nodules being apparent at the tip. Cracking around the graphite nodules is seen in Figure 6 and some separation at grain boundaries ahead of the crack is seen in Figure 5. Isolated grain boundary void formation was apparent near the crack tip. At 1 mm ahead of the tip very few voids are seen (Figure 7) while 3 mm from the tip there are no such features (Figure 8).

If the microstructure were examined for creep voids to determine creep damage, the conclusion would be that the material had degraded significantly but that there was no evidence of creep damage unless examination had been made at the cracks themselves. There are no signs of carbide decohesion except on a local scale adjacent to the cracks.

Figure 11 shows a metallographic section taken adjacent to the fracture on the outer surface of a CrMo tube that ruptured after 70,000 h at a nominal temperature of 490°C. In (a) the material was polished and etched to emphasize grain boundary decohesion or cavities: in (b) polishing and etching were made carefully to outline the grain boundaries. In (a) the conclusion could be drawn that the material was damaged considerably be creep mechanisms: in (b) the conclusion could be drawn that such damage was negligible. But the tube had failed adjacent to the region where this specimen was taken.

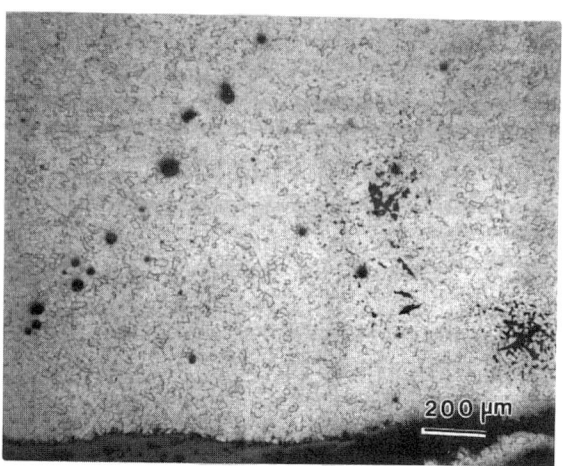

Figure 4. Microstructure of failed 15Mo3 tube, showing aligned graphite nodules. Nital etch.

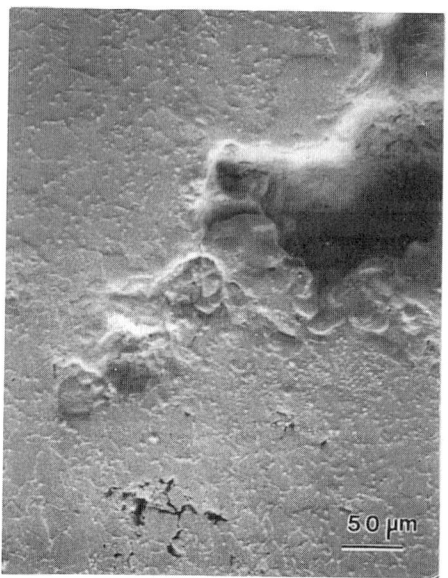

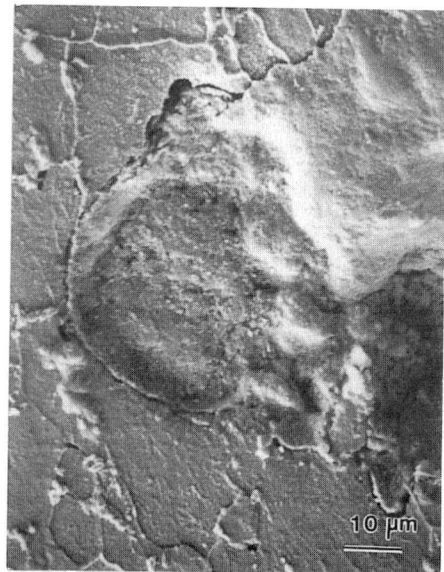

Figure 5. Tip of crack in 15Mo3 steel.   *Figure 6.* Cracking at graphite nodules.

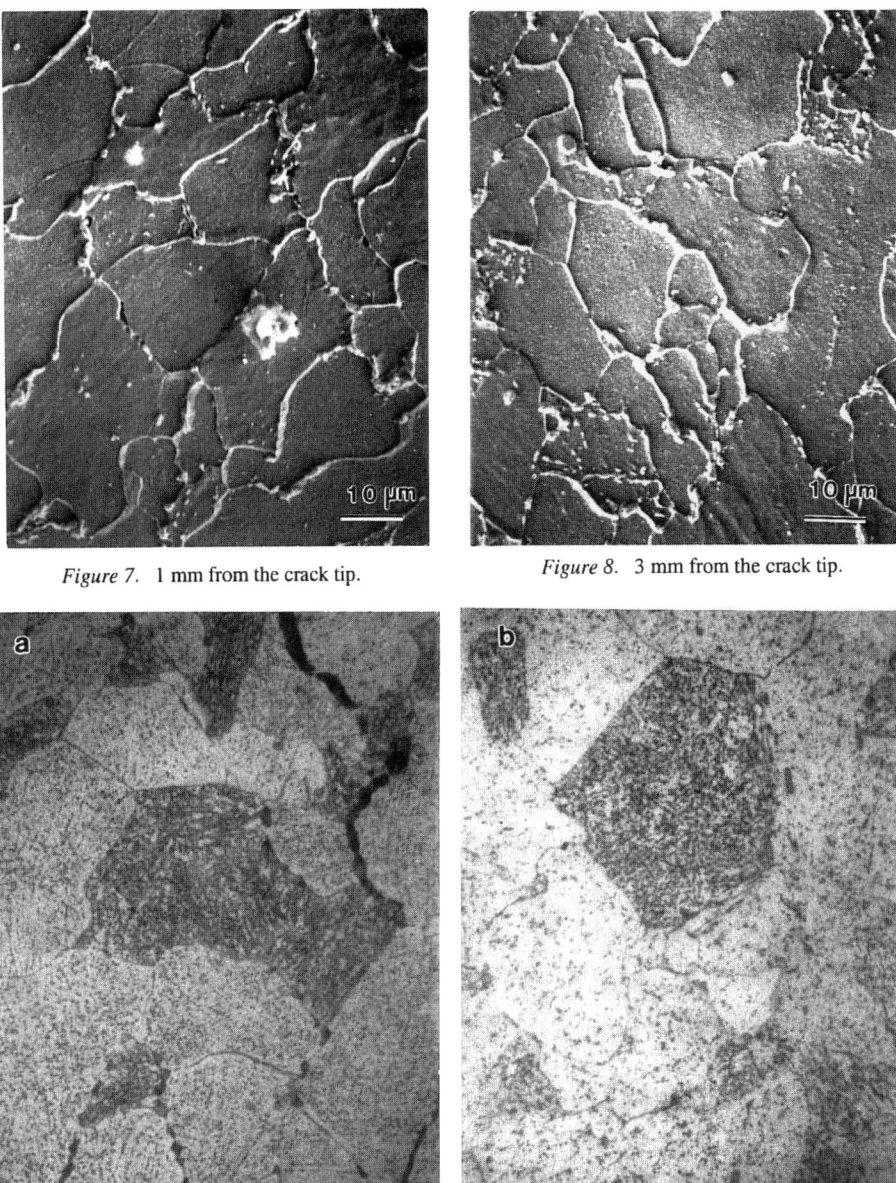

Figure 7.   1 mm from the crack tip.   Figure 8.   3 mm from the crack tip.

Figure 9.   Section near to the fracture and at the surface of CrMo tube that ruptured after 70,000 h at 490°C; (a) polished and etched to emphasize grain boundary cavities; (b) polished and etched carefully to show grain boundaries. Nital etch.

The value of the *A*-parameter was evaluated in the region of the tube wall from which the specimen shown in Figure 9 was prepared. The *A*-parameter was determined across the wall from the outside to the inner surface, and this was done using metallographic procedures (1) to emphasize voids and (2) to delineate grain boundaries without eating out carbides or otherwise removing grain boundary material. The results are shown in Figure 10. The *A*-parameter determined has a large variation across the wall and there is a large difference in its value at any depth as determined by the two polish/etch procedures. As hardness, or its loss, is a useful measure of the deterioration of a component at high temperature, Figure 11 shows hardness values taken through the wall thickness. The significant variation in hardness through the wall is apparent. The observed differences in *A*-parameter and hardness through the wall are not surprising in that creep damage and consequent cracking generally develop from the outside surfaces of boiler tubes and headers.

It was interesting to note that, when microhardness indentations were made in the creep damaged CrMo steel, fine grain boundary microcracks were detected frequently at the edges of the indentations (Furtado and Le May, 1997). Thus, it appears that the damaged grain boundaries may have lost cohesion and strength although they may not have separated physically to any detectable extent when examined metallographically.

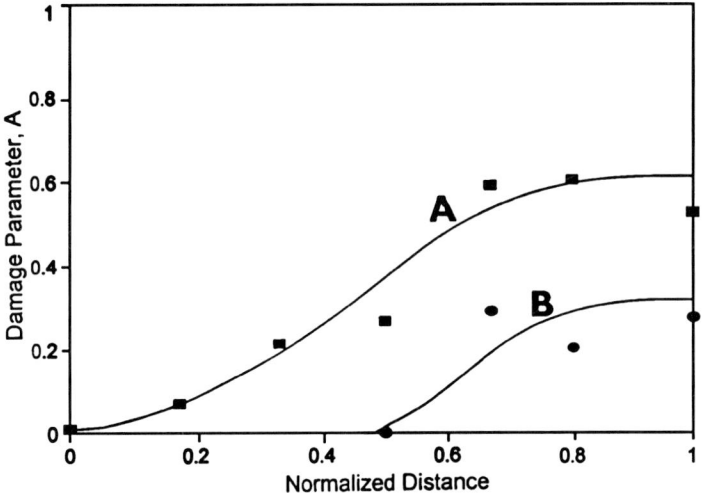

Figure 10. Plots of the *A*-parameter across the wall of the CrMo tube of Figure 11. The plots are for heavy etching (A) and careful etching (B). The outer surface is at the right.

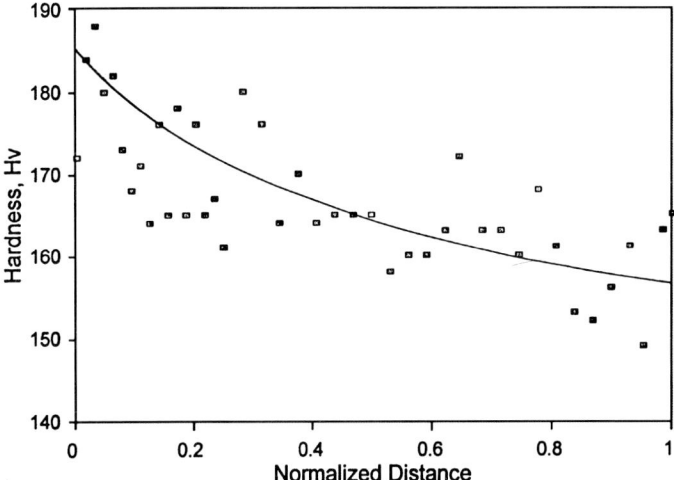

Figure 11.   Microhardness across the wall of the material of Figure 12. The outer surface is at the right.

4.   Damage Assessment Methodology

It is relevant to ask what should be the preferred method or methods for the evaluation of creep damage and remaining life of steam generator components operating at elevated temperature and subject to creep damage, in the light of the uncertainties and difficulties discussed.

At the outset it must be emphasized that creep failures are often very localized and examination even a very small distance from the point where damage has concentrated and a crack originates may give an unreasonably optimistic picture. Bearing this in mind, metallographic methods are useful, and should be considered on two bases, at least. The first is in terms of grain boundary weakness or decohesion that can be revealed as apparent voids and microcracks through a standardized repeated polish/etch procedure, which results in a strong attack and eats out carbides where there is decohesion of these at grain boundaries. Evaluation can then be made on a qualitative basis or a semi-quantitative one, for which the *A*-parameter method at least provides a number, although a number that should not be taken as being of great exactness. Semi-quantitative procedures such as that of VGB (1992) can also be used. The second basis for metallographic evaluation relates to the degree of spheroidization of the carbides. This is non-quantitative, but a qualitative evaluation procedure such as that of Toft and Marsden (1961) or Brear *et al.* (1997) can be used to provide an assessment. At the same time that metallographic examination is made, hardness measurements should be taken. This is straightforward when the surface has been prepared for replication and is clean and smooth. These measurements provide a useful check and correlation with the metallographic observations.

4.1.   EXTENSION OF THE KACHANOV APPROACH

Recently, there has been increased interest in using the approach of Kachanov (1958) to assess damage and remaining life. Two approaches have been proposed, one by Penny (1974, 1996), the other by the Materials Properties Council (MPC) (Prager, 1995). These methods do not depend on metallographic assessment of damage.

4.1.1   The Approach of Penny

Penny (1974) published an extension of the Kachanov (1958) and Rabotnov (1969) approach and more recently (Penny, 1996) developed the method further.

Integrating the Rabotnov relation,

$$\dot{\omega} = \dot{\omega}_0 (1 - \omega)^r \tag{7}$$

for temperature T = constant, where $\dot{\omega}_0 = B\sigma_0^k$, B and k being material constants, we obtain

$$(1-\omega)^{(1+r)} = 1 - B(1+r)\sigma_0^k t \tag{8}$$

The stress σ_t on area A_t at time t is

$$\sigma_t = \sigma_0 A_0 / A_t = \sigma_0 / (1 - \omega) \tag{9}$$

where ω represents a loss of effective area. At failure, $t = t_r$, $\omega = \omega_{cr}$ and the average stress is $\sigma_r (\equiv \overline{\sigma})$, reached between σ_y (yield) and σ_u (UTS). Thus

$$\sigma_0 / (1 - \omega_{cr}) = \overline{\sigma} = \sigma_y + \alpha(\sigma_u - \sigma_y) \tag{10}$$

where α is a proportion above σ_y, and

$$\overline{\sigma} = \delta\sigma_y \tag{11}$$

where

$$\delta = \left[1 + \alpha(\sigma_u / \sigma_y - 1)\right] \tag{12}$$

For nonhardening materials (the extreme case) $\delta = 1$, but a reasonable estimate of a bounding value for ductile materials is $\delta = m/(m + 1)$ (Penny, 1996). The failure condition is bounded by $m/(m + 1) \leq \delta \leq \sigma_u/\sigma_y$. In all cases $\omega_{cr} < 1$. Substituting from Eqn. (10) in Eqn. (8),

$$t_r = \left[1 / B(1+r)\sigma_0^k \right]\left[1 - (\sigma_0 / \overline{\sigma})^{(1+r)} \right] \qquad (13)$$

or

$$t_r = \beta t_b \qquad (14)$$

where

$$t_b = \left[1 / B(1+r)\sigma_0^k \right] \qquad (15)$$

which is the result for "brittle" failure, and

$$\beta = \left[1 - (\sigma_0 / \overline{\sigma})^{(1+r)} \right] \qquad (16)$$

β is a factor that depends on the level of σ_0 with respect to $\overline{\sigma}$, the distance $\log\beta$ being shown in Figure 12.

In Penny's (1974) approach to creep strain versus creep damage, the term $(1 - \omega)$ from Eqn. (8) is substituted into the Norton equation in the form:

$$\dot{\varepsilon} = K\sigma_0^m / (1 - \omega)^p \qquad (17)$$

where p is a material constant relating to microstructural damage. Integrating,

$$\varepsilon / \varepsilon_0 t_b = \eta \left[1 - (1 - t / t_b)^{1/\eta} \right] \qquad (18)$$

where

$$\eta = (1 + r) / (1 + r - p) \qquad (19)$$

Substituting Eqn. (14) in Eqn. (18),

$$\varepsilon / \dot{\varepsilon}_0 t_r = (\eta / \beta)\left[1 - (1 - \beta t / t_r)^{1/\eta} \right] \qquad (20)$$

At $t = t_r$,

$$\varepsilon_r / \eta\dot{\varepsilon}_0 t_r = \left[1 - (1 - \beta)^{1/\eta} / \beta \right] \qquad (21)$$

Penny (1996) provided an alternative derivation based on strain variation during creep that leads to a useful result. From constant volume considerations (with A_t and L_t being area and length, respectively, at time t),

$$dL_t / L_t = -dA_t / A_t = -d(1-\omega)/(1-\omega) \tag{22}$$

or

$$(1+\varepsilon)(1-\omega) = 1 \tag{23}$$

Based on true strain $e = \ln(1 + \varepsilon)$, and from Eqns. (8), (14) and (15),

$$e = -\ln(1 - \beta t / t_r)^{1/(1+r)} \tag{24}$$

Differentiating,

$$[(t_r / \beta) - t] de / dt = 1/(1+r) \tag{25}$$

Eqn. (15) indicates that remaining life times the current (true) strain rate is a constant. If the strain rate can be measured in service, the remaining life can be computed, as r can be determined from creep rupture tests and β can be estimated, being ~1 at low stresses.

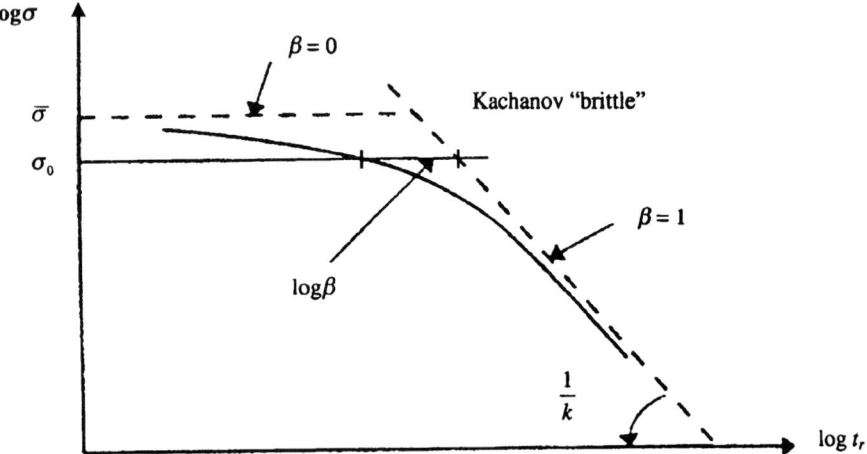

Figure 12. The modified Kachanov rupture curve. After Penny (1996)

4.1.2 The Omega Method

The Omega method was developed from the Kachanov-Rabotnov approach to assess components in high temperature service (Prager, 1995). The Kachanov-Rabotnov equation in strain rate form is:

$$\dot{e} = \dot{e}_0 \left(\sigma / \sigma_0\right)^m \left[1 / \left(1 - \omega\right)^p\right] \tag{26}$$

where e is the true strain. Substituting for $\left(\sigma / \sigma_0\right) = exp(e)$ from constant volume considerations,

$$\dot{e} = \dot{e}_0 \, exp(me) \left[1 / \left(1 - \omega\right)^p\right] \tag{27}$$

As Eqn. (27) is not easily integrated, it is written as an exponential function including three factors on which strain rate depends, namely increasing stress, increasing damage, and microstructural changes that are not related to damage. Hence,

$$\dot{e} = \dot{e}_0 \, exp(me) \left[1 / exp(-pe)\right] \left[1 / exp(-ce)\right] \tag{28}$$

where c is a constant obtained from experimental data that accounts for deficiencies in the Norton exponent and other microstructural factors associated with stress change. Thus,

$$\dot{e} = \dot{e}_0 \, exp(me) \left[(m + p + c)e\right] \tag{29}$$

Integrating,

$$\left[1 / \dot{e}(m + p + c)\right]\left\{1 - exp\left[-(m + p + c)e\right]\right\} = t \tag{30}$$

$$e = -\left[1 / (m + p + c)\right] \ln\left[1 - \dot{e}_0 (m + p + c)t\right] \tag{31}$$

For values of $(m + p + c)e \gg 2$ or 3, $exp[m + p + c]e$ is negligible; thus at failure

$$1 / \dot{e}_0 (m + p + c) = t_r = 1 / e_0 \Omega_p \tag{32}$$

where

$$\Omega_p = m + p + c \tag{33}$$

From Eqn. (29)

$$d(\ln \dot{e}) / de = m + p + c = \Omega_p \qquad (34)$$

The value of Ω_p can be determined by plotting $\ln \dot{e}$ versus e and taking the slope. From Eqn. (30), at fracture

$$t_r = (1 / \dot{e}\Omega_p)\left[1 - \exp(-\Omega_p e_r)\right] \qquad (35)$$

and the remaining life is

$$t_r - t = (1 / \dot{e}\Omega_p)\left[\exp(-\Omega_p e) - \exp(-\Omega_p e_r)\right] \qquad (36)$$

Now,

$$\dot{e} = \dot{e}_0 \exp(\Omega_p e) \qquad (37)$$

Neglecting e_r in Eqn. (36) and substituting from Eqn. (37),

$$t_r - t \cong 1 / \dot{e}\Omega_p \qquad (38)$$

or

$$(t_r - t)de / dt = 1 / \Omega_p \qquad (39)$$

Eqn. (39) provides the same prediction as with Penny's analysis, namely that the product of remaining life and (true) strain rate are constant during service.

5. Concluding Remarks

It is apparent that, for some high temperature alloys at least, grain boundary cavitation is minimal during creep in service and that its detection or otherwise cannot be a sufficient method for evaluation of the extent of damage and remaining life. For other alloys and for weld metal, cavitation appears to be more prevalent. Questions remain concerning the damage mechanisms and the modelling of these in many high temperature alloys, including the matter of the effect of environment, which is known to promote cavity formation in some alloys. For example, intergranular oxygen penetration can react with carbon or carbides at grain boundaries in nickel to form CO_2 bubbles, which act as nuclei for creep cavitation (Woodford, 1992).

Metallographic methods remain an important tool in assessing in-service components operating in the creep regime, but it is important that preparation procedures are used in a standardized manner and that dependence is not placed solely on the detection of apparent grain boundary cavities. Quantitative measurements of apparent grain boundary cavities by means of methods such as the A-parameter need to be carefully considered and not used blindly. Metallographic methods should, wherever possible, be supplemented by others, and the use of the continuum damage approach provides a very appropriate tool in this regard, as it does not depend on metallographic evaluation of voids or of other microstructural changes.

Testing to determine the true strain rate of aged materials may be possible *in situ* in some cases, or specimens can be removed for laboratory testing. At low stress levels there is virtually no primary creep in exposed boiler steels (Prager, 1995), so testing is relatively simple to determine the value of remaining life, $t_{rem} = (t_r - t)$, provided a data base of materials properties is available.

There is a need for further studies on the mechanisms of damage and crack initiation in a number of high temperature alloys of practical importance. Effects of environment are still a major uncertainty, and one may speculate that the decohesion observed at grain boundary carbides near the surface of CrMo steels may relate to mechanisms similar to those occurring in nickel alloys. Laboratory creep testing of such alloys in controlled atmospheres would be a valuable exercise. Without more detailed physical studies, modelling of the processes thought to be occurring may be an exercise in futility.

6. References

Auerkari, P. (1983) Remanent creep life estimation of older power plant steam piping systems, in D.A. Woodford and T.R. Whitehead (eds.), *ASME International Conference on Advances in Life Prediction Methods*, ASME, New York, pp. 353-356.

Beere, W. (1981) Theoretical treatment of creep cavity growth and nucleation, in J. Gittus (ed.), *Cavities and Cracks in Creep and Fatigue*, Applied Science, London, pp. 1-27.

Brear, J.M., Auerkari., P. and de Araujo, C. (1997) Metallographic techniques for condition assessment and life prediction in SP249 guidelines, in A. Jovanovic and L. Verelst (eds.), Proceedings of the SP249 Final Project Workshop, Linkebeek, Belgium, pp. 2.3.1-2.3.30.

Cane, B.J., and Shammas, M.S. (1984) *A Method for Remanent Life Estimation by Quantitative Assessment of Creep Cavitation on Plant: Report TRPD/L/2645/N84*, CEGB, Leatherhead.

Cocks, A.C.F., and Ashby, M.F. (1982) On creep fracture by void growth, *Progress in Matls. Sci.*, **27**, 189-244.

da Silveira, T.L. and Le May, I. (1992) Effects of metallographic preparation procedures on creep damage assessment, *Materials Char.* **28**, 75-85

da Silveira, T.L., Silva, T.C.S., and Le May, I. (1988) Boundary decohesion during high temperature creep deformation, in P.O. Kettunen, T.K., Lepistö and M.E. Lehtonen (eds.), *Strength of Metals and Alloys*, Pergamon, Oxford, pp. 947-952.

Dyson, B.F. (1976) Constraints on diffusional cavity growth rates, *Metal Science*, **10**, 349-353.

Furtado, H.C. and Le May, I. (1997) Modelling of creep damage to estimate remaining life, *Matls. Sci. and Engg.*, **A234-236**, 87-90.

Kachanov, L.M. (1958) Rupture time under creep conditions, *Izv. Akad. Nauk SSSR, Otd. Tekhn. Nauk*, **8**, pp. 26-31 (in Russian).

Hull, D., and Rimmer, D.E. (1959) The growth of grain boundary voids under stress, *Phil Mag.*, **4**, 673-687.

Liu, Y., Murakami, S., and Sugita, Y. (1994) Identification of creep damage variable from A-parameter by a stochastic analysis, *Int. J. Pres .Ves. & Piping*, **59**, 149-159.

Neubauer, B., and Wedel, V. (1983) Restlife estimation of creeping components by means of replicas, in D.A. Woodford and T.R. Whitehead (eds.), *ASME International Conference on Advances in Life Prediction Methods*, ASME, New York, pp. 307-313.

NORDTEST (1991) *Remanent Lifetime Assessment of High Temperature Components in Power Plants by Means of Replica Inspection*, Nordtest NT 010.

Penny, R.K. (1974) The usefulness of engineering damage parameters during creep, *Metals and Materials*, **8**, 278-283.

Murakami, S., Liu, Y., and Sugita, Y. (1992) Interrelation between damage variables of continuum damage mechanics and metallographic parameters in creep damage, *Int. J. Damage Mechanics*, **1**, 172-190.

Penny, R.K.. (1996) The use of damage concepts in component life assessment, *Int. J. Pres. Ves. & Piping.*, **66**, 263-280.

Prager, M. (1995) Development of the MPC omega method for life assessment in the creep range, *J. Pres. Ves. Tech.*, **117**, 95-103.

Rabotnov, Yu. N. (1969) *CreepProblems in Structural Members*, North Holland, Amsterdam.

Samuels, L.E., Coade, R.W., and Mann, S.D. (1992) Precracking structures in a creep-ruptured low-carbon Cr-Mo steel, *Materials Char.* **29**, 343-363.

Shammas, M.S. (1988) Metallographic methods for predicting the remaining the remanent life of ferritic coarse-grained weld heat affected zones subject to creep cavitation, in *International Conference on Life Assessment*, EPRI, Palo Alto, pp. 238-244.

Toft, L.H. and Marsden, R.A. (1961) in *Structural Processes in Creep: Special Report No. 70*, Iron and Steel Inst., London, pp. 276-294.

VGB (1992) *Guideline for the Assessment of Microstructure and Damage Development of Creep Exposed Materials for Pipes and Boiler Components: VGB-TW 507*: VGB Technical Association of Large Power Plant Operators, Essen.

Woodford, D.A. (1992) Creep strength and creep damage considerations for life cycle engineering, in D.A. Woodford, C.H.A. Townley and M. Ohnami (eds.), *Creep: Characterization,Damage and Life Assessment, Proc. Fifth Int. Conf. On Creep of Matls.*, ASM, Materials Park, OH, pp. 23-32.

CREEP DAMAGE ACCUMULATION AND FAILURE IN NARROW REGIONS OF STEEL WELDS

D.J. Smith, N.S. Walker and S.T. Kimmins
Department of Mechanical Engineering,
University of Bristol,
Bristol, BS8 1TR, UK.

1. Introduction

When structural components are joined together by fusion welding the thermal cycle introduces changes in the microstructure between the base (or parent) material and the weld filler metal, [Easterling, 1983]. A typical weld consists of base, weld and heat-affected-zone (HAZ) metals, as shown in Figure 1. The HAZ is a transition region, which can be subdivided into a number of zones, which depend on the weld and base metals. Typical HAZ zones, shown in Figure 1, for a ferritic steel weld with matching weld and base metal compositions include a coarse grained (CGHAZ) zone adjacent to the weld metal, fine grained (FGHAZ) and intercritical (ICHAZ) zones.

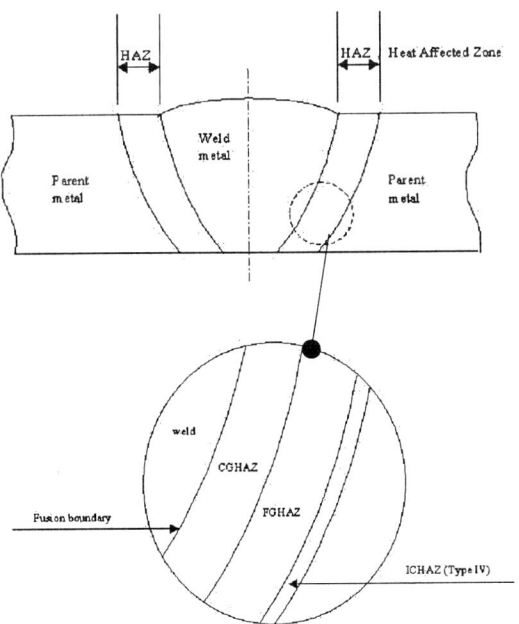

Figure 1 Schematic of section through a weld

257

S. Murakami and N. Ohno (eds.), IUTAM Symposium on Creep in Structures, 257–266.
© 2001 Kluwer Academic Publishers. Printed in the Netherlands.

The variations in the microstructural characteristics of each region of a typical steel weld also give rise to marked variations in creep properties. The mismatch in creep properties leads to highly complex mechanical and material behaviour. Operational experience of pressurised components, such as steam pipes, reveals that many types of failure mechanisms can arise [Shuller et al, 1974]. The failure location and mechanism depends on factors such as the creep strength, ductility and width of each zone in the weld.

In ferritic steel pipes, the predominant problem during long-term service is creep damage accumulation in the ICHAZ zone. Following the classification scheme developed by Shuller et al [1974], this type of weld damage is called Type IV. Recent literature reveals that Type IV failure is a world-wide problem in power generation systems operating at high temperature. There are extensive reviews of the features associated with this failure type, [Gooch and Kimmins, 1987, Kimmins et al, 1993, Kimmins and Smith, 1998, Ellis and Viswanathan, 1998, and Nishimura et al, 1999]. The cause of Type IV failure is principally associated with a microstructural region that has low creep strength. This region is surrounded by materials that are stronger in creep.

There is only limited research has been reported that has examined creep damage in the narrow region of the Type IV zone. Recent studies [Walker et al, 1996, Walker, 1997] examined creep damage, in the form of grain boundary cavities within the Type IV zone, and using specimens interrupted at various fractions of their creep life. The study included both cross-weld and simulated Type IV materials. Most damage studies in the form of grain boundary cavitation explore behaviour of simulated HAZ were there is a large grain size, [Cane, 1979 and Chuman et al, 2000]. The unique structure of the Type IV zone makes estimates of cavity growth rate and area fraction difficult to quantify. This paper examines creep damage accumulation and failure for this type of weld damage.

2. Experiments

Three main sets of experiments were carried out. Cross-weld and simulated Type IV specimens were used to obtain information about creep damage accumulation in the Type IV zone. Some of the cross-weld results were reported by Walker et al [1996]. Multiaxial behaviour was explored using notched bar and shear pin tests from cross-weld and simulated Type IV samples. Results for these tests are given by Kimmins et al [1996]. Finally, a series of creep cracking tests were carried out using cross-weld and simulated Type IV samples.

The base material was a 1/2CrMoV ferritic steel. This is the same material used in an earlier study by Gooch and Kimmins [1987]. Three material conditions were tested; parent steel with a grain size of 19μm, a simulation of the Type IV zone with a fine grain size (5μm) together with a lesser density of coarser ferritic grains (19μm), and finally, a MMA weld and associated HAZ. The Type IV region for these welds was characterised by a refined microstructure with a width of the order of 1mm, and grain size of about 5-10μm.

From the MMA welds, threaded cross-weld specimens, with total length 190mm, were extracted. The specimen design was such that there was a change of section so that the two Type IV regions were subjected to two different stresses. The low stress section had a square cross-section of 25mm by 25mm, and the high stress section had a cross-section of 20mm by 25mm.

Shear pin (8mm diameter) specimens and round notched bar specimens were also extracted from the MMA welds. At the centre of the gauge length of the notched bar circumferential notches, with radius 3.4mm, were machined. The diameter of the bar at the root of the notch was 10.2 mm. The major axes of the shear pins and notched bars were approximately perpendicular to one side of the weld fusion boundary.

Two types of creep crack growth specimens were extracted from the MMA welds, double edge notch tension (DENT) and compact tension (CT). Cracks were introduced by spark erosion using a wire diameter of 0.15mm. For the DENT geometry two crack depths were used, shallow and deep cracks, with a_0/W=0.08 and 0.5, where a_0 is the initial crack length and W is the specimen width. For the CT geometry, a_0/W was 0.5.

Specimens were also extracted from simulated HAZ material. To obtain damage accumulation in this material 15mm by 15mm section bars were used. Shear pin, notched bar, and creep crack growth specimens were also machined from the simulated material.

All tests were carried out in air, over a range of temperatures from 600°C to 700°C, using dead-load creep machines. For plain bar specimens such as cross-weld and simulated Type IV specimens, displacements were measured using transducers attached to extensometers located on specimens. For shear pin tests displacements were measured from the load train. The details of the procedure for the shear pin tests are reported by Kimmins et al [1996].

The detailed test procedure for measuring damage accumulation in the cross-weld tests, simulated Type IV and notched bar tests is given by Walker et al [1996] and Walker [1997]. It is suffice to indicate here that tests were interrupted at different proportions of their lives. Cavity measurements were obtained using plastic replicas. It was always found that the region of damage in the Type IV region was very localised, and confined to a width less than about 1mm, [Walker, 1997].

For crack growth tests, cracking was monitored remotely using a conventional potential drop technique [Webster and Ainsworth, 1994]. Displacements of the loading pins for the CT specimens and on the load line of the DENT specimens were measured.

3. Uncracked Creep Rupture and Damage Accumulation

Two features of the uncracked tests are presented here. An assessment is made of the failure times for creep rupture, and results are shown for creep damage accumulation in the form of cavity measurements. Tests were carried out at various temperatures and

D.J. SMITH et al.

stresses.   A time-temperature parameter, T_{PR}, is used for the horizontal axis of the stress rupture diagram, where

$$T_{PR} = -1000(log(t_R)-20.01)/(T + 104) \tag{1}$$

where t_R is the lifetime (hours) of the specimen and T is the test temperature (°C).

Test results in terms of applied stress as a function of T_{PR} for uniaxial cross-weld, plain bar simulated Type IV tests, and multiaxial tests are shown in Figure 2. Also shown are rupture curves for the base 1/2CrMoV steel and for wrought 1Cr1/2Mo steel. Similar to earlier results [Gooch and Kimmins, 1987] for Type IV failure, the test results show that at high stresses the strengths of the Type IV and base material converge. The Type IV cross-weld results are combined with earlier results from Gooch and Kimmins [1987]. The failure times for the simulated Type IV tests were similar to the cross-weld tests.

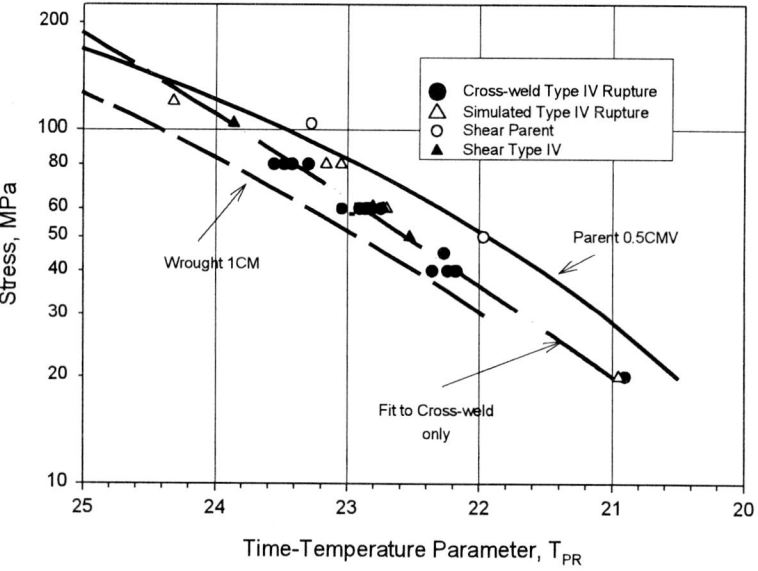

Figure 2. Creep Rupture of Uncracked Parent and Type IV Materials

The relationship between equivalent stress and creep rupture life of the shear pin and notched bar tests for base material, cross-weld and simulated Type IV tests are shown in

Figure 2. The equivalent stress for shear tests is $\sqrt{3}\tau$, where τ is the applied shear stress. For equivalent stress for the notched is the average von-Mises stress across the throat of the notch, [Hayhurst and Webster, 1986], and is 1.34 time the net section stress.

For the multiaxial tests, there are a number of choices for the equivalent stress. For example, maximum principal, equivalent stress and combination of these two. Kimmins et al [1996] suggest that for these tests the equivalent stress is the best choice. Results in Figure 2 confirm this. The shear pin tests using base (parent) material agree extremely well with the best-fit line for uniaxial tests from the ISO database. The cross-weld shear pin and notched bar results expressed in terms of equivalent stress agree with the best-fit line to uniaxial cross-weld tests.

The failure times for the simulated Type IV shear pin and notched bar tests agree well with the equivalent cross-weld tests. These results together with the earlier uniaxial tests suggest that the Type IV zone can be tested independently of adjacent material.

Damage accumulation obtained from cross-weld tests is reported by Walker et al [1996] and Walker [1997]. A compilation of results of cavity measurements, as a function of life fraction, t/t_R, from cross-weld specimens, is shown in Figure 3.

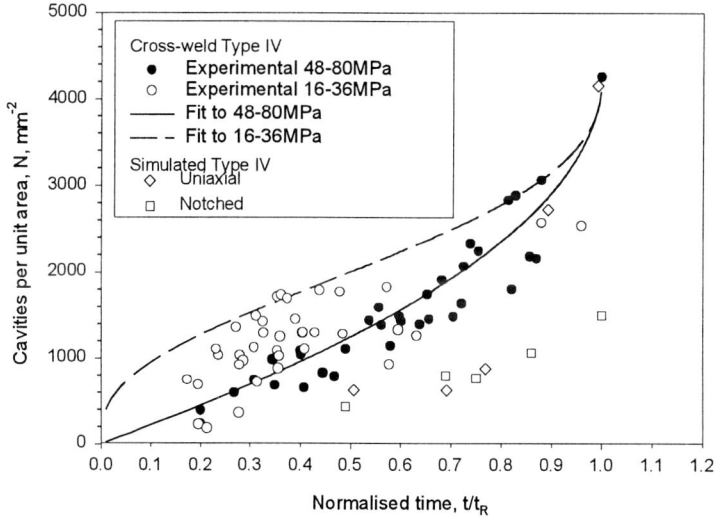

Figure 3 Damage accumulation in cross-weld and simulated Type IV material

Irrespective of temperature, the measurements show that damage accumulation is insensitive to stresses in the range 48 to 80MPa. Similarly, for a lower range of stresses,

16 to 36MPa. However, damage accumulation is more pronounced at the lower stresses and for life fractions less than about 0.7. An empirical relationship between N and t/t_R is

$$N = N_R \left(1 - \left(\frac{t}{t_R} \right)^m \right)^\psi \tag{2}$$

where N_R, m and ψ are constants. N_R ($=4260\text{mm}^{-2}$) is the maximum measured number of cavities per unit area. For stresses in the range 48 to 80MPa, $m=1$, $\psi=0.5$ and for stresses in the range 16 to 36MPa, $m=1/3$ and $\psi=0.4$. The curves corresponding to equation 2 are shown in Figure 3.

In addition to cavity measurements from cross-weld specimens, measurements were also obtained from plain round bars of simulated Type IV material. Cavity measurements for 60MPa and 640°C are shown in Figure 3. At rupture N_R for the simulated Type IV material was very similar to that for the cross-weld specimen. However, the rate of damage accumulation for the simulated Type IV material was about half that from cross-weld tests.

Measurements were made using an SEM to determine the cavity size distribution for various life fractions. Weibull distributions from cross-weld and simulated Type IV are compared in Figure 4.

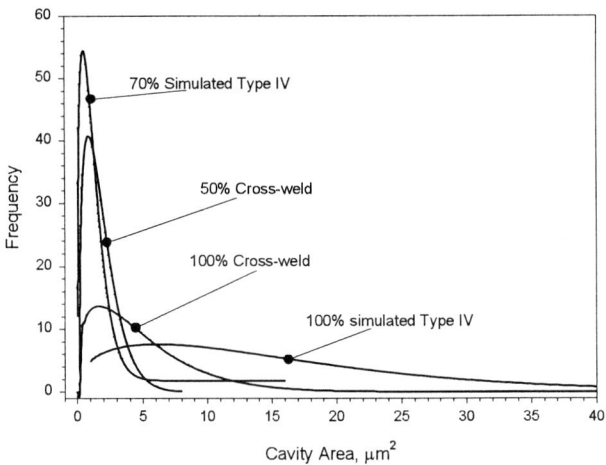

Figure 4 Cavity area distribution at various life fractions

There are very similar cavity area distributions for the two cases at life fractions between 50 and 70%. However, it is important to note that the number of cavities per unit area, N, for cross-weld specimens was greater than for simulated Type IV specimens (Figure 2). In contrast, N_R was very similar, but the peak cavity areas are distinctly different, Figure 4.

The majority of cavities retained a circular shape in the cross-weld tests compared with the simulated Type IV tests where cavities were more lenticular (normal to the applied stress). This suggests that the cavity growth in the cross-weld specimens was more constrained than in the simulated Type IV tests. There was no evidence of cavity coalescence in either the cross-weld or simulated Type IV tests.

4. Cracked Creep Rupture and Crack Growth

In practise, it is important to distinguish between failure resulting from discrete crack growth and creep rupture across the remaining ligament of the cracked specimen. In this section, results from the creep cracking experiments are presented.

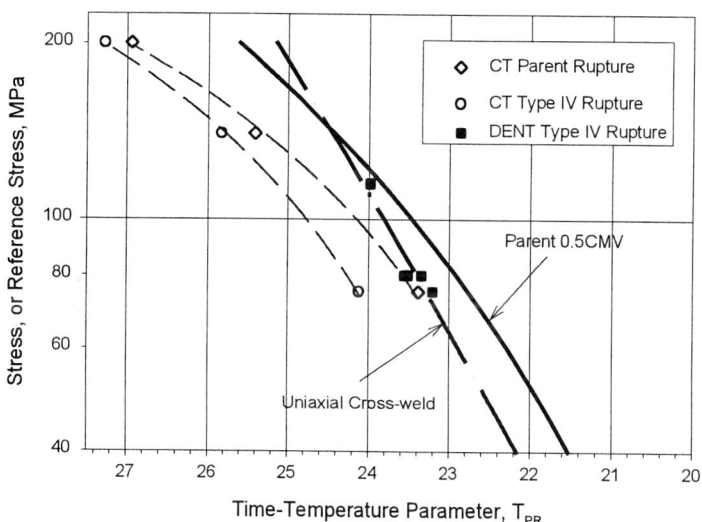

Figure 5 Creep rupture of cracked and uncracked parent and Type IV materials

The creep rupture times were evaluated using the reference stress, σ_{ref}. Equations for σ_{ref} for the DENT and CT specimens are given by Webster and Ainsworth [1994]. The reference stress was evaluated for plane stress conditions at the initial crack length, a_0.

As with the creep rupture tests for uncracked samples, the failure times at various temperatures are presented using the time-temperature parameter, T_{PR}, equation 1. Test results are shown in Figure 5. Also shown in Figure 5 are curves corresponding to best-fit lines to base material and cross-weld Type IV creep rupture.

For CT specimens, both parent and cross-weld creep rupture tests had substantially shorter lives than the uncracked material. For example, in the parent steel and the Type IV materials, the rupture time of a CT specimen was approximately 3 times less than the rupture life of a uncracked specimen. This is with $a_0/W=0.5$ at a reference stress of 75MPa. At higher reference stresses (up to 200MPa) the failure time of the cracked specimen was about or greater than 10 times less than the uncracked specimens.

In contrast for the DENT specimens, the rupture lives of the cross-weld and simulated Type IV tests were similar to the rupture lives of uncracked specimens. This is exemplified in Figure 5 by experimental values for this cracked geometry lying on or close to the best-fit curve of the creep rupture lives of cross-weld specimens. There was no distinction in this figure between results from pre-cracked cross-weld and simulated Type IV tests.

5. Discussion and Concluding Remarks

Conventional finite element analysis [Perrin and Hayhurst, 1999, and Hyde et al, 1998], based on transverse strain compatibility give rise to significant constraint of creep deformation in the weak zone. This is particularly the case for the Type IV zone in a ferritic steel weld adjacent to the parent metal. Continuum damage analyses of the type developed in recent finite element analyses [Perrin and Hayhurst, 1999, and Hyde et al, 1998], include models where damage accumulation is enhanced by multiaxial stress states that arise due to constraint. The experimental results summarised in this paper do not indicate there is a strong influence of the maximum principal stress on damage accumulation. For example, creep rupture results from shear and notched bar tests, using cross-weld and simulated Type IV materials, indicate that the von-Mises effective stress is sufficient to describe failure.

Kimmins and Smith [1998] recently emphasised that, with the refined grain sizes in the Type IV region, there is the potential for a high degree of grain boundary accommodation through nucleation of cavities. In Figure 4 the peak in the frequency of the cavity area remains at small values, even close to failure. This illustrates that there is continuous cavity nucleation, together with cavity growth, throughout the life of the Type IV zone.

In all types of specimens the failure time of cross-weld and simulated Type IV were similar. However, the rate of increase in number of cavities in cross-weld tests, both plain and notched, were greater than in the simulated Type IV tests. As suggested by Kimmins and Smith [1998] relaxation of the constraint in the cross-weld tests could occur through grain boundary sliding, which in turn leads to continued cavity nucleation and growth.

The results from pre-cracked specimens revealed that creep rupture lives were a strong function of geometry. The DENT specimens containing shallow cracks failed at times similar to those for the uncracked cross-weld tests. This is consistent with the failure mechanism dominated by ligament damage. Discrete crack growth had a relatively minor influence on the overall failure time. The CT specimens exhibited the other extreme, with the failure time considerably less than the failure time from ligament damage.

Acknowledgements

This work was supported by the Health and Safety executive of the UK government and Nuclear Electric (now British Energy).

References

Cane, B.J., [1979], "Interrelationship between creep deformation and creep rupture in 2.25CrMo Steel", Metal Science, Vol. 10, 287, 1979

Chuman, Y, Yamauchi, and Hiroe, T., [2000], "Study on Evaluation Procedure of Multiaxial Creep Strength of Low Alloy Steel", Key Engineering Materials, Vols, 171-174, 2000, 305-312

Easterling, K, [1983], "Introduction to the Physical Metallurgy of Welding", Butterworths Monographs in Metals.

Ellis, F.V. and Viswanathan, R, [1998], "Review of Type IV cracking in pipe welds", in Proc. Int. Conf. on Integrity of High-Temperature Welds, PEP, London, 1998, 125-134

Gooch, D.J. and Kimmins, S.T., [1987], 'A study of Type IV cracking in ½%CrMoV/2¼CrMo weldments', Proc. 3rd Int. Conf. Creep and Fracture of Eng. Mats. and Structs., 1987, 689-703.

Hayhurst, D.R. and Webster, G.A., [1986], "An overview of studies of stress state effects of circumferentially notched bars", in Techniques for Multiaxial Creep Testing, ed. D.J. Gooch and I.M. How, 1986, 137-176

Hyde, T.H., Tang, A., and Sun, W., [1998], "Analytical and Computational Stress Analysis of Welded Components under Creep Conditions", in Proc. Int. Conf. on Integrity of High-Temperature Welds, PEP, London, 1998, 285-307

Kimmins, S.T., Coleman, M.C. and Smith, D.J., [1993], 'An overview of creep failure associated with heat affected zones of ferritic weldments', Proc. 5th Int. Conf. Creep and Fracture of Eng. Mats. and Structs., 1993, 681-694.

Kimmins, S.T., Smith, D.J. and Walker, N.J., [1996], "Creep deformation and rupture of low alloy ferritic weldments under shear loading", J. Strain. Analysis, 1996, Vol 31, No 2, 125-133

Kimmins, S.T. and Smith, D.J., [1998], "On the relaxation of interface stresses during creep of ferritic steel weldments", J. Strain. Analysis, 1998, Vol 33, No 3, 195-206

Nishimura, N, Iwamoto, K, Yamauchi, M, Masuyama, F, Imamoto, T and Yokoyama, T, [1999], "Development of Life Assessment System for High Energy Piping in Fossil Power Boilers, in Proc. 4th Int. Conf. on Reliability, Maintenance and Safety, Shanghai, May 1999.

Perrin, I.J. and Hayhurst, D.R., [1999], "Cotinuum damage mechanics analyses of type IV creep failure in ferritic steel crossweld specimens", Int. J. Pres. Ves. and Piping, 76, 599-617

Shüller, H.J., Hagn, L. and Woitscheck, A., [1974], 'Cracks in the weld area of formed parts in superheated steam lines - material analysis', Der Maschinenschaden, 1974, 1, 1-13.

Shron, R.Z. and Korman, A.I., [1975], 'Evaluation of the supporting power of welds of steam pipes fabricated of Cr-Mo-V steels', Thermal Engineering (Teploenergetika (USSR)), 1975, 22 (10), 33-36.

Walker, N.S., [1997], "Type IV Creep Cavitation in Low Alloy Ferritic Steel Weldments", PhD Thesis, University of Bristol, 1997

Walker, N.J Kimmins, S.T., and Smith, D.J, [1996], "Type IV creep cavitation in ferritic steel welds", in Creep and Fatigue, Design and Life Assessment at High Temperature, Proc. *I.Mech E Conf*, Paper C494/056/96, pp341-350, 1996

Webster G.A. and Ainsworth, R., "High Temperature Component Life Assessment", Chapman and Hall, London, 1994

LONG-TERM CREEP LIFE PREDICTION BASED ON UNDERSTANDING OF CREEP DEFORMATION BEHAVIOR OF FERRITIC HEAT RESISTANT STEELS

K. YAGI, F. ABE, K. KIMURA and H. KUSHIMA
National Research Institute for Metals
1-2-1, Sengen, Tsukuba, Ibaraki 305-0047, Japan

1. Introduction

An understanding of long-term creep properties is important for safe designing and reliable assessment of structural components operating at high temperature. However, it is difficult to predict the long-term creep rupture life by extrapolating the short-term rupture data. The rupture life is a reflex of creep deformation behavior. The shape of creep curves is dependent on tested material, stress and temperature conditions [1], because the creep deformation is affected by the change in microstructure during creep. Therefore, the long-term creep life must be predicted based on the understanding of creep deformation behavior.

It is necessary to characterize a creep curve for investigating the creep deformation behavior. The long-term creep curve consists of the primary creep and tertiary creep regions without steady-state creep region [1]. The equation based on θ projection concept [2] can well reproduce such a long-term creep curve. For the present work, the equation based on modified θ projection concept [3] is applied to characterize the creep curve.

In this paper, the creep curves of various types of ferritic heat resistant steels are analyzed by the modified θ projection concept, and the creep rupture life is predicted from the creep deformation behavior. Then, the relationship between creep rupture life and creep deformation behavior is investigated. Moreover, the limit in creep life prediction is discussed.

2. Characterization of Creep Curve

For the modified θ projection concept, a creep curve is described by the following

S. Murakami and N. Ohno (eds.), IUTAM Symposium on Creep in Structures, 267–276.

equation;

$$\varepsilon = \varepsilon_0 + A\{1 - \exp(-\alpha t)\} + B\{\exp(\alpha t) - 1\} \tag{1}$$

where ε is the true strain, t is the time (h) , and the parameters ε_0, A, B and α are determined so that equation (1) gives a best fit to a measured creep curve. From the equation (1), creep rate, $d\varepsilon/dt$, is expressed as;

$$d\varepsilon/dt = A\ \alpha\exp(-\alpha t) + B\ \alpha\exp(\alpha t) \tag{2}$$

The equations (1) and (2) can express mostly the curves of creep strain and creep rate in the vicinity of minimum creep rate region, except the early stage of primary creep and the final stage of tertiary creep.

The minimum creep rate, $d\varepsilon_{min}/dt$, and the creep rupture life, t_r, are expressed by the following equations;

$$d\varepsilon_{min}/dt = 2\alpha\ (AB)^{0.5} \tag{3}$$

$$P = (1/\alpha)\ln\{(\varepsilon_r - \varepsilon_0 - A)/B\} \tag{4}$$

$$t_r = CP^q \tag{5}$$

where ε_r is the rupture strain and P is the rupture parameter. C and q are the constants which are obtained by best fitting between P and the measured rupture life.

3.   Creep Deformation Behavior and Rupture Life of Ferritic Steels

The creep deformation behavior of 1.3Mn-0.5Mo-0.5Ni, 1Cr-0.5Mo, 2.25Cr-1Mo, Mod.9Cr-1Mo, and 12Cr steels is analyzed in this work. The detailed data of these steels are obtained in NRIM Creep Data Sheet, Nos.18B, 1B, 3B, 43 and 13B, respectively.

3.1. 12Cr STEEL [4]

Figure 1 shows the relationship between stress vs. Larson-Miller parameter (LMP). Heat A has the highest creep rupture strength, and Heat C shows the lowest creep rupture strength. On the other hand, Heat B has the same strength at higher stresses as Heat A, but the strength of Heat B becomes equal at lower stresses to that of Heat C. The creep curves for this steel were analyzed using the equations (1) and (2), and the

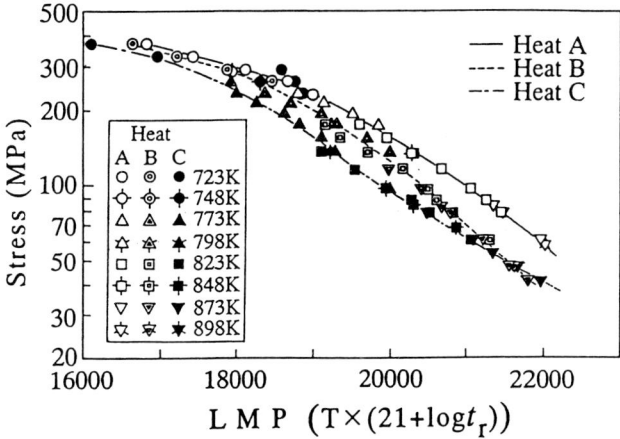

Figure 1.   Relationship between stress and Larson-Miller Parameter of 12Cr steel.

dependence of temperature and stress on 4 parameters was examined. The heat to heat variation in rupture life was recognized only in the magnitude of α. The stress dependence of α showed the same trend as that of rupture life. The creep deformation behavior of 12 Cr steel can be characterized by the magnitude of α. For the heat with small α value, the minimum creep rate was low, the increase in tertiary creep rate was mild and the rupture life was long. This heat-to-heat variation in the magnitude of α corresponded to that in the stability of microstructure, and the creep deformation behavior and rupture life for this steel depended on the initial morphology and stability of microstructure.

3.2. 1Cr-0.5Mo STEEL [5]

The creep rupture curves for 1Cr-0.5Mo steel show an inverse sigmoidal bending as shown in Figure 2. In order to clarify the origin of appearance of the inverse sigmoidal bending, the creep and creep rate curves were analyzed using the equations (1) and (2) . Figure 3 shows the creep and creep rate curves at the testing condition at which the inverse sigmoidal bending occurred. Two minima appear in the creep rate curve. Because two minima appeared in the creep rate curve, the overall creep process was divided in two processes; Type I for the initial process and Type II for the later process. The measured creep and creep rate curves are represented well by the combination of the calculated Type I and Type II processes. For higher stresses, the minimum creep rate and rupture life were controlled by the Type I process. For lower stresses, the minimum creep rate and rupture life were controlled by the Type II process. On the other hand, for the middle stress range, the minimum creep rate was controlled by the Type I process and the rupture life was controlled by the Type II process.

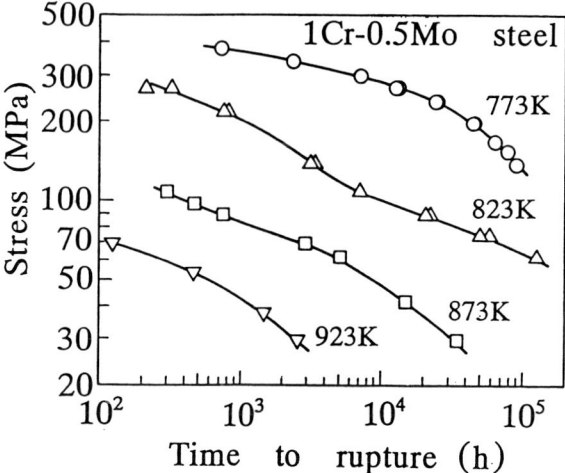

Figure 2. Creep rupture curves of 1Cr-0.5Mo steel.

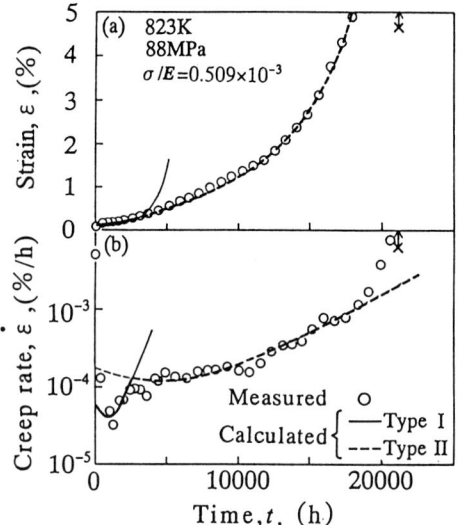

Figure 3. Comparison of experimental data with calculated creep and creep rate curves of 1Cr-0.5Mo steel.

Figure 4 shows the creep rupture curves calculated by the Type I and Type II processes. The change in controlling factor for creep deformation from the Type I to the Type II causes the inverse sigmoidal bending in the creep rupture curves, and the Type II process determines the long-term creep and rupture behavior.

The relationship between rupture life and minimum creep rate is well known to be

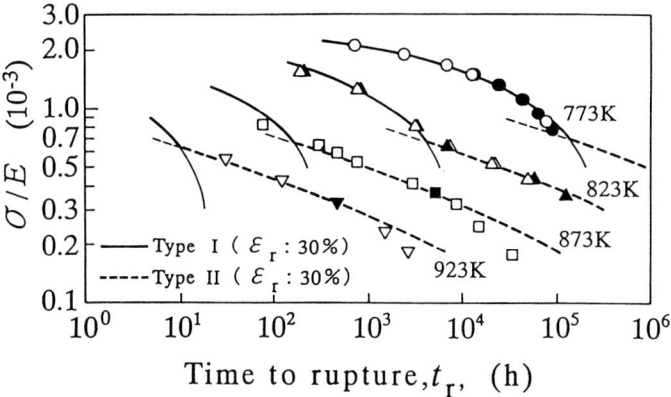

Figure 4. Comparison of measured rupture life with predicted one of 1Cr-0.5Mo steel. The creep curves whose rupture life is represented by the solid symbols were used for the analysis.

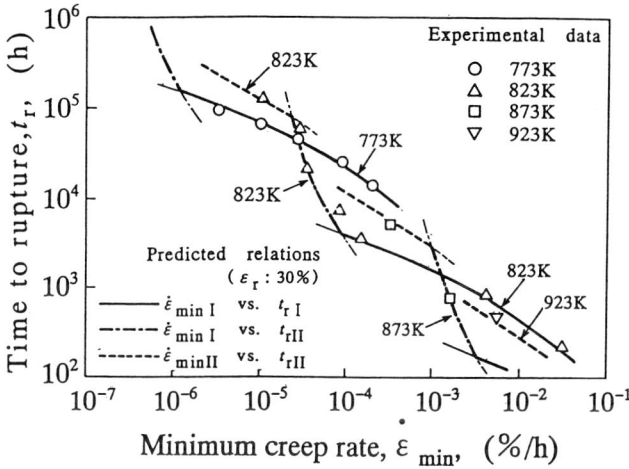

Figure 5. Comparison of measured results with predicted relations of time to rupture vs. minimum creep rate of 1Cr-0.5Mo steel.

linear in both logarithmic scale. For Cr-Mo ferritic heat resistant steels, however, this relation has large amount of scatter. Figure 5 shows the comparison of the experimental results with the relationship between rupture life and minimum creep rate calculated by the equations (3) and (5). The predicted lines agree well with the measured data. This relation is not linear, and the data scattering in this relation is caused by the change in controlling factor of minimum creep rate and rupture life with decreasing stress and increasing temperature.

3.3. 2.25Cr-1Mo STEEL [6]

The creep rupture curves for 2.25Cr-1Mo steel also shows the inverse sigmoidal bending. The creep curves of this steel were analyzed in the same way as 1Cr-0.5Mo steel. Figure 6 shows the relationship between rupture life and minimum creep rate. This relation has large amount of data scatter at longer rupture lives and lower minimum creep rates. The predicted lines agree well with the experimental data. Therefore, this scattering can be explained by the change in controlling factor of creep deformation and rupture life. Although the data scattering of Monkman-Grant relation is often interpreted from the decrease in creep rupture ductility, the investigation from the viewpoint of creep deformation controlling factor is necessary.

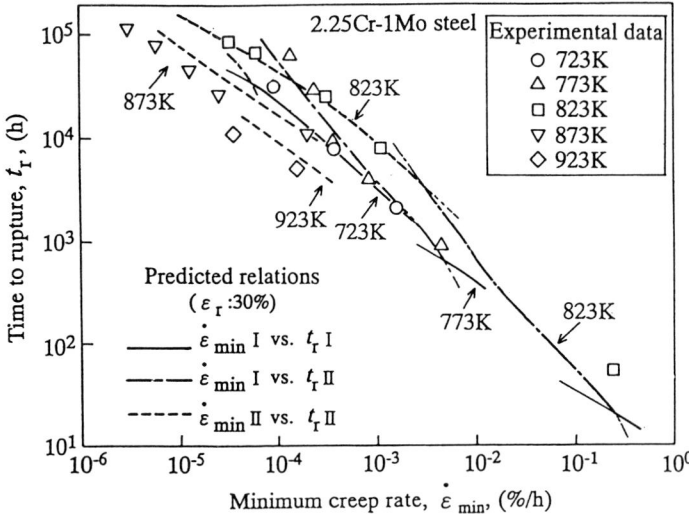

Figure 6. Comparison of measured results with predicted relations of time to rupture vs. minimum creep rate of 2.25Cr-1Mo steel.

3.4. 1.3Mn-0.5Mo- 0.5Ni STEEL [7]

The creep properties of two heats, La and Ld, were compared. Figure 7 shows the creep rupture curves and elongation. For higher stresses and shorter times, Heat La shows the same creep rupture strength as Heat Ld, but with decreasing stress, the rupture life of Heat La becomes shorter. The elongation of Heat La is lower at longer times, because the formation of creep cavities is remarkable for Heat La [8].

The creep curves of two heats were analyzed using the equations (1) and (2). The heat-to-heat variation in creep deformation behavior was characterized by the parameter B. The rupture life of both heats was calculated using the equation (5). The

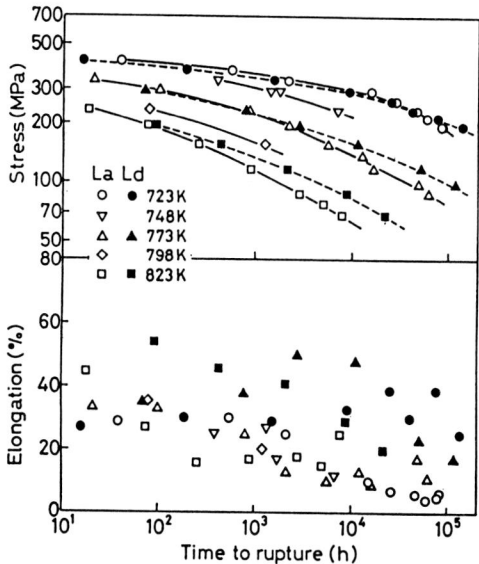

Figure 7. Creep rupture strength and elongation of 1.3Mn-0.5Mo-0.5Ni steel.

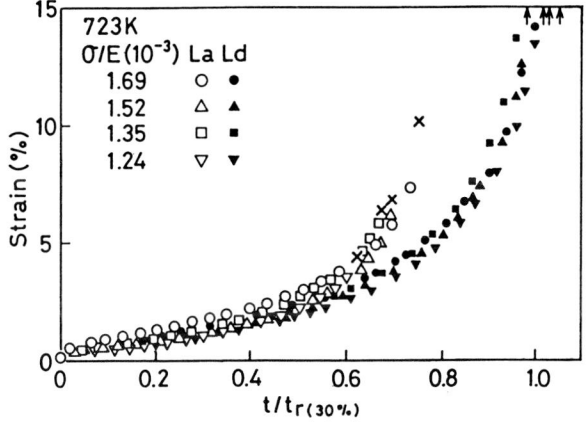

Figure 8. Relationship between creep strain and rupture life ratio, t/t_r.

calculated rupture curves agreed well with the experimental results. For Heat La, the rupture life was a little dependent on the value of ε_r, and with decreasing stress, the predicted rupture life agreed better with the measured one when the ε_r value was small.

For the steels whose creep damage formation is remarkable and rupture ductility is low, however, it should not be mistaken that the rupture life can be predicted by lowering the ε_r value. Figure 8 shows the relation of strain vs. time for both heats. The time is normalized using the t_r value of both heats calculated by the equation (5). The normalized creep curves of Heat La agree with those of Heat Ld except the final stage of tertiary creep region. The equations based on modified θ projection concept are obtained from the analysis of creep curves in the vicinity of minimum creep rate region as mentioned in Section 2. For the materials without creep damage formation, the rupture life can be predicted using the creep deformation behavior in the vicinity of the minimum creep rate region. However, for the steels with severe creep damage formation, not only the decrease in ductility but also the increase in creep rate due to the formation of creep damage must be considered.

3.5. Mod. 9Cr-1Mo STEEL [9]

The creep curves of Mod. 9r-1Mo steel (ASME P91) were analyzed using the equations (1) and (2), and the rupture life was calculated using the equation (5). Figure 9 shows the comparison of the predicted rupture curves with the experimental results. The predicted rupture life agrees well with the measured one except for the rupture life of longer than 10,000h at 873K and 923K. However, the rapid decrease in creep rupture strength at 10,000h can not be predicted by the temperature accelerating test method. The rupture elongation is not low and no formation of creep damage was observed.

The microstructural change with creep deformation was observed. For higher stresses at short term region less than 10,000h, homogeneous progress in recovery of tempered martensitic microstructure was observed. On the other hand, for longer than 10,000h region at 873 and 923K, the remarkable progress in the recovery of micro-

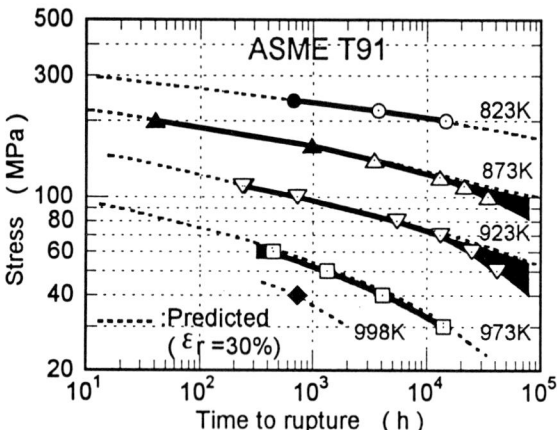

Figure 9.   Creep rupture curves of Mod. 9Cr-1Mo steel. The predicted lines were estimated based on the analysis of the creep curves indicated by solid symbols.

structure was observed in the vicinity of prior austenitic grain boundary. The tertiary creep rate was accelerated by such preferential recovery along a prior austenitic grain boundary.

The experimental result presents that the inhomogeneous microstructure which is formed during creep can promote the creep rate. That is, even if no creep damage such as cavity is formed, the creep rate in tertiary stage can be accelerated by local recovery of microstructure.

4. Long-term Creep Life Prediction

The microstructure in ferritic heat resistant steels changes during high temperature creep, and the creep deformation behavior is affected by the microstructural change. For the steels which occur homogeneous microstructural change, the rupture life could be predicted using the equation based on modified θ projection concept which characterizes the creep deformation behavior in the vicinity of the minimum creep rate region including tertiary creep. However, for the steels which occur the remarkable creep damage formation or inhomogeneous microstructural recovery, the rupture life was overestimated because the acceleration of tertiary creep rate due to local damage and recovery was not taken into consideration in modified θ projection concept.

Advanced ferritic heat resistant steels will have strong matrix such as tempered martensitic microstructure. After long service, preferential recovery of microstructure might occur in these steels. The creep damage might be also formed in restrained portion and welding joint. Therefore, the development of rupture life prediction method, in which the creep deformation due to creep damage formation and local microstructural recovery is taken into consideration, is necessary.

5. Conclusions

The creep curves of ferritic heat resistant steels was analyzed, and the creep deformation behavior of these steels was characterized using the equation based on modified θ projection concept. For the steels which occurred homogeneous microstructural change, the rupture life could be predicted well using the equation based on this concept. But, for the steels which occurred the creep damage formation and local recovery of microstructure, further investigation is necessary to develop a new life prediction procedure including these factors.

Acknowledgment

The authors wish to thank the members of NRIM Creep Data Sheet Project for performing the long-term tests.

References

1. Sakamoto, M., Yagi, K., Morishita, H., Kubo, K., Monma, Y., and Tanaka, C.: Classification of Creep Deformation Behaviour of 2.25r-1Mo Steel, 304 and 316 Steels, *J. Soc. Mater. Sci., Japan*, **39**(1990), 674-680.

2. Evans, R.W., Parker, J.D., and Wilshire, B.: *Recent Advances in Creep and Fracture of Engineering Materials and Structures*, ed.by Wilshire, B., and Owens, D.R., Pineridge Press, (1982), p.135.

3. Maruyama, K., and Oikawa, H.: A New Departure of Long Term Creep Curve Prediction up to the Tertiary Stage, *Tetsu-to-Hagane*, **73**(1987), 26-33.

4. Kushima, H., Kimura, K., Yagi, K., and Maruyama, K.: Long-term Creep Strength Property and Microstructural Stability of 12 Cr Steel, *Tetsu-to-Hagane*, **81**(1995), 214-219.

5. Kushima, H., Kimura, K., Yagi, K., Tanaka, C., and Maruyama, K.: Characterization of Creep Deformation Behaviour for Cr-Mo Steel, in Y. Hosoi, H. Yoshinaga, K. Oikawa and K. Maruyama (eds.), *Proc. of 7th JIM Intern. Symp. on Aspects of High Temperature Deformation and Fracture in Crystalline Materials*, The Japan Institute of Metals, (1993), 609-616.

6. Kushima, H., Kimura, K., Abe, F., Yagi, K., and Maruyama, K.: Assessment of Creep Strength Properties of 2.25Cr-1Mo Steel Based on Creep Deformation, *Report of 123rd Comm. on Heat-Resisting Metals and Alloys, JSPS*, **35**(1994), 261-274.

7. Kushima, H., Watanabe, T., Yagi, K., and Maruyama, K.: Evaluation of Creep Deformation and Rupture life of 1.3Mn-0.5Mo-0.5Ni Steel by Modified θ Projection Concept, *Tetsu-to-Hagane*, **78**(1992), 918-925.

8. Shinya, N., Kyono, J., Kushima, H., and Yokoi, S.: Creep Fracture Mechanism and Rupture Life of 1.3Mn-0.5Mo-0.5Ni Steel, *J.Soc. Mater. Sci., Japan*, **33**(1984), 441-446.

9. Kushima, H., Kimura, K., and Abe, F.: Degradation of Mod. 9Cr-1Mo Steel during Long-term Creep Deformation, *Tetsu-to-Hagane*, **85**(1999), 841-847.

NEAR-THRESHOLD FATIGUE CRACK GROWTH IN SUS304 STEEL AT ELEVATED TEMPERATURES

Shiro KUBO, Eiichi TAMURA, Nobuhiro TAGAMI,
and Kiyotsugu OHJI
Department of Mechanical Engineering and Systems,
Graduate School of Engineering, Osaka University,
2-1 Yamadaoka, Suita, Osaka, 565-0871, Japan
E-mail: kubo@mech.eng.osaka-u.ac.jp

1. Introduction

Many efforts have been devoted to the study of fatigue crack growth behavior at high temperatures [1-20]. It has been reported that the fatigue crack growth behavior at high temperatures can be classified into cycle-dependent and time-dependent ones. Threshold was found to exist in the cycle-dependent crack growth even in the creep regime.

The near-threshold fatigue crack growth and the threshold are important for evaluating the long-term lives of structures and their components for high temperature use. The near-threshold fatigue crack growth and the threshold at high temperatures are complicated due to many factors. The oxidation is more pronounced at high temperatures than at room temperature. In the near-threshold region the oxide-induced crack closure [21-23] is expected to be more significant in addition to plasticity-induced crack closure. The exclusion of crack closure or its measurements is required for the investigation of the near-threshold fatigue crack growth and the threshold [24]. The near-threshold crack growth depends on the loading history. When high range load excursions were introduced, crack grew even below the threshold obtained by conventional tests [11]. Under creep dominant conditions the presence of creep crack growth resulted in apparent loss of the threshold of cycle-dependent fatigue crack growth [9-10].

In this study fatigue crack growth experiments in the near-threshold region were conducted at 350, 450, 550, 650 and 700 (°C) for a type 304 austenitic stainless steel to investigate the character of the near-threshold fatigue crack growth and the threshold. The effects of stress ratio R and temperature were investigated. The laser-based interferometric strain-displacement gauge (I.S.D.G.) method [25-26] was used for the measurement of crack closure.

The behavior of fatigue crack growth in the near-threshold and subthreshold regions was investigated using K_{max}-constant ΔK-decreasing test method. By

S. Murakami and N. Ohno (eds.), IUTAM Symposium on Creep in Structures, 277–286.
© 2001 *Kluwer Academic Publishers. Printed in the Netherlands.*

increasing the value of ΔK or $K_{\max}$ for a crack in a threshold condition, the behaviors of subsequent fatigue crack growth in ΔK-decreasing tests were studied.

2. Material and Specimen

Material used for the investigation is a solution-treated SUS 304 stainless steel. The chemical composition and mechanical properties at room temperature are shown in Tables 1 and 2. Center-cracked specimens, which were sometimes called M(T) specimens, were used. The geometry of M(T) specimen is also shown in Fig. 1. Since Young's modulus of the material used was not obtained, Young's moduli at 350, 450, 550, 650 and 700 (°C) were interpolated from the ones at 29.4, 399, 482, 538, 593, 649 and 704 (°C) obtained for other lot of type 304 steels [27], considering that Young's modulus is almost the same for type 304 stainless steels.

Table 1   Chemical composition of SUS304 steel tested (wt.%)

C	Si	Mn	P	S	Ni	Cr	Fe
0.04	0.60	1.69	0.036	0.001	8.29	18.34	Bal.

Table 2   Mechanical properties of SUS304 steel at room temperature

Tensile Strength (MPa)	Elongation (%)	Reduction in Area (%)	Hardness (HB)
618	50.2	62.0	164

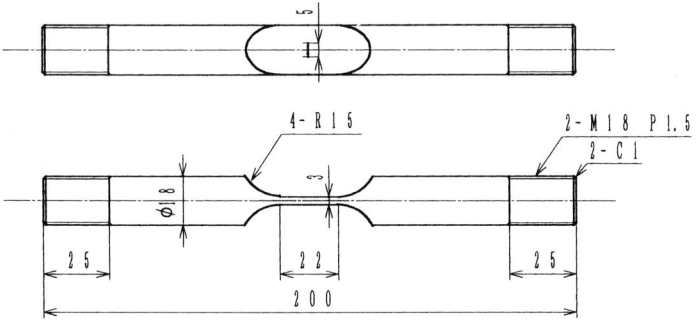

Fig. 1   Geometry of center-cracked specimen (in mm)

3. Experimental Procedures

For evaluating the near-threshold fatigue crack growth and the threshold value, fatigue crack growth experiments were conducted at 350, 450, 550, 650 and 700 (°C). Stress ratio, R was kept at -0.1 or 0.5. Loading waveform was sinusoidal and its frequency was set at 30 (Hz). ΔK-decreasing test method or load range ΔP-constant test method were applied for the fatigue crack growth tests. A.C. electric potential method was used for the measurement of crack length. The apparatus, the measurement of crack length and the procedures of fatigue crack growth test were detailed in the previous paper [24].

The laser-based I.S.D.G. method was used for the measurement of crack closure. Laser beam was emitted towards two Vickers indentations placed near the crack tip in symmetry with respect to the crack line. Two indentations reflected the laser beams, which formed interference fringe patterns. By measuring the movement of the fringe patterns, change in the distance between two Vickers indentations was calculated. In this study, I.S.D.G. was modified using platinum plate and optical filters. The details were given in the previous paper [24].

The behavior of fatigue crack growth in the near-threshold and subthreshold regions was investigated using K_{max}-constant ΔK-decreasing test method. Following the conventional ΔK-decreasing test, the values of ΔK and/or K_{max} were increased for a crack in a threshold condition and the behaviors of fatigue crack growth and its threshold were investigated.

4. Calculation of Fracture Mechanics Parameters

Crack opening ratio, U was determined by the unloading elastic compliance method from the measurement of crack opening displacement with the laser-based I.S.D.G. Effective stress intensity factor range, ΔK_{eff} was calculated by multiplying stress intensity factor range, ΔK and crack opening ratio, U. Since crack closure was not observed at $R=0.5$ and 650 (°C) in the previous paper [24], crack opening ratio, U was set to be 1 at $R=0.5$ or higher.

J-integral range, ΔJ_f consists of elastic and plastic components. It was shown in the previous paper [24] that non-elastic behavior in load-displacement curve was not significant in the near-threshold region. Then the value of effective J-integral range, $\Delta J_{f,eff}$ was given by the elastic component as follows with E denoting Young's modulus.

$$\Delta J_{f,eff} = \Delta K_{eff}^2 / E \qquad (1)$$

5. Near-Threshold Fatigue Crack Growth and Threshold

Fatigue crack growth experiments were conducted using the ΔK-decreasing tests and the ΔP-constant tests for stress ratio R of -0.1 and 0.5 at 350, 450, 550, 650 and 700

($^{\circ}$C). As an example relationship between da/dN with ΔK, ΔK_{eff} and $\Delta J_{f,eff}$ obtained at 550 ($^{\circ}$C) are shown in Figs. 2(a), (b), and (c), respectively. Solid symbols and open symbols in these figures show the results obtained at R = -0.1 and 0.5, respectively. Difference in symbols means difference in specimens tested.

In Fig. 2(a) the relationship between crack growth rate da/dN and ΔK for R= -0.1 deviated from that for R=0.5. In the relationships between da/dN versus ΔK_{eff} and da/dN versus $\Delta J_{f,eff}$ shown in Figs. 2(b) and 2(c), R dependence of da/dN is reconciled. The square open symbols deviate from other data. This may be due to difficulties in accurate determination of U for low U values. These figures show that crack growth rate, da/dN can be expressed as a function of ΔK_{eff} or $\Delta J_{f,eff}$. The solid lines in Figs. 2(b) and (c) are approximate lines for the relationship of da/dN with ΔK_{eff} and $\Delta J_{f,eff}$. Similar results were obtained for the other test temperatures.

Approximate equations of da/dN expressed in terms of $\Delta J_{f,eff}$ are given as follows with $\Delta J_{f,eff}$ and da/dN expressed in [kJ/m^2] and [m/cycle], respectively.

In the case of 350 ($^{\circ}$C)
$\quad \Delta J_{f,eff} > 0.122$(kJ/m^2)
$$da/dN = 3.60 \times 10^{-8} \, \Delta J_{f,eff} \qquad (2)$$
$\quad \Delta J_{f,eff} < 0.122$(kJ/m^2)
$$da/dN = 1.17 \times 10^{-4} \, \Delta J_{f,eff}^{4.84} \qquad (3)$$
In the case of 450 ($^{\circ}$C)
$\quad \Delta J_{f,eff} > 0.140$(kJ/m^2)
$$da/dN = 6.07 \times 10^{-8} \, \Delta J_{f,eff} \qquad (4)$$
$\quad \Delta J_{f,eff} < 0.140$(kJ/m^2)
$$da/dN = 1.17 \times 10^{-4} \, \Delta J_{f,eff}^{4.84} \qquad (5)$$
In the case of 550, 650 and 700 ($^{\circ}$C)
$\quad \Delta J_{f,eff} > 0.167$(kJ/m^2)
$$da/dN = 1.20 \times 10^{-7} \, \Delta J_{f,eff} \qquad (6)$$
$\quad \Delta J_{f,eff} < 0.167$(kJ/m^2)
$$da/dN = 1.17 \times 10^{-4} \, \Delta J_{f,eff}^{4.84} \qquad (7)$$

Equation (6) is coincident with the characteristics of cycle-dependent crack growth rate in the intermediate da/dN region of M(T) specimen of a type 304 steel obtained in the previous paper [9]. The other equations were determined by the least squares method assuming that the power law was applicable in the respective regions. It is seen from Eqs. (6) and (7) that crack growth rate, da/dN at 550, 650 and 700 ($^{\circ}$C) can be expressed using only two equations expressed in terms of $\Delta J_{f,eff}$. From Eqs. (3), (5) and (7) crack growth rate, da/dN at 350, 450, 550, 650 and 700 ($^{\circ}$C) for $\Delta J_{f,eff}$ lower than certain values can be expressed by the same equation of $\Delta J_{f,eff}$. The crack growth rate da/dN at 350 and 450 ($^{\circ}$C) is a little lower than those at the other temperatures for large values of $\Delta J_{f,eff}$. Thus the temperature dependence of crack growth characteristics is small when da/dN is correlated with $\Delta J_{f,eff}$.

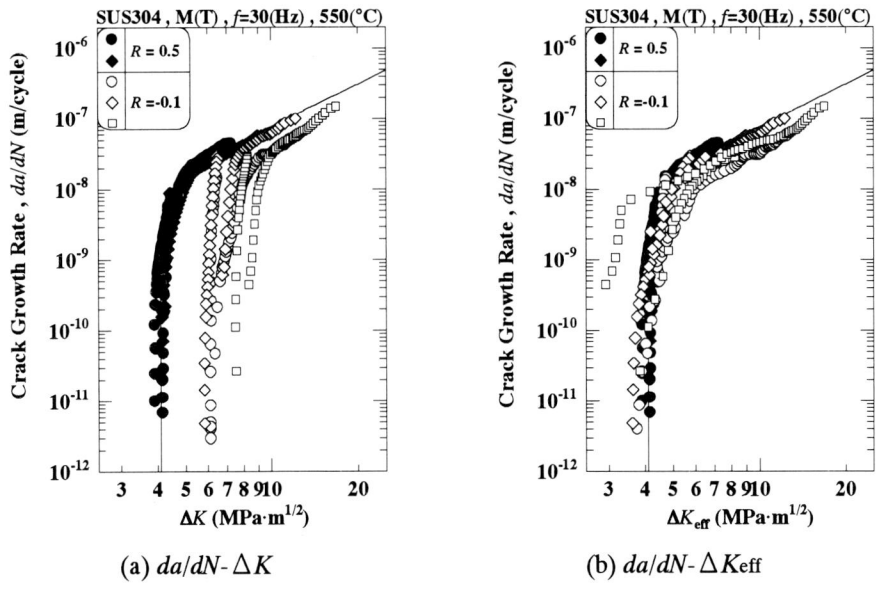

(a) da/dN-ΔK

(b) da/dN-ΔK_{eff}

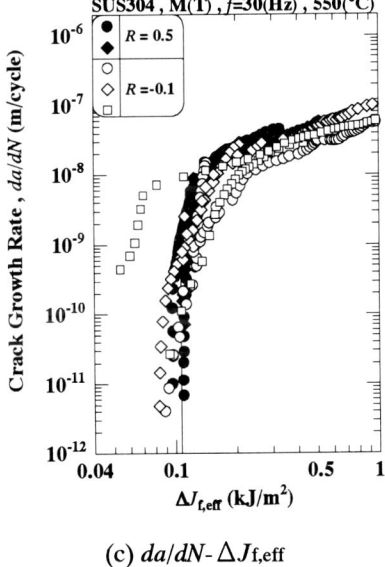

(c) da/dN-$\Delta J_{f,eff}$

Fig. 2   Correlation of da/dN with ΔK, ΔK_{eff} and $\Delta J_{f,eff}$ at 550 (℃) obtained using ΔK-decreasing tests and ΔP-constant tests

The value of ΔK_{eff} or $\Delta J_{f,eff}$ corresponding to da/dN of 10^{-10} (m/cycle) was taken as the threshold of fatigue crack growth. The threshold values obtained at test temperatures investigated are shown in Table 3. This table shows that the threshold value at 550 (°C) are the highest. It has been reported [12, 13] that the value of threshold of fatigue crack growth of a type 304 steel at room temperature is in the range of $\Delta K_{eff,th}$ = 1.5 - 2.5 (MPa·m$^{1/2}$), which are much lower than that at high temperatures obtained in this study. These results indicate that the threshold of the type 304 steel reaches a maximum at a temperature between 450 and 550 (°C).

Table 3   Values of fatigue crack growth threshold at 350, 450, 550, 650 and 700 (°C).

	350(°C)	450(°C)	550(°C)	650(°C)	700(°C)
$\Delta K_{eff,th}$ (MPa·m$^{1/2}$)	3.5	4.0	4.1	3.24	3.208
$\Delta J_{f,eff,th}$ (kJ/m^2)	0.07	0.097	0.106	0.07	0.07

6. Fatigue Crack Growth below Conventional Threshold

The behavior of fatigue crack growth below the threshold value, which was obtained by using the conventional ΔK-decreasing test method, was investigated using K_{max}-constant ΔK-decreasing test method and single-edge cracked (SEC) specimens. Threshold condition was established by decreasing the value of ΔK and keeping K_{max} constant. Then the value of K_{max} and/or ΔK were increased for a crack in a threshold condition, and the fatigue crack growth behavior in the following ΔK-decreasing test was investigated. To exclude the effect of crack closure, stress ratio R was set to be higher than 0.5.

As an example Fig. 3 shows the da/dN-ΔK relation obtained by the tests at 350 (°C). In the test shown in Fig. 3(a) the value of K_{max} was increased in the conventional threshold condition. The value of ΔK at the restart of the ΔK-decreasing test is equal to the conventional threshold value. As is seen in the figure the crack grew even in the subthreshold region.

In the test shown in Fig. 3(b) the value of ΔK together with K_{max} was increased at the restart of the ΔK-decreasing test. The da/dN-ΔK relation is not so different from that obtained in the conventional test.

These results show that the crack grew in the subthreshold region when ΔK at the restart of the test is equal to or less than the conventional threshold value, while the subthreshold crack growth was negligible when the initial value of ΔK is larger than the threshold value.

Then the value of K_{max} was successively increased at the restart of the ΔK-decreasing tests in the newly established threshold region. Finally a ΔK-decreasing test with the initial value of ΔK larger than the conventional threshold value was

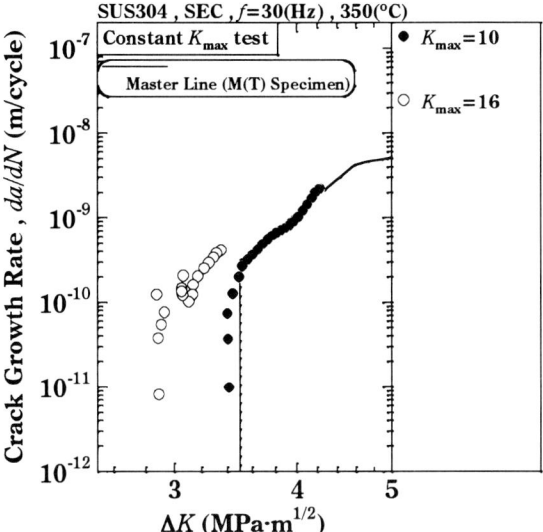

(a) When initial value of ΔK at restart of test was equal to conventional threshold

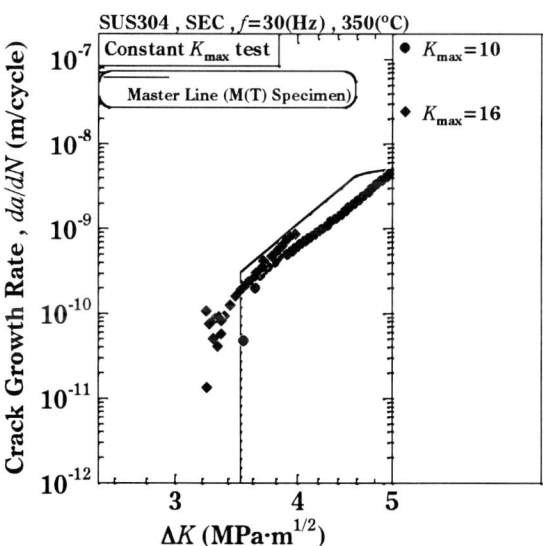

(b) When initial value of ΔK at restart of test was higher than conventional threshold

Fig. 3   Fatigue crack growth observed after increase in K_{max} and/or ΔK for a crack in threshold condition at 350 (°C)

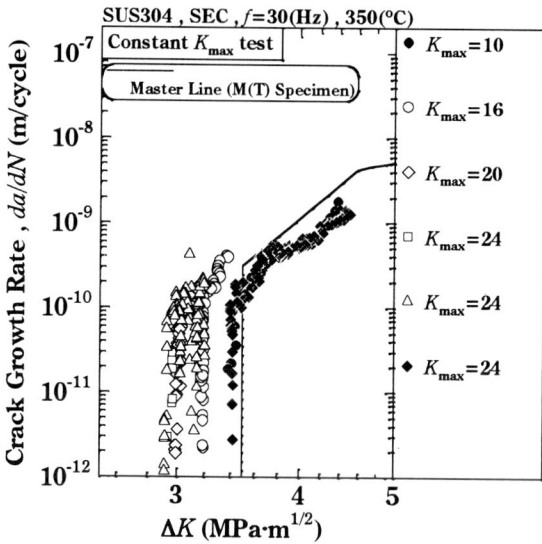

Fig. 4   Subthreshold fatigue crack growth after repeated increase in K_{max} for a crack in threshold condition at 350 (°C)

conducted.   Open symbols in Fig. 4 shows the da/dN-ΔK relations obtained in the tests when ΔK at the restart of the tests is equal to or less than the conventional threshold value.   It is seen that crack grew in the repeated ΔK-decreasing tests in the subthreshold region.   Even after repeated increase in K_{max} the crack no longer grew for ΔK smaller than a certain value.

Diamond-shaped solid symbols in Fig. 4 show crack growth rate da/dN obtained for the last ΔK-decreasing test with the initial value of ΔK larger than the threshold. The da/dN-ΔK relation of the test coincides with that obtained in the first ΔK-decreasing test even though R of the last test is the highest in the series of tests.   This suggests that the effect of stress ratio R is small and that the initial value of ΔK at the restart of the ΔK-decreasing test has a major influence on the following fatigue crack growth behavior.

Similar results were obtained at 550 and 650 (°C).   Those results are not shown in this paper for the sake of space.

7. Conclusions

Fatigue crack growth behavior in the type 304 austenitic stainless steel was investigated at elevated temperatures, i.e. 350, 450, 550, 650 and 700 (°C).   Crack growth rates for stress ratio R=-0.1 plotted against stress intensity factor range, ΔK, deviated from those for R=0.5.   Using the effective stress intensity factor range, ΔK_{eff} and effective J-integral range, $\Delta J_{f,eff}$, which were calculated using crack opening ratio, the stress ratio dependence in crack growth behaviors for R = -0.1 and 0.5 was reconciled.   The value of fatigue crack growth threshold at elevated temperatures expressed in terms of ΔK_{eff} or $\Delta J_{f,eff}$ was much higher than that at room temperature.   It was found that the threshold value reached a maximum at a temperature between 450 and 550 (°C).

The behavior of fatigue crack growth was investigated using the K_{max}-constant ΔK-decreasing test method.   By increasing the values of K_{max} and/or ΔK for a crack in a threshold condition, the behaviors of fatigue crack growth were investigated.   It was found that the crack grew even in the region below the threshold value obtained by using the conventional ΔK-decreasing test.   Crack growth was observed by successively increasing the K_{max} value in newly established threshold regions.   In the region below a certain value of ΔK, however, the crack no longer grew.

References

1. James, L.A.: The Effect of Frequency upon the Fatigue-Crack Growth of Type 304 Stainless Steel at 1000 F, *Proceedings of the 1971 National Symp. on Fracture Mechanics, Part I*, ASTM STP 513 (1972), 218-229.
2. Haigh, J.R., Skelton, R.P., and Richards, C.E.: Oxidation-Assisted Crack Growth during High Cycle Fatigue of a 1% Cr-Mo-V Steel at 550°C, *Materials Science and Engineering* **26** (1976), 167-174.
3. Ellison, E.G., and Harper, M.P.: Creep Behavior of Components Containing Cracks - A Critical Review, *Journal of Strain Analysis* **13** (1978), 35-51.
4. Smith, D.J., and Ellison, E.G.: Modelling Crack Growth for Creep and Fatigue Loading, *International Journal of Pressure Vessels and Piping* **50** (1992), 231-241.
5. Ohtani, R., Yamada, K., Kashiwagi, T., and Matsubara, H.: Crack Propagation of 304 Stainless Steel in Low Cycle Fatigue at Elevated Temperature, *Transactions of the Japan Society of Mechanical Engineers* (in Japanese) **48** A (1982), 1378-1390.
6. Okazaki, M., Shiraiwa, F., Hattori, I., and Koizumi, T.: Effect of Strain Wave Shape on Low-Cycle Fatigue Crack Propagation of SUS 304 Stainless Steel at Elevated Temperatures, *Journal of the Society of Materials Science, Japan* (in Japanese) **32** (1983), 645-650.
7. Sadananda, K.: Crack Propagation under Creep and Fatigue, *Nuclear Engineering and Design*, **83** (1984), 303-323.
8. Kato, Y., Hasegawa, N., Matsushima, M., and Nakashima, H.: Effects of Cyclic Strain Aging and Crack Surface Oxidation on Threshold Stress Intensity Factor Range ΔK_{th} in Low Carbon Steel at Elevated Temperatures, *Transactions of the Japan Society of Mechanical Engineers* (in Japanese) **52** A (1986), 903-907.
9. Ohji, K., and Kubo, S.: Fracture Mechanics Evaluation of Crack Growth Behavior Under Creep and Creep-Fatigue Conditions, *Current Japanese Materials Research - Vol.3 High Temperature Creep-Fatigue*, Ohtani, R., Ohnami, M., and Inoue, T. ed., Elsevier Appl. Sci., Soc. Mat. Sci., Japan (1988), 91-113.
10. Ohji, K., and Kubo, S.: Crack Growth Behaviour and Fractographs Under Creep-Fatigue Including Near-Threshold Regime, *Current Japanese Materials Research - Vol.6 Fractography*, Koterazawa, R., Ebara, R., and Nishida, S. ed., Elsevier Appl. Sci., Soc. Mat. Sci., Japan (1990), 105-123.
11. Ohji, K., Kubo, S., and Nakai, Y.: Near-Threshold Fatigue Crack Growth Behavior at High Temperatures, Creep: *Characterization, Damage and Life Assessment*, Proc. 5th Int. Conf. on Creep of Materials, Woodford, D.A., Townley, C.H.A., and Ohnami, M. ed., ASM (1992) 379-388.
12. Ogura, K., Miyoshi, Y., and Nishikawa, I.: Fatigue Crack Growth and Threshold Behavior at Elevated

Temperatures, *Transactions of the Japan Society of Mechanical Engineers* (in Japanese) **52** A (1986), 89-98.

13. Takeuchi, E., Matsuoka, S., and Nishijima, S.: Near-Threshold Fatigue Crack Growth Properties for SB42 and 304 Steels at Elevated Temperature, *Transactions of the Japan Society of Mechanical Engineers* (in Japanese) **52** A (1986), 2155-2161.

14. Nikbin, K.M., and Webster, G.A.: Prediction of Crack Growth under Creep-Fatigue Loading Conditions, *Low Cycle Fatigue*, ASTM STP 942 (1988), 281-292.

15. Webster, G.A.: Lifetime Estimates of Cracked High Temperature Components, *International Journal of Pressure Vessels and Piping*, **50** (1992), 133-145.

16. Smith, H.H., Kullen, P.S., and Michel, D.J.: Fatigue Crack Propagation Behavior of Titanium Alloys 6242S and 5621S at Elevated Temperature, *Metallurgical Transactions* **19A** (1988), 881-885.

17. Kobayashi, H., and Park, K.: Effect of Oxidation on Fatigue Crack Growth Threshold of Alloy-and Carbon-Steels at Elevated Temperature, *Journal of Japan High Pressure Institute* (in Japanese) **30** (1992), 14-23.

18. Makhlouf, K., and Jones, J.W.: Near-Threshold Fatigue Crack Growth Behavior of a Ferritic Stainless Steel at Elevated Temperatures, *International Journal of Fatigue* **14** (1992), 97-104.

19. Ghonem, H., Nicholas, T., and Pineau, A.: Elevated Temperature Fatigue Crack Growth in Alloy 718 - Part II : Effects of Environmental and Material Variables, *Fatigue and Fracture of Engineering Materials and Structures* **16** (1993), 577-590.

20. Ogura, K., Nishikawa, I. and Kimura, Y.: Crack Closure Measurement at Elevated Temperatures Using Laser Interferometric Gage and Fatigue Threshold, *Transactions of the Japan Society of Mechanical Engineers* (in Japanese) **56** A (1990), 526-531.

21. Stewart, A.T.: The Influence of Environment and Stress Ratio on Fatigue Crack Growth at Near Threshold Stress Intensities in Low-Alloy Steels, *Engineering Fracture Mechanics* **13** (1980), 463-478.

22. Ritchie, R.O., Suresh, S., and Moss, C.M.: Near-Threshold Fatigue Crack Growth in 2 1 / 4 Cr-1Mo Pressure Vessel Steel in Air and Hydrogen, *Journal of Engineering Materials and Technology* **102** (1980), 293-299.

23. Suresh, S., Parks, D.M. and Richie, R.O.: Crack Tip Oxide Formation and Its Influence on Fatigue Thresholds, *Fatigue Thresholds* **1** (1982), 391-408.

24. Tamura, E., Kubo, S., Ohji, K., and Nakai, Y.: An Application of Interferometric Strain-Displacement Gauge to Measurement of Fatigue Crack Closure at High Temperatures, *Proc. of Int. Conf. on Advanced Technology in Experimental Mechanics*, JSME (1997), 55-60.

25. Sharpe, W.N. Jr.: Applications of the Interferometric Strain/Displacement Gage, *Optical Engineering* **21** (1982), 483-488.

26. Sharpe, W.N. Jr., and Su, X.: Closure Measurements of Naturally Initiating Small Cracks, *Engineering Fracture Mechanics* **30** (1988), 275-294.

27. Simmons, W.F. and Cross, H.C.: Data Sheets: 18 percent Chromium - 8 percent Nickel, *Report on the Elevated -Temperature Properties of Stainless Steels*, ASTM STP 124 (1952), 60-70.

APPROXIMATE VISCOPLASTIC NOTCH ANALYSIS

G. Härkegård and H.-J. Huth
Norwegian University of Science and Technology
NO-7491 Trondheim, Norway

Abstract
Approximate methods for the calculation of notch stress and strain in elastic-plastic and elastic-viscoplastic bodies have been reviewed. Equations based on Neuber's rule, total strain energy density (TSED), strain energy density (SED) and the contour integral J have been formulated within a common framework. Common and individual features of the different models have been pointed out. The TSED model agrees with Neuber's rule for an incompressible solid. Under certain simplifying assumptions, SED and J based models become equivalent. Whereas the stress concentration factor is fairly insensitive to the choice of model, strain concentration factors differ considerably between models. Nevertheless, under fully (visco)plastic conditions, all models yield identical asymptotic predictions of stress *and* strain concentration factors.

1. Introduction

Life prediction of mechanical components requires the assessment of stress and strain peaks, which occur due to sudden changes in geometry, material and load distribution. The analysis of high-temperature components involves non-linear material behaviour in the form of plastic deformation and creep. Calculations are routinely carried out by means of a (commercially available) finite element code on a desktop computer. In particular, this is true for geometric stress raisers or 'notches', such as grooves, holes, fillets etc. At first glance, therefore, modern computerised tools have completely replaced approximate methods for the non-linear analysis of notches. However, a complete non-linear analysis requires comprehensive modelling effort, and computing time may become prohibitive, if the designer needs to consider a number of design alternatives. Therefore, a well-validated approximate method of non-linear notch analysis is a valuable *complement* to finite element analysis. Moreover, it gives the designer an improved *understanding* of the development of notch stress and strain under plastic deformation and creep conditions.

At the last IUTAM Symposium on Creep in Structures, Kubo and Ohji [1] presented a method for estimating the stress and strain history at the rounded tip of a narrow, flat-surfaced notch under *large-scale creep* conditions based on the path-independence of the contour integral J. In 1994, Moftakhar [2] put forward a dissertation on the calcula-

S. Murakami and N. Ohno (eds.), IUTAM Symposium on Creep in Structures, 287–296.

tion of notch stress and strain under *small-scale* creep postulating the invariance of strain energy density and total strain energy density. Last year, Sørbø [3] presented his doctoral thesis on viscoplastic notch analysis under *large-scale* creep based on Neuber's rule. Although different models have been used, authors report good agreement between predictions and finite element results. In the following, formulations and predictions of the different models will be presented and compared within a common framework. Both elastic-plastic and elastic-viscoplastic solids have been considered. In order to keep the mathematics simple, the presentation has been restricted to a two-dimensional notched body under *simple tension* and *plane stress* conditions.

2. Neuber's rule

Härkegård and Sørbø [4] recently presented a formulation of Neuber's rule for notch analysis permitting multiaxial stress and large-scale plastic deformation or creep. The application of the model to a notched plate in plane stress will be demonstrated in the following. For details as well as applications to a variety of configurations, the reader is referred to the original work [3, 4].

Based on the analysis of a notched body under transverse shear, Neuber [5] postulated the invariance of the product of *equivalent* stress and strain concentration factors under general, multiaxial loading

$$K_{\sigma_e} \cdot K_{\varepsilon_e} = K_{te}^2 \tag{1}$$

where $K_{\sigma_e} = \sigma_e / \sigma_{ne}$ and $K_{\varepsilon_e} = \varepsilon_e / \varepsilon_{ne}$. In the following, *nominal* stress and strain will be defined as *reference* stress and strain (cf. [4]). An alternative formulation of Eq. (1) is in terms of the product of local equivalent stress and strain (= 'total' deviatoric strain energy density):

$$\left(\sigma'_{ij}\varepsilon'_{ij} =\right)\sigma_e\varepsilon_e = K_{te}^2\sigma_{ne}\varepsilon_{ne}\left(= K_{te}^2\sigma'_{n,ij}\varepsilon'_{n,ij}\right) \tag{2}$$

For a power hardening *elastic-plastic* solid, *equivalent* strain is given by

$$\varepsilon_e = \varepsilon_e^e + \varepsilon_e^p = \sigma_e/3G + A \cdot \sigma_e^n \tag{3}$$

Under *simple tension* $\sigma_e = \sigma$, i.e. Eq. (2) becomes

$$\sigma \cdot \left(\sigma/3G + \varepsilon^p\right) = K_t^2 \cdot \sigma_n \cdot \left(\sigma_n/3G + \varepsilon_n^p\right) \tag{4}$$

and the strain in the direction of the (uniaxial) stress component can be written as

$$\varepsilon = \varepsilon^e + \varepsilon^p = \sigma/E + A \cdot \sigma^n \tag{5}$$

By introducing suitably defined normalised uniaxial stress and strain (cf. [4]), we can re-formulate Eq. (5) as

$$e = e^e + e^p = s + s^n \tag{6}$$

With $s = K_\sigma s_n$, the dimensionless formulation of Eq. (4) becomes

$$\frac{E}{3G} K_\sigma^2 s_n^2 + K_\sigma^{n+1} s_n^{n+1} = K_t^2 \left(\frac{E}{3G} s_n^2 + s_n^{n+1} \right) \tag{7}$$

Having solved Eq. (7) for K_σ, we obtain the strain concentration factor as

$$K_\varepsilon = \frac{K_\sigma s_n + K_\sigma^n s_n^n}{s_n + s_n^n} \tag{8}$$

For large plastic strains, i.e. $s_n^n \gg 1$, stress and strain concentration factors approach

$$K_{\sigma\infty} = K_t^{\frac{2}{n+1}} \tag{9}$$

$$K_{\varepsilon\infty} = K_{\sigma\infty}^n = K_t^{\frac{2n}{n+1}}$$

For a power hardening *elastic-viscoplastic* solid, *equivalent* strain *rate* is given by

$$\dot{\varepsilon}_e = \dot{\varepsilon}_e^e + \dot{\varepsilon}_e^p = \dot{\sigma}_e / 3G + A \cdot \sigma_{ne}^n \tag{10}$$

Under *simple tension* the strain rate in the direction of the (uniaxial) stress component can be written as

$$\dot{\varepsilon} = \dot{\varepsilon}^e + \dot{\varepsilon}^p = \dot{\sigma} / E + A \cdot \sigma^n \tag{11}$$

By introducing suitably defined normalised stress, strain and time (cf. [4]), we can re-write Eq. (11) as

$$\dot{e} = \dot{e}^e + \dot{e}^p = \dot{s} + s^n \tag{12}$$

Under nominal creep conditions, i.e. σ_n = constant, it is natural to introduce $s = \sigma/\sigma_n$ (= K_σ), $e = E\varepsilon/\varepsilon_n$ and $\tau = EA\sigma_n^{n-1}t$, i.e. $s_n = 1$ and $e_n = 1 + \tau$. Differentiating Eq. (4) with respect to time and eliminating ε^p, we obtain a first order differential equation for the stress concentration factor K_σ

$$\frac{dK_\sigma}{d\tau} = -K_\sigma \frac{K_\sigma^{n+1} - K_t^2}{\dfrac{E}{3G}K_\sigma^2 + K_t^2\left(\dfrac{E}{3G} + \tau\right)} \tag{13}$$

with the initial condition $K_\sigma(0) = K_t$. The strain concentration factor is

$$K_\varepsilon = \frac{e}{e_n} = \frac{K_\sigma + \int_0^\tau K_\sigma^n d\tau}{1 + \tau} \tag{14}$$

3. Invariance of total strain energy density

By substituting stress and strain for the corresponding deviatoric quantities into Eq. (2), we obtain for the 'total' strain energy density [6]

$$\left(W + \overline{W} =\right)\sigma_{ij}\varepsilon_{ij} = K_{tTSED}^2 \sigma_{n,ij}\varepsilon_{n,ij} \tag{15}$$

This is a generalised version of the TSED model as described by Moftakhar [2]. The corresponding 'stress concentration factor' is given by the linearly elastic solution (denoted by *) through

$$K_{tTSED} = \sqrt{\frac{\sigma_{ij}^* \varepsilon_{ij}^*}{\sigma_{n,ij}^* \varepsilon_{n,ij}^*}} \tag{16}$$

Under *simple tension*, $K_{tTSED} = K_t$, and the TSED equations can be formally obtained from the corresponding 'Neuber' equations by setting $E/3G = 1$. Thus, Eq. (7) for the stress concentration factor of an *elastic-plastic* solid turns into

$$K_\sigma^2 s_n^2 + K_\sigma^{n+1} s_n^{n+1} = K_t^2\left(s_n^2 + s_n^{n+1}\right) \tag{17}$$

Similarly, for an *elastic-viscoplastic* solid, we obtain from Eq. (13)

$$\frac{dK_\sigma}{d\tau} = -K_\sigma \frac{K_\sigma^{n+1} - K_t^2}{K_\sigma^2 + K_t^2(1 + \tau)} \tag{18}$$

4. Invariance of strain energy density

Assuming *small-scale* plasticity, Molski and Glinka [7] postulated the strain energy density, $W = \int \sigma_{ij} d\varepsilon_{ij}$, at the root of a notch to be independent of the material behaviour at a given external load. Sørbø and Härkegård [6] suggested a formulation including the case of *large-scale* plasticity (cf. Eqs. (15) and (16)):

$$\left(\frac{W}{W_n}\right)_{\varepsilon_{n,ij}}^{\varepsilon_{ij}} = \frac{\int_0^{\varepsilon_{ij}} \sigma_{ij} d\varepsilon_{ij}}{\int_0^{\varepsilon_{n,ij}} \sigma_{n,ij} d\varepsilon_{n,ij}} = \frac{\int_0^{\varepsilon_{ij}^*} \sigma_{ij}^* d\varepsilon_{ij}^*}{\int_0^{\varepsilon_{n,ij}^*} \sigma_{n,ij}^* d\varepsilon_{n,ij}^*} = K_{tSED}^2 \tag{19}$$

Under *simple tension*, the right hand side is given by K_t^2, and Eq. (19) can be written as

$$\left(W =\right)\int_0^\varepsilon \sigma d\varepsilon = K_t^2 \int_0^{\varepsilon_n} \sigma_n d\varepsilon_n \left(= K_t^2 W_n\right) \tag{20}$$

With uniaxial *elastic-plastic* strain given by Eq. (5), integration of Eq. (20) yields

$$\frac{\sigma^2}{2E} + \frac{n}{n+1}\sigma \cdot \varepsilon^p = K_t^2\left(\frac{\sigma_n^2}{2E} + \frac{n}{n+1}\sigma_n \cdot \varepsilon_n^p\right) \tag{21}$$

or in dimensionless form

$$K_\sigma^2 s_n^2 + \frac{2n}{n+1}K_\sigma^{n+1}s_n^{n+1} = K_t^2\left(s_n^2 + \frac{2n}{n+1}s_n^{n+1}\right) \tag{22}$$

Asymptotic stress and strain concentration factors are again given by Eq. (9).

For the analysis of an *elastic-viscoplastic* solid, we differentiate Eq. (20) with respect to time:

$$\dot{W} = \sigma \cdot \dot{\varepsilon} = K_t^2 \sigma_n \cdot \dot{\varepsilon}_n \tag{23}$$

Substitution of the strain rate of Eq. (11) into Eq. (23) gives us a first order differential equation for K_σ:

$$\frac{dK_\sigma}{d\tau} = -\frac{K_\sigma^{n+1} - K_t^2}{K_\sigma} \tag{24}$$

5. Path independence of the contour integral *J*

In his much quoted paper on the contour integral

$$J = \int_{\Gamma} \left(W dy - \sigma_{ji} n_j \frac{\partial u_i}{\partial x} ds \right) \tag{25}$$

Rice [8] demonstrated its use in the approximate analysis of the strain concentration at smooth-ended notch tips. In the spirit of Rice, we shall study a straight slot in a plate under simple tension and plane stress conditions. Two different integration paths Γ are considered: Γ_{notch} along the surface of the notch (including the notch tip) and Γ_{far} away from the notch tip. The path independence of *J* for a (non-linearly) elastic solid requires

$$J_{notch} = J_{far} \tag{26}$$

Since $\sigma_{ji} n_j = 0$ on the *stress-free* surface of the notch, and $dy = 0$ on the *flat* surfaces of the notch

$$J_{notch} = \int_{\Gamma_{tip}} W dy = \int_{-\pi/2}^{\pi/2} W \cdot \rho \cos\theta \cdot d\theta \tag{27}$$

where ρ denotes the (constant) radius of the notch tip, θ the angle between the radius and the *x*-axis. The left-hand side of Eq. (21) gives the strain energy density of a power hardening elastic-plastic solid, i.e.

$$W(\theta) = \frac{\sigma(\theta)^2}{2E} + \frac{n}{n+1} \sigma(\theta) \cdot \varepsilon^P(\theta) \tag{28}$$

With $\sigma(\theta)^2 = \sigma_{max}^2 \cdot f^e(\theta)$ and $\sigma(\theta) \cdot \varepsilon^P(\theta) = \sigma_{max} \varepsilon_{max}^P \cdot f^P(\theta)$, we obtain

$$J_{notch} = \rho \cdot \left(\frac{\sigma_{max}^2}{2E} \cdot \phi^e + \frac{n}{n+1} \cdot \sigma_{max} \varepsilon_{max}^P \cdot \phi^P \right) \tag{29}$$

where ϕ^e and ϕ^P denote the integrals

$$\phi^e = \int_{-\pi/2}^{\pi/2} f^e(\theta) \cos\theta \cdot d\theta, \quad \phi^P = \int_{-\pi/2}^{\pi/2} f^P(\theta) \cos\theta \cdot d\theta \tag{30}$$

The contour integral J_{far} should be accurately estimated by the *crack* solution provided that $\rho \ll a$. Unfortunately, available (approximate) solutions for *J* in an elastic-plastic solid [9-11] do not admit a straightforward, consistent generalisation to an elastic-visco-plastic solid. In particular, for a given hardening exponent n, the stress under steady-state creep does not agree with that in the fully plastic condition, as required by Hoff's analogy. In order to eliminate this inconsistency, we suggest that the far-field J be expressed as

$$J_{far} = \kappa \cdot W_n a = \kappa \cdot a \int_0^{\varepsilon_n} \sigma_n d\varepsilon_n \qquad (31)$$

For a power hardening *elastic-plastic* solid, described by Eq. (5), integration of Eq. (31) yields

$$J_{far} = \kappa \cdot a \cdot \left(\frac{\sigma_n^2}{2E} + \frac{n}{n+1} \sigma_n \cdot \varepsilon_n^p \right) \qquad (32)$$

Under linearly elastic conditions

$$\left(J_{notch} = \right) \phi^e \cdot \rho \cdot \frac{K_t^2 \sigma_n^2}{2E} = \frac{K^2}{E} \left(= J_{far} \right) \qquad (33)$$

$$\left(J_{far} = \right) \kappa \cdot a \cdot \frac{\sigma_n^2}{2E} = \frac{K^2}{E} \qquad (34)$$

According to Shin et al. [12]

$$K_t \approx 2F \sqrt{\frac{a}{\rho}} \qquad (35)$$

where F is the geometry factor in the expression for the stress intensity factor

$$K = F \cdot \sigma_n \sqrt{\pi a} \qquad (36)$$

Substitution of Eqs. (35) and (36) into Eq. (33) yields $\phi^e = \pi/2$. Lacking theoretically well-founded expressions for f^e and f^p of a power-hardening solid, we assume that $f^e = f^p$, and that f^e retains its elastically calculated angular dependence at all stress levels, i.e. $\phi^e = \phi^p = \pi/2$. Further, by substituting Eq. (36) into Eq. (34), we obtain $\kappa = 2\pi F^2$. Combination of Eqs. (26), (29), (32) and (35) yields equations for σ (subscript *max* has been dropped) and K_σ identical to Eq. (21) and Eq. (22), respectively, of the strain energy density model. Similarly, for a power-hardening *elastic-viscoplastic* solid, differentiation of Eqs. (26), (27) and (31) with respect to time reproduces Eq. (24). However, this equation is at variance with the formulation by Kubo and Ohji [1], who used the *elastic-plastic* model, cf. Eq. (21), as a starting point for formulating the *viscoplastic* model.

6. Comparison between approximate models and finite element analysis

Following Kubo and Ohji [1], we study a uniaxially loaded plate of width $2w$ and height $2h = 4w$ with a straight centre slot of length $2a = 0.6w$ and tip radius $\rho = 0.02w = a/15$. Fig. 1 shows a quadrant of the plate, which was modelled in ABAQUS using 6500 eight-node elements. The element size at the tip of the slot is $0.05\rho = 0.001w$. According to a linearly elastic finite element calculation, the (net-section) stress concentration factor $K_t = 6.76$. Results from stress-strain analyses of power-hardening elastic-plastic and elastic-viscoplastic solids with $n = 5$ are briefly presented below.

Eqs. (7), (17) and (22) are solved for K_σ of the *elastic-plastic* plate by means of a quasi-Newton procedure, and K_ε is calculated by means of Eq. (8). The finite element model has been analysed by means of ABAQUS. Fig. 2 shows the results. The *stress* concentration factor of the finite element analysis is in close agreement with predictions by the approximate models. The finite element *strain* concentration factor falls below that of the TSED and Neuber models for all levels of nominal stress and below that of the SED model for large nominal stresses.

For the *elastic-viscoplastic* plate, Eqs. (13), (18) and (24) have been solved for K_σ by means of a Runge-Kutta procedure, and K_ε has been calculated by means of Eq. (14), where Simpson's rule has been used to evaluate the integral. Fig. 3 shows the results. Again, there is good agreement between the *stress* concentration factors predicted by the various methods. The stress concentration factors are also in good agreement with that calculated by Kubo and Ohji except for short times ($\tau < 0.001$). The main reason for this is their (lower) value of the elastic stress concentration factor, viz. $K_t = 5.48$. The *strain* concentration factor of the finite element analysis falls below that of the TSED and Neuber models for all times and below that of the SED model for long times.

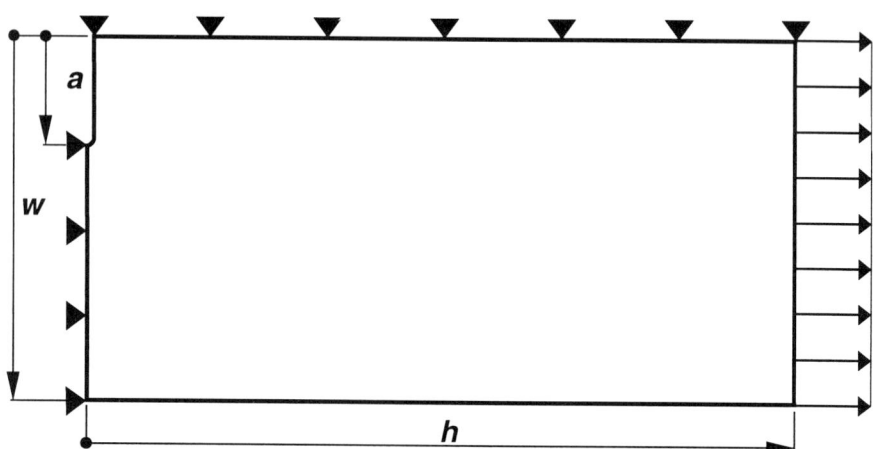

Figure 1. Quadrant of a uniaxially loaded plate ($h = 2w$) with a central slot ($a = 0.3w$, $\rho = a/15$).

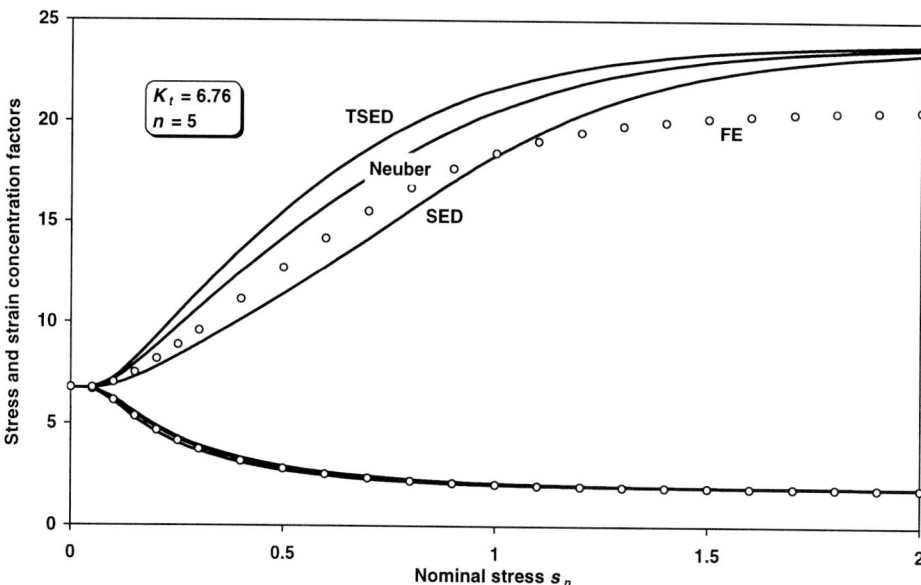

Figure 2. Stress and strain concentration factors as functions of nominal stress for notched elastic-plastic body under simple tension and plane stress conditions.

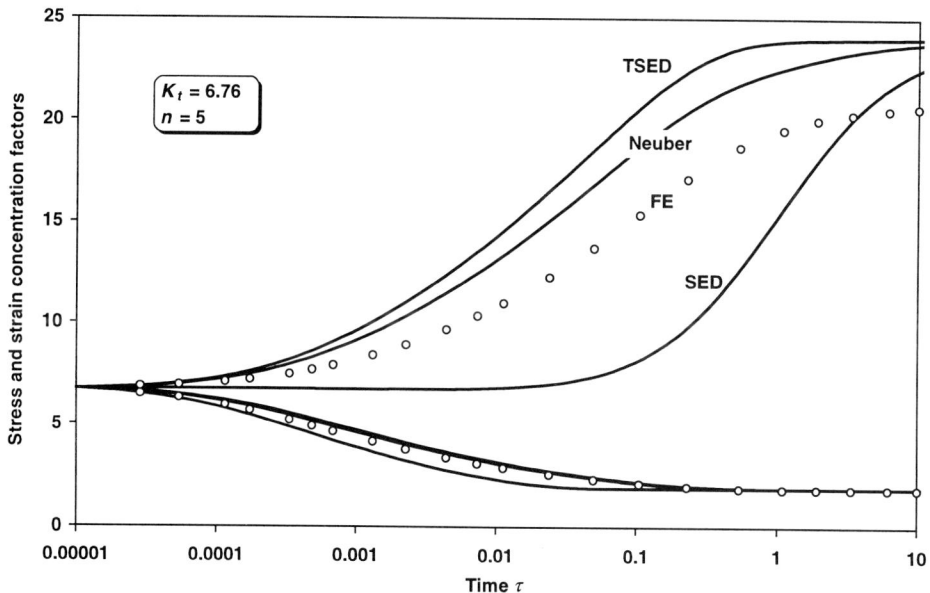

Figure 3. Stress and strain concentration factors as functions of nominal stress for notched elastic-visco-plastic body under simple tension and plane stress conditions.

References

1. Kubo, S. and Ohji, K.: Small-scale and large-scale creep predictions of notch stress and strain under creep conditions, M. Zyczkowski (ed.), *Creep in Structures*, IUTAM Symposium Cracow 1990, Springer-Verlag, Berlin, 1991, pp. 363-370.
2. Moftakhar, A.: *Calculation of time independent and time dependent strains and stresses in notches*, Dissertation, University of Waterloo, Canada, 1994.
3. Sørbø, S.: *Visco-Plastic Notch Analysis*, Dissertation, Norwegian University of Science and Technology, Trondheim, 1999.
4. Härkegård, G. and Sørbø, S.: Applicability of Neuber's rule to the analysis of stress and strain concentration under creep conditions, ASME *Journal of Engineering Materials and Technology* **120** (July 1998) 224-229.
5. Neuber, H.: Theory of stress concentration for shear strained prismatical bodies with arbitrary nonlinear stress-strain law, ASME *Journal of Applied Mechanics* **28** (Dec. 1961) 544-550.
6. Sørbø, S. and Härkegård, G.: Evaluation of approximate methods for elastic-plastic analysis of notched bodies, in M.H. Aliabadi et al. (eds.), *Localized Damage III*, Computational Mechanics Publications, Southampton, 1994, pp. 471-478.
7. Molski, K. and Glinka, G.: A method of elastic-plastic stress and strain calculation at a notch root, *Mat. Sci. Engng.* **50** (1981) 93-100.
8. Rice, J.R.: A path independent integral and the approximate analysis of strain concentration by notches and cracks, ASME *Journal of Applied Mechanics* **35** (1968) 379-386.
9. Webster, G.A. and Ainsworth, R.A.: *High Temperature Component Life Assessment*, Chapman & Hall, London, 1994.
10. He, M.Y. and Hutchinson, J.W.: Bounds for fully plastic crack problems for infinite bodies, *Elastic-Plastic Fracture, Volume I-Inelastic Crack Analysis*, ASTM STP 803, American Society for Testing and Materials, 1983, pp. 277-290.
11. Dowling, N.E.: J-integral estimates for cracks in infinite bodies, *Engineering Fracture Mechanics* **26** (1987) 3: 333-348.
12. Shin, C.S., Man, K.C. and Wang, C.M.: A practical method to estimate the stress concentration at notches, International Journal of Fatigue, **16** (1994) 242-256.

THE REFERENCE STRESS METHOD IN CREEP DESIGN

A Thirty Year Retrospective

J. T. BOYLE
Department of Mechanical Engineering
University of Strathclyde
Glasgow
Scotland, United Kingdom

R. SESHADRI
Faculty Engineering
& Applied Science
Memorial University
of Newfoundland
St. John's
Newfoundland, Canada

1. Introduction

In the 1970 IUTAM Symposium on Creep in Structures, Marriott [1] reviewed an astonishing new and simple technique for predicting creep deformations – the *reference stress method*. Originally devised by Anderson, Gardner and Hodgkins in 1963, in the form suggested by Mackenzie in 1968 the method maintains that steady state creep deformations in a component could be related to the result of a single creep test conducted at the so-called reference stress. In 1968 Sim had demonstrated that the reference stress itself could be derived from the limit load for the component. At the same time several researchers also noted that the reference stress was related to the stress at the skeletal point – a location in a creeping structure where the stress scarcely varied during creep – this was further discussed by Anderson [2] in the 1980 IUTAM Symposium. During the 1970's and early 1980's there was considerable activity in the United Kingdom to verify and extend the capabilities of the reference stress method beyond steady creep; this was summarized at the time [3,4]. Eventually a robust procedure for high temperature life assessment was derived – this is elegantly summarized in the monograph by Webster & Ainsworth [5]. Apart from this, interest in the reference stress method has declined somewhat during the past decade – it was barely mentioned in the 1990 Symposium.

Nevertheless the authors believe that the reference stress method continues to offer the only real simplified and robust alternative to creep design other than detailed inelastic stress analysis. Indeed even if initial design can be based on inelastic stress analysis, it is the responsibility of a professional engineer to provide verification – the use of reference stress, and related concepts, is the simplest way to achieve this. However it has often been argued that wider acceptance of the reference stress method, and its further development, was held back due to the need, at the very least, to obtain limit loads or the location of skeletal points. Over the past decade several developments in

S. Murakami and N. Ohno (eds.), IUTAM Symposium on Creep in Structures, 297–309.

simplified methods for inelastic behavior have removed these difficulties. These developments include the notion of the *r-node* [6] which can be used to evaluate skeletal points, limit loads and the reference stress, and be directly used in inelastic design assessment [7], and the related *elastic compensation method* [8] which can be used to estimate limit and shakedown loads for large complex structures using linear elastic finite element analysis alone [9,10].

The aim of this paper is to summarize developments since IUTAM 1970 in simplified methods for creep, plasticity and fracture, which can be used to enhance the reference stress approach for creep design. This is not a review (this can be found in [1-5]), simply a retrospective: with increasing computing capacity there is a renewed need to consider robust design rules for structures and components subject to creep – the reference stress method could have a guiding role.

2. The Reference Stress Method for Steady Creep

We begin with a brief introduction to the essential concepts of the reference stress method as it was originally formulated for simple steady creep described by a power-law:

Consider the simple plane truss composed of pin-jointed bars of uniform cross-sectional area A subject to a vertical load Q as shown in Figure 1.

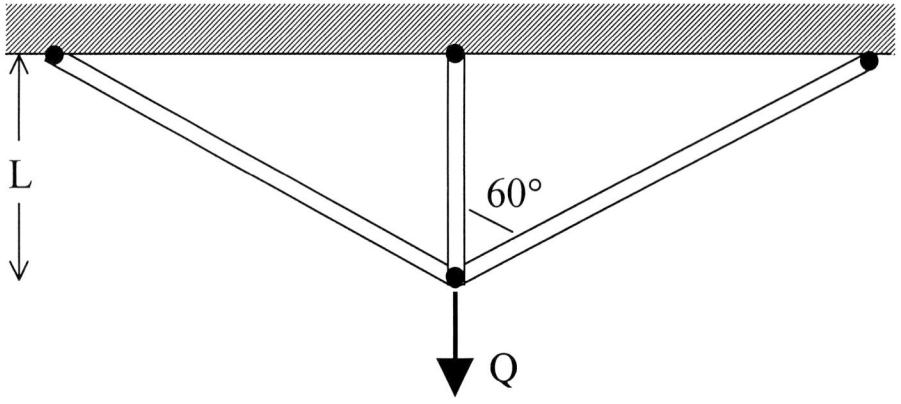

Figure 1: Simple three-bar truss

If we assume a power law material $\dot{\varepsilon} = B\sigma^n$ in the usual notation where B and n are material parameters then it can be shown that the vertical displacement rate of the common joint where the load is applied is given by

$$\frac{\dot{q}}{L} = \frac{B(Q/A)^n}{\left(1+(1/2)^{2/n}\right)^n}$$

The reference stress method starts by rewriting this is the specific form

$$\dot{q} = \delta \times \dot{\varepsilon}(\sigma_R)$$

where σ_R is the *reference stress* and δ is the *scaling factor*,

$$\frac{\delta}{L} = \frac{(1/\alpha)^n}{\left(1+(1/2)^{2/n}\right)^n}$$

where we have written the reference stress in the form $\sigma_R = \alpha \dfrac{Q}{A}$ for some arbitrary scalar α. If we now examine the variation of the scaling factor with n for various values of α then if $\alpha < 1/2$ the scaling factor $\delta/L \to \infty$ as $n \to \infty$ whereas if $\alpha > 1/2$ the scaling factor $\delta/L \to 0$ as $n \to \infty$. However if $\alpha = 1/2$ the scaling factor tends to a limit, $\delta/L \to 2$ as $n \to \infty$.

Therefore

$$\dot{q} \cong (\lim_{n\to\infty} \delta) \times \dot{\varepsilon}(\sigma_R)$$

to a good approximation provided we take the reference stress $\sigma_R = \dfrac{1}{2}\dfrac{Q}{A}$.

Expressed in this way, *the scaling factor and reference stress are independent of the material properties*. Further, and this was the original reference stress concept, we can interpret $\dot{\varepsilon}(\sigma_R)$ as the steady creep rate from a uniaxial creep test held at the reference stress. That is, we don't need to know the material parameters, we just need to perform *one* creep test!

At the time the basic reference stress concept was verified for many simple structures and components; it was also shown to be valid for steady creep material models other than power-laws [11]. More significantly, it was also noticed by Sim [12] that the reference stress itself could be written in a very revealing, but perhaps puzzling, way:

$$\sigma_R = \frac{Q}{Q_L} \sigma_y$$

where Q_L is the *limit load* for an equivalent structure composed of a perfectly plastic material with yield stress σ_y.

A partial explanation of this was given by Leckie and Ponter [13] who demonstrated also that the scaling factor could be bounded as

$$\delta \le \frac{V\sigma_R}{Q}$$

where V is the volume of the structure. A more complete explanation was later given by Boyle [14], using Calladine & Drucker's [15] *Theorem of Nesting Surfaces for Power-Law Materials*, which also extended the result to multiple loads – we will return to this later.

As a final point, equally remarkable, when reference stresses were evaluated for common components with a distribution of stress, it was found that the equivalent stress at the so-called *skeletal point* could also be related to the reference stress in an approximate way.

In studies of the creep stress redistribution through time of a structure under constant load, Marriott & Leckie [16] noted that there existed points in the structure where the initial elastic stress remained virtually constant: these locations were called the *skeletal points*. It was soon noticed that the value of the stress at the skeletal point was a good approximation to the reference stress, although Marriott [1] recommended some caution with this interpretation at the time. Indeed if a structure, such as a beam in bending, was analyzed in steady creep using a power-law material then it was found also that the stress at the skeletal point was almost independent of the exponent n, similar to the reference stress concept. When this was first recognized, it seemed to offer some insight into the relationship between references stress for creep, limit load and skeletal point stress – with hindsight it probably deepened the mystery. But, as we will see later, even more mysterious behavior appeared.

3. The Reference Stress Method in High Temperature Design

The simplicity of the whole reference stress concept gave direction to a design and life assessment philosophy for high temperature components [5], more specifically, and naturally, in the context of pressure vessels. In the context of steady creep under constant load, for simplicity, according to the reference stress approach, the rate of work being done by the applied loads in deforming the component must be less than that due to the reference stress:

$$\dot{W}_R = V \sigma_R \dot{\varepsilon}(\sigma_R)$$

Then [5], from a design point of view, if section thickness can be chosen so that the creep strain corresponding to the reference stress is limited during the design life, excessive creep deformation of the component should be avoided. Of course under constant load the stress will redistribute from the elastic through to the steady state, but it is found that the additional creep strain at the skeletal point due to stress redistribution is approximately σ_R / E, where E is Young's Modulus. Combining this with the steady creep strain at the reference stress the total accumulated creep strain under constant load may be estimated, and appropriate section thickness may be chosen.

Of course in design consideration must also be given to estimating the lifetime due to creep failure. It has been shown [5] that the creep failure time due to damage propagation in a complex component cannot be greater than, but is often quite close to, the lifetime of a uniaxial creep specimen at the reference stress. Further, the rate of damage propagation is approximately proportional to the creep strain rate at the reference stress and inversely proportional to the material creep ductility.

The preceding brief description of a simple high temperature design philosophy based on reference stress concepts has been implemented and extended in the UK *R5: Assessment Procedure for High Temperature Response of Structures* [17] to cracked structures and components. Beginning with an elastic-plastic fracture assessment, fracture is considered to be avoided if

$$K_r = \frac{K}{K_{mat}} \leq f\left(\frac{\sigma_R}{\sigma_y}\right)$$

where K is the elastic stress intensity factor and K_{mat} is the fracture toughness; the function $f( )$ is prescribed by the boundary of a *failure assessment diagram* [5,17]. The design boundary of the FAD is determined by the fracture toughness and the ratio of flow stress to yield (0.2% proof) stress through three options: (i) a general purpose curve which is biased to a lower bound for a range of materials and is independent of material, geometry, loading and crack size, (ii) an explicit curve where, from an examination of *J*-integral estimates from experimental and finite element data the geometry dependence can be eliminated and material dependence expressed in terms of the stress-strain curve, or (iii) an explicit curve based directly on an equivalence of the FAD to a *J*-integral; analysis which is specific to geometry, loading, crack size, shape and the material stress strain relation which may be derived from inelastic finite element analysis.

This philosophy may then be extended into the creep range through the creep crack-growth parameter C* in its simplest form for steady creep

$$C^* = \sigma_R \dot{\varepsilon}(\sigma_R) R'$$

where R' is an elastic-plastic defect size parameter [5]. Further the creep crack-growth rate can then be estimated as

$$\dot{a} = \frac{0.3 \left(C^* \right)^{0.85}}{\varepsilon_f}$$

where ε_f is the percentage ductility. It has further been shown that this approach can be used in more general primary and tertiary creep [5].

4. The Elastic Compensation Method

It would be apparent from the above that one possible difficulty in the design philosophy based on reference stress would be *the need to calculate the limit load*, not only for uncracked components, but also for those containing defects. Over the past decade a technique – called the *elastic compensation method* - has been developed as a simple means of simulating various plastic failure mechanisms in complex structures using conventional linear *elastic* finite element analysis. The fundamental concept is that the elastic modulus of each element in a finite element model is adjusted according to the ratio of the stress in the element to a nominal stress in order to redistribute the stress field away from highly stressed regions. The resulting stress and strain field can then be applied to the bounding theorems of plasticity to obtain limit (and shakedown loads). The technique has been demonstrated to give useful results for a range of structural problems. The details of the method will not be repeated here (they can be found in references [8-9] and the review paper [10]), rather a simple summary is provided:

In elastic compensation analysis an initial linear elastic finite element analysis is performed for an *arbitrary* load set Q and the elastic stress field is obtained. This forms iteration zero in a series of linear elastic finite element analyses, in which the elastic modulus, E, of each element is adjusted according to the rule:

$$E_i = E_{(i-1)} \left(\frac{\sigma_n}{\sigma_{i-1}} \right)$$

where subscript i is the sequence number of the next analysis, $(i$-1) represents the current analysis, σ_n is a nominal stress value and $\sigma_{(i-1)}$ the maximum (unaveraged) nodal stress (von Mises) associated with the element from the ith solution. The value of the nominal stress is not significant: it is the relative adjustment of the elastic modulii, and the resulting stress field, which is important. In fact the elastic compensation method can cause the maximum stress to increase or decrease. But generally over a number of iterations there is a net decrease in the maximum stress with respect to the initial solution. The iterative procedure is carried out for a few iterations, usually between six and ten.

A lower bound limit load can then be calculated by invoking the lower bound limit load theorem, which states that if a statically admissible stress field in which the stress nowhere exceeds yield for a given component under given loading, the loading is a lower bound limit load. The elastic compensation solution meets the first requirement of the lower bound theorem in that it is statically admissible. Therefore to satisfy the second condition, the best value for the lower bound limit load possible for a given stress distribution is one in which the maximum nodal unaveraged stress is at yield. For example for a single load, since the iteration solutions are linear elastic, the applied load giving such a maximum stress can be calculated through simple proportionality. Hence the best lower bound limit load is given by considering the solution in which the stress has the lowest value, say iteration k, in which the maximum nodal unaveraged stress is σ_k. Therefore the lower bound limit load is given by the simple relation

$$Q_L \simeq Q \frac{\sigma_y}{\sigma_k}$$

From the same sequence of analyses an upper bound limit load can also be derived from the upper bound theorem since the strain and displacement field in each iteration are compatible and hence kinematically admissible. Without going into detail, the upper bound limit load for a single load can be derived as:

$$Q_L \simeq Q \frac{D_1}{U_1}$$

where the strain energy U_1 and plastic dissipation D_1 for the finite element model must balance for the best upper bound.

This apparently simple procedure, which requires only a few elastic finite element analyses has been verified on a large range of problems: the recent review [10] runs to over seventy related publications. Further Ponter & Carter [18] have demonstrated the use of an extension of the simple method described above to deal with the evaluation of limit loads in *cracked* structures, and to provide a more rigorous theoretical basis (see [10]).

In terms of the reference stress it can be seen from the above that *the reference stress* σ_R *can be taken as the maximum nodal unaveraged stress* σ_k *from each iteration of the elastic compensation procedure* – that is, it can be evaluated directly from the maximum component stress after a few elastic analysis! Interestingly this new development in the mystery of the reference stress could possibly have been predicted. Webster & Ainsworth [5] made the observation that "… essentially, the reference stress is a useful quantity because creep strain is usually strongly stress dependent. Then, creep tends to make stresses uniform in a structure in a similar manner to the uniform stresses that occur in the plastically deforming regions of a structure at plastic collapse …". Recall from the above, the aim of the elastic compensation procedure is to adjust the elastic stiffness of each element of a finite element model according to the ratio of the stress in the element to a nominal stress in order *to redistribute the stress field away from highly stressed regions* – creep has the same effect. Webster & Ainsworth also commented that "… the identification of the skeletal point where the stress remains essentially constant, and equal to the reference stress, provides further insight into the reference stress method … where a skeletal point can be identified, there will be greater confidence in the accuracy of the technique …". The existence of a skeletal point in the application of the elastic compensation procedure, which derives the reference stress directly from the maximum redistributed stress, now brings us to the pivotal notion of a *redistribution node*:

5. The Redistribution Node (r-node) Concept

Over the past decade Seshadri and his collegues [6,19] have developed the concept of redistribution nodes in structures. The redistribution node concept can be used to link the concepts of reference stress, skeletal points and elastic compensation procedures [7]. The *r-node method* was proposed [20] as a simple technique which *directly* determined the reference stress (at the skeletal points) without prior knowledge of the limit load – indeed *the limit load could be evaluated from the reference stress*, rather than the other way round! Further only two elastic finite element analyses are required to evaluate the reference stress.

Skeletal points can be thought of as 'nodes of redistribution of stresses': when inelastic stress redistribution occurs, involving say the entire cross-section of a component or structure, the statically indeterminate stresses undergo a redistribution throughout the component except at the r-nodes (skeletal points) which are essentially statically determinate and therefore would be expected to be independent of the constitutive relation. Thus the stresses at the r-nodes are induced to preserve equilibrium with the externally applied load. In the context of creep the r-nodes are then the skeletal points, whereas in the context of plasticity the r-nodes are where plastic hinges form (or there is nucleation of plasticity).

Thus, if the limit load is eliminated from the reference stress definition and the reference stress is related directly to the stress at the r-node, there is no need to establish successive re-analyses in the elastic compensation procedure (to obtain better bounds with the theorems of plasticity). As a consequence only *two* elastic finite element analyses are required. The basic procedure is then similar to elastic compensation, except that the elastic modulii are modified only once. Then the redistribution nodes are locations where the stress values are the same in the two analyses. If there is only a single r-node location, then the effective stress at this node can be taken as the reference stress. If there are multiple r-nodes (as in the case of the formation of multiple plastic hinges before plastic collapse) then an average can be taken for the reference stress. However a more consistent approach makes use of the Theorem of Nesting Surfaces [14,15] cited earlier:

According to the Theorem [14], the reference stress can be defined as the limit

$$\sigma_R = \lim_{n \to \infty} Q_e$$

where

$$Q_e = \left(\frac{1}{V} \int_V \sigma_e^{n+1} dV \right)^{\frac{1}{n+1}}$$

for a power-law of steady creep, where σ_e is the effective stress and V the volume of the structure.

As $n \to \infty$ the surfaces Q_e 'nest' inside each other bounded on the outside by $n = 1$ and on the inside by $n \to \infty$. For a finite element discretisation the integral is replaced by a summation over the elements.

In the case of multiple r-nodes it is suggested that the reference stress is taken as the approximation:

$$\sigma_R \simeq \left(\frac{1}{V} \int_V \sigma_e^2 dV \right)^{\frac{1}{2}}$$

after the first elastic modulus adjustment. Indeed it is possible [19, 21], but so far unproven, that the sequence of elastic analyses from an elastic compensation analysis gives rise to a sequence of nesting surfaces of reference stress as defined above, tending in the limit to that corresponding to the limit load (that is, $n \to \infty$).

Using these ideas, together with Mura's extended lower bound limit load theorems [22], Seshadri & Mangalaramanan [21] developed the m_α-method of lower bound limit analysis. In the m_α-method, a statically admissible load multiplier, m_α, is calculated on the basis of two linear elastic finite element analyses. The new concept of a 'reference volume', based on the Theorem of Nesting Surfaces, is used to derive the limit load. In other words a limit load theorem can be expressed directly in terms of the reference stress.

It seems there are several advantages to working directly with the reference stress as a basis for design rather than limit load concepts. As an example, in the context of pressure vessel design by analysis using the elastic route, the need to identify primary stress is removed if the primary stress limit is rewritten using the r-node stress:

$$\left[\sigma_{e,\text{r-node}}\right]_{\max} \le S_m$$

where S_m is the allowable for primary membrane stress. In effect, the above stated stress limit takes into consideration both primary membrane and primary bending loads.

A final observation by Mangalaramanan & Seshadri [23] on the relationship between reference stress, limit loads, skeletal points and redistribution nodes added the notion of *minimum weight design* to the mystery:

Consider the simple pin-jointed two-bar structure under combined load shown in Figure.2 which was analyzed in [14]. The bars have equal length, cross sectional area A and are at right angles.

The 'nesting' surfaces Q_e for steady power-law creep for are shown in Figure 3. The nesting surfaces shown have points where the surfaces corresponding to $n = 1$ and $n \rightarrow \infty$ are coincident. This can be attributed to the situation where all the material in a structure have the same value of absolute stress. This is precisely the condition for minimum weight design under a given load. Minimum weight is both necessary and sufficient for all the nesting surfaces to coincide [15] – the entire volume undergoes plasticity at collapse.

Mangalaramanan & Seshadri [23] have suggested the *iso-r-node* procedure for minimum weight design. After an initial pair of elastic analyses to establish the location of the r-nodes, material (or, elements in the overall finite element mesh) are successively removed to equalize the r-node stresses. Minimum weight design is achieved by modifying the geometry such that the entire structure would undergo concurrent yielding. As well, the minimum weight structure would be subjected to primary stresses only.

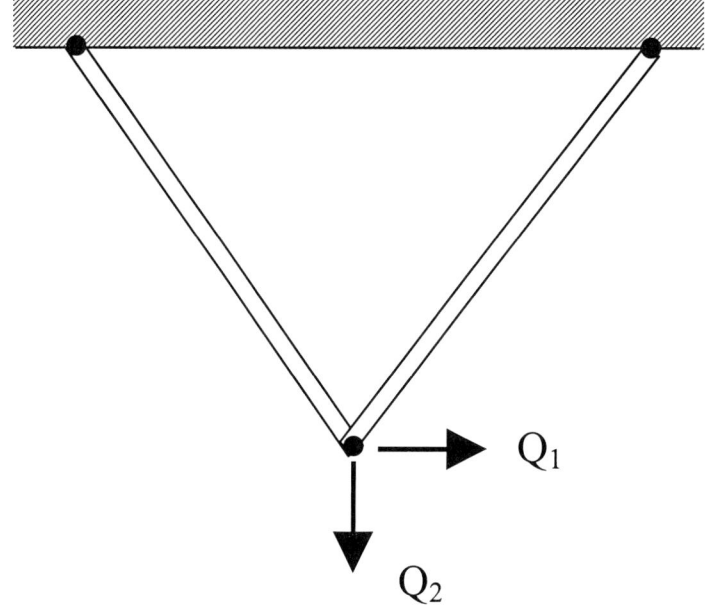

Figure 2: Simple two-bar structure

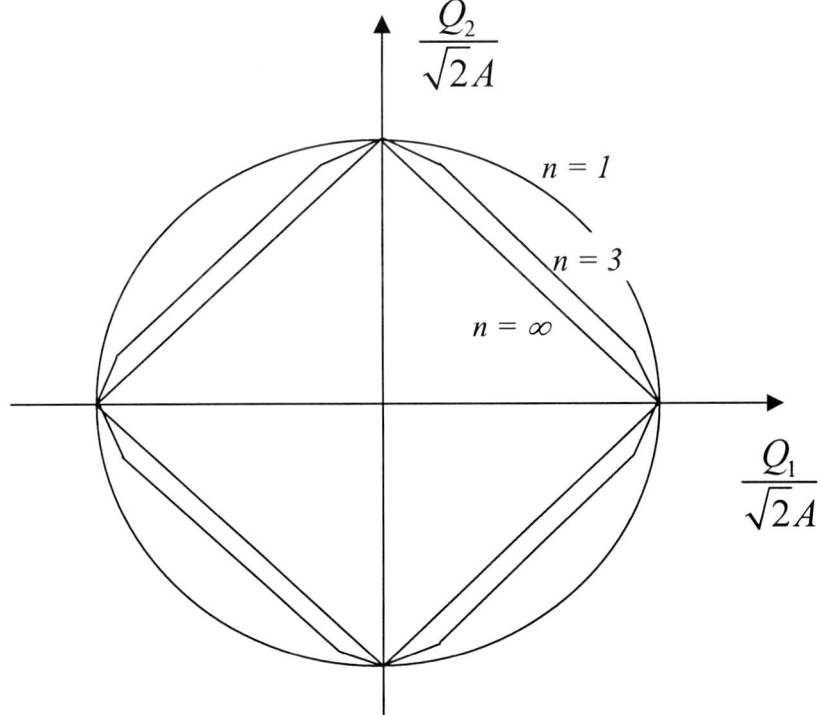

Figure 3: Nesting surfaces for two-bar structure

6. Conclusions

We have mentioned several times that there is something mysterious in the relationship between reference stress, limit load and in particular plastic collapse at plastic hinges, skeletal point and redistribution nodes in inelastic (plastic and creeping) structures. While this apparent simplicity in complex inelastic structural behavior has always been recognized [1,2,7] it was certainly not expected at the time that the whole situation would be simplified even further through the simple device of carrying out a couple of elastic finite element analyses where the relative elastic modulii of the individual elements were adjusted to redistribute the stress. For the historically curious, it is perhaps interesting to note that the suggestion to modify the elastic modulus in this way in fact came from Marriott [24]! Indeed the authors often wonder what research in inelastic and high temperature creep design could have achieved if this simple technique had been known at the 1970 [1] or 1980 [2] *Creep in Structures* Symposium. In 1983, the first author wrote of the reference stress method [3]: "... it is one of few results in creep mechanics which contains an element of surprise – if all else fails it should be treasured for this ...". It seems the surprises continue – perhaps the structures and components are prompting us to keep looking for something even more profound.

References

1. DL Marriott: A review of reference stress methods for estimating creep deformations. Proc. Second IUTAM Symposium on *Creep in Structures*, Gothenburg, Springer Verlag, 1970

2. RG Anderson: Some observations on reference stresses, skeletal points, limit loads and finite elements, Proc. Third IUTAM Symposium on *Creep in Structures*, Leicester, Springer Verlag, 1980

3. JT Boyle: The reference stress method and its role in high temperature design. In *Engineering Approches to High Temperature Design*, Ed B Wilshire & DRJ Owen, Pineridge Press, 1983

4. JT Boyle & J Spence: *Stress Analysis for Creep*, Butterworths, 1982

5. GA Webster & RA Ainsworth: *High Temperature Component Life Assessment*, Chapman & Hall, 1994

6. R. Seshadri and C.P.D. Fernando: Limit loads of mechanical components and structures using the GLOSS R-Node method, Transaction of the ASME: Journal of Pressure Vessel Technology, 114, pp. 201-208, 1992

7. R. Seshadri and D.L. Marriott: On relating the reference stress, limit load and the ASME stress classification concepts, International Journal of Pressure Vessels and Piping, 56, pp. 387-408, 1993

8. D Mackenzie, JT Boyle, C Nadarajah, J Shi : Simple bounds on limit loads by elastic finite element analysis, Trans ASME Jrn Pressure Vessel Technology, vol 115, no 1, pp27-31, 1994

9. D Mackenzie, JT Boyle: A method of estimating limit loads by iterative elastic analysis I, II & III, Int Jrn Pres Ves and Piping, vol 53, no 1, pp77-142, 1992.

10. D Mackenzie, JT Boyle & R Hamilton: The elastic compensation method in limit and shakedown analysis: a review. Journ. Strain Analysis (In press)

11. JT Boyle: Approximations in the reference stress method for creep design. Int. Journ. Mech. Sci., vol.22, pp73-81, 1980

12. RG Sim: PhD Thesis, University of Cambridge, 1968

13. FA Leckie & ARS Ponter: Deformation bounds for bodies which creep in the plastic range. Trans ASME, Journ. Applied Mechanics, vol.37, pp426-430, 1970

14. JT Boyle: The theorem of nesting surface in steady creep and its application to generalised models and limit reference stresses. Res Mechanica, vol.4, pp275-294, 1982

15. CR Calladine & DC Drucker: Nesting surfaces of constant rate of energy dissipation in creep. Quart. Appl. Math., vol.20, pp79-86, 1962

16. DL Marriott & FA Leckie: Some observations on the deflections of structures during creep. Proc. Inst. Mech. Engnrs. Vol.178(3L0, pp115-125, 1963

17. RA Ainsworth, DW Dean &PJ Budden: Developments in creep fracture assessments within the R5 procedure. Proceeding *Creep in Structures 2000*, Nagoya, 2000

18. ARS Ponter & KF Carter: Limit state solutions based on linear elastic solutions with spatially varying elastic modulus. Computer Methods in Applied Mechanics & Engineering, vol.140, pp237-258, 1997

19. R Seshadri: Simplified methods in plasticity, creep and fracture – some recent developments. Trans CSME, vol.22, pp419-433, 1998

20. R Seshadri: The generalised local stress strain (GLOSS) analysis – theory and applications. Trans ASME , Journ of Pres Vess Techn., 25th Anniversary Volume, pp219-227, 1991

21. R Seshadri & SP Mangalaramanan:, Lower bound limit loads using variational concepts: the m_α - method, Int. J. Pres. Ves. & Piping, vol.71, pp93-106, 1997

22. T Mura et al: Extended theorems of limit analysis. Quart Appl Math, vol.23, pp171-179, 1965

23. SP Mangalaramanan & R Seshadri: Minimum weight design of pressure components using r-nodes". Trans ASME, J. Pres. Ves. Tech. vol.119, 224-231, 1997

24. DL Marriott: Evaluation of deformation or load control of stresses under inelastic conditions using elastic finite element stress analysis, Proc. ASME PVP Conf., Pittsburgh, 136, pp3-9, 1988

STUDY ON CREEP-FATIGUE LIFE PREDICTION METHODS BASED ON LONG-TERM CREEP- FATIGUE TESTS FOR AUSTENITIC STAINLESS STEEL

Y. TAKAHASHI
Research Fellow
Materials Science Department
Central Research Institute of Electric Power Industry
2-11-1 Iwado Kita, Komae-shi, Tokyo 201-8511, JAPAN

Abstract

Low-carbon, medium-nitrogen 316 stainless steel is considered as a main structural material of future fast breeder reactor plants in Japan. A number of long-term creep-fatigue were conducted for several products of this steel and two representative creep-fatigue life prediction methods, i.e., time fraction rule and ductility exhaustion method were applied. Stress relaxation behavior during strain holding was simulated by an addition of a viscous strain term to the conventional creep strain but only the latter was counted in the evaluation of creep damage in the ductility exhaustion method. The ductility exhaustion method showed good accuracy in creep-fatigue life prediction for all materials tested, while the time fraction rule tended to overpredict failure.

1. Introduction

Development of fast breeder reactor (FBR) plants is one of the targets of highest priority for the electric utilities in Japan and large R&D efforts are being taken under the sponsorship of electric companies as well as the government. In FBR plants and some of other plants, many components have been and will be made of austenitic stainless steels because of their superior high-temperature strength and ductility.

At present, type 316 stainless steel with low carbon and medium nitrogen contents called 316FR (Fast Reactor) is regarded as a principal structural material in the future FBR plants. This steel has considerably better creep strength than the 316 steel with conventional chemical composition by reducing the amount of precipitation of chromium-carbides along grain boundaries which promote initiation of creep cavities. However, principal loading mode of fast reactor components is creep-fatigue interaction rather than simple creep, because thermal stresses are dominant compared to primary stresses due to low coolant pressure and large temperature variation. Therefore accurate prediction of creep-fatigue failure is quite important for sound operation of the

311

S. Murakami and N. Ohno (eds.), IUTAM Symposium on Creep in Structures, 311–320.

plants for a design life as long as 40 years.

The author has been conducting creep as well as creep-fatigue tests for several products of this steel for several years (Takahashi, 1998; Takahashi, 1999). Evaluation of two representative creep-fatigue failure prediction methods practically used in structural design or life assessment of power generation plants has been made based on these test results. This paper summarizes the updated results obtained by this study.

2. Outline of Experiment

Four kinds of products of 316FR were tested in this study. Two of them, hereafter called plate A and plate B, were produced by hot-rolling and their thickness was 50 mm. The third material was produced by forging and its thickness was 400mm. The remaining material was made as a form of a pipe with an outer diameter of 165.2mm and thickness of 18.2mm. Solution heat treatment was given by keeping them, for 30minutes for the plates and the pipe and for 7.5 hours for the forging, at 1050°C, followed by water quenching. Chemical compositions of these four products are shown in Table 1 with recommended range of this steel. No significant difference can not be observed between them. However, grain size differed much and the average grain size number according to ASTM standard of each material was 4.0, 5.5, 2.5 and 6.0, respectively.

A number of trapezoidal-wave, uniaxial creep-fatigue tests were conducted mainly at 550°C and 600°C. Axial strain was kept constant for a given hold time at the tensile peak of each cycle. Strain rate during ramping periods was fixed at 0.1%/s. Several strain ranges were chosen between 0.35% and 1.0% while changing the hold time from 6 minutes to 50 hours. The longest test duration was about 35,000 hours and most of the creep-fatigue tests conducted in the present study took longer than those reported in the past for this steel (Ueta et al., 1995) and the conventional 316 stainless steel (e.g. Wareing, 1981) as well.

3. Estimation of stress relaxation

Whichever life prediction method is used, estimation of stress relaxation during strain hold period is essential. It is traditional in the design of fast reactor plants to use creep strain equations representing primary and secondary creep deformation under the constant load and temperature generalized by the strain hardening law. However, the rapid drop of the stress after the initiation of strain hold after the ramping period with relatively high strain rates can not be predicted well by this method. Therefore, the author proposed the following equation to cope with this problem (Takahashi, 1998):

$$\dot{\varepsilon}_{in} = \dot{\varepsilon}_c + \dot{\varepsilon}_v \tag{1}$$

where $\dot{\varepsilon}_c$ is the "conventional" creep strain rate obtained from the uniaxial creep strain equation and $\dot{\varepsilon}_v$ is the "viscous" strain rate evaluated as

$$\dot{\varepsilon}_v = \dot{\varepsilon}_{v0} \exp(\frac{\sigma - \sigma_0}{\sigma_v}) \tag{2}$$

Here, $\dot{\varepsilon}_{v0}$ is the inelastic strain rate before strain hold, σ_0 is the stress at the onset of the holding. σ_v is a constant which was determined from the comparison of the simulated stress relaxation behaviors with actual ones. By using this equation, it can be expressed that the inelastic strain rate at the onset of holding is equal to the prior inelastic strain rate and also that it decreases rapidly with stress relaxation.

Examples of comparison between the simulated stress relaxation behavior with the actual data are shown in Figure 1. Reasonable agreement between the predictions with the viscous strain and the test data for two prior strain rates can be seen. From the above-mentioned nature, the viscous strain develops and saturates very rapidly whereas the creep strain rate does not change so much in spite of stress relaxation as well as strain hardening.

4. Creep-Fatigue Life prediction

4.1. OUTLINE OF LIFE PREDICTION METHODS

4.1.1. *Evaluation of Creep Damage*
Time fraction rule for the estimation of the creep damage has been employed in the design for high-temperature components of fast reactors (e.g. American Society for Mechanical Engineers, 1995). According to this rule, the creep damage during a hold period is calculated by the following equation:

$$D_c^{TF} = \int_0^{t_H} \frac{dt}{t_R(\sigma, T)} \tag{3}$$

where t_H is the hold time, t_R is creep rupture time at the stress, σ and the temperature, T during the hold period.

At the same time, various creep-fatigue life prediction methods based on inelastic strains rather than stress have been proposed in the last two or three decades. Among them, ductility exhaustion method developed mainly in the United Kingdom (Priest and Ellison, 1980, Hales, 1983, Clayton, 1988) has been receiving much attention from the viewpoint of application to long life components and incorporated in the assessment procedure for high-temperature components in power generation plants (Nuclear Electric, 1995). According to this method, creep damage per cycle in creep-fatigue tests is calculated as

$$D_c^{DE} = \int_0^{t_H} \frac{\dot{\varepsilon}_c}{\delta} dt \tag{4}$$

where $\dot{\varepsilon}_c$ is creep strain rate, δ is the so-called "creep ductility", which may or may not be a function of $\dot{\varepsilon}_c$. It is considered that the inelastic strain faster than a certain value does not contribute to creep damage because diffusion of atoms required for the growth of the grain boundary cavities is insufficient. Moreover, in the former investigation based on the experimental stress relaxation behavior (Takahashi, 1995), creep damage is considerably overestimated when all of the inelastic strain is assumed to contribute to the creep damage, especially at small strain ranges where the ratio of initial rapid inelastic strain is large. Based on these facts, we employed a simple assumption that only the strain predicted by the conventional "creep" equation gives creep damage.

As a value of δ, various quantities such as rupture elongation, reduction of area and true rupture strain have been used depending on researchers and/or materials concerned. Moreover, its dependency on the creep strain rate is taken into account in some cases but not in others. In this study, the minimum values of rupture elongation obtained in the creep rupture tests for each combination of the material and the temperature were used regardless of the creep strain rate during stress relaxation.

4.1.2. Evaluation of Fatigue Damage

Conventionally, fatigue damage per cycle, D_f, is calculated as a reciprocal of the pure fatigue life, N_{f0}, obtained at the same strain range, temperature and strain rate as those in the creep-fatigue tests as

$$D_f = \frac{1}{N_{f0}(\Delta\varepsilon)} \tag{5}$$

4.1.3. Prediction of Failure Life

In the application of the time fraction rule, a bilinear failure criterion connecting (1,0), (0.3,0.3) and (0,1) firstly proposed by Campbell (1971) is widely used for type 304 and 316 stainless steel in all existing design codes for fast reactors and thus the same criterion was used in this study. According to this criterion, failure life is calculated as

$$N_f = \frac{1}{max(D_f, D_c^{TF}) + \frac{7}{3}min(D_f, D_c^{TF})} \tag{6}$$

In the case of the ductility exhaustion approach, the simplest linear damage summation criterion represented by a straight line connecting (1,0) and (0,1) in the interaction diagram was employed. Thus failure life is simply estimated by

$$N_f = \frac{1}{D_f + D_c^{DE}} \tag{7}$$

4.2 RESULT OF LIFE PREDICTION

Failure lives predicted by the above two methods for the plate A are plotted against the hold time for several strain ranges with the experimental results in Figure 2. It can be seen that life reduction by introducing hold time was larger at the smaller strain range and this behavior was well captured in the prediction by the ductility exhaustion method. On the other hand, life reduction predicted by the time fraction rule was considerably smaller than the test results at the smaller strain ranges.

Direct comparison of the predicted cycles to failure for each test condition and the actual experimental data is shown in Figure 3. It can be seen failure lives were predicted by the present ductility exhaustion approach within a factor of 2 for all test conditions. On the other hand, failure lives tended to be overpredicted by the time fraction rule with a maximum factor being about 30. The only exception is the results of the tests conducted at highest strain range (1.0%) and 600°C, for which predictions by the time fraction rule agreed with the test data well .

To investigate the reason for the above results, accumulated creep and fatigue damage at failure are plotted in Figure 4 with failure criteria assumed for each life prediction method. In the case of the ductility exhaustion method, a reasonable amount of creep damage was estimated and the linear damage summation assumption constitutes a fairly good approximation of the data trend. On the contrary, creep damage estimated by the time fraction rule was much smaller and no synthetic correlation can be found between the creep damage and fatigue damage. Providing an alternative failure criterion as done for different materials in the ASME Code (American Society for Mechanical Engineers, 1995) is quite difficult because there are a bunch of data close to zero creep damage and two of them very close to the origin.

5. Discussion

It is well known that the product of the minimum (secondary) creep rate and the time to rupture in the creep tests is approximately constant for a wide range of stress as

$$\dot{\varepsilon}_m t_R = k \tag{8}$$

where $\dot{\varepsilon}_m$ is the minimum creep rate and k is the constant usually called the Monkman-Grant coefficient. Having this assumption, the time fraction creep damage can be rewritten using the minimum creep strain rate as follows:

$$D_c^{TF} = \frac{\int_0^{t_H} \dot{\varepsilon}_m \, dt}{k} \tag{9}$$

Using eq. (4) and eq. (9) and assuming constant δ, we obtained the following relationship between two creep damages:

$$\frac{D_c^{TF}}{D_c^{DE}} = \frac{\delta}{k} \frac{\int_0^{t_H} \dot{\varepsilon}_m \, dt}{\int_0^{t_H} \dot{\varepsilon}_c \, dt} \tag{10}$$

The constant k took the value of about 5 and 10% at 550°C and at 600°C, respectively for this steel.   Because k is smaller than the rupture elongation, δ, which is at least 22 %, the time fraction creep damage becomes larger than the ductility exhaustion damage if there is no primary creep strain, i.e. $\dot{\varepsilon}_c = \dot{\varepsilon}_m$, or relatively small compared to total (primary plus secondary) creep strain.   On the contrary, if the total creep strain rate is much larger than the secondary creep strain rate, the ductility exhaustion damage becomes larger than the time fraction damage.   The latter is the case of the present study because the total creep strain rate is larger in up to three order of magnitude than the secondary creep strain rate even after 10,000 hours hold.   These investigations, as well as the results of life prediction described above, clearly suggests the importance of considering the primary creep strain in assessing the creep damage if its ratio to total creep strain is large.   The time fraction rule could not take account of the effect of primary creep strain generated each cycle of creep-fatigue tests and consequently underestimated the creep damage (In the case of pure creep tests, primary creep deformation occurs only once).

6. Conclusion

Long-term creep-fatigue tests were conducted for four products of low-carbon, nitrogen-added 316 stainless steel (316FR) developed for use in fast breeder reactors and the applicability of the two representative creep-fatigue life estimation methods was investigated.   Results of the study can be summarized as follows:
(1) A simple approach of ductility exhaustion method where only the "creep" (primary and secondary) strain is assumed to give the creep damage and the minimum rupture elongation is taken as the ductility, was found to predict failure lives with very good accuracy for a wide range of strain range and holding time, without respect to material and test temperature.   It is therefore recommended to use this method as a basis of creep-fatigue damage assessment in the design of fast breeder reactor components made of 316FR.
(2) The time fraction rule considerably overpredicted creep-fatigue life for many tests, especially for small strain ranges, and thus was found to be not suitable for long-term integrity assessment of the high-temperature components made of this steel.
(3) This contrasty difference in life predictions by the two methods can be attributed to the difference in the way of dealing with the effect of primary creep strain generated every cycle in creep-fatigue tests.

7. References

American Society for Mechanical Engineers (1995) *Boiler and Pressure Vessel Code, Section III, Division 1 – Subsection NH.*

Campbell, R. D.(1971) Creep/fatigue interaction correlation for 304 stainless steel subjected to strain-controlled cycling with hold times at peak strain", *Journal of Engineering for Industry*, 887-892.

Clayton, A. M. (1988) Creep-fatigue assessment procedures for fast reactors, *Recent Advances in Design Procedures for High Temperature Plant*, Institution of Mechanical Engineers, pp. 49-54.

Hales, R. (1983) A method of creep damage summation based on accumulated strain for the assessment of creep-fatigue endurance, *Fatigue of Engineering Materials and Structures* **6**, 121-135.

Nuclear Electric plc (1995) *An Assessment Procedure for High Temperature Response of Structures, R5*, Issue 2.

Priest, R. H. and Ellison, E. G. (1980) A combined deformation map –ductility exhaustion approach to creep-fatigue analysis, *Material Science and Engineering* **49**, 7-17.

Takahashi, Y. (1995) Long-term high temperature strength of 316FR steel, *ASME PVP-Vol. 315*, 421-427

Takahashi, Y.(1998) Evaluation of creep-fatigue life prediction methods for low-carbon nitrogen-added 316 stainless steel", *J. Engineering Materials and Technology* **120**, 119-125.

Takahashi, Y.(1999) Further evaluation of creep-fatigue life prediction methods for low-carbon nitrogen-added 316 stainless steel", *J. of Pressure Vessel Technology*, **121**, 142-148.

Ueta, M. Nishida, T., Koto, H., Sukekawa, M. and Taguchi, K. Creep-fatigue properties of advanced 316-steel for FBR structures, *ASME PVP-Vol. 313-2*, 423-428.

Wareing. J (1981) Creep-fatigue behaviour of four casts of type 316 stainless steel, *Fatigue of Engineering Materials and Structures* **4**, 131-145.

Table 1 Chemical composition of test materials (weight percent)

Material (G.S.)	C	Si	Mn	P	Ni	Cr	Mo	S	N
Plate A (4.0)	0.009	0.57	0.86	0.025	11.25	16.85	2.06	0.005	0.0766
Plate B (5.5)	0.008	0.54	0.84	0.027	11.16	16.83	2.1	0.004	0.0754
Forging (2.5)	0.008	0.48	0.82	0.025	11.41	16.57	2.03	0.003	0.0935
Pipe (6.0)	0.01	0.5	0.84	0.023	11.29	16.78	2.08	0.004	0.089
Target Range	<0.02	0.30 - 0.75	0.50 - 1.50	0.02 - 0.03	10.5 - 12.5	16.0 - 18.0	2.05 - 2.55	<0.010	0.06 - 0.11

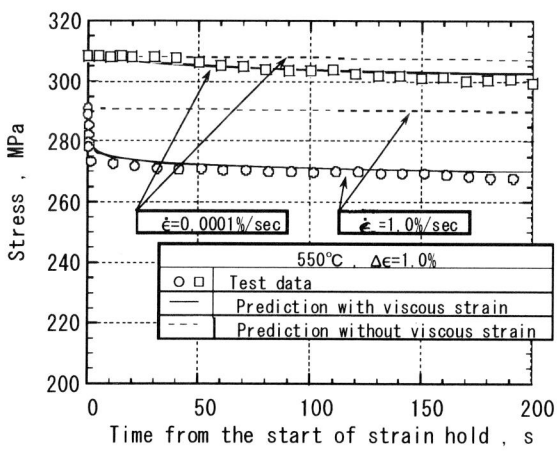

Figure 1   Comparison of predicted and experimental stress relaxation behavior in creep-fatigue tests

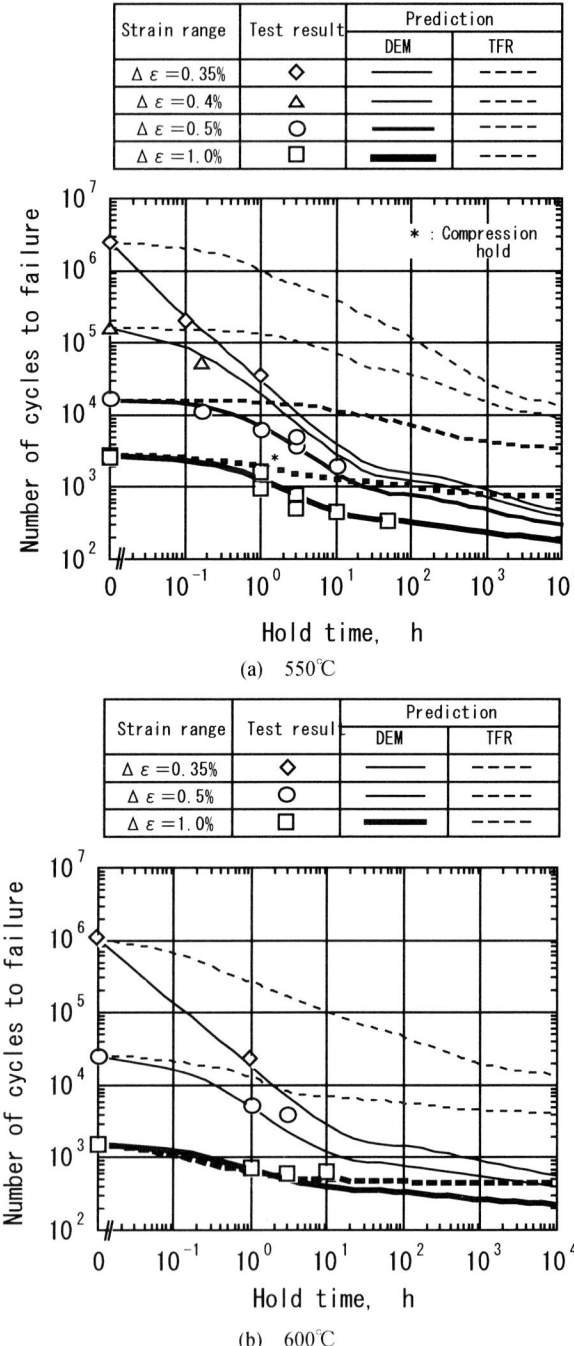

(a)   550°C

(b)   600°C

Figure 2   Relation between hold time and number of cycles to failure

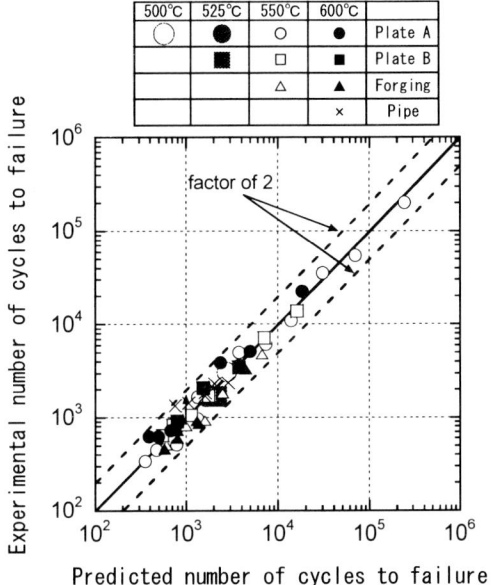

(a) Ductility exhaustion method

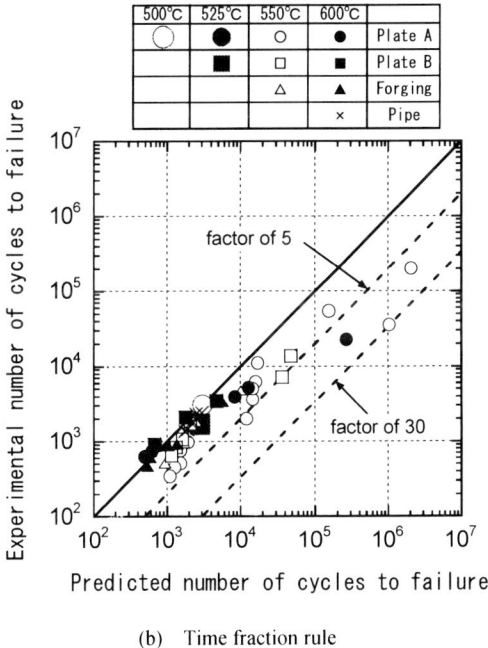

(b)   Time fraction rule

Figure 3   Comparison between predicted and experimental failure life

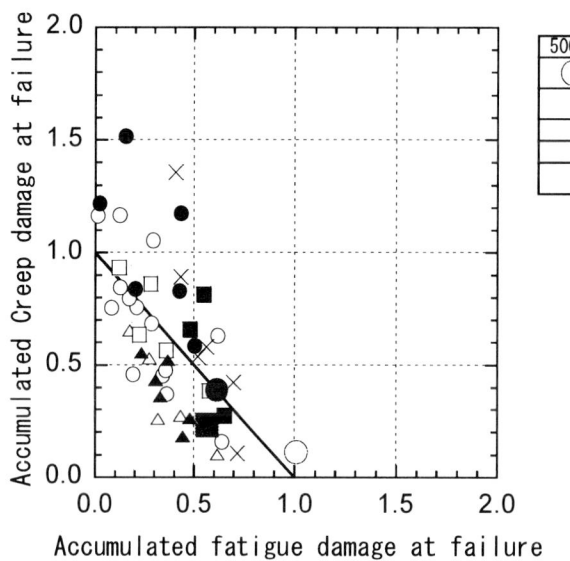

(a) Ductility exhaustion method

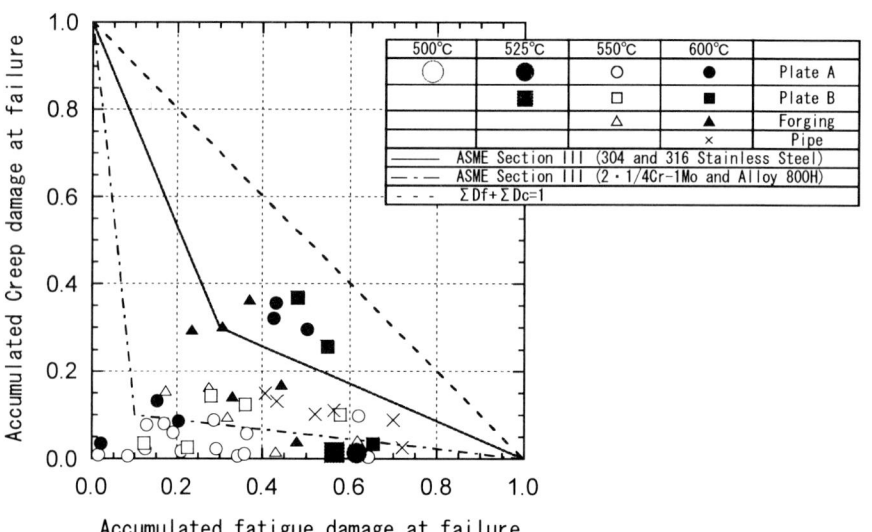

(b) Time fraction rule

Figure 4   Accumulated damages at failure

DEVELOPMENTS IN CREEP FRACTURE ASSESSMENTS WITHIN

THE R5 PROCEDURE

R. A. AINSWORTH, D. W. DEAN AND P. J. BUDDEN
British Energy Generation Ltd
Barnwood, Gloucester, GL4 3RS, UK

Abstract. The R5 procedures have been developed within the UK power generation industry to provide a comprehensive assessment approach for components operating at high temperatures. Within R5, there are specific procedures for assessing components containing defects. These are largely based on approximate reference stress techniques and are continuing to be developed. This paper describes developments in the creep fracture parts of R5, including supporting background information.

1.   Introduction

At the time of the third IUTAM Conference on Creep in Structures in 1990, the first issue of the R5 high temperature assessment procedures had just been released [1], although the defect assessment parts of R5 had been available for several years [2]. During the past 10 years, the defect assessment procedures have been incorporated into a British Standards Published Document PD6539 [3] with an ASTM standard [4] being available for generating the required creep crack growth data.

Confidence in the basic R5 procedures has now been established to the extent that they are routinely used within British Energy for assessments of operating high temperature plant. The procedures have also been extended and developed in a number of areas. Following a brief description of the basic defect assessment procedures in Section 2, this paper describes these developments in four selected areas: failure assessment diagram methods; probabilistic creep crack growth; crack growth in weldments; and, use of an approximate σ_d approach.

2.   Basic Defect Assessment Procedure

The background to the defect assessment procedures in R5 and PD6539 has been set out in [5]. Here, only the basic procedures for steady loading are briefly presented to aid the presentations on new developments in Sections 3-6. The procedures use approximate techniques based on a reference stress related to the applied load, F, by

$$\sigma_{ref} = F\sigma_Y / F_L(\sigma_Y, a) \qquad (1)$$

where F_L is the plastic collapse load for a yield stress σ_Y and crack size, a.

S. Murakami and N. Ohno (eds.), IUTAM Symposium on Creep in Structures, 321–330.
© 2001 *Kluwer Academic Publishers. Printed in the Netherlands.*

Having evaluated the reference stress, three times are estimated. First the time, t_{CD}, for creep damage to propagate through a structure and lead to failure is taken as

$$t_{CD} \simeq t_r[\sigma_{ref}(a_o)] \tag{2}$$

where $t_r(\sigma)$ is the rupture time at stress, σ, from conventional stress/time-to rupture data and the reference stress is calculated for the initial crack size, a_o.

Next, the initiation time, t_i, during which the initial crack blunts without any significant crack extension is estimated. At present, the testing standard [4] does not include methods for obtaining initiation data but recent recommendations [6] propose that initiation is defined for engineering purposes as corresponding to 0.2mm crack extension. The method of representing initiation data then depends on observed specimen response. For steady state creep conditions with an essentially constant displacement rate, the initiation time in test specimens is correlated with experimental estimates of the crack tip parameter C* through

$$t_i C^{*q} = B \tag{3}$$

where B and q are constants. More generally, initiation times can be related to measurements of crack opening displacement. Given data in the form of eqn (3), the initiation time in a component is obtained from an estimate of C* given by

$$C^* = \sigma_{ref} \dot{\varepsilon}^c_{ref} R \tag{4}$$

where $\dot{\varepsilon}^c_{ref}$ is the creep strain rate from uniaxial data at the reference stress of eqn (1) calculated for the initial defect size, a_o. The length R may be estimated from the numerical results or approximately from stress intensity factor solutions [5].

The third time required is the time, t_g, for the crack to propagate by an amount Δa. For steady state creep this is obtained from creep crack growth data in the form

$$\dot{a} = AC^{*q} \tag{5}$$

where A and q are constants, using eqn (4) to estimate C* for crack sizes between a_o and $a_o + \Delta a$. More generally, transient creep can be accommodated by allowing for the increased amplitude of the crack tip fields at short times [5]. Note, here the constant q has been taken the same in the eqns (3) and (5) representing initiation and growth data but generally these exponents may differ.

3.   Failure Assessment Diagram Methods

Recently, a high temperature failure assessment diagram (FAD) method has been developed and added to R5 [7]. This is consistent with R6 at lower temperatures. The high temperature method requires two parameters to be calculated

$$L_r = F / F_L(\sigma^c_{0.2}, a_o) \tag{6}$$

$$K_r = K^P(a_o) / K^c_{mat} \tag{7}$$

The definition of L_r uses similar quantities to those in eqn (1) but $\sigma^c_{0.2}$ is the stress corresponding to 0.2% inelastic strain on the isochronous stress-strain curve for the temperature and assessment time of interest. With the FAD, only assessments for initiation or small amounts of crack growth, Δa, are considered so the calculation is performed for the initial crack size a_o.

In eqn (7), K^P is the stress intensity factor corresponding to the load F and K^c_{mat} is a 'creep toughness'. This is obtained from constant load tests using

$$K^c_{mat} = [K^2 + \frac{n}{n+1} \frac{EF\Delta_c}{B_n(W-a)} \eta]^{\frac{1}{2}} \tag{8}$$

where η is the factor from [4] used to obtain C*, B_n, W, a are net specimen thickness, width and crack size, K is the stress intensity factor, E is Young's modulus, n is the creep exponent and Δ_c is the load-line displacement due to creep at the time for which the crack extension is Δa; K^c_{mat} is then the creep toughness for that time and that crack extension.

Having evaluated L_r and K_r, the point (L_r, K_r) is plotted on the FAD, Figure 1. If it lies within the curve then the crack extension does not exceed Δa and creep rupture is avoided. This is achieved by defining the curve by a cut-off at

$$L^{max}_r = \sigma_r / \sigma^c_{0.2} \tag{9}$$

where σ_r is the stress to rupture in the time of interest and, for $L_r < L^{max}_r$, by

$$K_r = \left[\frac{E\varepsilon_{ref}}{L_r\sigma^c_{0.2}} + \frac{L^3_r\sigma^c_{0.2}}{2E\varepsilon_{ref}} \right]^{-\frac{1}{2}} \tag{10}$$

where ε_{ref} is the total strain from the isochronous stress-strain curve at the reference stress. The limit $L_r = L^{max}_r$ is equivalent to using eqn (2) to estimate rupture time. The curve of eqn (10) is equivalent to using the C* approach of Section 2 for steady state creep with R in eqn (4) taken as $(K^P / \sigma_{ref})^2$, but also incorporates transient creep effects. As briefly summarised here, the method only covers constant primary loading but it may be extended to cover thermal and residual stresses, fatigue crack growth and variable primary loads [7].

Currently, there is effort to generate creep toughness data as historically data have been used to define relationships of the form of eqns (3) and (5). Some typical data are shown in Figure 2. In general, creep toughness is a function of time and crack extension Δa, reducing with increasing time and reducing crack extension. This simply reflects a requirement to use a lower toughness to avoid small crack extensions in long times than to avoid large crack extensions in short times.

The FAD approach is illustrated in [7] and in Section 4 below, the latter in the context of probabilistic fracture mechanics calculations.

4. Probabilistic Creep Crack Growth

The methods of Sections 2, 3 are straightforward to apply and, therefore, amenable to probabilistic methods. Recently, an Appendix has been added to R5 describing application of such methods using either the basic procedures or the FAD method. These methods are illustrated by considering a worked example of a pressurised pipe of mean radius R_m=243mm and thickness t=25mm. The pipe is assumed to have a surface defect of depth a_o=3mm and length ℓ_o=30mm detected after operating for 5 years. The probability of further crack extension of 2mm prior to inspection after a further year is required. This is referred to here as 'failure'. Creep rupture is not limiting in this case; as the crack is detected in service it is assumed to be growing so that initiation is not relevant; therefore, the assessment has focussed on creep crack growth described by eqn (5). Note, the pipe is subjected to a small number of fatigue cycles so that the allowable creep crack growth is slightly less than 2mm.

To apply the procedures of Section 2, eqn (5) is taken to hold with q = 0.67. For the 2¼Cr weld where the defect is located, the constant A has a mean value of 2.8 and 95 percentile value of 14.6 for $\dot{a}$ in mmh^{-1} and C* in MPamh^{-1}. A log-normal distribution defines the distribution of creep crack growth rates. A secondary-tertiary mean creep strain equation defines the strain rate to calculate C* from eqn (4). To allow for variability, the mean is multiplied by a factor f which has a mean value of unity, a 95 percentile value of 5 and a log-normal distribution.

Standard stress intensity factor and limit load solutions are used and enable the procedures of Section 2 to be applied. Using Monte Carlo techniques, repeating the deterministic calculations 5000 times, the probability of failure was obtained as 0.11, i.e. an 11% probability of crack growth exceeding 2mm in one further year of operation.

Models suggest that creep crack growth rates are inversely proportional to ductility [5]. This would lead to high values of A being associated with low values of f and vice versa. Therefore, the calculations were repeated with negative correlation between A and f and this led to a reduced failure probability of 0.02. For completeness, a positive correlation was also considered and this increased the failure probability to 0.14. The probability of failure is shown as a function of time in Figure 3; the effects of correlations are clearly visible. For the low failure probabilities of practical interest, negative correlations have a significant effect.

The same example was used with the FAD approach of Section 3. The assessment point is shown on Figure 4 based on mean properties. This lies inside the assessment curve indicating that failure does not occur for mean properties within one year. However, variability in L_r arises from variability in the creep strain data which leads to variability in $\sigma^c_{0.2}$; variability in K_r arises from variability in creep crack growth data which leads to variability in K_{mat}^c. The margin on creep toughness, F^K, was calculated as the ratio AB/AC in this figure. If the only variability were in creep toughness, then the probability of failure would be

$$P_f^K = \Phi \left[\frac{-1.645 \quad \log F^K}{\log (\overline{K}_{mat}^c / K_{mat(5\%)}^c)} \right]$$ (11)

where Φ is the standard normal cumulative distribution; the log-normal distribution of A leads to a log-normal distribution of creep toughness with mean $\overline{K}_{mat}^c$ and 5 percentile value $K_{mat}^c(5\%)$. A similar expression is obtained for P_f^σ, the failure probability assuming that only $\sigma_{0.2}^c$ is variable. Then, based on earlier work relating to R6, an approximate estimate of failure probability is [8]

$$P_f = P_f^K + P_f^\sigma$$ (12)

This leads to a value of 0.10, within 10% of the value from Monte Carlo methods. This illustrates the power of the FAD since the estimate of eqn (12) is obtained by one deterministic calculation and straightforward evaluation of margins, without the need for repeated calculations involving crack growth. However, the method of eqn (12) does not currently enable correlations to be treated and, in view of the significance of these in creep, is a limitation of the method.

5. Crack Growth in Weldments

Crack growth assessment methods for similar welds have been in R5 for some years and applications of the approach to heat-affected zone (HAZ) and Type IV cracking in full-size welded 0.5CrMoV pipes are reported in [9, 10]. For such welds, global redistribution of stress occurs under predominantly hoop stressing in butt welded pressurised pipes [11] due to the differing creep deformation behaviour of the base metal, weld, and HAZ. This leads to the R5 k-factor, where k<1 or k>1 for zones relatively weak or strong in creep compared with the parent material. For loading predominantly transverse to the weld, such overall stress redistribution cannot occur and k=1. The homogeneous reference stress of eqn (1) is multiplied by k for each characteristic material zone within the weld and eqn (2) is then used to assess rupture of each zone using the appropriate local rupture data. Crack growth follows from eqns (4) and (5), again using the relevant factored reference stress and creep crack growth and deformation data appropriate to the cracking location. Strain compatibility under hoop loading, however, enables the creep strain rate in eqn (4) to be set to the parent rate in that case.

In dissimilar metal welds (DMWs), such as between 2.25Cr1Mo and Type 316 steels, cracking is usually associated with the ferritic to weld metal interface region. Methods for assessing interfacial crack growth in DMWs have recently been added to

R5. Creep crack growth data in the form of eqn (5) are required with the crack in the appropriate interface region. Such data have been obtained [12, 13] for Inconel and 316 weld metals using standard test procedures with bi-material compact tension (CT) specimens. It was shown [13], firstly, that the experimental η-factor for the CT specimen is equally applicable for interface cracking in DMWs and, secondly, that the measured growth rates in both cases lay below reported upper bound parent 2.25Cr behaviour [14]. To estimate C* in an assessment, R5 recommends that eqns (1) and (4) are used but based on the strain rate properties of the faster-creeping material in the region of the crack, typically the ferritic HAZ. This approach to calculating C* has been shown, by comparison with detailed bi-material finite element analysis of CT [12, 13], single edge cracked [15] and circumferentially-cracked cylinder [16] specimens, to be conservative. Similarly, the overall assessment approach has been shown to be conservative by application to a test on a circumferentially-notched tube under pure bending [17] and by unreported assessments of both Inconel and Type 316 weld pressurised pipe tests. The R5 methods avoid the need to both characterise and obtain data for the various microstructural regions across a typical DMW; for instance those associated with the ferritic buttering layer and its associated HAZ in 2.25Cr/Type 316/Type 316 English Electric Mk.III welds [13]. The level of conservatism in the approach depends on the constraint of the specimen and in some cases it can be argued that the parent 2.25Cr data suffice.

6.   The σ_d Approach

The σ_d approach provides an alternative method to that of Section 2 for predicting creep crack initiation and has recently been included as a new appendix to R5. This type of approach was originally proposed in the French design code RCC-MR [18] as a practical method for estimating initiation of fatigue cracking from notches or pre-existing defects.   Subsequently, the method was extended to encompass initiation under creep and creep-fatigue conditions [19].

     The σ_d procedure set out in R5 is currently restricted to the assessment of creep crack initiation in austenitic stainless steels. This reflects the current status of experimental validation of the approach, although proposed future developments will consider creep-fatigue conditions and ferritic materials.

     The method is illustrated schematically in Figure 5 and is based on evaluation of a stress, σ_d, a short distance, $d = 50\,\mu m$, ahead of the crack tip. The linear elastic stress, σ_{de}, at a distance d from the crack tip is calculated from

$$\sigma_{de} = K / \sqrt{2\pi d} \tag{13}$$

where K is the linear elastic stress intensity factor.   The plastic strain, ε_2, at the reference stress, σ_{ref}, is then estimated from the monotonic stress/strain curve at the temperature of interest.   This allows the elastic-plastic stress, σ_d, to be estimated using Neuber's construction [20] together with the monotonic stress/strain curve based on σ_{de} and an initial strain of $\sigma_{de}/E' + \varepsilon_2$, where $E' = 3E/2(1+\nu)$ and ν is

Poisson's ratio. The initiation time, t_i, is then estimated using

$$t_i = t_r[\sigma_d] \qquad (14)$$

where $t_r(\sigma_d)$ is the rupture time at stress, σ_d.

7. Closing Remarks

This paper has described some recent developments in R5. The procedure continues to be developed in order to expand the scope of the procedures, to reduce conservatisms and to include simplified methods which enable practical defect assessments to be performed rapidly.

8. Acknowledgement: This paper is published with permission of British Energy.

9. References

1.   Goodall, I W and Ainsworth, R A , An assessment procedure for the high temperature response of structures, Proc. IUTAM Conf on Creep in Structures (Ed.Zyczkowski, M), 303-311, Springer-Verlag, Berlin (1991).

2.   Ainsworth, R A, Chell, G G, Coleman, M C, Goodall, I W, Gooch, D J, Haigh, J R, Kimmins, S T and Neate, G J: CEGB Assessment procedure for defects in plant operating in the creep range, *Fatigue Fract Engng Mater Struc*, **10** (1987) 115-127.

3.   PD6539:1994, Guide to methods for the assessment of the influence of crack growth on the significance of defects in components operating at high temperatures, BSi, London (1994).

4.   E1457-98, Standard test method for measurement of creep crack growth rates in metals, American Society for Testing and Materials, Philadelphia (1998).

5.   Webster, G A and Ainsworth, R A: *High Temperature Component Life Assessment*, Chapman & Hall, London (1994).

6.   Schwalbe, K-H, Ainsworth, R A , Saxena, A and Yokobori, T: Recommendations for a modification of ASTM E1457 to include creep-brittle materials, *Engng Fract Mech*, **62** (1999) 123-142.

7.   Ainsworth, R A, Hooton, D G, and Green, D. Failure assessment diagrams for high temperature defect assessment, *Engng Fract Mech*, **62** (1999) 95-109.

8.   Wilson, R, Mitchell, B J and Ainsworth, R A, Relationship between conditional failure probabilities and corresponding reserve factors derived from the R6 failure assessment diagram, ASME PVP Conference, Vol 323 (1996) 401-404.

9.   Budden, P J, Analysis of the type IV creep failures of three welded ferritic pressure vessels, *Int. J. Press. Vess. and Piping*, **75**, (1998) 509-519.

10.   Jones, M R and Coleman, M C, The assessment of creep crack growth in a welded pressure vessel, Proc. 4th. Int. Conf. Creep and Fracture of Engineering Materials and Structures, Swansea, Inst. Metals, (1990) 605-619.

11.   Coleman, M C, Parker, J D and Walters, D J, The behaviour of ferritic weldments in thick section 1/2Cr1/2Mo1/4V pipe at elevated temperature, *Int. J. Press. Vess. and Piping*, 18, (1985) 277-310.

328   R.A. AINSWORTH et al.

12.   Budden, P J and Curbishley, I, Assessment of creep crack growth in dissimilar metal welds, ASME
      PVP, Vol. 349 (1997). *To appear in Nuc. Engng. Design* (2000).

13   Ainsworth, R A, Characterisation of creep fracture at interfaces in weldments, Proc. ICF9, Vol.1,
      143-153 (1997).

14   Saxena, A, Han, J and Banerji, K, Creep crack growth behaviour in power plant boiler and steam pipe
      steels, *ASME J. Press. Tech.*, **110**, (1988) 137-146.

15   O'Dowd, N P, Budden, P J and Griffiths, E R J, Finite element analysis of a bimaterial SENT
      specimen under elastic-plastic loading, IUTAM Symposium on Nonlinear Analysis of Fracture, Ed.
      Willis, J R, Kluwer Academic Publishers, (1997) 33-42.

16   O'Dowd, N P and Budden, P J, The effect of mismatch on J and C* for interfacial cracks in plane and
      cylindrical geometries, 2nd. Intnl. Symposium on Mis-matching of Interfaces and Welds, Eds.
      Schwalbe, K H and Kocak, M, GKSS, (1997) 221-232.

17.   Curbishley, I and Hooton, D G, Assessment of creep and creep-crack growth of
      21/4Cr1Mo/Inconel/Type 316 dissimilar metal weld features in pure bending, p.273-282, Proc. Conf.
      "Integrity of High Temperature Welds", Nottingham, PEP (1998).

18.   RCC-MR, Design and Construction Rules for Mechanical Components of FBR Nuclear Islands,
      AFCEN, Paris (1985).

19.   Moulin, D, Drubay, B, and Acker, D, A Practical method based on stress evaluation (σ_d criterion)
      to predict initiation of cracks under creep and creep-fatigue conditions, PVP, Vol 223, Pressure
      Vessel Fracture, Fatigue and Life Management, ASME (1992).

20.   Neuber, H, Theory of stress concentration for shear strained prismatic bodies with arbitrary nonlinear
      stress strain law, Trans ASME *Jnl Appl Mech*, **28** (1961), 544-550.

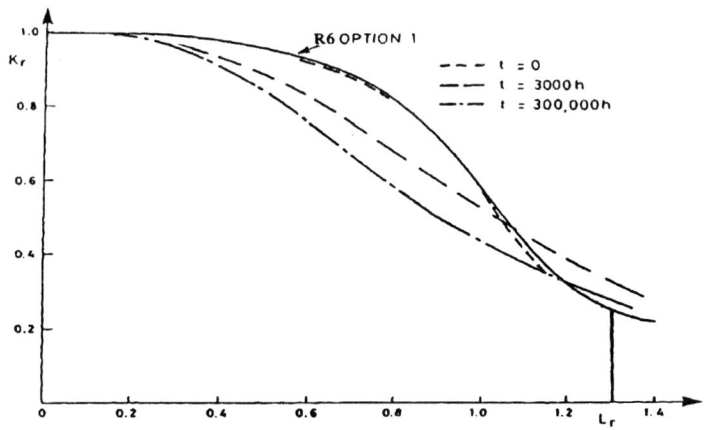

Figure 1   High Temperature Failure Assessment Diagram

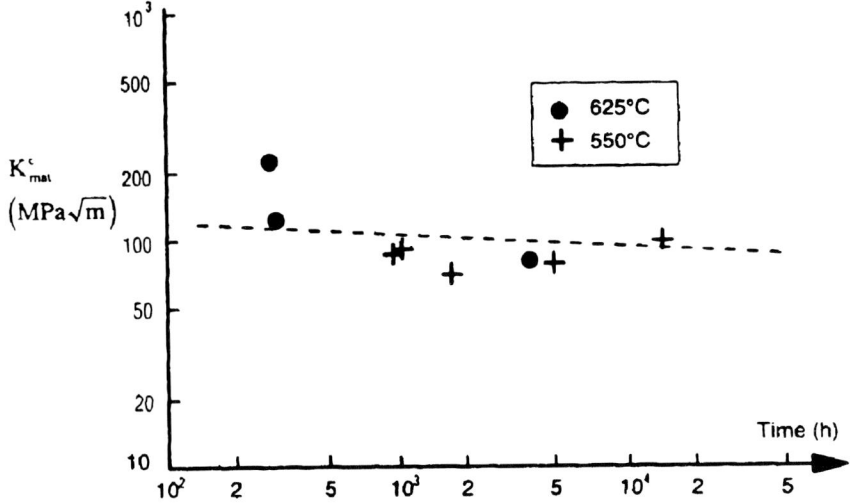

Figure 2   **'Creep Toughness' Data**

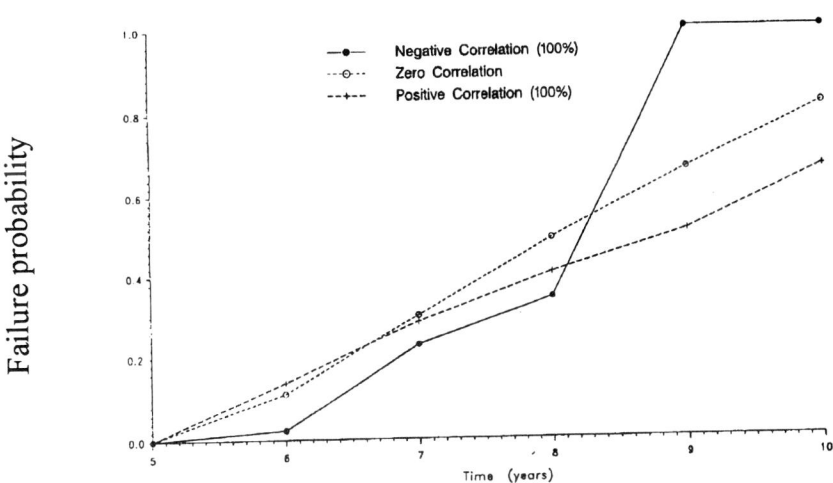

Figure 3   **Probability of Crack Extension of 2mm**

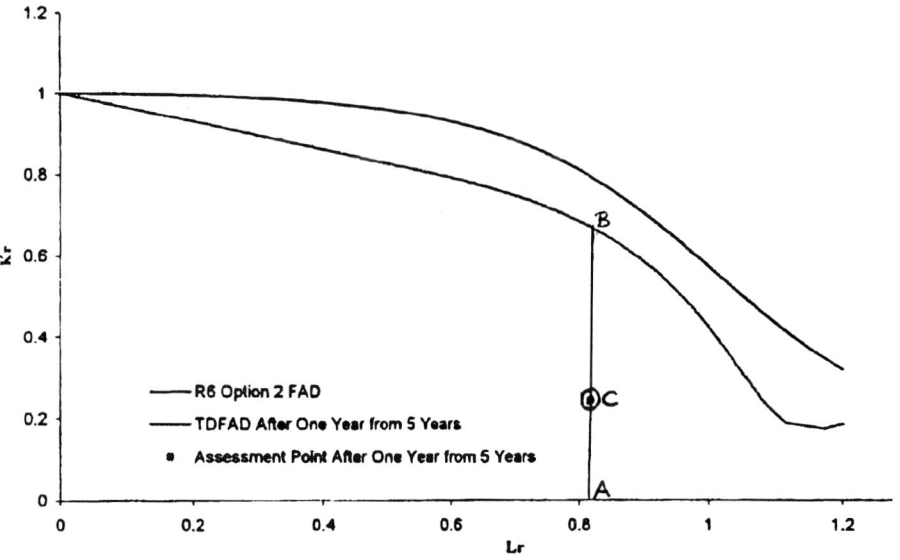

Figure 4   Assessment Using the FAD

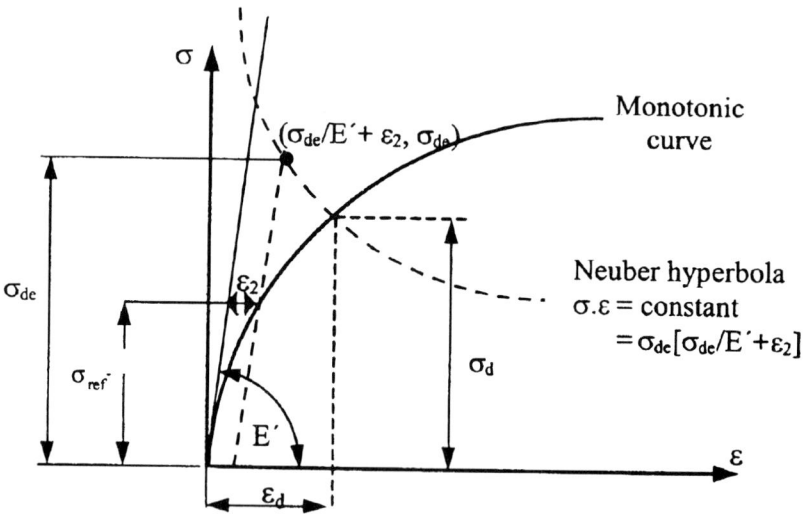

Figure 5   Illustration of the Approach Used to Estimate σ_d

ON GLOBAL APPROACHES TO SOME PROBLEMS INVOLVING PLASTICITY AND VISCOSITY EFFECTS

K. Dang Van
Laboratoire de Mécanique des Solides, UMR7649
Ecole polytechnique, 91128, Palaiseau, France.
dangvan@lms.polytechnique.fr

1 Introduction

In spite of extensive research done these last two decades (in particular for jet engines and nuclear applications), the prediction of the resistance of structures undergoing thermomechanical loadings which may induce creep is still a difficult problem. This is particularly true for problems involving creep fatigue interaction under anisothermal thermomechanical loadings which are frequently encountered in mechanical industries. In the automotive industries, this problem is also present and important: many hot structures exists (see for instance cylinder heads, exhaust manifolds, or brake discs...), which must be designed very carefully. However until very recently, because of the lack of efficient computational methods only empirical design methods are used: many tests must be done on prototypes which necessitate a long and expensive development time.

To be able to compute the resistance of such structures to a definite number of thermomechanical service loading cycles, it is necessary to overcome a certain number of difficulties which are summarised in the chart represented hereafter [Figure 1].

Because of non linear behaviour, one of the crucial points is the choice of the constitutive equations representing the material behaviour in the thermomechanical loading range. From this choice depends the computational scheme and strategy, which is also very important since, most of the time, discretisation of industrial structures necessitates a great number of variables. One hundred thousand DOF is common and applications require to have reasonable computation time to derive the pertinent parameters controlling the damage phenomena. The choice of damage evolution laws, and finally of failure criterion, are also other crucial points.

Thus, reliable and robust thermoanelastic constitutive equations and associated computational schemes are needed. Moreover, these equations must be known for a large range of temperature for which the material behaviour differs considerably. At low temperature, this behaviour is rather elastoplastic and at high temperature it is more of viscous type.

S. Murakami and N. Ohno (eds.), IUTAM Symposium on Creep in Structures, 331–340.

Most of the proposed models, in particular those based on internal variable formalism in thermodynamics of irreversible processes are efficient when these two phenomena are separated, (see for instance [1]). However, when they occur simultaneously, the obtained results are not so good. Indeed unified viscoplastic models have been used for different high temperature applications to metals, metal alloys and composites. However, in many cases, these types of approach are not fully adequate, as is shown on the example presented hereafter. It is of course possible to consider that anelastic deformation derives from two potentials the first one corresponding to plastic strain rate, the second to viscoplastic strain rate. But from the standpoint of practical utility, it is very difficult to separate the evolutions of these potential corresponding strain hardening effects. A crucial applicability factor of these models is related to the difficulty of identifying the (generally numerous) material constants from experimental data due in particular to the lack of direct physical significance of these parameters. Moreover, it is often necessary (in structures operating at a large range of varying temperature for example) to be able to account simultaneously for the reversible and irreversible time- dependant deformation.

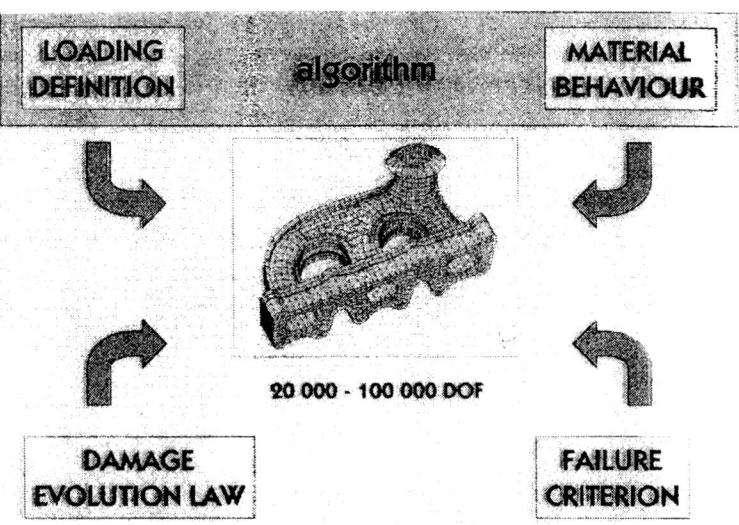

Figure 1. Chart of a global approach in thermomechanical design

This is the reason why we propose another type of model, based on a systematic way to introduce the desired effects with the help of rheological elements. The model presented here depends only on a small number of material parameters which make the identification problem easier. However as it is a key question for practical utility of the proposed method, we also develop optimisation procedures to estimate these parameters.

Another fundamental aspect refers to the stress–strain integration scheme in computational algorithm. As will be shown, this model leads to very simple implementation in most of the available computational codes.

2 Typical behaviour of the polyethylene material

Most of the existing models are based on a strain decomposition : total strain is a sum of elastic, plastic and viscoplastic strain.

We are working on another type of model that we call the viscoelastic and elastoplastic double layer model. In this model, the plastic and the viscous effects are distinguished using the stress. It corresponds for instance to a rheological model composed of an elastoplastic element in parallel with a non linear viscoelastic element. This model was proposed when studying polyethylene material which is very sensitive to loading rate and to temperature (J. Kichenin, A. Ouakka, K. Dang Van, et al., see for instance References 2 and 3.).

Typical behaviour of this type of material for tension-relaxation-recovery tests (TRR tests) is shown hereafter
Figure 2 a-b represents the stress and strain history; solid line is the imposed loading, dashed line is the corresponding response.
Figure 3 - Stress strain curves for two different loading rates the temperature being the same; the highest curve corresponds to a loading rate ten times greater than the lower one.
Figure 4 shows the influence of the temperature.

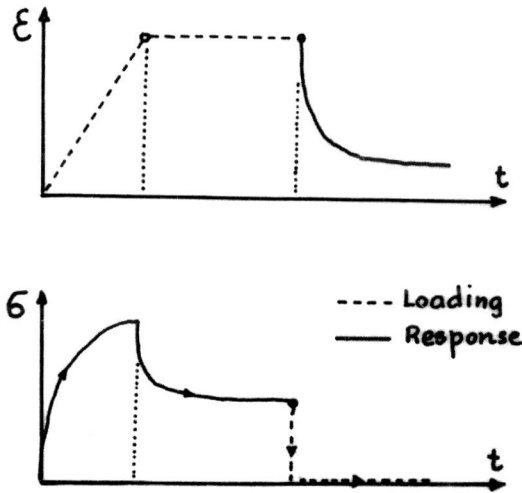

Figures 2-a, 2-b: stress –strain history: imposed loading and response

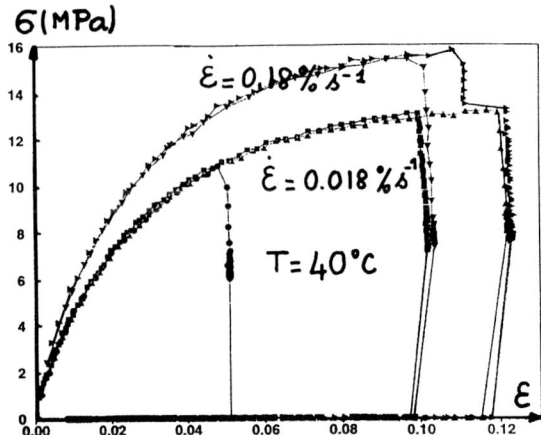

Figure 3: Influence of loading rate on stress-strain curve

It is clear that the classical anelastic models are not adequate to describe such behaviour which exhibits simultaneously varying apparent modulus, viscous effects and plastic effects. For instance, elastoplastic or viscoplastic models imply the existence of an elastic domain for which the young modulus is rate independent; at the same time, a linear or non linear solid viscoelastic model is not pertinent since the recovery and the stress fading phenomenon is not complete.

This is why we propose an other approach combining two rheological branches in parallel and represented in figure 5. The first branch corresponds to the introduction of the viscous effect, and the second one to the plastic effect. It is clear that this model is able to represent qualitatively the observed behaviour, i.e. the dependence of the apparent modulus and "yield" limit with the loading rate, and the existence of incomplete recovery and stress fading.

This type of model is similar to overlay models introduced quite a long time ago by Prof. Bessling [4] and then developed and used by Zienkiewicz, Nayak and Owen [5], and later by Meijers et al.[6].

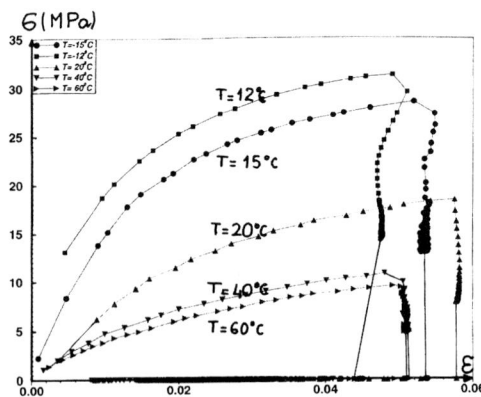

Figure 4: Influence of temperature on stress-strain curves.

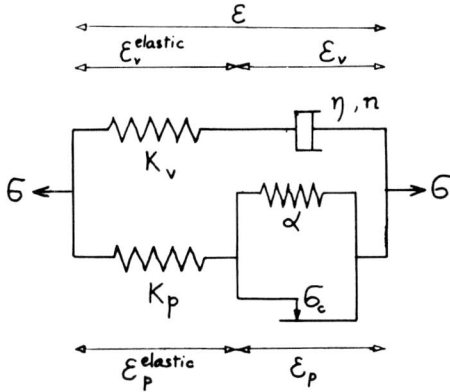

Figure 5: The "double layer" model

In the proposed model, the total strain can be decomposed as:

$\varepsilon = \varepsilon_v^{\text{elastic}} + \varepsilon_v$ corresponding to the viscous branch

$\varepsilon = \varepsilon_p^{\text{elastic}} + \varepsilon_p$ corresponding to the plastic branch

The corresponding stresses are:

$\sigma_v = K_v : \varepsilon_v^{\text{elastic}}$

$\sigma_v = K_p : \varepsilon_p^{\text{elastic}}$

$\sigma = \sigma_v + \sigma_p$

In this model, the total stress is the sum of a viscous part and a plastic part. Moreover, we allow non linear viscosity for a better quantitative description of the material behaviour. Finally, the standard model depends on six material parameters: $(K_v$, $K_p)$, respectively the elastic modulus of the viscous branch and the plastic branch, $(\eta$, $n)$, the viscous parameter and exponent, and (σ_y, α) the initial yield limit and a kinematic hardening parameter characterising the plastic branch.

Identification of the material parameters for the studied polyethylenes are done thanks to isothermal T.R.R. tests, and the use of a special software developed in our laboratory. It allows us to determine automatically the numerical values of the rheological model . A difficult problem finds its origin in the sensitivity of the determination of the set of parameters. This important particular point will be discussed in the paper of L. Verger et al. presented in the same conference.

Comparison between tests results and model for a typical polyethylene is shown in figure 6. Tests are done at two different temperature, T= 15°C and 40°C.

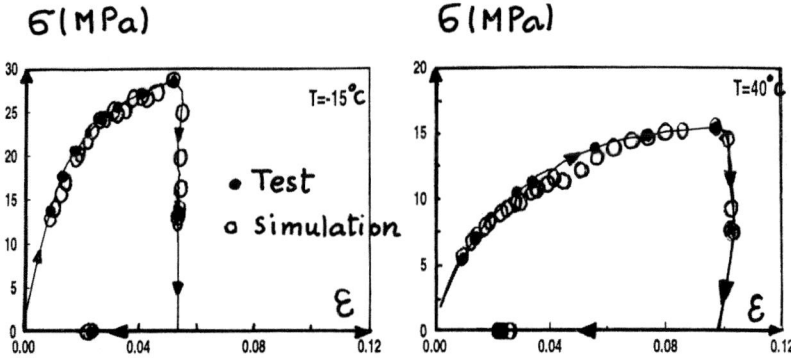

Figure 6: Comparison between test at 2 temperatures and numerical simulation

3 Generalisation: 3D constitutive equations and potential functions

For practical applications, generalisation in 3D can be done easily. One has:

$$\varepsilon_v^{\text{elastic}} = (1+v).\sigma_v/\ K_v - v.\text{trace}(\sigma_v)\ /\ K_v = (C_v)^{-1}.\sigma_v$$

$$\varepsilon_p^{\text{elastic}} = (1+v).\sigma_p\ /\ K_p - v.\text{trace}(\sigma_p)\ /\ K_p = (C_p)^{-1}.\sigma_p$$

$$\dot{\varepsilon}_p = \lambda\,\frac{s_p - X}{J_2(\sigma_p - X)}$$

$$\dot{\varepsilon}_v = \frac{3}{2}\left[\frac{J_2(\sigma_v)}{\eta}\right]^n \frac{s_v}{J_2(\sigma_v)}$$

$$\dot{\alpha} = \dot{\varepsilon}_p$$

$$\alpha = \frac{3\,X}{2h}$$

a dot on a parameter means time derivative; λ is the plastic multiplier. The meaning of the other (6) different parameters are shown in figure 5.

This model is compatible with an approach based on so called internal variable formalism in thermodynamics of irreversible processes. The material behaviour is characterised by two potentials, the free energy and the dissipation potential.

More precisely, let $\psi(\sigma_v,\sigma_p,\ X,\ T)$ and $\phi(\sigma_v,\sigma_p,\ X,\ T)$ be their complementary form, and T the temperature. The following assumptions are made:

Free energy can be decomposed into three terms:

$$\psi(\sigma_v, \sigma_p, \mathbf{X}, \mathbf{T}) = \psi_{viscous}(\sigma_v, T) + \psi_{plastic}(\sigma_p, T) + \psi_{hardening}(\mathbf{X}, \mathbf{T})$$

Dissipation potential is the sum of a viscous potential and a plastic potential:

$$\psi_{viscous} = \frac{1}{2\rho}\left[\frac{1+v}{K_v}.tr(\sigma_v : \sigma_v) - \frac{v}{K_v} tr(\sigma_v)^2\right]$$

$$\psi_{plasc} = \frac{1}{2\rho}\left[\frac{1+v}{K_p} tr(\sigma_p : \sigma_p) - \frac{v}{K_p} tr(\sigma_p)^2\right]$$

$$\psi_{hardening} = \frac{3}{4h} X : X$$

$$\varphi(\sigma_v, \sigma_p, \mathbf{X}, \mathbf{T}) = \varphi_{viscous}(\sigma_v, T) + \varphi_{plastic}(\sigma_p, \mathbf{X}, \mathbf{T})$$

$$\varphi_{iscous} = \frac{\eta}{n+1}\left[\frac{J_2(\sigma_v)}{\eta}\right]^{n+1}$$

$\varphi_{plastic}$ is the indicator of the convex function defined by the yield surface

One has classically for isothermal loadings:

$$\varepsilon_v^{elastic} = \rho\frac{\partial\psi}{\partial\sigma_v}, \varepsilon_p^{elastic} = \rho\frac{\partial\psi}{\partial\sigma_p}, \alpha = \rho\frac{\partial\psi}{\partial X}$$

$$\dot{\varepsilon}_v = \frac{\partial\varphi}{\partial\sigma_v}, \dot{\varepsilon}_p = \frac{\partial\varphi}{\partial\sigma_p}, \dot{\alpha} = \frac{\partial\varphi}{\partial\dot{X}}$$

or more precisely:

$$\dot{\varepsilon}_p = \lambda\frac{S_p - \frac{2}{3}h.\varepsilon_p}{J_2(\sigma - \frac{2}{3}h.\varepsilon_p)} = \overset{\bullet}{\alpha}$$

$$\dot{\varepsilon}_v = \frac{3}{2}\left[\frac{J_2(\sigma_v)}{\eta}\right]^n \frac{S_v}{J_2(\sigma_v)}$$

The temperature dependence is simply obtained by assuming that the coefficients K_v, K_p, η, h, σ_y, n are a function of the temperature T.

It can be proven by classical arguments of the thermodynamic of irreversible processes that the intrinsic dissipation D is the sum of a plastic part D_p and a viscous part D_v

$$D = \sigma_v : \dot{\varepsilon}_v + \sigma_p : \dot{\varepsilon}_p - X : \dot{\alpha}$$

$$D_p = \sigma_p : \dot{\varepsilon}_p - X : \dot{\alpha}$$
$$D_v = \sigma_v : \dot{\varepsilon}_v$$

4- Numerical implementation

The proposed model can be implemented in most of the existing numerical programs as is shown later. As one has the same total strain rate for the viscous branch and the plastic branch, the displacement rate field minimises the functional

$$J_1(\dot{u}) = \int_\Omega \left[\tfrac{1}{2}\varepsilon(\dot{u}) : C_v : \varepsilon(\dot{u}) - \dot{\varepsilon}_v : C_v : \varepsilon(\dot{u}) \right] d\Omega - \int_{\partial\Omega} \dot{F}_1 \dot{u}\, ds$$

for a solid of purely viscous type (corresponding to the viscous branch) submitted to the force $\dot{F}_1, \dot{\varepsilon}_v$ being given

and

$$J_2(\dot{u}) = \int_\Omega \left[\tfrac{1}{2}\varepsilon(\dot{u}) : C_p : \varepsilon(\dot{u}) - \dot{\varepsilon}_p : C_p : \varepsilon(\dot{u}) \right] d\Omega - \int_{\partial\Omega} \dot{F}_2 \dot{u}\, ds$$

for a solid of purely plastic type (corresponding to the plastic branch) submitted to the force $\dot{F}_2, \dot{\varepsilon}_p$ being given.

Consider the same solid submitted to the force $\dot{F} = \dot{F}_1 + \dot{F}_2, \dot{\varepsilon}_v, \dot{\varepsilon}_p$ being given, then the displacement rate minimises the following functional:

$$J(\dot{u}) = \int_\Omega \left[\tfrac{1}{2}\varepsilon(\dot{u}) : C : \varepsilon(\dot{u}) - \dot{\varepsilon}_v : C_v : \varepsilon(\dot{u}) - \dot{\varepsilon}_p : C_p : \varepsilon(\dot{u}) \right] d\Omega - \int_{\partial\Omega} \dot{F}$$

Where $C = C_v + C_p$

Thus in this manner, it is very simple to adapt this constitutive equation to existing numerical codes. Classical F.E.M. programs can be used; the rigidity matrix is calculated using the elastic matrix C, sum of the viscous branch and the plastic branch; the contribution of body forces induced by viscous deformation or plastic deformation is evaluated by using the corresponding elastic modulus.

We observe that for anisothermal loadings, it is more convenient to use implicit algorithms with the constitutive equations presented before to avoid difficulties arising from variation of parameters with temperature increments.

Another important remark is related to the important time duration required for such computations on industrial structures. it is difficult to imagine to simulate a great number of loading cycles. This is why, stabilised cycles are researched. This important point is developed in the paper presented by L; Verger, A. Constantinescu and E. Charkaluk at this symposium. See also reference [8].

5- Damage and failure criteria.

The definition of damage and failure must be examined when analysing structures from the point of view of a design engineer. The traditional distinction and discussions among scientists between initiation and propagation phase are difficult to transpose for studying the creep fatigue resistance of structures to varying thermomechanical loadings . We chose a simplified approach in considering that the failure has occurred when the structure lost its function. In the case of the exhaust manifold represented in figure 1; this stage corresponds to formation of a visible crack which is at the origin of its resistance decrease and the corresponding characteristic volume is the thickness of this structure.

Let us come back to the choice of damage parameter and criterion. The approaches deriving from L.C.F; and based on the concept of strain amplitude, are not pertinent for anisothermal problems since the stress varies with the temperature, for a given plastic strain. We prefer to characterise the material by the dissipated energy which can be computed without any ambiguity even for varying thermomechanical conditions.

This methodology is successfully applied to the design of some complex industrial structures like exhaust manifolds. Prediction of the number of fatigue life and locus of failure is in reasonable good agreement with experimental results [8].

6- Conclusions

A global methodology for studying the creep fatigue life of structures submitted to thermomechanical loading cycles is proposed. It is based on an uncouple approach between plastic / viscous behaviour (thermomechanical constitutive equations) and damage. In this modelling, we try to have a reasonable compromise between simplicity and precision of the prediction by using a systematic way to introduce viscosity and plasticity with the help of rheological elements, the parameters of which are function of the temperature; for creep fatigue the damage evolution is then characterised by the dissipated energy. The efficiency of this approach is proved by good prediction on industrial structures [7,8]. It is now in operation in some industries in France. Other kind of applications are also performed, for instance to predict slow creep crack growth in polyethylene gas pipes by local approach [9].

References

1. Lemaitre J., Chaboche J.L., *Mécanique Des Matériaux Solides*, Dunod, 1985.
2. Ouakka A., Dang Van K., *Identification et optimisation Des paramètres du modèle rhéologique du polyéthylène*, Protocole expérimentale & Validation. Rapport GDF/ LMS juin 1995
3. Ouakka A., Dang Van K. : *Identification et optimisation Des paramètres du modèle rhéologique du polyéthylène*, Proposition d'une loi de viscosité non linéaire, Rapport GDF/ LMS mai1997
4. Besseling J.F., A theory of elastic, plastic and viscoplastic deformation of an initially isotropic material, J. of Applied Mech., Vol. 25, 1958, pp. 612-627.
5. Zienkiewicz O.C., Nayak G.C., Owen D.R.J., *Composites and overlays models in numerical analysis of elastoplastic continua*, Int. Symp.Foundations of Plasticity, A. Sawczuk ed., 1972 Warsaw.
6. Meijers P., Roode F., *Experimental verification of constitutive equations for creep and plasticity based on overlay models*, J. Pressure Vessel Technology, Vol. 105, 1983, pp.277-284.
7. Charkaluk E., *Dimensionnement Des Structures à la Fatigue Thermomécanique*, Thèse de Doctorat de l'Ecole Polytechnique, 1999.
8. Lederer G., Charkaluk E., Verger L., *Constantinescu A., Numerical Lifetime Assessment of Engine Parts Submitted to Thermomechanical Fatigue, Application to Exhaust Manifold Design*, SAE Technical Paper Series 2000-01-0789, SAE 2000 World Congress, Detroit, March 6-9, 2000
9. D. Gueugnaut, P. Blouet, A. Ouakka, K.Dang Van : *Criteria for Natural and Mechanical Damages to Polyethylene Gas Pipes*, International Gas Research Conference, San Diego, USA, pp.281-291, 1998.

ON THE SIMULATION OF LARGE VISCOPLASTIC STRUCTURES UNDER ANISOTHERMAL CYCLIC LOADINGS *

L. VERGER[1,2] (verger1@mpsa.com),
A. CONSTANTINESCU[1] (constant@lms.polytechnique.fr),
E. CHARKALUK[3] (charkalu@mpsa.com)

[1] *Laboratoire de Mécanique des Solides (CNRS UMR 7649), Ecole Polytechnique, 91128 Palaiseau, France*
[2] *P.S.A. Peugeot-Citroën – Direction des Techniques et Achats, 18 rue des Fauvelles, 92250 La Garenne-Colombes, France*
[3] *P.S.A. Peugeot-Citroën – Direction de la Recherche et de l'Innovation Automobile, Chemin de la Malmaison, 91570 Bièvres, France*

Abstract. The optimal design of parts submitted to thermomechanical loadings is a key issue for the safety and quality assessments of structures. In this context, the present paper discusses the choice of a behavior model and the determination of its parameters. For cast iron and aluminum alloy applications, two constitutive laws are compared : one is based on a classical unified viscoplastic model and the other is based on a two-layer plastic-viscous rheological model. The following discussion shows that the use of anisothermal experiments is very important for the identification of material parameters and that simple models can be successfully applied for the predictive lifetime assessments of large 3D structures.

1. Introduction

Non-linear constitutive material behaviors have been the subject of extensive researches from the theoretical and numerical point of view over the last decades.

A large number of phenomenological models are devoted to the behavior of metals at high temperatures. They usually attempt to describe the metallurgical phenomena precisely [3,5]. However, they are not always suited for industrial design processes. Indeed, when computing cyclic multiaxial ther-

---

* Presented at IUTAM'2000 Creep in Structures, Nagoya, Japan

S. Murakami and N. Ohno (eds.), IUTAM Symposium on Creep in Structures, 341–350.

momechanical loadings on large structures, design engineers have to fulfill demands of predictive lifetime assessments within short design periods.

Then a complete computational design approach consists of three principal steps [1]:

- *The thermomechanical loading.* The mechanical analysis is based on numerical computations depending on the precise knowledge of the temperature history and boundary conditions. The temperature history is obtained from a combustion modeling and its following thermal transfer analysis.
- *The constitutive model.* The model should describe the behavior over the whole range of operating temperatures of the structure in isothermal as well as in anisothermal loading conditions. It has to be implemented in standard FEM computer codes using efficient and robust integration algorithms.
- *The lifetime prediction criterion.* The criterion is a low-cycle fatigue criterion in a multiaxial and anisothermal context. The criterion should also take into account the standard deviation due to the industrial manufacturing process.

The present paper focuses on the choice of a constitutive model in an anisothermal context and its underlying questions. Two important *a priori* assumptions are made concerning the model. We suppose on one hand that the cyclic softening/hardening behavior can be neglected and on the other hand that damage does not affect the material behavior during the main part of its life. In spite of the errors potentially induced by such assumptions, this leads to both shorter computation times and predictive final results [2,10].

Two metallic alloys will be discussed in the sequel: a spheroïdal graphite *cast iron* (SiMo SG) with Mo and Si additions, used for exhaust manifolds in a $0 - 800^{o}C$ temperature range, and an *aluminum-silicon alloy* (ASME A356) with Mg addition, used for cylinder heads in a $0 - 300^{o}C$ temperature range.

For these alloys, both a unified model and a two-layer elastoviscoplastic model are presented, together with a summary of their numerical integration scheme.

A series of questions are discussed regarding the identification of the parameters of the models in an anisothermal context, the difficulty of choosing between the two models, the passage from 1D tension-compression tests to 3D structures and from isothermal loading conditions to anisothermal ones.

As a final conclusion, it is shown that complete FEM computations can be performed on structures in spite of some behavior-modeling uncertainties, and that subsequent results can be used for successful lifetime prediction.

2. Constitutive models

The choice of the constitutive model is a key point in the global computational approach. It should represent the material behavior in the temperature and strain-rate range of the structure and it should be identified from a series of simple experiments on specimens.

2.1. EXPERIMENTS

The following experiments were at our disposal: *isothermal uniaxial tests* at different temperature levels which highlight the mechanisms of the phenomenological model ; *anisothermal uniaxial tests* which describe the transient behavior from room temperature to maximum temperature ; *anisothermal tests with temperature gradient* which exhibit a local state comparable with that of the complete structure.

The *uniaxial isothermal tests* are both tension-relaxation-compression tests (TRC) at different strain rates and low-cycle fatigue tests (LCF) characterizing the cyclic behavior and lifetime at different strain levels. The relaxation parts of the TRC stress/time data show exponential shapes (Fig. 1*a*). Therefore a Norton-Hoff power-law can be used for viscosity modeling. LCF data exhibit small cyclic softening which allows us to neglect isotropic hardening. Finally, cyclic TRC data and LCF data show a quick stabilization of the stress response (Fig. 1*b*). As a consequence, we shall assume that the damage undergone by the specimen at each cycle does not significantly affect the cyclic behavior.

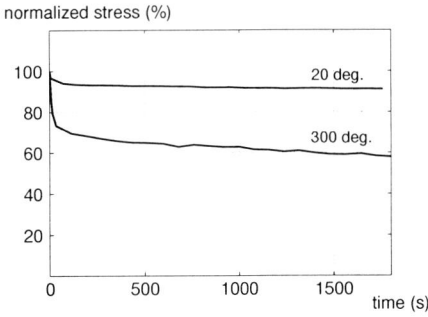
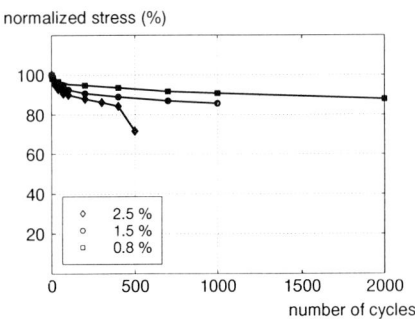

Figure 1.   Experimental observations on *aluminum alloy* : *(a)* exponential shape of relaxation in TRC tests, *(b)* negligible cyclic softening behavior in LCF tests

The *anisothermal uniaxial tests* are difficult to perform but they provide important data for identification [7]. The difficulties stem from the simultaneous need of a homogeneous temperature field in the specimen and a precise and rapid control of temperature and strain rates.

In the case of the *aluminum alloy*, two testing conditions were defined:

- an out-of-phase compression with temperature ranging from $50°C$ to $300°C$, which is used to fit the variation of characteristics with temperature,
- an out-of-phase compression with a pseudo-dwell at maximum strain and maximum temperature which enhances the influence of the viscosity at high temperature.

Finally, some *anisothermal tests with a temperature gradient along the specimen* were performed on a special machine [2]. The specimen was clamped between two cantilever beams and heated by Joule effect. The maximum temperature distribution along the *aluminum alloy* (respectively *cast iron*) specimen varies from $40°C$ at its ends to $300°C$ (respectively $700°C$) in its middle section.

2.2. PHENOMENOLOGICAL VISCOPLASTIC MODELS

Two viscoplastic models have been deduced from the previous experimental observations using classical phenomenological formulations. They both take into account plastic behavior with a linear kinematic hardening rule and viscous behavior with a Norton-Hoff power law.

The first model is a *unified viscoplastic* model *(UM)* [4,5]. Viscosity and plasticity are represented by a single internal variable : the inelastic strain. The corresponding rheological representation is presented on Fig. 2.

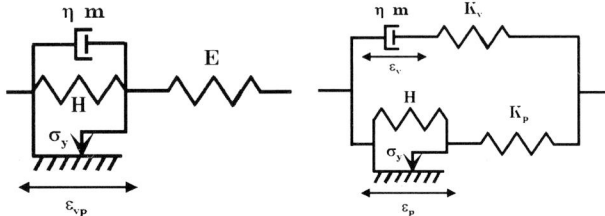

Figure 2.   Rheological representation of the unified model *(UM)*, and the two-layer model *(TLM)*

The second model is a *two-layer plastic-viscous rheological* model *(TLM)*, initially proposed by Kichenin [6]. The inelastic strain is represented by two independent internal variables.

The models are conditioned by temperature-dependent parameters which can be described as follows for 1D formulation: E, K_p and K_v are the elastic moduli of each layer, σ_y is the yield stress and H the hardening modulus, η is the viscosity coefficient and m the power coefficient of Norton-Hoff law.

2.3. CONSTITUTIVE LAWS

In the sequel, vectorial and tensorial variables are denoted by bold fonts. ε and σ stand for strain and stress tensors. In the *TLM* case, the subscripts *ev* and *ep* denote quantities related to viscous and plastic layers respectively.

The strain is supposed to decompose additively into elastic and inelastic parts:

$$
\begin{array}{ccl}
(UM) & & (TLM) \\
\varepsilon = \varepsilon^e + \varepsilon^{vp} & \quad \varepsilon \quad = & \varepsilon^e_{ev} + \varepsilon^v_{ev} \\
& = & \varepsilon^e_{ep} + \varepsilon^p_{ep}
\end{array}
\tag{1}
$$

The stresses and hardening variables are linked to the strain variables by the following equations:

$$
\begin{array}{cccc}
(UM) & & & (TLM) \\
\sigma & = & C : \varepsilon^e & \quad \sigma \quad = \quad \sigma_{ev} + \sigma_{ep} \\
X & = & -H : \alpha & \quad \sigma_{ev} \quad = \quad C_{ev} : \varepsilon^e_{ev} \\
& & & \quad \sigma_{ep} \quad = \quad C_{ep} : \varepsilon^e_{ep} \\
& & & \quad X_{ep} \quad = \quad -H_{ep} : \alpha_{ep}
\end{array}
\tag{2}
$$

C and H denote the elasticity andx hardening tensors. In this case, the hardening variable α is equal to the cumulated plastic strain.

The von Mises equivalent stress is used to define the plasticity criterion:

$J_2(s) = \sqrt{\dfrac{3}{2}(s : s)}$ where $s = \mathrm{dev}\sigma$.

Then the flow-rules can be expressed under the following form:

$$
\dot{\varepsilon}^{in} = \gamma \cdot \frac{A}{\sqrt{A : A}}
\tag{3}
$$

where:

$$
\begin{array}{ll}
(UM) & (TLM) \\
A = \mathrm{dev}(\sigma - X) & A_{ev} = \mathrm{dev}(\sigma_{ev}) \\
X = -\frac{2}{3}H : \varepsilon^{vp} & \gamma_{ev} = \sqrt{\frac{3}{2}}\left\langle \frac{J_2(A_{ev})}{\eta} \right\rangle^m \\
\gamma = \sqrt{\frac{3}{2}}\left\langle \frac{J_2(A)-\sigma_y}{\eta} \right\rangle^m & A_{ep} = \mathrm{dev}(\sigma_{ep} - X_{ep}) \\
& X_{ep} = -\frac{2}{3}H_{ep} : \varepsilon^p_{ep}
\end{array}
\tag{4}
$$

For theoretical details of the presented thermodynamical framework, see [5].

The constitutive equations have been integrated by an implicit Newton-Raphson integration scheme using a return-mapping algorithm and consistent tangent moduli [8,9].

3. Identification of material parameters

3.1. A CLASSICAL APPROACH

In a first stage, the parameters have been directly estimated from the stabilized isothermal TRC data. The identification was based on some simple interpretations of the physical meaning of the parameters: the elastic and hardening moduli are slopes read on stress/strain tension curves, the viscosity parameters fit the relaxation curve, the yield stress is found from the relaxed stress, etc. The temperature-dependent parameters were obtained by a linear interpolation between the TRC temperatures.

In the *cast iron* case, the parameters have been determined as described above, and afterwards they have been used to simulate the anisothermal clamped test. The local stress/strain state in the specimen is multiaxial due to the temperature gradient along its axis. The computed and measured axial stresses are in a reasonable accordance (Fig. 3*a*), but there is a difference in the radial stress/strain responses between the *(UM)* and the *(TLM)* models (Fig. 3*b*). All the same, the numerical results obtained for

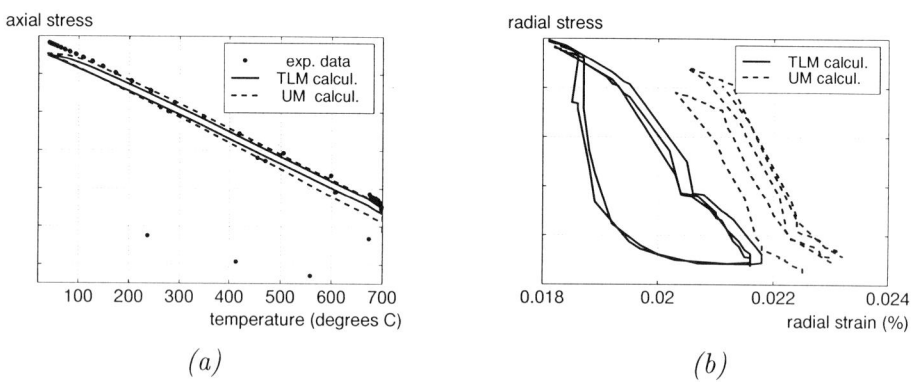

Figure 3. Anisothermal clamped test for *cast-iron:* (a) experimental/computed data comparison on a stabilized cycle (b) computational data comparison for two models

cast iron with the *(TLM)* model permitted correct lifetime predictions on exhaust manifolds [2,10].

The understanding of the differences between the behavior of the two models during the simulation of the clamped test is not an easy task. A first difficulty comes from the experiment, in which the stress and strain fields are controlled by a complex temperature distribution. The transient temperature field has to be computed very precisely in order to obtain accurate results for the mechanical fields. Moreover, due to the 3D nature of the fields, it is particularly difficult to read the sensitivities of the models and parameters with regard to the loading.

The second difficulty comes both from the models and the isothermal identification method. Indeed both models inherit an indetermination between the plastic and the viscous quantities (Fig. 4), which implies that the evolution of the parameters with temperature is badly managed. Therefore the prediction of the anisothermal mechanical response may be dubious.

In conclusion, we should find a simpler way to enhance identification. A better experiment should exhibit a homogeneous temperature field. Then the computed identification problem would be a 1D problem, much easier to analyze. This is the case of the anisothermal experiments without gradient described above. We should also dispose of a more general identification procedure permitting the combination of isothermal and anisothermal tests data.

3.2. AN OPTIMAL CONTROL APPROACH

In the sequel the identification of the material parameters is essentially based on the minimization of a cost functional defined as:

$$J(p) = \frac{1}{2} \int [\sigma_{exp}(t) - \sigma_{calc}(p,t)]^2 \, dt$$

where the time-integral measures the difference between the experimental and computed stresses (σ_{exp} and σ_{calc} respectively) over a complete experiment. In order to determine parameters from several experiments, the corresponding cost functionals can be added using weight factors.

The minimization procedure was based on a gradient computation using the adjoint state method and a BFGS gradient descent algorithm. This procedure can be combined with direct callibration of elastic and hardening moduli from the slopes of the isothermal curves as described in section 3.1.

The indetermination related to competition between plastic and viscous quantities is illustrated by plotting the cost functional as a function of the viscous parameters (η, m) (Fig.4). In the case of an isothermal tensile test, one can remark a long flat valley proving the coupling effect between the two parameters as well as the laborious identification induced.

Several identifications showed that a complete parameters set can be determined from a single anisothermal experiment (Fig. 5b). However, a single experiment will generally not assure the validity of the set in all test conditions. Therefore one should combine isothermal and anisothermal experiments in the identification program. This combination conducts to a difficult multiobjective minimization, where the identified parameters set will depend on the weight attributed to the different experiments.

In spite of these difficulties, the final parameters obtained by this method permitted several reasonably accurate computations on structures as presented in the next section.

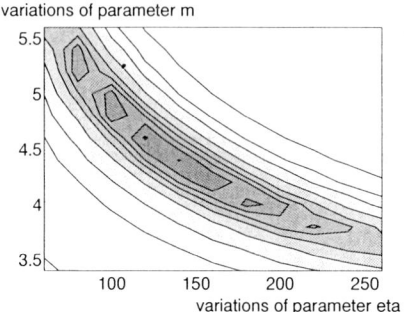

variations of parameter m

variations of parameter eta

Figure 4.   Cost functional distribution for three isothermal tests with aluminum alloy

4. Computations on structures

The computations on structures have been realized using the ABAQUS Standard FEM code after the determination of the parameters using the previous techniques.

The global computational approach is based on an important assumption: as for plastic shakedown phenomena, we assume that the response of the structure will rapidly stabilize under cyclic thermomechanical loadings. The stabilized computed cycle is generally obtained after a few cycles (< 10) and the corresponding mechanical fields are used as the input of the fatigue analysis.

The FEM meshes for exhaust manifolds or cylinder heads had $\approx 10^4 - 10^5$ degrees of freedom. The thermomechanical computations were completed within approximately 6 hours of CPU time on a HP-V class computer. The CPU time showed to be very sensitive both to nonlinearities of the constitutive model and to the contact boundary conditions.

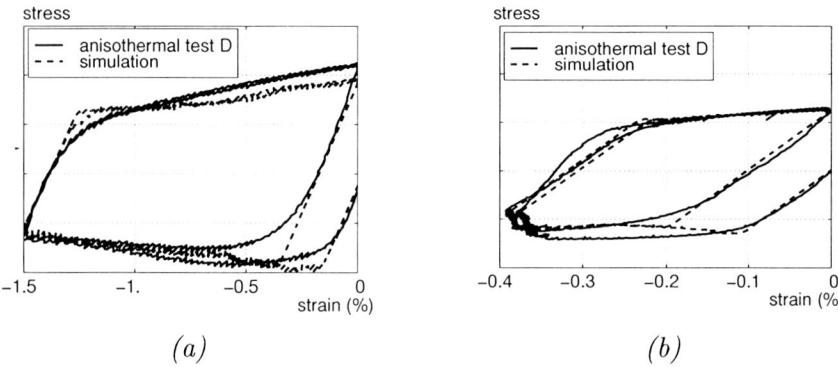

Figure 5.   Experimental and computed anisothermal uniaxial test with parameters identified from: *(a)* isothermal and anisothermal tests, *(b)* anisothermal tests

In the case of *cast-iron* exhaust manifolds, parameters of the constitutive models are identified using the classical method. Once more, there is a difference between the stress/strain responses of the *UM* and *TLM* models. Using an energy-based criterion for thermomechanical fatigue [2], the results from the *TLM* model permitted to determine the lifetime of several 3D structures: the clamped specimen under four test conditions and three exhaust manifold geometries (Fig.7). This shows the reliability of the presented global approach in thermomechanical fatigue.

In the case of *aluminum alloy* cylinder heads, parameters of the constitutive models are determined using the optimal control approach including anisothermal tests. 3D computations have been conducted on a one-cylinder head prototype for one particular test configuration.

A numerical comparison in the case of the clamped specimen shows a small difference in the mechanical response between the two models (Fig. 6). This proves the great importance of using anisothermal tests for the identification of the parameters. The lifetime results are not available yet to validate the global approach in the case of the *aluminum alloy* cylinder heads.

5. Conclusion

This paper discussed some problems related to the choice and the identification of a constitutive law for thermomechanical computations on real 3D structures. A series of difficulties have exhibited the importance of using anisothermal tests during the identification process. The encouraging results show that simple constitutive laws are sufficient for a robust lifetime prediction in an industrial context.

Acknowledgement to A. Bignonnet (PSA) and K. Dang Van (LMS) for encouring this project and G. Lederer (PSA) for continous support.

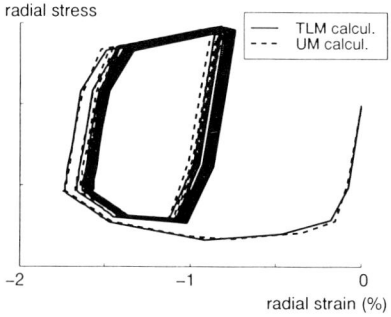

Figure 6.   Comparison of the stress/strain curves for the two models in the case of aluminum clamped specimen

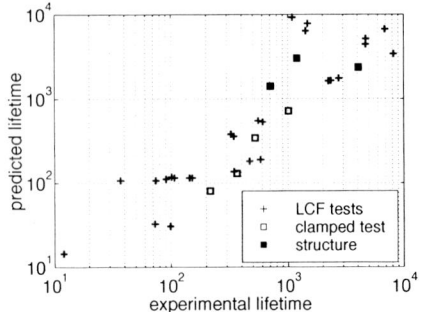

Figure 7. Lifetime assessments for cast-iron structures (exhaust-manifolds)

References

1. E. CHARKALUK, A. CONSTANTINESCU, A. BIGNONNET AND K. DANG VAN in *Low cycle fatigue and elasto-plastic behavior of materials*, ed. K.T. Rie et P.D. Portella, Elsevier, 1998, pp 815–820.

2. E. CHARKALUK, A. CONSTANTINESCU, An energetic approach in thermomechanical fatigue for silicon molybden cast iron, accepted for publication in J. Mat. High. Temp., 2000

3. D. FRANCOIS, A. PINEAU, A. ZAOUI in *Comportement mécanique des matériaux - vol II*, Hermès, Paris, 1994

4. J.L. CHABOCHE, Sur l'utilisation des variables d'état interne pour la description du comportement viscoplastique et de la rupture par endommagement, *Symposium Franco-Polonais de Rhéologie et de Mécanique*, 1977

5. J. LEMAITRE AND J. L. CHABOCHE, *Mécanique des matériaux solides*, Dunod, Paris, 1985

6. J. KICHENIN, *Comportement thermomécanique du polyéthylène - Application aux structures gazières*, PhD Thesis, Ecole Polytechnique (France), 1992

7. L. REMY Thermal and thermal-mechanical fatigue of superalloys - a challenging goal for mechanical tests and models, in *Low cycle fatigue and elasto-plastic behaviour of materials*, ed. K.T. Rie et P.D. Portella, Elsevier, 1998, pp 119–130

8. J.C SIMO and T.J.R. HUGHES, *Computational Inelasticity*, Springer Verlag, 1998

9. E. CHARKALUK, L. VERGER, A. CONSTANTINESCU, G. LEDERER AND C. STOLZ, Lois de comportement viscoplastiques anisothermes pour calculs cycliques sur structures, *Colloque Nat. en Calcul des Structures*, ed. Guédra-Degeorges and Co., 1999, pp. 575–580

10. G., LEDERER, E. CHARKALUK, L. VERGER, A. CONSTANTINESCU, Numerical life assessment of engine parts submitted to thermomechanical fatigue - Application to exhaust manifolds, *SAE Technical paper series*, 2000-01-0789

DESCRIPTION OF INELASTIC BEHAVIOR OF PERFORATED PLATES BASED ON EFFECTIVE STRESS CONCEPT

T. IGARI [*1], **T. TOKIYOSHI**[*1] and **Y. MIZOKAMI**[*2]
*1 *Nagasaki Research and Development Center*
*2 *Nagasaki Shipyard and Machinery Works*
   Mitsubishi Heavy Industries, Limited, Nagasaki, Japan

1. Introduction

High-temperature heat exchangers are well known as important components of high-temperature nuclear reactors such as High-Temperature Gas-Cooled Reactors and Fast Breeder Reactors.   Perforated plates in heat exchanges are one of the critical parts from a viewpoint of creep-fatigue damage.

Stress analysis based on the equivalent-solid-plate concept is often used in the structural design of high-temperature components containing perforated plates.   When adopting this concept, prediction methods of both macroscopic behavior and local stress-strain concentrations around a hole are required in order to obtain macroscopic deformation behavior and creep-fatigue damage around a hole of perforated plates.

Figure 1 shows two types of approach for describing the macroscopic behavior of perforated plates in case of tensile creep for example. The equivalent-solid-plate approach considers the section area neglecting the existence of circular holes and higher creep rate than base material, and the effective stress approach, on the other hand, considers the smaller cross section than the perforate plate and the same creep rate as the base material.

Previous studies adopt the former approach, and prediction methods of macroscopic and local behavior for elastic analysis are established by O'Donnell (1976), and are shown in ASME Boiler and Pressure Vessel Code Sec. III A-8000.   As for the elastic-plastic analysis, Uragami et al. (1981) proposed these methods by assuming the bilinear stress-strain behavior. As for the creep analysis, on the other hand, Uragami et al. (1981) and Porowski et al. (1980) proposed these prediction methods when assuming the

S. Murakami and N. Ohno (eds.), IUTAM Symposium on Creep in Structures, 351–360.

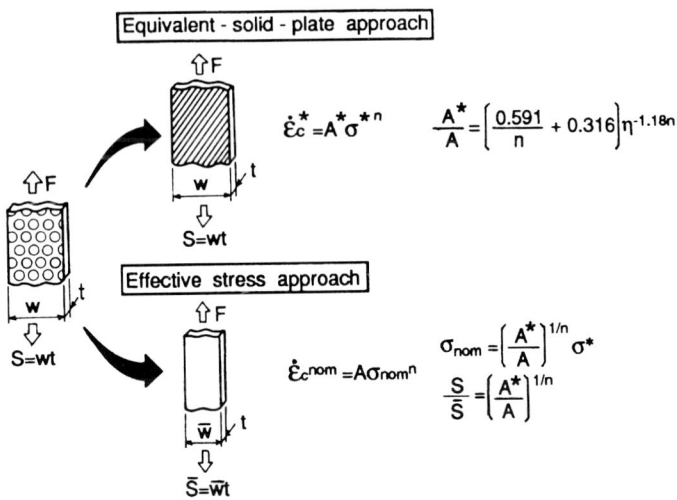

Figure 1. Equivalent-solid-plate approach and effective-stress approach

Norton's creep law. When assuming the Norton's creep law, Igari et al. (1985) proposed advanced prediction methods which are applicable to wider range of materials and temperatures than previous works.

In the previous IUTAM Symposium held at Cracow/Poland in 1990, Igari et al. (1990) proposed prediction methods applicable to the conditions involving both primary and secondary creeps. Fundamental idea for the proposal of macroscopic creep behavior of perforated plates is to consider the effective stress concept which plays a role of the reference stress for describing macroscopic deformation based on the properties of base material. Basic idea for predicting the local stress-strain based on the Neuber's rule is to focus on the exponent of stress in creep equation which governs the stress distribution in perforated plates.

This paper shows further extensions of our approach. Firstly an extension to the case of plasticity is proposed considering the analogy between creep and plasticity. Secondly the total procedure considering both plasticity and creep in multi-axial stress state is applied to the thermo-inelastic analysis and creep-fatigue life prediction of perforated cylinder subjected to cyclic thermal stress.

2. Proposal of Prediction Methods in Plasticity and Creep Regime

2.1. MACROSCOPIC BEHAVIOR OF PERFORATED PLATES

Macroscopic behavior of perforated plate under uniaxial loading in plastic and creep regime is expressed as follows in comparison with the behavior of the base material composing the perforated plates.

$$\varepsilon = \varepsilon_e + \varepsilon_p + \varepsilon_c \qquad \text{(Base material)} \tag{1}$$

$$\varepsilon^* = \varepsilon_e^* + \varepsilon_p^* + \varepsilon_c^* \qquad \text{(Equivalent solid plate)} \tag{2}$$

where the superfix * denotes the properties of the equivalent solid plate, and ε_e, ε_p and ε_c denote the elastic , plastic and the creep strains, respectively.

As for the basic case of creep, Igari et al. (1985) already proposed the prediction method as follows for the case assuming the Norton's creep law.

$$\dot{\varepsilon}_c = A\sigma^n \qquad \text{(Base material)} \tag{3}$$

$$\dot{\varepsilon}_c^* = A^*\sigma^n \qquad \text{(Equivalent solid plate)} \tag{4}$$

where σ, A and n are stress and material constants, respectively. The prediction equation of the equivalent solid plate property A^* of the perforated plates with triangular penetration pattern of circular holes as shown in Fig.1 is examined for three types of loading under plane stress condition. The equivalent solid plate property A^* can be expressed by the following equation.

$$A^* = (0.591/n + 0.316)\eta^{(-1.18n)}A \tag{5}$$

where η (=h/p) is a ligament efficiency which defines a penetration pattern of circular holes. Equation (5) is determined so that the predicted values of A^*/A correspond to E/E^* shown in A-8000 for the special case of "n=1", and also the predicted values of A^*/A correspond to the results of creep analysis by FEM for various values of exponent n. And so, eq. (5) covers wide range of exponent n corresponding to wide range of materials and temperatures.

Let us consider the creep behavior in the next form representing actual materials with both primary and secondary creep.

$$\varepsilon_c = f(\sigma, t) = P\sigma^q t^r + A\sigma^n t \qquad \text{(Base material)} \tag{6}$$

A prediction method for macroscopic behavior of perforated plates using the effective stress has been proposed (Igari et al., 1990).

$$\varepsilon_c^* = f(\sigma_{nom}, t) = P\sigma_{nom}^q t^r + A\sigma_{nom}^n t \tag{7}$$

where σ_{nom} is the effective stress which is defined by the next equation,

$$\sigma_{nom} = (A^*/A)^{(1/n)}\sigma^* \tag{8}$$

or simply defined as follows for the case of exponent n from 5 to 10.

$$\sigma_{nom} = (E/E^*)\sigma^* \tag{9}$$

The effective cross section of perforated plate is reduced by the existence of circular holes. The effective stress is a stress against this effective cross section. The basic concept of eq. (7) is to use σ_{nom} in the same function as eq. (6) in order to predict the

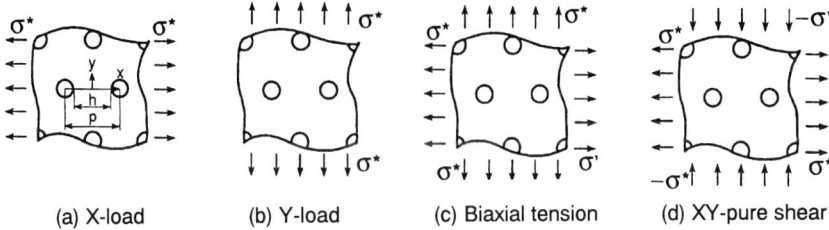

(a) X-load   (b) Y-load   (c) Biaxial tension   (d) XY-pure shear

Figure 2.   Loading conditions

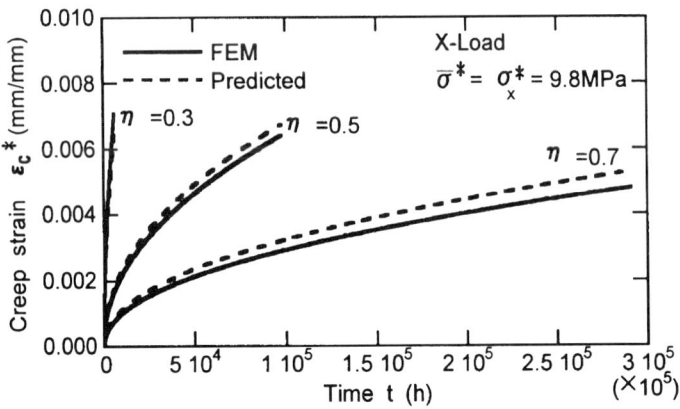

Figure 3.   Prediction of macroscopic creep behavior

macroscopic creep behavior of perforated plates. Propriety of this method is confirmed through comparing the prediction with numerical results by FEM for four types of loading pattern shown in Fig.2: (a)X-load, (b)Y-load, (c)Biaxial tension and (d)XY-pure shear. An example of comparison between prediction and numerical results by FEM is shown in Fig.3 (Igari et al., 2000).

Assume that the plastic strain of the base material and the equivalent solid plate is expressed as follows.

$$\varepsilon_p = B\sigma^m \qquad \text{(Base material)} \tag{10}$$

$$\varepsilon_p^* = B^*\sigma^{*m} \qquad \text{(Equivalent solid plate)} \tag{11}$$

where B and m are material constants.   When considering the analogy between two materials with the stress-strain behavior of eq. (3) and eq. (10), the above-mentioned results for creep are considered to be applicable to plasticity.   That is to say, by replacing A to B and n to m, respectively, the next relation is obtained.

$$B^* = (0.591/m + 0.316)\eta^{(-1.18m)}B \tag{12}$$

For different types of stress-strain equation from eq. (10), the concept of the effective

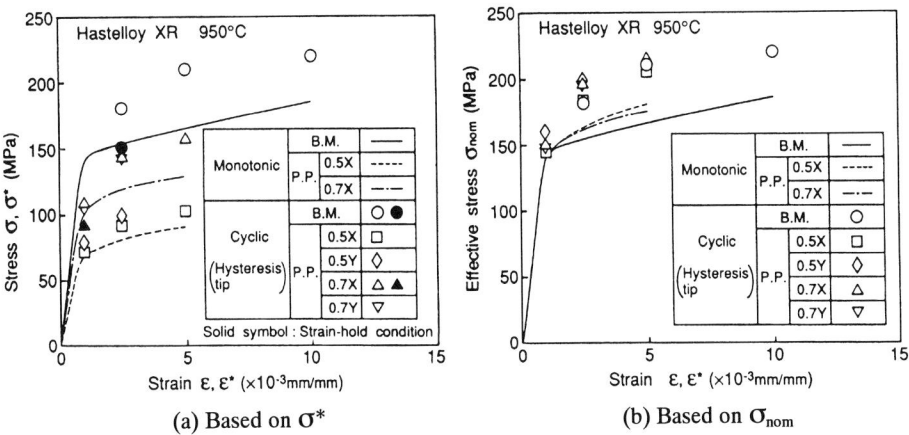

Figure 4. Monotonic and cyclic stress-strain behavior of base material (B.M.) and perforated plate (P.P.)

stress shown in eqs. (7) and (8) for the case of creep can be applied. That is to say, even if the plastic strain is expressed by arbitrary function of stress, the macroscopic plastic behavior of perforated plates can be predicted by using the following effective stress which is defined by considering the analogy between plasticity and creep.

$$\sigma_{nom} = \left(B^* / B\right)^{(1/m)} \sigma^* \tag{13}$$

Experimentally obtained monotonic and cyclic stress-strain curves of perforated plates made of Hastelloy X are successfully predicted from those of base material as shown in Fig.4 (Igari et al., 1993).

2.2. LOCAL STRESS-STRAIN CONCENTRATION OF PERFORATED PLATES

The maximum local strain around a hole ε_{max} can be expressed as follows.

$$\varepsilon_{max} = \varepsilon_{e\,max} + \varepsilon_{p\,max} + \varepsilon_{c\,max} = K_t\varepsilon_e^* + K\varepsilon_p\varepsilon_p^* + K\varepsilon_c\varepsilon_c^* \tag{14}$$

where ε_{emax}, ε_{pmax} and ε_{cmax} are maximum elastic strain, maximum plastic strain and maximum creep strain, respectively. As for the creep strain concentration factor $K\varepsilon_c$, the following equations have been proposed by the authors (1985) when considering the Norton's creep law.

$$K\varepsilon_c = \varepsilon_{c\,max} / \varepsilon_c^* = \left(CK_t\right)^{(2n/(n+1))} \tag{15}$$

where

$$C = \left(-1.359\log_{10}\eta - 0.551\right)\left(1/n - 1\right) + 1 \tag{16}$$

$$K_t = E^* Y_{max} / \left(E\eta\right) \tag{17}$$

where K_t is the elastic strain concentration factor and Y_{max} $(=\sigma_{max}/(\sigma^*/\eta))$ is the elastic stress multiplier shown in A-8000. Equation (15) is basically based on the Neuber's rule, and the modification factor C is introduced. The predicted results by eq. (15) coincide with the elastic strain concentration factor Kt by eq. (17) for special case of n=1, and also corresponds with numerical results by FEM for wide range of materials and temperatures for various values of exponent n from 1.0 to infinity. Furthermore the application of the method to the case with both primary and secondary creep as shown in eq. (6) is reported by the authors (1990), and propriety of this method is confirmed (Igari et al., 2000) through comparing the prediction with measured creep strains by Moire method and numerical results by FEM.

Plastic strain concentration factor $K\varepsilon_p$ is thought to be obtained when considering the analogy between plasticity and creep just as shown in the case of macroscopic behavior.

$$K\varepsilon_p = \varepsilon_{p\max} / \varepsilon_p^* = \left(CK_t\right)^{2m/(m+1)} \tag{18}$$

$$C = \left(-1.359 \log_{10} \eta - 0.551\right)(1/m - 1) + 1 \tag{19}$$

As for the stress concentration factor $K\sigma$ during creep of relaxation period can be described by the next equation.

$$K_\sigma = \sigma_{\max} / \left(\sigma^* / \eta\right) = \eta \left(CK_t\right)^{(2/(n+1))} \left(A^* / A\right)^{1/n} \tag{20}$$

where C, K_t and A*/A are shown in eqs. (16), (17) and (5), respectively.

Predicted stress-strain behavior at around a hole is successfully applied to the creep-fatigue life prediction of perforated plate made of Hastelloy X under strain-controlled cycling in uniaxial stress state at 950℃ in air (Igari et al., 1993).

3. Thermal Fatigue Tests of Perforated Cylinder

Thick-walled cylinder made of Hastelloy X with outer diameter of 57mm and thickness of 8mm is machined from a round bar with diameter of 60mm. Eighteen circular holes with diameter of 2.5mm is machined radially on the circumference of the cylinder. Penetration pattern of circular holes is triangular at the middle surface of the wall thickness and the ligament efficiency η is equal to 0.7. There are fifteen layers of holes in the axial direction. Figure 5 schematically shows the test method. Firstly the test specimen is heated up to the uniform temperature of 950℃ in the electric furnace, and then is cooled down by the compressed air flowing into the cylinder. In the cooling process, axial temperature distribution occurs because the perforated portion is cooled faster than other part. Test specimen is again heated up to the uniform temperature of 950℃ and is kept at this temperature for 30 minutes. Time history of the temperature at the center of perforated portion is schematically shown in Fig.5.

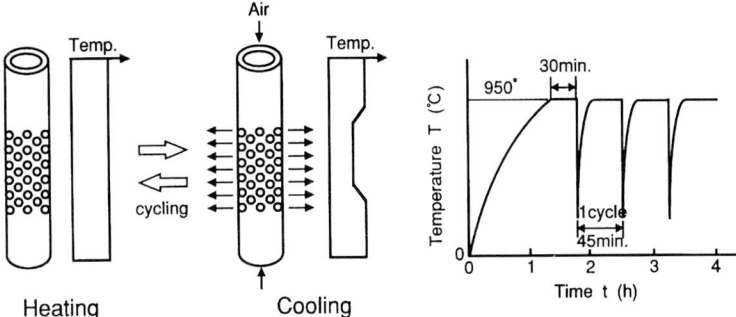

Figure 5.   Schematic view of test method

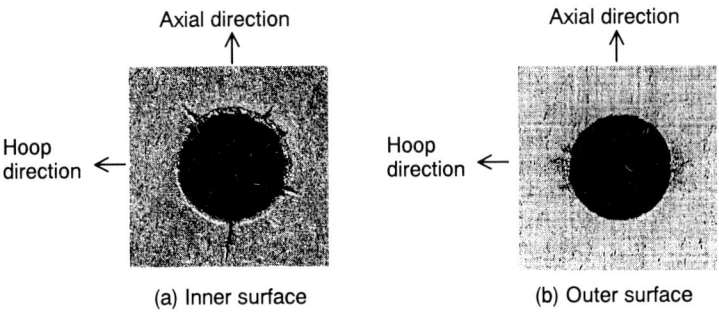

(a) Inner surface   (b) Outer surface

Figure 6.   Macroscopic view of damage around a hole

Number of cycles is 1200, and damage around a hole of the inner and outer surfaces is examined by using the replication method at 400, 800, 1000 and 1200 cycles at both the center of the cylinder and the rim-ligament boundaries which mean the edge of the perforated portion.   The view of the damage at both inner and outer surfaces of rim-ligament boundaries is shown in Fig.6. At the outer surface, the intergranular cracks perpendicular to the axial direction are found.   At the inner surface, on the other hand, transgranular cracks in radial direction of circular holes are found.   Micro cracks up to 0.2~0.5mm are found at 800 cycles at the inner surface and at 1000 cycles at the outer surface.

4.   Inelastic Analysis and Creep-Fatigue Life Prediction of Perforated Cylinder

Inelastic behavior of perforated plate in multiaxial stress state is known as transversally isotropic (Murakami et al., 1982).   The approach in A-8000 is to regard the perforated plate as isotropic in multiaxial stress state, if the thickness of perforated plate is larger than twice of pitch of circular hole.   The inelastic analysis in this paper adopts the assumption of initial isotropy following A-8000, because the hoop and axial

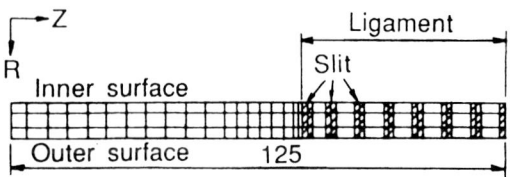

Figure 7.   Mesh division of analytical model

stresses occurring in the perforated cylinder are basically in-plane stresses. The volume change in multiaxial inelasticity is assumed to zero just as usual metallic material. Propriety of this assumption in describing the creep behavior of perforated plates under multiaxial in-plane loading is shown in the literature (Igari et al., 2000).

The model for numerical analysis by FEM is shown in Fig.7.   Half length of the specimen is modeled to the axisymmetric cylinder.   In the heat conduction analysis, circular holes are modeled to slits in which both heat conduction and heat transfer from the circular holes are considered.   Ligament portion in the figure is modeled to the equivalent solid cylinder.   Firstly, transient heat conduction analysis is performed by using the heat transfer coefficients that are determined from the measured temperature distribution. Secondly, elastic-plastic-creep analysis is performed up to three cycles based on the numerical results of the heat conduction.   Elastic-plastic behavior of the base material and the equivalent solid plate is modeled to bilinear shape with temperature dependence.   As for creep behavior of both materials, both primary and secondary creep are considered together with the concept of the equivalent stress σnom. The kinematic hardening rule for plasticity and strain hardening rule for creep are used in the inelastic analysis.

An example of results of the elastic stress analysis in the cooling procedure from the uniform temperature of 950℃ is shown in Fig.8.   Ligament side at rim-ligament boundary is the most critical part, and the axial stress-strain hysteresis loops at this portion from the inelastic analysis are shown in Fig.9 for both inner and outer surfaces. At the outer surface, stress state at the hold period at 950℃ is tensile as shown in Fig.9, and the hoop stress is negligible as can be seen from Fig.8.   At the inner surface, on the other hand, stress state at the hold period at 950℃ is compressive as shown in Fig.9, and the hoop stress has the same sign and about the same magnitude as the axial stress as shown in Fig.8.   This fact well corresponds to the microscopic damage in Fig.6.

Creep-fatigue life prediction based on the linear damage rule by Robinson and Taira is performed by using the analytical results.   Creep damage during both the heating procedure up to 950℃ and the hold period at 950℃ is calculated using the equivalent stress of von Mises.   Predicted lives for inner and outer surfaces are 305 and 323 cycles, respectively, which are conservative when compared with experimental ones, 800 and

1000 cycles, respectively. When compared with the case of creep-fatigue tests of perforated plates at uniform temperature of 950℃ (Igari et al., 1993), predicted lives are a little shorter than experimental ones.

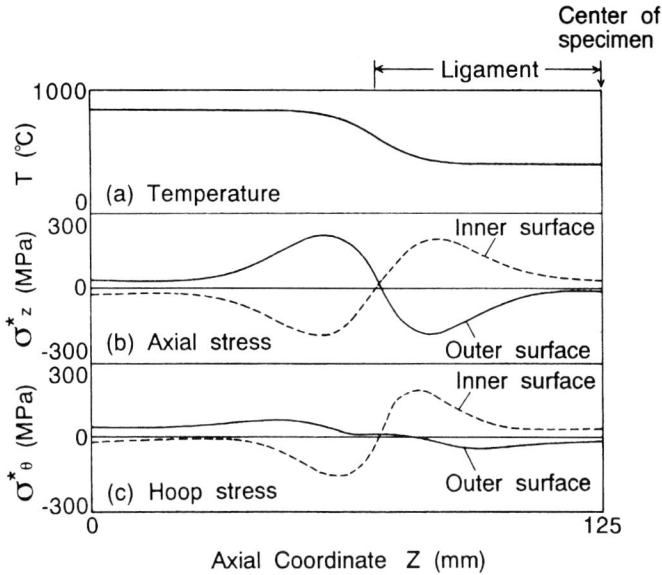

Figure 8.   Stress distribution by elastic analysis

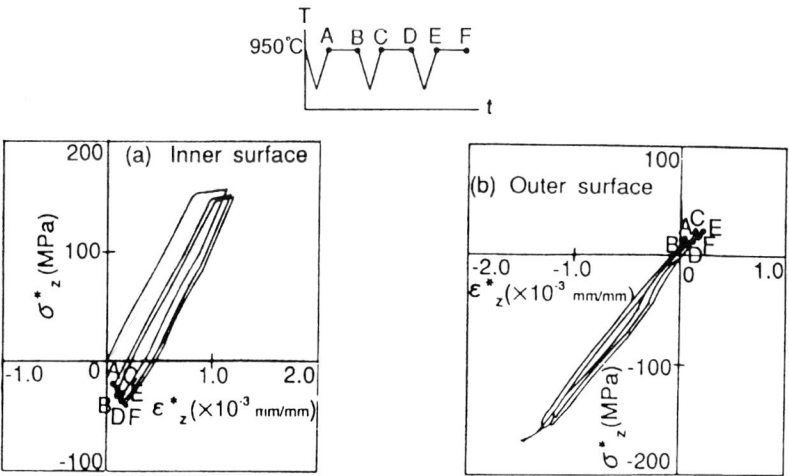

Figure 9.   Stress-strain hysteresis of rim-ligament boundaries

5. Conclusion

Prediction methods for both macroscopic and local stress-strain behavior of perforated plates in plastic and creep regime are proposed in this paper. Inelastic analysis based on this proposal is performed for thermal fatigue test of perforated cylinder, and predicted creep-fatigue lives based on the inelastic analysis are compared with experimental results. Results obtained are summarized as follows.

(1) Macroscopic behavior of perforated plates including cyclic stress-strain behavior and relaxation behavior is predictable based on the proposed method in this paper.

(2) Numerically obtained stress-strain behavior of perforated cylinder corresponds to the damage of inner and outer surfaces of perforated cylinder. Predicted lives are conservative when compared with experiment.

6. References

Igari,T., Setoguchi,K. and Nomura,S. (1985) Simplified creep analysis of perforated plates under steady creep condition, *Nuclear Engineering and Design*, **97**, 161-166.

Igari,T., Setoguchi,K. and Nomura,S. (1987) Experimental study on macroscopic creep behavior of perforated plates, *Trans. SMiRT-9*, L4/4, 219-224.

Igari,T., Setoguchi,K., Matsumura,N., Hamaue,Y. and Kawaguchi,K. (1988) Experimental study of macroscopic and local creep behavior of perforated plates, *Proc. ICPVT-6*, Beijing China, pp.1093-1100.

Igari,T., Setoguchi,K. and Nomura,S. (1990) Prediction of macroscopic and local creep behavior of perforated plates, *Proc. of IUTAM Symposium Creep in Structures IV*, Cracow Poland, pp.579-586.

Igari,T., Tokiyoshi,T., Mizokami, Y. and Nomura, S. (1993) Simplified inelastic analysis and creep-fatigue life prediction of perforated plates in elevated temperature, Proc. of 2nd ASME/JSME Nuclear Engineering, ASME Book No. 10343B-1993, pp.865-870.

Igari,T. Tokiyoshi,T. and Y. Mizokami (2000) Prediction of macroscopic and local creep behaviors of perforated plates under primary and secondary creep conditions, *J. Japan Society of Mechanical Engineers*, **66**, to appear.

Murakami, S. and Konishi, K. (1982) An elastic-plastic constitutive equation for transversely isotropic materials and its application to the bending of perforated circular plates, *Int. J. Mechanical Science*, **24**, 763-775.

O'Donnell,W.J. (1976) Perforated plates and shells, *Pressure Vessels and Piping : Design and Analysis, A Decade of Progress*, **2**, ASME, pp.1041-1101.

Porowski,J.S., O'Donnell,W.J., Tanaka,T. and Badlani,M. (1980) Stress and strain concentration in perforated structures under steady creep condition, *J.Pressure Vessel Technol.*, **102**, 419-429.

Uragami,K., Nakamura,K., Asada,K. and Kano,T. (1981) Simplified inelastic analysis method of ligament plate, *J.High Pressure Institute of Japan*, **19**, 57-65.

THE OVERSTRESS MODEL APPLIED TO NORMAL AND PATHOLOGICAL BEHAVIOR OF SOME ENGINEERING ALLOYS

ERHARD KREMPL AND KWANGSOO HO[1]

Mechanics of Materials Laboratory

Rensselaer Polytechnic Institute

Troy, NY 12180-3590

Abstract

The isotropic, small strain viscoplasticity theory based on overstress (VBO) is introduced. This version is applicable to low, high, constant and variable homologous temperature. It is shown that the overstress dependence enables the modeling of the difference in creep behavior found when a creep test is started at the same stress level upon loading and unloading. Adding a constant to the growth law of the state variable equilibrium stress enables the modeling of rate independence and of negative rate sensitivity that can be found in the dynamic strain-aging regime. It is shown that rate independence and relaxation found in 9Cr-1Mo steel at 400 °C can be modeled with a version of VBO that needs only seven constants. Theories dealing with the modeling of "pathological" behavior of engineering alloys are now beginning to appear and will be extended.

1. Introduction

The availability of economical computing power in the design office has made inelastic stress analyses possible. When the capabilities of the available and accepted material models such as creep and plasticity were evaluated they were found to be deficient (Pugh, Liu et al. 1972) and new material models were developed. Usually, state variable models do not have separate repositories for creep and plasticity but can model creep-plasticity interaction rather well. For a recent account of some of these models see (Krausz and Krausz 1996).

One of the new state variable models is the viscoplasticity theory based on overstress (VBO) developed by the authors and former students. VBO models positive, non-linear rate sensitivity (stress level increases with an increase in loading rate), creep and relaxation. Cyclic hardening or softening as well as cyclic neutral behavior can be

S. Murakami and N. Ohno (eds.), IUTAM Symposium on Creep in Structures, 361–373.

modeled. The overstress dependence of the inelastic strain rate is responsible for modeling many unusual or paradoxical material deformation behaviors, see (Krempl 1995).

VBO will be introduced in the first part of this paper followed by examples to highlight its predictive capability. Then a new term introduced in the growth law of the state variable equilibrium stress is discussed. It enables the modeling of rate independence and negative rate sensitivity. Rate independent behavior of a 9Cr-1Mo steel is then modeled with a VBO version that needs only seven constants.

2. Viscoplasticity theory based on overstress

2.1. THE VBO EQUATIONS

The simplified model for variable temperature and static recovery with modeling capability of positive and negative rate sensitivity and rate independence is introduced. The theory consists of the volumetric and deviatoric flow law and the growth laws for the state variables. Let σ and ϵ denote the stress and the strain tensor respectively, and let a superposed dot designate the material time derivative. With E and v as the elastic constants the volumetric relation is given by

$$tr\dot{\epsilon} = \frac{d}{dt}\left(\frac{1-2v}{E}tr\sigma\right) + 3\alpha\dot{T} \tag{1}$$

with α the coefficient of thermal expansion based on temperature rate and T the temperature in degree Kelvin. The terms in gray print in an equation are for variable temperature. Isothermal behavior can be obtained by specialization. The deviatoric relations are

$$\dot{e} = \dot{e}^{el} + \dot{e}^{in} = \frac{d}{dt}\left(\frac{1+v}{E}s\right) + \frac{3}{2}\frac{s-g}{Ek[\Gamma/D]}$$

$$\dot{e} = \dot{e}^{el} + \dot{e}^{in} = \frac{d}{dt}\left(\frac{1+v}{E}s\right) + \frac{3}{2}F[\Gamma/D]\frac{s-g}{\Gamma} \tag{2}$$

where the stress and strain deviators are s and e, respectively. The inelastic strain rate can be expressed in terms of the positive, decreasing viscosity function k with the dimension of time. It comes directly from the standard linear solid lineage of VBO, see (Krempl and Ho 2000). The flow function F, with dimension of reciprocal time, is positive, increasing and $F[0]=0$. The power function and the hyperbolic sine functions are examples of F and are frequently used in modeling of metallic materials. The drag stress is D and can vary with deformation to become a state variable. In this paper the drag stress will be constant. The overstress invariant is

$$\Gamma = \sqrt{\frac{3}{2}tr\left((s-g)(s-g)\right)}$$

and the effective inelastic strain rate is

$$\dot{p} = \sqrt{\frac{2}{3} tr\left(\dot{e}^{in}\dot{e}^{in}\right)} = F\left[\Gamma/D\right] = \frac{\Gamma}{Ek\left[\Gamma/D\right]}.$$

The equilibrium stress deviator is g and its growth law is

$$\dot{g} = \frac{\partial}{\partial T}\left(\frac{\psi[\,]}{E}\right)\dot{T}s + \frac{\psi}{E}\left(\dot{s} + \frac{s-g}{k[\,]} - \frac{\Gamma}{k[\,]}\frac{(g-f)}{A+\beta\Gamma}\right) + \left(1 - \frac{\psi[\,]}{E}\right)\dot{f} - R[\overline{g}]g \qquad (3)$$

The arguments *[Γ/D]* of the functions $k[\,]$ and $\psi[\,]$ are not written for simplicity; $\overline{g} = \sqrt{3/2 tr(gg)}$ is the effective equilibrium stress.

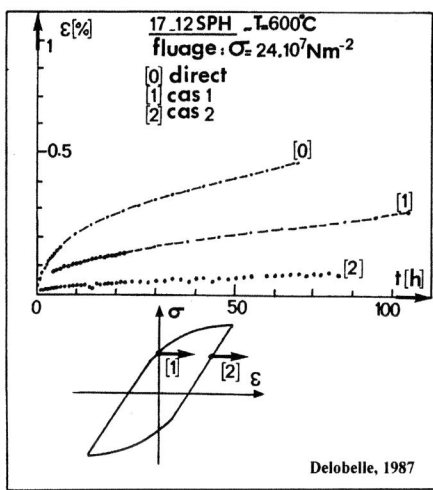

ε[%]

17_12SPH _T_600°C
fluage: σ= 24.10⁷Nm⁻²
[0] direct
[1] cas 1
[2] cas 2

Delobelle, 1987

Figure 1. Creep at the same stress level on loading and on unloading

The first term on the left hand side ensures the path independence of the elastic growth of g for variable temperature while the next two terms are positive and contribute to hardening and are referred to as hardening terms. The term preceded by the negative sign is the dynamic recovery term that depends on the difference $g - f$ where f is the kinematic stress whose growth law is given below. The term $(\beta\Gamma)$ in the denominator is the repository for the modeling of rate-independence and negative rate sensitivity. The state variable A is called the isotropic stress since it has the same function as its counterpart in plasticity: It models cyclic hardening or softening when it is formulated in a rate independent fashion. With an appropriate growth law it can model softening in monotonic loading as well, see (Majors and Krempl 1994; Tachibana and Krempl 1998). The term multiplied by the time derivative of the kinematic stress f is needed so that an asymptotic solution can be modeled, see below. For high homologous temperature the static recovery term $-R[\overline{g}]g$ has to be added as shown. The growth law for the equilibrium stress g is then in a Bailey-Orowan format of hardening and softening. A steady state of deformation is then possible. This steady state can be achieved since all the terms in Eq. (3), except the static recovery term, are homogeneous of degree one in the rates. (This property is hard to see in the stress formulation given in Eq. (3). The growth law for the equilibrium stress can be written in terms of strain rates to make this property obvious, see (Krempl 1996, Eq. (3) and Eq. (4)).

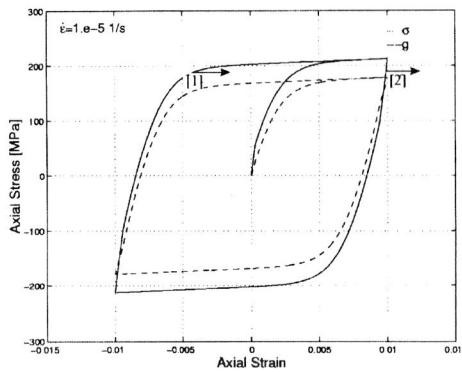

Figure 2. Computed hysteresis loops. Material data from Table 1

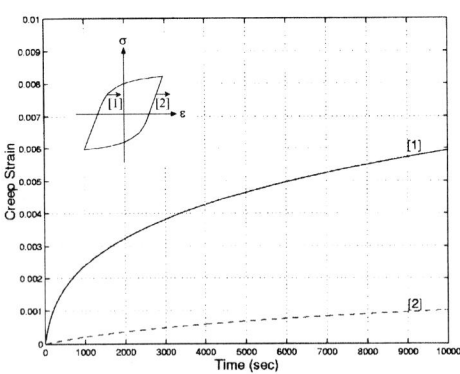

Figure 3. Computed creep curves at the same stress level of 190 MPa on loading [1] and on unloading [2].

The transition from the initially quasi-elastic region to the fully developed inelastic flow is influenced by the decreasing, non-zero shape function ψ with the dimension of stress. It will be shown that it is instead possible to use a constant, see (Maciucescu, Sham et al. 1999) and (Tachibana and Krempl 1998).

The kinematic stress is the repository for modeling the Bauschinger effect and sets the slope of the stress and of the equilibrium stress at the maximum strain of interest. It has the form of the Prager kinematic hardening law

$$\dot{f} = \frac{2}{3}\hat{E}_t\dot{e}^{in} = \frac{\hat{E}_t}{E}\frac{s-g}{k[\Gamma/D]} \tag{4}$$

The tangent modulus at the maximum strain is Et with the corresponding slope based on inelastic strain $\hat{E}_t$. The two are related by the equation

$$E_t = \hat{E}_t / \left(1 + \hat{E}_t / E\right)$$

2.2. THE EFFECT OF OVERSTRESS

In this paper the drag stress D is kept constant. If this is the case then the flow law, Eq. (2), shows pure overstress behavior, i.e. the inelastic strain rate depends only on the current overstress. Only one state variable, the deviatoric equilibrium stress g is used in the flow law. Frequently the question arises as to why a state variable is needed and what effect the state variable has on the modeling. Considering a specific example best conveys an answer.

It is not so well known that during cyclic loading at high homologous temperature the creep rate at the same stress level is larger on loading than on unloading. In Fig. 1 taken from (Delobelle 1987), a French stainless steel was loaded cyclically at 600 °C and creep tests at the same stress level were performed upon loading and unloading as shown in Fig. 1. It is seen that the creep strain accumulation in about 80 hr. was much larger on loading than on unloading. This behavior is not an isolated instance but is a general property and well known to the experts. Materials scientists explain this behavior by stating that in going from point [1] to [2] in Fig. 1 additional inelastic deformation is taking place and therefore the internal make-up at [2] is different from that at [1] in Fig. 1. While this observation is very important for modeling it is not, however, directly useful to the modeler, since the observation was not reported in a form that can be used in a differential equation.

The Norton law is a special form of the overstress theory and is given by

$$\dot{\varepsilon}^{in} = \dot{\varepsilon}^{creep} = F\left[\frac{\bar{\sigma}_0}{D}\right]\frac{\sigma_0}{\bar{\sigma}_0} \tag{5}$$

where $\bar{\sigma}$ is the effective stress and the subscript $_0$ indicates that the stress is constant. Since the stress level is the same upon loading and unloading the Norton law will predict equal creep curves. It cannot model the observation given in Fig. 1. A modification of Eq. (5) by introducing a state variable in addition to the effective stress can in principle model the behavior. It is, however, not clear how it could be done.

TABLE 1. Material Constants used for

the graphs in

Properties	Figs. 2 and 3	Figs. 7 and 8
Moduli	$E = 195\ 000$ MPa $E_t = 1\ 000$ MPa	$E = 190\ 000$ MPa $E_t = 3\ 000$ MPa
Viscosity function $k = k_1\left(1 + \Gamma/k_2\right)^{-k_3}$	$k_1 = 3.142\times10^{-5}$ s $k_2 = 60$ MPa $k_3 = 21.98$	$k_1 = 10000$ s $k_2 = 85$ MPa $k_3 = 21.0$
ψ and A are constant (simple VBO) $\beta = 0$	$\bar{\psi} = \psi/E = 0.435$ $A = 170$ MPa	$\bar{\psi} = \psi/E = 0.8$ $A = 523$ MPa

Introduction of the overstress changes the picture. Since the observation given in Fig. 1 is a general property, i.e. it should be observed for any metal or alloy, a model should be capable of reproducing this behavior in principle and not only through a choice of specific constants. This has been done in Fig. 6 of (Krempl 1995) where only a sketch was used that depended on the evolution of the equilibrium stress relative to the stress during one cycle. It was shown that the overstress on loading was larger than on unloading, thus explaining the observation. Here a numerical experiment was performed using the simplified overstress model with material constants that do not model a specific alloy, see Table 1.

The computed hysteresis loop is depicted in Fig. 2 and the creep curves are shown in Fig. 3. The simplified model can reproduce the observed behavior without any difficulty. The reason lies in the overstress dependence of the inelastic strain rate.

Although the stress level is the same in both cases the overstress is different and because of this property the model can reproduce the behavior shown in Fig. 1. At point 1 the overstress is larger than at point 2 and this fact gives rise to the different creep curves.

It is possible to assume that the equilibrium stress incorporates the effect of the changing microstructure since it is the repository for modeling the observed paradoxical behavior. The overstress dependence is also capable of reproducing special relaxation behaviors, see (Krempl 1995) and (Krempl and Nakamura 1998). Fig. 2 shows that the possible strain accumulation near the zero stress axis is limited (assuming that the evolution of the equilibrium stress during constant stress is approximately the same as during unloading). At zero stress there is recovery induced by the gap between zero stress and equilibrium stress. This gap is small and corresponds to what is generally called the anelastic strain. This property has been investigated elsewhere and it appears that VBO models anelastic strain naturally, see (Krempl 1999).

2.3. LONG-TIME SOLUTIONS AND DIFFERENT RATE SENSITIVITIES

2.3.1. General properties of VBO

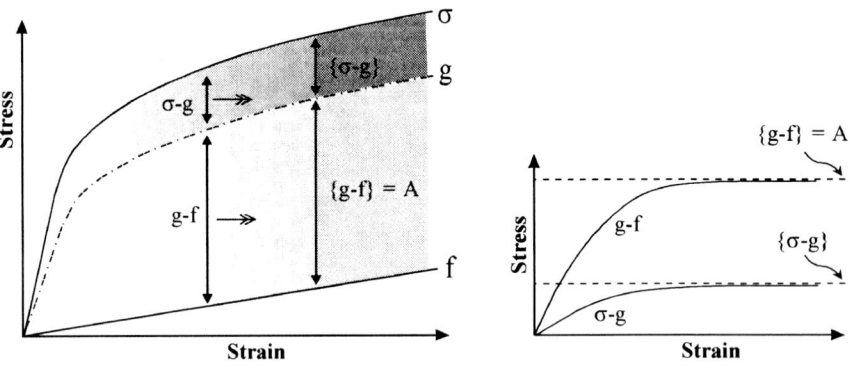

Figure 4. The evolution of the stress, equilibrium and the kinematic stresses for constant strain rate. The graphs on the right show that the differences will become constant ultimately.

It can be shown that the above set of coupled differential equations admits a long-time asymptotic solution that is schematically shown in Fig. 4 for the uniaxial case. (The { } indicate asymptotic solution or the value when the asymptotic solution is reached.) The difference $(\sigma - g)$ and its asymptotic value $\{\sigma - g\}$ are shown together with $(g - f)$ and the asymptotic value $\{g - f\}$. The graph on the right shows how these differences reach the asymptotic value and the shading in the graph on the left mirrors this approach. It is important to realize that the differences reach a constant value although the stress does not.

The asymptotic value of the overstress can be calculated as

$$\{s-g\} = \frac{2}{3}\left\{\frac{\Gamma}{F[\Gamma/D]}\right\}\left\{\left(I - \frac{1+v}{E}\frac{dg}{de}\right)\dot{e}\right\} = \frac{2}{3}\{k[\Gamma/D]E\}\left\{\left(I - \frac{1+v}{E}\frac{dg}{de}\right)\dot{e}\right\} \quad (6)$$

and is nonlinearly related to the strain rate, see (Krempl 1996 and Ho 1998) either through the viscosity function k or the flow function F. The second term on the right involves the slope of g. In the uniaxial case this term equals $\left\{1 - \frac{dg}{d\varepsilon}/E\right\}\dot{\varepsilon} = \dot{\varepsilon}^{in}$. In the asymptotic region when $\{dg/d\varepsilon\} = E_t$ and $E_t \ll E$ the second factor may be approximated by I. The rate-independent contribution is $\{g-f\} = \{A\}\frac{\{s-g\}}{\Gamma}$, which is also in the direction of the overstress. When the asymptotic limit is reached $\{\dot{f}\} = \{\dot{g}\} = \{\dot{\sigma}\}$ as shown in Fig. 4. The slope of f determines the slope of the stress and of the equilibrium stress at the maximum strain of interest. The model then exhibits: Viscous hardening; the rate-independent hardening by $\{A\}$ and the rate-independent contribution given by f. Details can be found in (Cernocky and Krempl 1979; Krempl 1996; Ho 1998). The VBO model allows a continuously changing stress while the overstress and the difference $(g-f)$ reach a constant value.

After some calculations it can be shown that, see (Ho 1998),

$$\{s-f\} = (\{A\}+(1+\{\beta\})\{\Gamma\})\frac{\{s-g\}}{\{\Gamma\}}. \quad (7)$$

TABLE 2. Rate Sensitivity

Factor β	Rate Sensitivity
$\beta < -1$	Negative
$\beta > -1$	Positive
$\beta = -1$	Zero
$\beta = 0$	Normal VBO; positive

This equation demonstrates that the difference $\{s-f\}$, see Fig. 4, is in the direction of the overstress and that it consists of the asymptotic value of A (in case A has a growth law) plus the term with $\{\beta\Gamma\}$ that represents the viscous contribution. Note that $\beta \neq 0$ is the normal VBO model. The augmented model can reproduce different rate sensitivities by merely changing β as shown in Table 2.

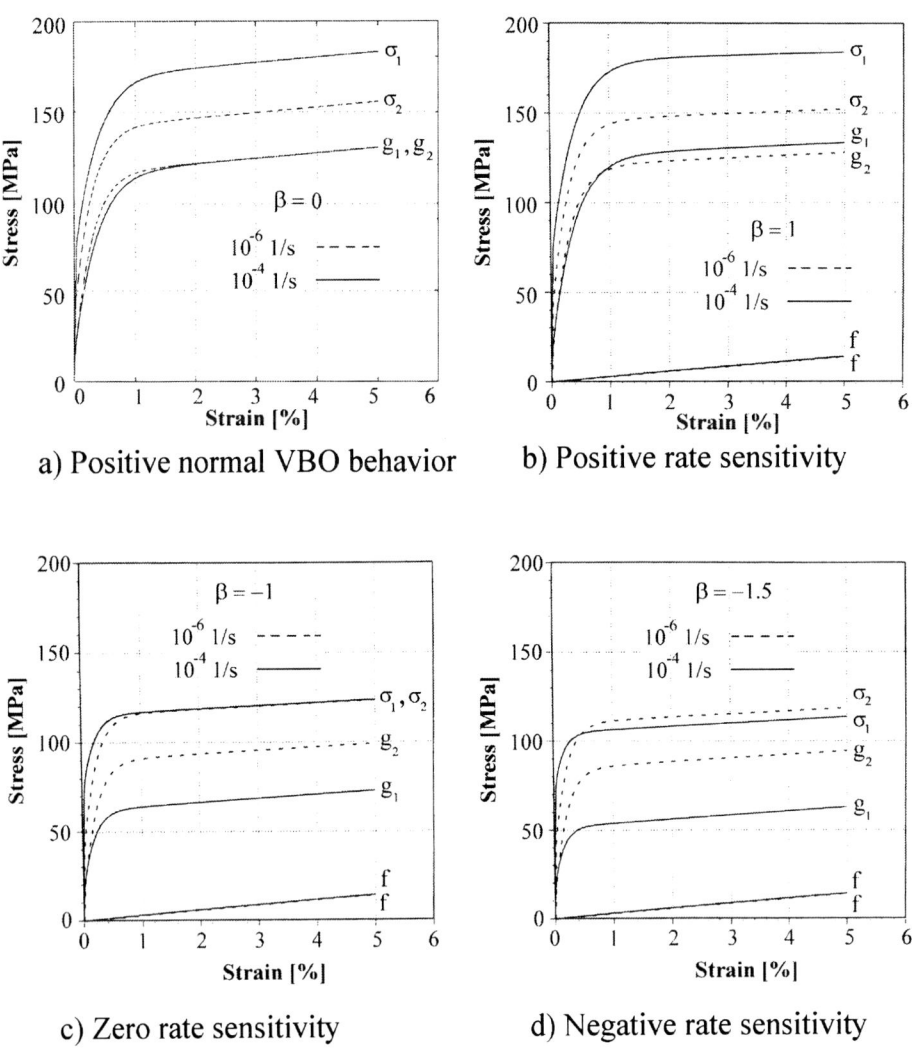

a) Positive normal VBO behavior   b) Positive rate sensitivity

c) Zero rate sensitivity   d) Negative rate sensitivity

Figure 5. The influence of the factor β on the stress-strain diagrams

Numerical stress-strain diagrams for four values of β are plotted in Fig. 5. In each figure two strain controlled tests with the same two strain rates (10^{-6} and 10^{-4} 1/s) are plotted together with the equilibrium stress g and the kinematic stress f. The figure in the upper left corner represents normal behavior. In the asymptotic region, when the inelastic flow is fully established the equilibrium stresses are identical and the stresses show a nonlinear spacing that is maintained all the time. Fig. 5b is for positive rate sensitivity and the equilibrium curves do not coincide in the inelastic region. Zero rate sensitivity is shown in Fig. 5c. In this case the two stress-strain curves coincide whereas the g-curves are separated. Note that the g_2-curve, which pertains to the slow strain rate, is above the g_l curve. This is also the case for the negative rate sensitivity shown in Fig. 5d.

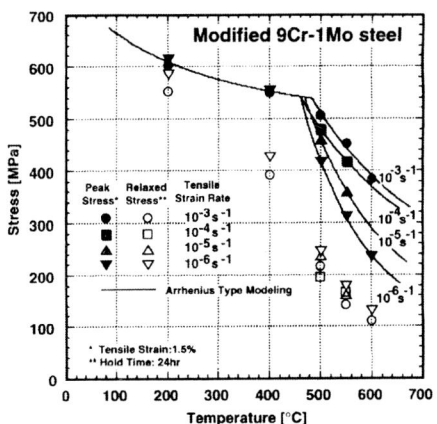

Figure 6. Results of tensile and relaxation tests as a function of temperature for a 9 Cr-1 Mo steel.

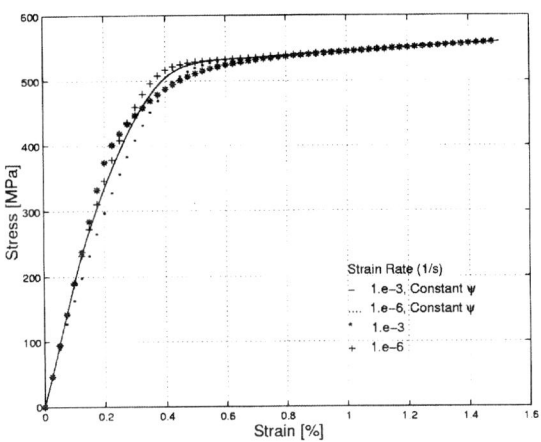

Figure 7. Tensile tests on 9 Cr-1 Mo steel at 400 °C at the indicated strain rates. The curves labeled constant ψ are for the simplified version of VBO.

Eq. (6) does not depend on the augmentation factor β and the overstress is not affected by this quantity. This can be easily verified by measuring the respective overstresses in every diagram of Fig 5. However, Eq. (7) shows that the difference $\{\sigma - f\}$ depends on the factor β. This changes the asymptotic stress when β is changed.

Since the overstress is not affected by β, the creep and relaxation behavior will not change when the factor β changes.

The graphs of Fig. 5 demonstrate that the introduction of the factor β has no deleterious influence on the stress-strain curves. At the origin the inelastic strain rate has very little influence and the full effect of the factor β is only realized after the quasi-elastic region. Although the results of Eq. (6) are strictly valid only at infinite time, and therefore at infinite strain, Fig. 5 shows that the asymptotic values are reached at strains as low as 2% within the accuracy of a ruler.

2.3.2. Modeling of modified 9 Cr 1 Mo steel

Fig. 6 shows the results of comprehensive tensile tests performed by (Yaguchi and Takahashi 1999) at different strain rates including 24 hr relaxation tests between 200 and 550 °C. At 400 °C rate independence is observed which changes to a small negative rate sensitivity at 200 °C. It is surprising to observe that relaxation continues to exist even if there is rate independence or negative rate sensitivity. At all temperatures except at 550 °C identical relaxation properties are found. The end point of the 24 hr relaxation test is lowest for the fastest prior strain rate. For a discussion of these findings, see (Krempl and Nakamura 1998).

(Ho and Krempl 2000 b) modeled this behavior using a simple version of VBO. The results were very encouraging as VBO proved capable of modeling regular behavior (positive rate sensitivity) and unusual behaviors including the different rate sensitivities and the associated relaxation behavior. Included are regular rate sensitivity, rate insensitivity and negative rate sensitivity, rate changes and creep and relaxation curves inside and outside the dynamic strain-aging regime, which is below 400 °C. In these simulations the shape function ψ was used which needs three constants

The already compact VBO was simplified further by replacing the shape function ψ by a constant, see (Tachibana and Krempl 1998; Maciucescu, Sham et al. 1999). A further reduction in the constants would have precluded the modeling of essential phenomena. The modeling results using the simplified VBO that needs a minimum number of constants were very encouraging. It was applied to alloys as varied as a Pb-Sn alloy and alloy 800H. All modeling was at high homologous temperatures and softening and time-dependent effects were very important. At the temperatures considered the behavior of these alloys was normal, i. e. they showed positive rate sensitivity, creep and relaxation and softening in cyclic loading.

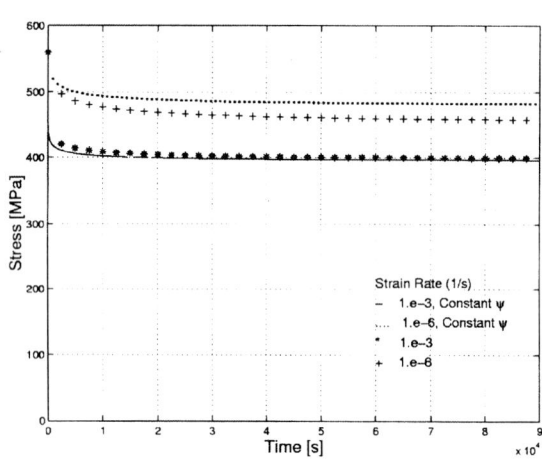

Figure 8. Simulation of relaxation tests at 400° C. The curves labeled constant ψ are computed using the simplified version.

The VBO version employed by (Ho and Krempl 2000 b) was simplified further by assuming $\bar{\psi} = \psi/E =$ *constant*. This constant is restricted to $1 > \bar{\psi} > E_t/E$, see (Krempl 1996).

This so modified VBO model was used to simulate the rate insensitive and relaxation behavior of the modified 9Cr-1Mo steel at 400 °C. After some trial numerical integrations the factor $\bar{\psi} = 0.8$ was selected to simulate the behavior, which now has two constants less than the model used by (Ho and Krempl 2000 b).

In Figs. 7 and 8 the results shown in (Ho and Krempl 2000 b) for 400 °C are plotted together with the curves obtained here with $\bar{\psi} = 0.8$. Fig. 7 depicts the tensile tests at the strain rates of 10^{-3} and 10^{-6} 1/s. It can be seen that the rate insensitivity is correctly modeled at the initial portion and beyond 0.8 % strain. As expected, some discrepancies are encountered in between. It should be noted that beyond the initial elastic region creep and relaxation will be experienced even if the stress-strain curves overlap. Other than the deviations around the knee of the stress-strain curve there is not much difference between the two cases

Two relaxation tests of 24 hr duration are shown in Fig. 8. Again the old and the new results are plotted together. It is seen that the relaxation curves are changed, especially for the slow prior strain rate. Overall the simplified version produces acceptable results and should be investigated further.

3. Conclusion

An isotropic, small strain version of the viscoplasticity theory based on overstress (VBO) is introduced. This model consists of a set of three coupled, nonlinear, stiff ordinary differential equations that have to be integrated for a given history. This theory reproduces normal rate sensitivity, i.e. the stress level is non-linearly increasing with strain rate. By introducing a factor in the dynamic recovery term of the growth law for the equilibrium stress, K. Ho has extended the theory to the rate independent case and to negative rate sensitivity, see (Ho 1998).

Negative rate sensitivity and rate independence are found in many engineering alloys in some temperature region where so-called dynamic strain aging takes place. In the dynamic strain aging regime jerky motion or the Portevin Le Chatelier effect is observed. This phenomenon is said to be caused by the alternate attachment and breaking loose of dislocations. The jerky motions and the underlying dislocation motions cannot be described by VBO. However, it is completely competent at modeling the simultaneously observed negative rate sensitivity and rate insensitivity. The last phenomenon is not identical to the rate independence of classical plasticity that does not allow creep and relaxation motions. In the dynamic strain aging case stress-strain curves at different rates happen to coincide. In contrast to plasticity, however, there is creep and relaxation, see (Irizarry-Quiñones 1999) and (Ho and Krempl 2000 a).

In Fig. 7 the stress-strain curves do not coincide around yielding. An interpretation of this fact could be that regions of non-overlapping stress-strain diagrams imply rate dependence whereas coinciding portions show classical plasticity. However, rate dependence is prevalent everywhere except close to the origin!

One example of the usefulness of the overstress is given in the case of creep. Data from (Delobelle 1987) show that the creep rate at the same stress level is higher on loading than on unloading. This result is not an isolated instance, rather it is generally observed. It has been demonstrated that the overstress dependence can easily model such behavior with a single state variable, the equilibrium stress and its appropriate growth law. It is theoretically possible that another state variable could be found that can also match the data in a simple way as the overstress does, but it is highly unlikely.

Other possibilities of modeling unusual phenomena via the augmentation factor or function β are reported in (Ho and Krempl 2000 a; Ho and Krempl 2000 b; Ho and Krempl 2000 c). Applications include modeling of stress increase and dynamic strain aging as the temperature increases, a gradual change from positive rate sensitivity to negative rate sensitivity and vice versa with accumulated inelastic effective strain.

The present paper shows a possibility of extending the model from the realm of normal behavior in monotonic loading (positive, nonlinear rate sensitivity, creep and relaxation) to pathological behavior (rate-independence, nonlinear, negative rate sensitivity with creep and relaxation). The latter behavior is observed in the dynamic strain aging region of engineering alloys, see Fig. 6. The behavior shown in Fig. 6 has been modeled by (Ho and Krempl 2000 b). The predictions of a further simplified VBO are depicted in Figs. 7 and 8. This theory has 2 constants less than the theory used in (Ho and Krempl 2000 b). The predictive capability for the other cases treated in (Ho and Krempl 2000 b) taken from (Yaguchi and Takahashi 1999) will be investigated further.

(Yaguchi and Takahashi 1999) generated the unique experimental results that have been used by (Ho and Krempl 2000 b). (Yaguchi and Takahashi 2000) and (Ho and Krempl 2000 b) have applied their respective theories to the same data and both approaches can model most of the observed behavior but with a different number of constants.

(Nakamura 1998) used an independently modified VBO for modeling the cyclic deformation of 316 FR Steel. This steel is rate insensitive at 923 K for the initial cycles and normal rate sensitivity develops as cycling progresses. At the 20th cycle normal rate sensitivity has returned. These pathological features were modeled satisfactorily.

The modeling of pathological deformation behavior is just starting. Interesting results can be expected when constitutive equations capable of modeling these pathological behaviors are applied to the stress analyses of structures.

4. Acknowledgement

The early support of the Department of Energy Grant DE-FG02-96ER14603 is acknowledged. O. U. Colak and F. J. Khan helped in the preparation of the manuscript.

5. References

Cernocky, E. P. and E. Krempl (1979). "A nonlinear uniaxial integral constitutive equation incorporating rate effects, creep and relaxation." International Journal of Non-Linear Mechanics 14: 183-203.

Delobelle, P. (1987). Sur lois de comportment viscoplastique a variables internes. Ou, peut-on modeliser simplement des phenomenes complexes? Besancon, France, Laberatoire de Mecanique Appliquee-- LaBouloie, Route Gray: 110 pages, 34 figures.

Ho, K. (1998). Application of the viscoplasticity theory based on overstress to the modeling of dynamic strain aging of metals and to solid polymers, specifically Nylon 66. PhD thesis, Mechanical Engineering, Aeronautical Engineering & Mechanics. Troy, N.Y., Rensselaer Polytechnic Institute.

Ho, K. and E. Krempl (2000 a). "Modeling of positive, negative and zero rate sensitivity using the viscoplasticity theory based on overstress (VBO)." Mechanics of Time-Dependent Materials, in print.

Ho, K. and E. Krempl (2000 b). "The modeling of unusual rate sensitivities inside and outside the dynamic strain aging regime." J. of Engineering Materials and Technology, in print.

Ho, K. and E. Krempl (2000 c). "Extension of the viscoplasticity theory based on overstress (VBO) to capture non-standard rate dependence in solids," International Journal of Plasticity, in print

Irizarry-Quiñones, H. (1999). Rate-dependent properties of Al-Mn and Al-Mg alloys during dynamic strain aging , and of OFHC-Copper and 304 Stainless Steel. PhD thesis, Mechanical Eng., Aeronautical Eng. and Mechanics. Troy, NY, Rensselaer Polytechnic Institute: 377.

Krausz, A. and K. Krausz, Eds. (1996). Unified Constitutive Laws of Plastic Deformation. San Diego, Academic Press.

Krempl, E. (1995). "The overstress dependence of the inelastic rate of deformation inferred from transient tests." Materials Science Research International 1(1): 3-10.

Krempl, E. (1996). A small strain viscoplasticity theory based on overstress. Unified Constitutive Laws of Plastic Deformation. A. Krausz and K. Krausz. San Diego, Academic Press: 281-318.

Krempl, E. (1999). Creep-Plasticity Interaction. Modeling of Creep and Damage Processes in Materials and Structures. J. J. Skrzypek and H. Altenbach. Udine, Italy, Springer Verlag.

Krempl, E. and K. Ho (2000). An overstress model for solid polymer deformation behavior applied to Nylon 66. Time dependent and nonlinear effects in polymers and composites. R. A. Schapery and C. T. Sun. West Conshohocken , PA, American Society for Testing and Materials. STP 1357: 118-137.

Krempl, E. and T. Nakamura (1998). "The influence of the equilibrium growth law formulation on the modeling of recently observed relaxation behaviors." JSME International Journal, Series A 41: 103-111.

Maciucescu, L., T.-L. Sham, et al. (1999). "Modeling the deformation behavior of a Pn-Pb solder alloy using the simplified viscoplasticity theory based on overstress (VBO)." Journal of Electronic Packaging 121: 92-98.

Majors, P. S. and E. Krempl (1994). "The isotropic viscoplasticity theory based on overstress applied to the modeling of modified 9Cr-1Mo steel at 538C." Materials Science and Engineering A186: 23-34.

Nakamura, T. (1998). "Application of viscoplasticity theory based on overstress (VBO) to high temperature cyclic deformation of 316 FR steel." JSME International Journal, Series A 41: 539-546.

Pugh, C. E., K. Liu, C,, et al. (1972). Currently recommended constitutive equations for inelastic design analysis of FFTF components. Oak Ridge , TN, Oak Ridge National Laboratory.

Tachibana, Y. and E. Krempl (1998). "Modeling of high homologous temperature deformation behavior using the viscoplasticity theory based on overstress (VBO): Part III-A simplified model." Journal of Engineering Materials and Technology 120: 193-196.

Yaguchi, M. and Y. Takahashi (1999). "Unified inelastic constitutive model for modified 9Cr-1Mo steel incorporating dynamic strain aging effect." JSME International Journal, Series A: Solid Mechanics and Materials Engineering 42: 1-10.

Yaguchi, M. and Y. Takahashi (2000). "A viscoplastic constitutive model incorporating dynamic strain aging effect during cyclic deformation conditions." International Journal of Plasticity 16: 241-262.

---

[1] Now at Yeungnam University, Korea

CREEP STRAIN UNCERTAINTIES ASSOCIATED WITH TESTPIECE EXTENSOMETER RIDGES: THEIR IDENTIFICATION AND REDUCTION

D. R. HAYHURST
Department of Mechanical Engineering, UMIST,
Manchester, M60 1QD, U.K.
J. LIN
Department of Manufacturing and Mechanical Engineering,
The University, Birmingham, B15 2TT, U.K.
Z. L. KOWALEWSKI
Institute of Fundamental and Technological Research,
Polish Academy of Sciences, Warsaw, Poland.
B. F. DYSON
Department of Materials,
Imperial College of Science, Technology and Medicine,
Prince Consort Road, London, SW7 2BP, UK.

Abstract

This paper reviews recent research concerned with identifying and subsequently reducing sources of uncertainty in strain measurement associated with the ubiquitous use of extensometer ridges in the design of high-temperature creep testpieces. Both experimental and theoretical procedures have been employed. Three principal sources of strain uncertainty have been identified: firstly, the interaction of deformation fields due to the hoop constraints associated with each extensometer ridge; secondly, the notch effect caused by any differences in bar diameter between the gauge length side and loading shank side of the ridge; and thirdly, time-variation of the testpiece temperature during the test. It is shown by theory and experiment that an efficacious means of eliminating the first source of uncertainty is to machine multiple axial slits within each extensometer ridge. This is shown to be particularly advantageous for the short gauge lengths used in LCF testing. The uncertainty in measured creep strains due to the "notch" effect has been reduced by displacing the step outwards and away from the gauge section side of the ridges. To reduce the uncertainty in strain measurement due to temperature variability, it has been proposed that temperature histories throughout each test should be accurately measured and recorded. These data, along with measured strain histories, can then be used to calibrate appropriate constitutive equations to enable more accurate strain/ time histories to be computed.

1. Introduction

There are four factors known to significantly influence the accuracy and repeatability of experimental data collected from high-temperature creep tests. They are: material variation; the level of superimposed bending stress; temperature control; and strain

S. Murakami and N. Ohno (eds.), IUTAM Symposium on Creep in Structures, 375–390.
© 2001 Kluwer Academic Publishers. Printed in the Netherlands.

measurement technique. Material variations are usually minimised by cutting all testpieces from the same block of material. Effects of superimposed bending stresses have been studied by Hayhurst [1974] who showed that the testpiece percentage bending determined from:

$$\text{Bending } (\%) = \{(\varepsilon_1 - \varepsilon_2)/(\varepsilon_1 + \varepsilon_2)\} 100$$

should not exceed 6%. Where ε_1 and ε_2 are uni-axial surface strains measured at diametrically opposed surface points. To achieve initial bending levels below 6% the universal block specimen gripping system discussed by Hayhurst [1974] has been used.

The effects of temperature variation can be significant [Hayhurst, 1974] and precautions are usually taken to employ temperature measurement techniques and controllers which maintain temperatures close to constant for the very long test durations. In this paper, an alternative approach is taken which measures temperature histories and then interprets the strain history using a knowledge of the temperature history and of the constitutive equations. The interaction between testpiece design, extensometer ridges, and measured strains and lifetimes is the main theme of this paper.

The measurement of uni-axial creep strain in laboratory tests is frequently carried out using cylindrical bar testpieces on which ridges have been machined to delineate the gauge length over which strain is to be measured. Mechanical extensometers are fitted to these ridges which are used to transfer the displacements occurring during creep to a location outside of the high-temperature furnace where low cost transducers can accurately measure displacements at ambient temperatures. The measured displacements are then used to compute the variation of strain with time. This type of testpiece and extensometer will be studied within this section.

A previous theoretical study carried out by Lin, Hayhurst and Dyson [1993a] has shown that the uni-axial creep strains measured in tensile testpieces using ridged testpieces and extensometers do not agree with the true strains in the parallel section of the testpiece. The error levels have been shown by Lin, Hayhurst and Dyson [1993b] to be dependent upon the applied stress level, but principally upon the size of the gauge length. For Nickel superalloy testpieces of diameter 7.65 mm, with gauge lengths of 51 mm, errors have been predicted in excess of 10%. For gauge lengths of 10 mm, the error increases to typically 30%. These figures are for the low stress level of 118 MPa. They can be expected to double for the stress level of 250 MPa. Whilst gauge lengths of 10 mm are not frequently used in creep testing they are used in combined cyclic plasticity and creep testing. These phenomena are conveniently displayed in Figure 1 which is due to Lin, Hayhurst and Dyson [1993a] for a Nickel-based superalloy at different life fractions. As $R(=\varepsilon/\bar{\varepsilon})$, where ε is the gauge area strain and $\bar{\varepsilon}$ is the measured strain, tends to unity, this reflects no extensometer ridge effect; and, as the gauge length Y reduces, the extensometer ridge shows a pronounced effect.

The reason for these high errors has been shown by Lin, Hayhurst and Dyson [1993b] to be due to the circumferential reinforcement of the testpiece provided by the extensometer ridges. The ridges perturb the stress, strain, and damage fields above and below the ridge, with the perturbations extending a distance of typically 1.5 times the diameter of the parallel sided region of the testpiece. The circumferential stress generated in the extensometer ridge is predominantly compressive and Lin, Hayhurst and Dyson [1993b] have shown how these stresses may be relieved by the introduction of slits into the ridges; and how the errors in measured creep strains can, as a consequence, be reduced by a factor of typically two. The same circumferential reinforcement effect has been demonstrated experimentally by Ohashi et al [1975] in tubular specimens subjected to internal pressure. They tested a range of different specimens in which the geometry of the extensometer ridge was varied and shown to strongly affect the measured strains.

The theoretical work of Lin, Hayhurst and Dyson [1993b] on slitted extensometer ridged testpieces has been studied experimentally by Kowalewski, Lin and Hayhurst [1994], who tested gauge lengths of 50, 30 and 10 mm with extensometer ridges in both the slitted and unslitted states. In addition, they performed finite element analyses of the testpieces using the finite element solver DAMAGE XX. They obtained close agreement between the results of experiments and theory; and also showed how the

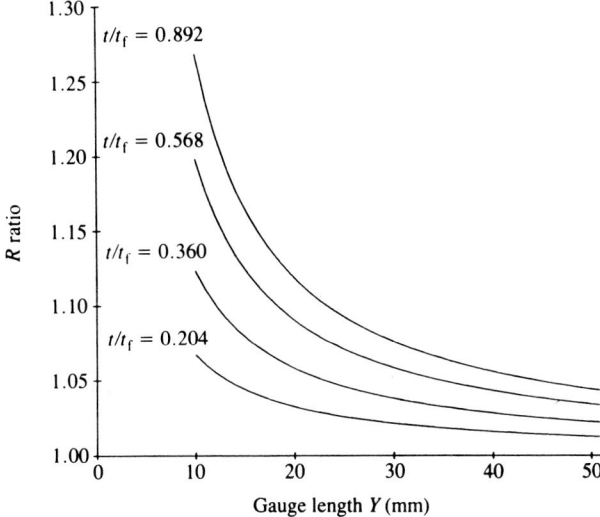

FIGURE 1. Extensometer ridge reinforcement effect. Variation of R ratio with gauge length Y for different life fractions t/t$_f$, where t$_f$ is the failure of lifetime

accuracy of the measured creep behaviour can be greatly improved by using slitted extensometer ridges. The same research has also been interpreted by Kowalewski, Lin and Hayhurst [1995].

The above computations were made using small-displacement methods and predicted that fracture would only take place at the extensometer ridges, whereas large-displacement analyses subsequently predicted that fracture should take place within the parallel gauge section, as observed experimentally. The transition locus between the two fracture fields has been computed by Dyson, Hayhurst and Lin [1996] to be a function of stress level, the material's uniaxial creep ductility and, most importantly, the damage criterion chosen for material fracture under the statically indeterminate stress field of the ridge region. As a consequence of constraints imposed by the ridges, strain inhomogeneities and associated multi-axial stress fields are formed within the gauge section and can cause two problems for materials metrologists: (i) a monotonically increasing error in the average gauge length strain 'measured' by an axial extensometer as strain accumulates; (ii) an even larger discrepancy between 'measured' strain at fracture and the material's uni-axial ductility input into the model, due to effects of stress state on ductility. Hence the need for improved testpiece/extensometer ridge design.

The investigations which have been summarised so far, relate principally to creep testpieces having relatively long gauge lengths. However, the implications have been considered for shorter gauge lengths of the magnitude encountered in low cycle fatigue testing. It is for these shorter gauge lengths that the need to reduce the extensometer ridge reinforcement effect is most acute. Those aspects of testpiece design responsible for errors in cyclic plasticity experiments have been studied by Lin, Dunne and Hayhurst [1999]. Theoretical CDM studies were made using the Finite Element programme DAMAGE XX for copper testpieces subjected to cyclic plasticity and to creep at temperatures of 20°C and 500°C, and were compared with experimental results. It was shown that inhomogeneous fields of strain and stress are generated in the gauge length region of the testpiece, which result from standard cyclic plasticity testpiece features, namely, the blend radius linking testpiece shank and gauge length, and the extensometer ridges. The influence of these features on the inhomogeneity of the stress and strain fields is exacerbated by the short gauge length which is necessary in cyclic plasticity testpieces to prevent buckling. The results obtained clearly demonstrate the dependence of cyclic plasticity data on testpiece design and suggests that the range of scatter observed experimentally in cyclic plasticity testing, for nominally identical testing conditions, may result from variations in testpiece design.

The latter investigation has been developed further by Lin, Hayhurst and Dunne [1996]; they examined two testpiece designs using a combined creep-cyclic plasticity damage interaction theory, and compared theoretical results with plain bar data. The testpieces were selected to examine two geometrical features: firstly, the extensometer ridges, and secondly, the loading shank-gauge section blend radii.

For copper at 500°C, undergoing creep-cyclic plasticity damage interaction, and at 20°C, undergoing cyclic plasticity damage only, the dominant geometrical feature was shown to be the blend radii.

Techniques were reported which enable errors in strain range to be identified and quantified using simple relations. These relations can be used to correct data obtained from the different testpiece geometries to achieve unified data from homogeneous uni-axial stress conditions.

In the next section a strategic approach is outlined for a combined experimental/theoretical investigation aimed at the identification of a slitted extensometer ridged testpiece design which introduces negligible errors, even for short gauge lengths.

2. Strategic Approach for New Testpiece Design

This paper will address theoretically and experimentally three causes of strain measurement uncertainty in testpiece design:

(i)   The gauge length-dependent interaction of deformation fields caused by hoop constraints associated with each extensometer ridge.

(ii)   The notch effect caused by differences in bar diameter on either side of each extensometer ridge.

(iii) Time variations in testpiece temperature during testing.

To assess the effect of gauge length in the experimental programme three gauge lengths will be considered of 50 mm, 30 mm and 10 mm. By way of establishing a benchmark which can be used to assess other designs, tests have been carried out on the unslitted ridged testpiece of Figure 2a.

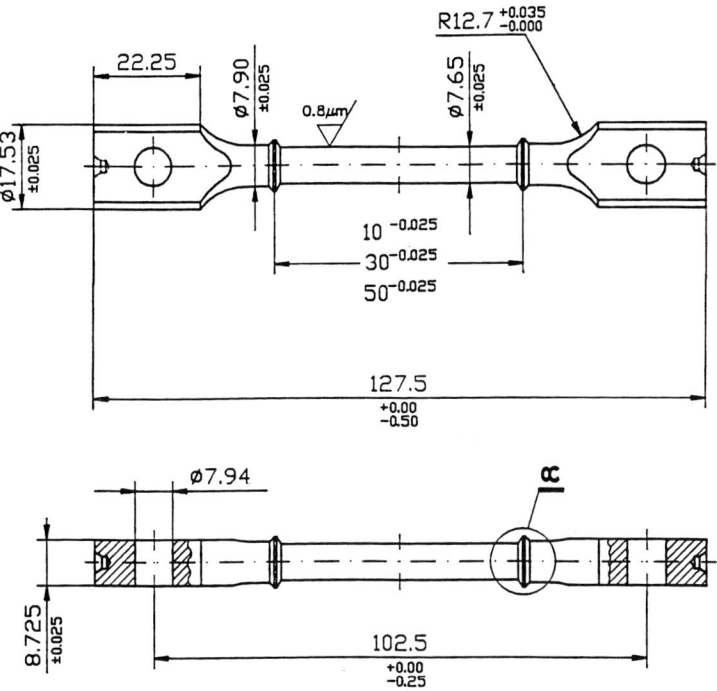

FIGURE 2a. Engineering drawing of the testpiece showing plan and side elevation views; scrapview detail α is given in Figures 2(b), 2(c) and 2(d) for different slit designs.

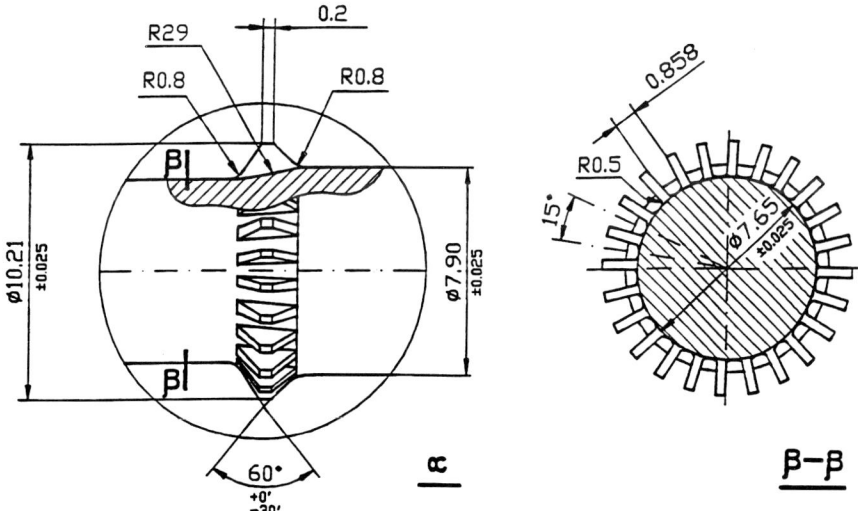

FIGURE 2b. Engineering drawing of the testpiece showing details of a slitted extensometer ridge. The scrapview detail is given in Figure 2(a).

To investigate (i) two arrangements of slitted extensometer ridge will be considered. Firstly, that due to Kowalewski, Lin and Hayhurst [1994], shown in Figure 2(b), and that shown in Figure 2(d) where the number of slits has been reduced from 24 to 6. This latter design removes almost entirely both the radial and circumferential constraints provided by the extensometer ridge.

To investigate (ii), two arrangements for the blend radii between the gauge section and the loading shanks have been selected. The first is that due to Kowalewski, Lin and Hayhurst [1994], shown in Figure 2(b), and the second is shown in Figure 2(c) where the toe of the bland radii have been extended 3 mm outwards away from the ridge and the gauge section. This same feature has also been included in the design given in Figure 2(d).

In what follows, tests will be reported on the four testpiece designs, carried out on an Aluminium alloy at the same nominal temperature of 150°C and at a stress level of 250MPa. The material selected is that used by Kowalewski, Hayhurst and Dyson [1994], namely a precipitation hardened Aluminium alloy manufactured to BS 1472. The testing techniques and specimen manufacturing methods are those specified by Kowalewski, Hayhurst and Dyson [1994].

In the next section is presented the constitutive equations and their calibration.

3.   Physics-Based Creep Damage Constitutive Equations

In the following section is summarized the set of constitutive equations, recently proposed by Kowalewski, Hayhurst and Dyson [1994], to carry out theoretical computations and to predict the behaviour of the slitted and unslitted testpieces. The stress level dependence of creep rate is described by a sinh function, and two damage state parameters are used to model tertiary creep softening caused by grain boundary nucleation and growth, and ageing of the particulate microstructure. Also a primary creep period is described by the model.

3.1   THE FORM OF THE CONSTITUTIVE EQUATIONS FOR UNI-AXIAL CONDITIONS

The form of the constitutive equations proposed by Kowalewski, Hayhurst and Dyson [1994], for uni-axial conditions, is given by the following set of equations:

$$\frac{d\varepsilon}{dt} = \frac{A}{(1-\omega_2)^n} sinh\left(\frac{B\sigma(1-H)}{1-\Phi}\right), \tag{1}$$

$$\frac{dH}{dt} = \frac{h}{\sigma}\frac{d\varepsilon}{dt}\left(1 - \frac{H}{H*}\right), \tag{2}$$

$$\frac{d\Phi}{dt} = \frac{K_c}{3}(1 - \Phi)^4, \tag{3}$$

$$\frac{d\omega_2}{dt} = \frac{DA}{(1 - \omega_2)^n} sinh\left(\frac{B\sigma(1 - H)}{1 - \Phi}\right), \tag{4}$$

where A, B, $H*$, h, K_c, D are material constants, and n is given by

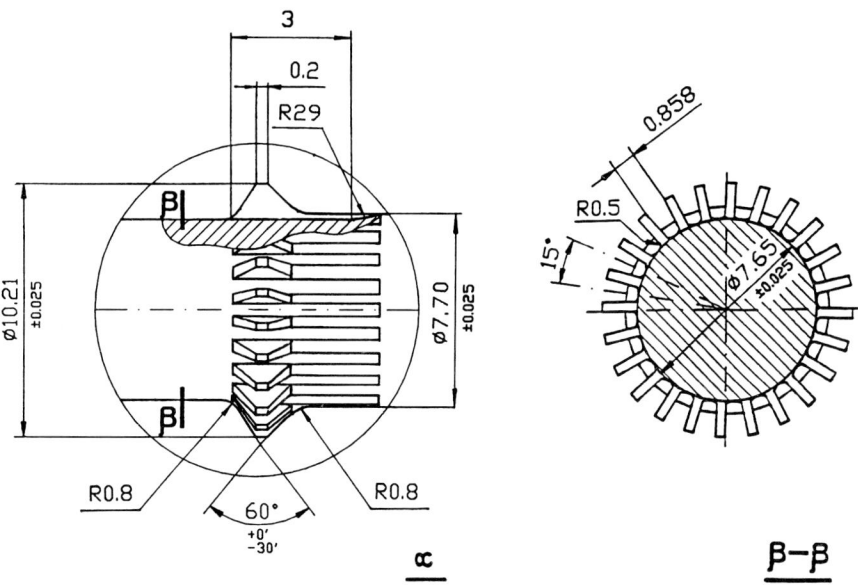

FIGURE 2c. Engineering drawing of the testpiece showing details of a slitted extensometer ridge with modifications based on Figure 2(b): (i) slits extended to the testpiece shanks; (ii) decrease in the shank diameter. The scrapview detail is given in Figure 2(a).

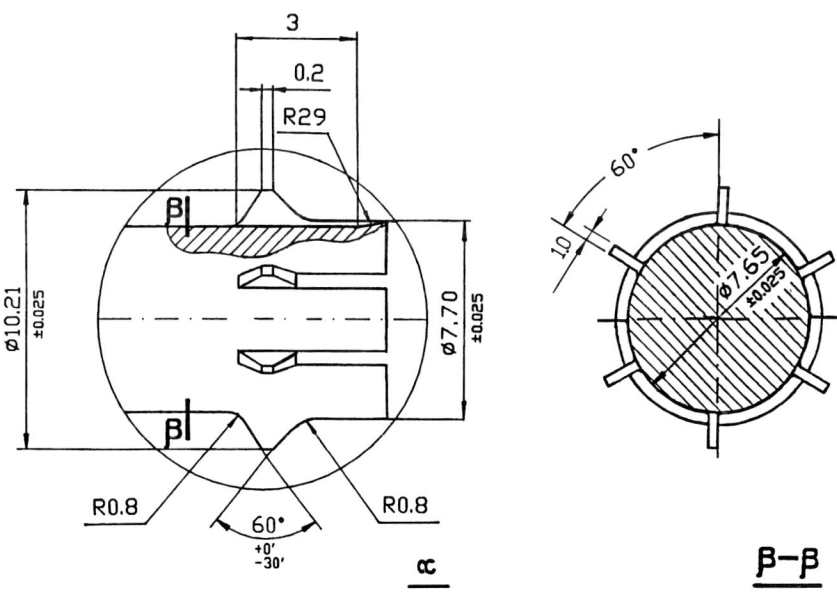

FIGURE 2d. Engineering drawing of the testpiece showing details of a slitted extensometer ridge with a further modification based on Figure 2 (c) : reduction of the number of slits to 6. The scrapview detail is given in Figure 2(a).

$$n = \left(\frac{B\sigma(1-H)}{1-\varPhi}\right) coth\left(\frac{B\sigma(1-H)}{1-\varPhi}\right). \tag{5}$$

Material parameters which appear in this model may be divided into three groups, i.e.
(i)   the constants h, H^* which describe primary creep;
(ii)   the constants A and B which characterise secondary creep;
(iii)   the constants K_c and D responsible for damage evolution and failure.

Equation (1) describes primary creep using variable H, which varies from 0 at the beginning of the creep process to H^*, where H^* is the saturation value of H at the end of primary period and subsequently maintains this value until failure.

As shown in previous papers by Kowalewski, Hayhurst and Dyson [1994], Hayhurst, Dimmer and Morrison [1984], Hayhurst, Brown and Morrison [1984], scalar parameters may be used to describe with sufficient accuracy damage evolution of material during creep, and therefore, such parameters have also been used in the creep constitutive model proposed. Equations (1)-(5) contain two scalar damage state variables used to model the tertiary softening mechanisms:
(i) $\varPhi$, which is described by Equation (3), is defined from the physics of ageing to lie within the range $0-1$ for mathematical convenience;
(ii) ω_2, which is defined by Equation (4), describes nucleation controlled grain boundary creep constrained caviation, the magnitude of which is strongly sensitive to alloy composition and to processing route.

3.2   TEMPERATURE-DEPENDENT PARAMETERS

The values of the parameters A, B and K_c in the constitutive Equations (1)-(5) increase with the increment of temperature and their temperature dependence can be defined by the following equations:

$$X_i = Y_i \exp - (Q_i / RT) \qquad\qquad i = 1,2,3 \tag{6}$$

where X_i ($i = 1, 2, 3$) represent the parameters A, B and K_c respectively; Y_i ($i = 1, 2, 3$) represents the corresponding constants A_1, B_1 and K_{c1}, Q_i ($i = 1, 2, 3$) is the activation energy; R (=2 cals/mole deg^{-1}) is the universal gas constant; and T is the absolute temperature in K. The determination of the constants Y_i and Q_i within the temperature-dependent equations will be discussed in the next Section.

The parameter D in the Equation (4) is inversely related to the ductility of the material. The value of D decreases with the increment of temperature since high ductility is observed from the experiments at a higher temperature. Thus, the temperature dependence of parameter D can be defined by

$$D = D_1 \exp(Q_4 / RT) \tag{7}$$

The determination of the constants D_1 and Q_4 will be introduced in the next Section.

The parameters h and H^* in the constitutive equations are used to model the primary hardening of the material under high temperature creep. The temperature

TABLE 1. Material parameters at 150°C and 160°C

T (K)	423	433
A (h^{-1})	4.040x10^{-15}	9.945x10^{-15}
B (MPa^{-1})	0.1126	0.1165
H (MPa)	2.95x10^4	2.95x10^4
H^* (-)	0.1139	0.1139
K_c (h^{-1})	1.820x10^{-4}	4.924x10^{-4}
D (-)	2.75	2.40

dependence of the primary hardening rate largely depends on the temperature-dependent creep rates. Thus the temperature effects of the parameters h and H^* on the hardening rate are ignored in this work.

4. Determination of temperature-dependent parameters

4.1 MATERIAL PARAMETERS AT 150 °C

The task of determining the values of the material parameters in the constitutive Equations (1)-(5) is not trivial [Dunne and Hayhurst 1991], since they involve the optimization of the six parameters using non-linear objective functions. Optimization methods and the technique for correcting experimental creep data to eliminate the effects of extensometer ridges have been reported elsewhere in the work of Kowalewski, Lin and Hayhurst [1994]. The true material constants in the constitutive equations have been determined from the corrected experimental data at temperature 150°C (423°K) by Kowalewski, Lin and Hayhurst [1994], which are listed in Table 1.

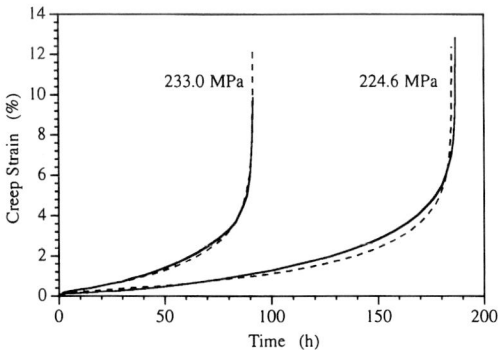

FIGURE 3. Comparison of experimental (solid) and predicted (dashed) creep curves for stress levels of 233.0 and 224.6 MPa at the mean temperature of 160°C.

4.2   MATERIAL PARAMETERS AT 160°C

In order to determine the constants in the temperature-dependent Equations (6) and (7), the values of the parameters in constitutive Equations (1)-(5) need to be defined at another constant temperature. Tests with two stress levels, 224.6 and 233.0 MPa, were therefore carried out at the temperature of 160°C in which each entire creep curve was measured. Both tests were carried out using the unslitted testpieces with gauge lengths of 50 mm. The creep test data have been corrected using the technique introduced Kowalewski, Lin and Hayhurst [1994] and the true creep curves for the tests are shown by the continuous lines in Figure 3. The activation energy for the minimum creep rate parameter A is $Q_1 = 33,000$ cals/mole (Equation (6)) and the value of the parameter A at 150°C is $A = 4.04 \times 10^{-15}$ h^{-1} (Table 1). Thus, Equation (6), when $i = 1$, can be solved for A_1 (Y_1), giving $A_1 = 3.76 \times 10^2$ h^{-1}. Once the constants A_1 and Q_1 have been determined in Equation (6), when $i = 1$, then the value of the parameter A at 160°C (433°K) can be calculated as, $A = 9.945 \times 10^{-15}$ h^{-1}.

It was assumed that the parameters h and H^* are temperature independent. Thus the values of the parameters A, h and H^* are known at 160°C. Optimization techniques are used to determine the other material parameters, B, K_c and D, from the corrected experimental creep data at 160°C, shown by the continuous lines in Figure 3. The material parameters determined at 160°C are given in Table 1. The broken lines in Figure 3 are the predicted creep curves. Close agreements between the corresponding experimental and computed creep curves are obtained.

4.3   DETERMINATION OF TEMPERATURE-DEPENDENT PARAMETERS

From the material parameters B, K_c and D listed in Table 1 at temperatures of 150°C and 160°C, the constants in the temperature-dependent Equations (6) and (7) can be determined. Thus creep curves at any temperature within the temperature range can be predicted. Table 2 summarises the constants determined for the temperature-dependent Equations (6) and (7), which are used together with:

$$A = A_1 \{exp(-Q_1/RT)\},\ B = B_1 \{exp(-Q_2/RT)\},$$
$$K_c = K_{c1}\{exp(-Q_3/RT)\},\ \text{and}\ D = D_1\{exp(Q_4/RT)\}.$$

TABLE 2. Constants for the temperature-dependent parameter equations

A_1 (h^{-1})	3.76×10^2
Q_1 (cals/mole)	33,000.0
B_1 (MPa^{-1})	4.91×10^{-1}
Q_2 (cals/mole)	1,246.4
K_{c1} (h^{-1})	1.02×10^{15}
Q_3 (cals/mole)	36,524.0
D_1 (-)	3.968×10^{-3}
Q_4 (cals/mole)	5,534.0

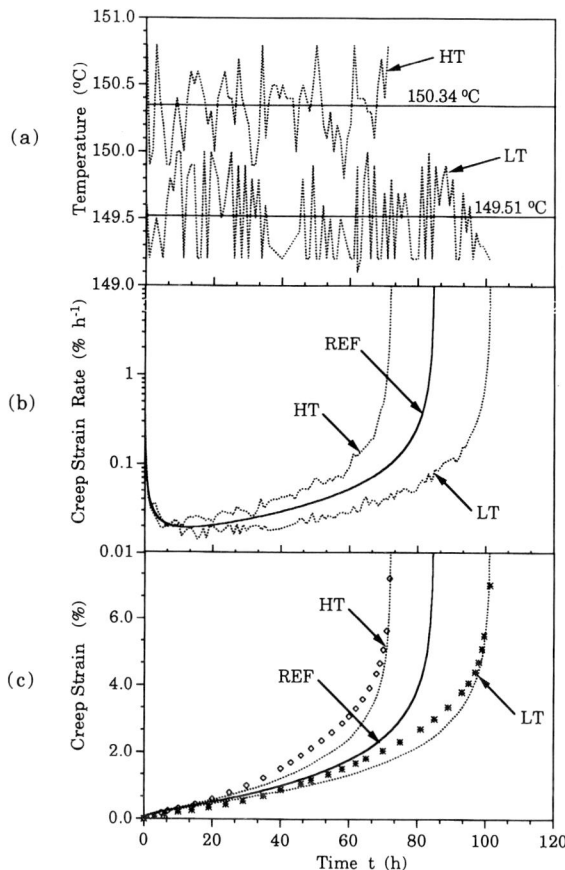

FIGURE 4. Effects of experimental temperature variations on creep rates, creep strains and lifetimes. Two test results are used: one has lower temperatures (LT) with mean value of 149.51°C, the other has higher temperatures (HT) with the mean value of 150.34°C. (a) Temperature histories for the two tests LT & HT. (b) Predicted strain rates using temperature histories HT & LT shown in (a), and the continuous curve (REF) denotes the creep rates predicted at the reference temperature of 150°C. (c) Comparison of creep curves predicted using the temperature histories shown in (a) with experimental results and the predicted curve at the reference temperature, the symbols denote experimental results.

5.   Effect of Temperature on the Creep Curve

Experimental data for unslitted ridged testpieces at a nominal temperature of 150°C using a 50 mm gauge length testpiece is presented in Figure 4. Figure 4a gives two temperature histories: one having a higher mean temperature HT of 150·34°C and the second having a lower mean temperature LT of 149·51°C. These temperature histories have been used as input to calculate the creep curves for a uni-axial stress of 250 MPa using the temperature dependent Equations (1) to (5), and the data given in Table 2. The coupled differential equations were solved numerically using a fourth order Runge Kutta method. The predicted strain rate behaviour is given in Figure 4b for both high and low temperature histories and is compared with that for a constant temperature of 150°C. Presented in Figure 4c are the corresponding creep curves. Despite the small temperature errors of less than 0·3% the differences in the curves are significant. Figure 4c clearly shows that the approach is capable of accurately predicting experimental results.

These theoretical techniques have been used to interpret the experimental results obtained on the three different testpiece designs.

TABLE 3. Mean temperature histories °C, for the creep tests, carried out at a stress level of 250 MPa with a target temperature of 150°C, on the four specimen designs

Gauge Length	Unslitted Ridges	SRDI	SRD2	SRD3
50mm	150.136	149.777	150.268	150.016
30mm	149.967	149.775	150.357	150.098
10mm	149.855	149.849	150.049	150.116

6.   Interpretation of Experimental Results and Assessment of Testpiece Designs

Presented in Figure 5 are experimental and theoretically derived results for the different testpiece designs, shown in Figure 2, for three different gauge lengths; the dotted line denotes 50 mm gauge length; the chain line denotes 30 mm gauge length; and the regularly broken line denotes 10 mm gauge length. In order to interpret the results more precisely, the mean temperatures for the twelve tests, referred to in Figure 5, are given in Table 3.

The mean temperature of each test, T_{mean} , has been calculated using the Equation:

$$T_{mean} = \sum_{i=1}^{N} T_i(t_i - t_{i-1})/t_f \tag{8}$$

where N is the number of data points recorded for a creep test, T_i is the temperature of the ith data point and t_f is the time at failure.

The testpieces have been manufactured using the techniques outlined by Kowalewski, Lin and Hayhurst [1994]; and methods have not been developed for volume production. For this reason, coupled with the limited volume of material available from the same block of raw material, only a limited number of specimens were available for use with each testpiece design.

The results presented in Figure 5 are now discussed in turn for each extensometer ridge design.

6.1   UNSLITTED RIDGED DESIGN, Figure 2(a)

These results are presented in Figure 5(a).   The solid line denotes the theoretical prediction for 150°C; and the broken lines denote experimental results for the three

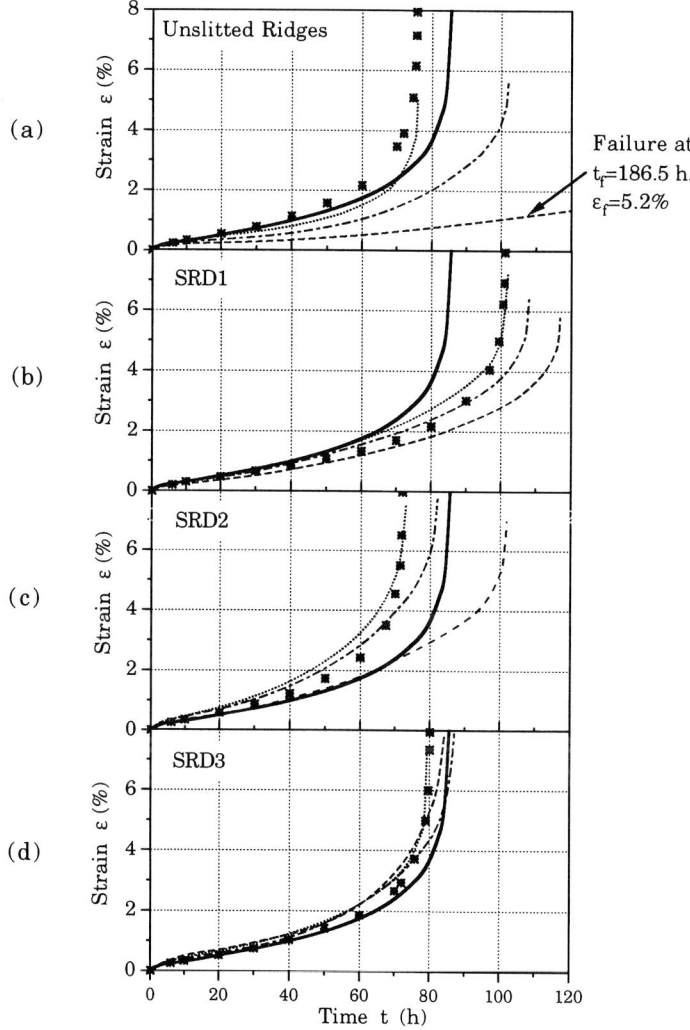

FIGURE 5   Comparison of experimental and theoretical predicted creep curves at 250 MPa for (a) unslitted ridged testpieces (c.f. Figure 2(a)), and, slitted ridge design: (b) SRDI (c.f. Figure 2(b)); (c) SRD2 (c.f. Figure 2(c)); (d) SRD3 (c.f. Figure 2(d)). The symbols denote theoretical predictions made using individual experimental temperature histories for 50mm gauge length testpieces, and the solid lines denote the theoretical prediction for the same testpiece at the reference temperature of 150°C. The experimental creep curves have been obtained for 50mm gauge lengths ············ ; 30mm gauge lengths ----------- ; and for 10mm gauge lengths -·--·--·--·--·-.

different gauge lengths of 50, 30 and 10 mm. Note the large errors for the 10 mm gauge length shown by the broken line. The symbols denote theoretical predictions made using the experimentally measured temperature history; very good agreement has been achieved with the experimental result given by the dotted line for the 50 mm gauge length.

6.2   SLITTED RIDGE DESIGN, SRD1 Figure 2(b)

In Figure 5(b) the same nomenclature has been used as for Figure 5(a). Note the close agreement between theory and experiment for the 50 mm gauge length testpiece. The creep curves for the three different gauge lengths are close. The slightly longer lifetimes than that for the theoretical result, denoted by the solid line, are due to slightly lower mean temperatures for the three slitted ridged testpieces. Here the average mean temperature over the tests carried out on the three different gauge lengths is 149.80°C; and, the mean temperatures vary by 0.074°C or 0.05%. Since the latter variation is small, the differences between the three curves can be attributed to the different designs. The slitted extensometer ridges very clearly have the effect of reducing the hoop constraint provided by the ridge.

6.3   SLITTED RIDGE DESIGN WITH DISPLACED SHANK BLEND RADII, SRD2 Figure 2(c)

The results are presented in Figure 5(c). Unfortunately the mean temperatures for the tests carried out on the three different gauge lengths are widely separated, 150.049 to 150.268°C; and hence, the figure does not show to full effect the advantage of the displaced shank blend radii. The average mean temperature is 150.225°C and the variation is 0.308oC or 0.2%. This variation is of the same order as for the test results on the unslitted specimens, c.f. 5(a); comparison of these two figures dramatically shows the effect of the slitted ridges and the displaced shank blend radii. It is unfortunate that the temperatures could not be better controlled to demonstrate the full effect of the displaced shank blend radii. However, this will be partly demonstrated in Figure 5(d).

6.4   HEAVILY SLITTED RIDGE DESIGN WITH DISPLACED SHANK BLEND RADII, SRD3 Figure 2(d)

The results are presented in Figure 5(d). The mean temperature of the tests on the testpieces of gauge length 50, 30 and 10 mm are very close in this case, showing a variation of 0.018°C or 0.01% with an average mean of 150.08°C; and, the effect of the new design is evident. For all gauge lengths, extremely close agreement has been achieved with the theoretical (solid line) curve, and with the theoretical prediction for the experimental temperature history for the 50 mm gauge length specimen.

6.5   EFFECTIVENESS OF TESTPIECE DESIGN

It is clear from these results that the specimen design SRD3, in which the ridge has been almost completely removed, and in which the toe of the shank blend radius has been moved away from the gauge section, has been successful in overcoming the proximity effects of the ridges and blend radii for even the short gauge length of 10 mm.

7. Effect of Fluctuating Temperature History

As shown explicitly in Figure 4 it is possible, with a knowledge of the constitutive equations, their temperature sensitivity, and the temperature history during the test, to predict very accurately the resultant creep curves. It is therefore proposed that a new approach to constitutive modelling be adopted in which temperature histories, as well as strain histories, are logged in all tests. From the mean temperatures, derived from Equation (8), and the initial stress levels it is proposed that these be used to fit the constants in a physics-based, temperature dependent constitutive equation. The constitutive equation for the desired temperature may then be obtained with little error.

It will be necessary to use optimisation techniques of the type used by Dunne, Othman, Hall and Hayhurst [1990] and Kowalewski, Hayhurst and Dyson [1994] and, fast digital computers to achieve this. The advantage of the approach is that the errors due to temperature variation are quantified, however there are inevitably a large number of parameters to be handled in the optimisation process. Of great importance is that the approach will help reduce the unwanted effects of poor temperature control over the long periods encountered in creep testing.

8. Conclusions

A combined experimental and theoretical approach has been used to show how the constraint effects associated with creep testpiece extensometer ridges, and of the stress raising effect of the gauge length-shank blend radii may be substantially reduced by careful testpiece design and manufacture.

Further, it has been shown how both temperature and strain histories may be recorded and used to interpret test results. It is suggested that this procedure be used, together with physically-based constitutive equations to calibrate the equations by inclusion of the temperature dependency.

References

Dyson, B.F., Hayhurst, D.R. and Lin, J. (1996) The ridged uniaxial testpiece: creep and fracture predictions using large-displacement finite-element analyses, *Proc. R. Soc. London* A, **452**, 655–676.

Dunne, F.P.E. and Hayhurst, D.R. (1991) An expert system for the determination of creep constitutive equations based on continuum damage mechanics, *J. Strain Analysis* **26**, 185-191.

Dunne, F.P.E., Othman, A.M., Hall, F.R. and Hayhurst, D.R. (1990) Representation of uniaxial creep curves using continuum damage mechanics, *Int. J. Mech. Sci.* **32**, *11*, *945–957*.

Hayhurst, D.R. (1974) The effect of test variable on scatter in high-temperature tensile creep-rupture data, *Int.J. Mech. Sci* **16**, 829–841.

Hayhurst, D.R., Dimmer, P.R. and Morrison, C.J. (1984) Development of continuum damage in the creep rupture of notched bars, *Phil. Trans. R. Soc. Lond.*, **A311**, 103-129.

Hayhurst, D.R., Brown, P.R. and Morrison, C.J. (1984) The role of continuum damage in creep crack growth, Phil. Trans. R. Soc. Lond., **A311**, 130-158.

Kowalewski, Z.L., Hayhurst, D.R. and Dyson, B.F. (1994) Mechanisms-based creep constitutive equations for an aluminium alloy, *J. Strain Analysis* **29**, 309–316.

Kowalewski, Z.L., Lin, J. and Hayhurst, D.R. (1994) Experimental and theoretical evaluation of a high-accuracy uniaxial creep testpiece with slit extensometer ridges, *Int. J. Mech. Sci.* **36**, 8, 751–769.

Kowalewski, Z.L., Lin, J. and Hayhurst, D.R. (1995) Investigation of a high accuracy uni-axial creep testpiece with slit extensometer ridges, *Arch. Mech.* **47**, 261–279.

Lin, J., Dunne, F.P.E. and Hayhurst, D.R. (1999) Aspects of testpiece design responsible for errors in cyclic plasticity experiments, *Int. J. Damage Mechanics* **8**, 109–137.

Lin, J., Hayhurst, D.R. and Dyson, B.F. (1993a) The standard ridged uniaxial testpiece: computed accuracy of creep strain, *J. Strain Analysis* **28**, 2, 101–105.

Lin, J., Hayhurst, D.R. and Dyson, B.F. (1993b) A new design of uniaxial creep testpiece with slit extensometer ridges for improved accuracy of strain measurement, *Int. J. Mech. Sci.* **35**, 1, 63–78.

Lin, J., Hayhurst, D.R. and Dunne, F.P.E. (1996) Errors in creep-cyclic plasticity testing: their quantification and correction for obtaining accurate constitutive equations, *UMIST, Department of Mechanical Engineering, Research Report* DMM.96.8.

Ohashi, Y., Tokuda, M. and Yamashita, H. (1975) Effect of third invariant of stress deviator on plastic deformation of mild steel, *J. Mech. Phys. Solids* **23**, 23-31.

EQUIVALENCE OF BACK STRESS DURING PLASTIC AND CREEP DEFORMATION

H. ISHIKAWA, K. SASAKI AND T. MAYAMA
Division of Mechanical Science, Hokkaido University
N13, W8, Kita-ku, Sapporo, 060-8628 Japan

Abstract- Back stresses caused by both cyclic and creep deformations are discussed in this paper. The equivalence of the back stress due to cyclic and creep deformation are shown by a series of experiments. Different creep curves due to cycle number in preloading are simulated by a constitutive model based on dislocation density (Estrin et al. 1996), considering the equivalence of the back stresses.

1. Introduction

Back stress plays an important role to explain plastic deformation, in particular in cyclic plasticity, and almost all constitutive models of cyclic plasticity utilize back stress. Constitutive models of cyclic plasticity were originally developed for time independent deformation. However, to apply the constitutive models to estimate fatigue failure, including creep-fatigue interaction, a constitutive model must also describe viscoplastic deformation such as creep and stress relaxation.

Detailed investigation for the interaction between plasticity and creep has been made (Kujawski, et al., 1980, Ikegami and Niitsu, 1985, Ruggles and Krempl, 1989, Ohno, et al., 1990, Wu and Ho, 1993, Wu and Ho, 1995, Freed and Walker, 1993, Kawai, 1994, Ishikawa and Sasaki, 1994, Sasaki and Ishikawa, 1995). Wu and Ho (1993) observed back stress due to both plasticity and transient creep, and pointed out that the value of the back stress developed by plastic deformation almost coincided with that of transient creep. They also proposed an endochronic type of constitutive model to simulate both plasticity and creep (Wu and Ho, 1995). Freed and Walker (1993) employed viscoplastic and plastic surfaces separately, and developed a constitutive model to describe short-term plastic strain, long-term creep strain, and the interaction between the two. Ishikawa and Sasaki (1994) conducted intermittent creep tests during cyclic loading and showed that back stress plays an important role in explaining the intermittent creep. They also showed correlation between plasticity and viscoplasticity with uniaxial and biaxial ratchetting tests using Type 304 stainless steel (Sasaki and Ishikawa, 1995).

According to the literatures a back stress develops during creep deformation in the transient creep region (Wu and Ho, 1993, Blum, et al., 1989, Davies, et al., 1992, Orlova 1993). One question is whether the back stress developed by creep deformation

391

S. Murakami and N. Ohno (eds.), IUTAM Symposium on Creep in Structures, 391–400.
© 2001 *Kluwer Academic Publishers. Printed in the Netherlands.*

is of the same magnitude as that developed by the plastic deformation. Wu and Ho (1993) showed that the back stress developed by creep deformation is equal to that by plastic deformation. However, they only compared monotonic but no cyclic loading.

This paper reports observations on the correlation between back stresses affecting plasticity and viscoplasticity. First, a series of intermittent creep tests during monotonic and cyclic loading were carried out, and the equivalence of back stresses due to the plasticity and creep are discussed. The other creep tests, so-called continuous creep tests, were also carried out at the maximum stress after several cycles of stabilized stress-strain loops caused by cyclic tension-compression preloading with a constant strain amplitude and zero mean strain. Different continuous creep curves were obtained though the stress-strain loop does not change after the stabilization. Then, if the back stresses due to the plastic or creep deformation are equivalent, the difference in continuous creep curves must come from other factors caused by cyclic loading after the stabilization. In this paper, it is assumed that dislocation mechanisms are one of the factors of the different continuous creep curves. Therefore, a constitutive model based on dislocation density, which is proposed by Estrin (1996), was employed to simulate the continuous creep curves after cyclic loading. All tests were conducted with thin tubular specimens of Type 304 stainless steel at room temperature.

2. Experimental Procedure

2.1 SPECIMENS AND TESTING MACHINE

The specimens used in this work have a gauge length of 50mm, and an inner and outer diameter of 20 and 23mm, respectively. The specimens were made of Type 304 stainless steel subjected to a solution heat treatment at 1070 °C for 5 minutes; its chemical compositions were 0.05C, 0.36Si, 1.13Mn, 0.034P, 0.004S, 9.08Ni, 18.15Cr and Fe balance in weight percent. Young's modulus at room temperature was equal to E=198GPa, while shearing modulus G=73.5GPa.

The servo-controlled axial-torsional machine (Shimazu EHF-EB10), together with both the Shimazu 4825 controller and the computer (NEC PC-9801DX), were used for computerized testing and data acquisition. Strain was measured using two strain gauges (Kyowa KFG-2-120-C1-16) applied on diametrically opposite side of the specimens. The axial force was measured using the load cell incorporated in the machine (Ishikawa and Sasaki, 1994).

2.2 TEST CONDITION

The following tests were conducted: (1) Intermittent creep during pure tension. (2) Intermittent creep during pure tension included partial unloading. (3) Cyclic tension-compression loading with strain amplitude 0.25% after pure tension and creep. (4) Intermittent creep during cyclic tension-compression loading with strain amplitude of 0.5%. (5) Continuous creep tests conducted at the maximum stress over the several cycles of stabilized stress-strain loops caused by cyclic tension-compression preloading with strain amplitude 0.25% under constant strain rate of 0.01 %/s, and this cyclic preloading was conducted with zero mean strain. The tests were carried out for time interval up to 20000 seconds after different number of cycles of 10, 30 and 100.

3. Equivalence of Back Stress

3.1 INTERMITTENT CREEP DURIN TENSION

Intermittent creep tests for the duration of 300 seconds were conducted at stress levels of 300, 350, 400, 450, and 500MPa during tension controlled by stress rate of 10MPa/s. The solid line in Fig. 1 shows the stress-strain curve caused by the intermittent creep tests, and the broken lines show the stress-strain curve caused by tension. The stress-strain curve of reloading after the intermittent creep at each stress levels coincides with the stress-strain curve without the intermittent creep. This phenomenon suggests that the strain hardening occurs during the intermittent creep, and that the value of the strain hardening due to the intermittent creep is equal to that due to the pure tension. Then, it is supposed that the back stresses due to the pure tension and creep should have a same value.

Figure 2 shows the stress-strain curve of the intermittent creep during tension included a partial unloading. The test was controlled by stress rate of 10MPa/s. The specimen is subjected to tension to the stress level of 240MPa (point a) and the intermittent creep for the duration of 300 seconds. After the intermittent creep the specimen is reloaded to 280MPa (point c) and unloaded to the stress level of 240MPa (point d). At the stress level (point d) the intermittent creep is performed. The broken lines in Fig. 2 show the stress-strain curve of tension as well as Fig. 1. At the stress level of 240MPa after partial unloading (point d) no creep strain is observed because the effective stress at point d is inside of the yield surface. The result shows the equivalence of the back stresses due to pure tension and creep.

3.2 CYCLIC LOADING AFTER TENSION AND CREEP

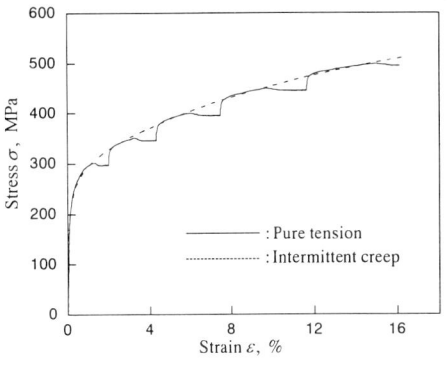

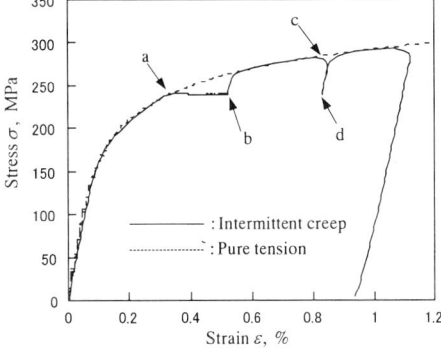

Figure 1 Stress-strain relation of creep during pure tension

Figure 2 Stress-strain relation of creep during pure tension included partial unloading

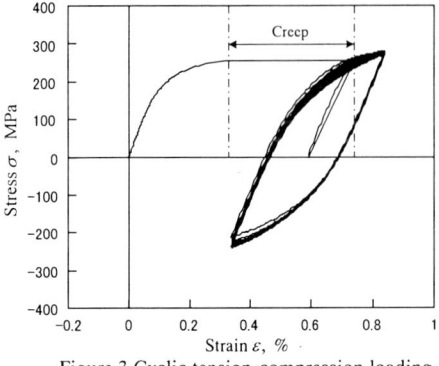

Figure 3 Cyclic tension-compression loading
after creep

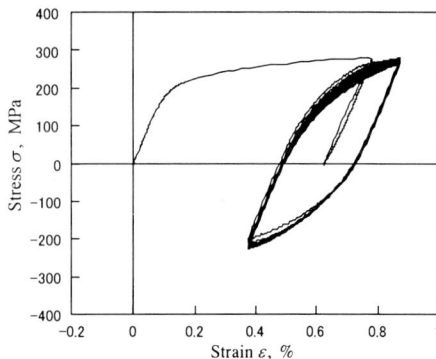

Figure 4 Cyclic tension-compression loading
after pure tension

The cyclic tension-compression loadings with strain amplitude of 0.25% after creep at 250MPa for the duration of 10000 seconds and after tension are shown in Fig. 3 and 4, respectively. Both cyclic loadings were started at a same strain of 0.59%. The hysteresis loops caused by each cyclic loading were almost the same. Figure 5 shows the relationship between the peak stress during the cyclic loading and number of cycle. The peak stresses due to both cyclic loadings are almost same. If the back stresses caused by creep

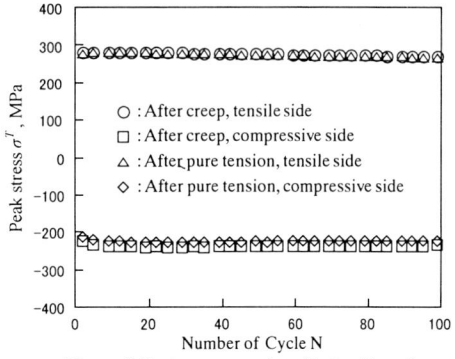

Figure 5 Peak stresses of cyclic loading after creep and pure tension

and that by pure tension are different, the hysteresis loop and the peak stresses should not be same. Then, the results show the equivalence of the back stresses.

3.3 INTERMITTENT CREEP DURING CYCLIC LOADING

Figure 6 shows intermittent creep test for the duration of 300 seconds during the cyclic tension-compression loading with strain amplitude of 0.5%. The specimen was subjected to cyclic tension-compression loading with constant strain amplitude of 0.5 % under constant stress rate of 10MPa/s. After the stress-strain relation stabilized the intermittent creep tests were conducted at the stress level indicated by a ~ g in Fig. 6 and Table 1. The tests were conducted by using one specimen. From the tests, following results were obtained (Ishikawa and Sasaki, 1994):

(1) At zero stress at points a and f there are very small but noticeable creep strains.

(2) There was no creep strain at point e, corresponding to the back stress developed by the cyclic preloading. Then, at points a and f the effective stresses are larger than the yield stress, and the back stress exists between the maximum (minimum) stress and the zero stress.

(3) Considering the equivalence of the back stresses due to creep and cyclic plasticity,

Table 1 Stress levels for intermittent creep during cyclic loading

Symbol	a	b	c	d	e	f	g
Stress (MPa)	0	160	240	240	160	0	318

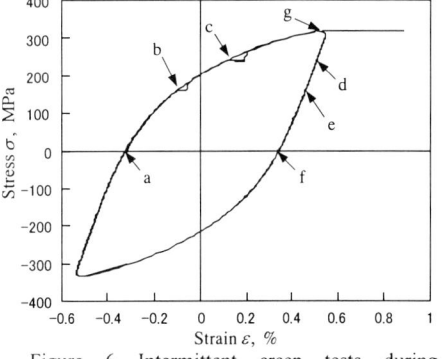

Figure 6 Intermittent creep tests during tension-compression loading

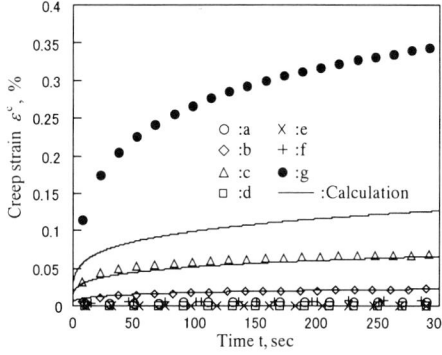

Figure 7 Creep curves of intermittent creep during cyclic tension-compression loading

the creep strains can be explained by the Bailey-Norton's transient creep law incorporating the back stress developed by the cyclic loading, except for the creep strain at point g, the maximum stress of the stress-strain loop. The conventional Bailey-Norton's transient creep law can be expressed by

$$\dot{\varepsilon}^c = 5.5 \times 10^{-16} |\sigma - \sigma_b|^{4.93} t^{-0.78} sign(\sigma - \sigma_b) \qquad (1)$$

where σ_b is the back stress at the stabilized cyclic loop and t is time in second, and $sign(x)$ is positive or negative sign of x. Figure 7 shows the experimental and calculated creep curves by Eq. (1). Equation (1) is the modified Bailey-Norton law obtained from the test results in Fig. 7, and should be noticed only to present an experimental fitting.

These results led to the conclusion that the back stress due to plasticity is important in describing the intermittent creep, and also that the creep strain at point g is of a different kind from the intermittent creep. This second point is further studied in this report, and the creep at point g is here termed continuous creep to distinguish it from intermittent creep.

4. Simulation of Continuous Creep

4.1 CONTINUOUS CREEP TEST

Figure 8 shows the stress-strain relation of the continuous creep tests after the 30th cycle of tension-compression preloading with the strain amplitude of 0.25 percent. First, the specimen is subjected to cyclic tension-compression loading with the strain amplitude of 0.25% under strain rate of 0.01%/s. After the stress-strain relation is stabilized, the specimen is unloaded to point A in Fig. 8, and reloaded to the peak stress

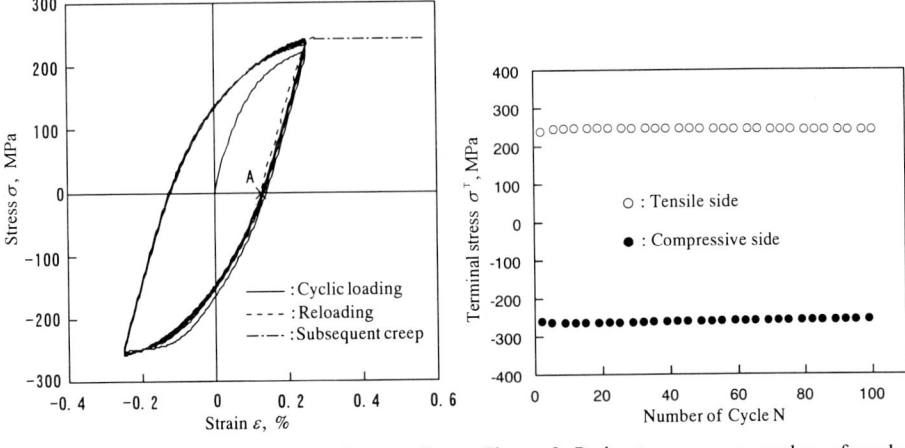

Figure 8 Continuous creep after cyclic tension-compression loading (strain amplitude 0.25%)

Figure 9 Peak stress versus number of cycle during cyclic preloading for continuous creep

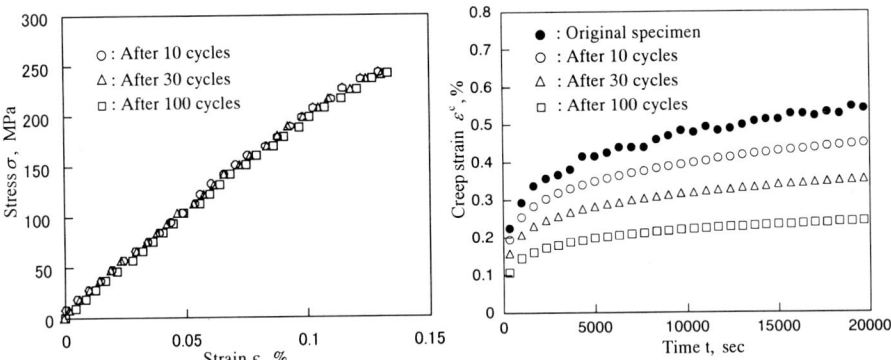

Figure 10 Stress-strain relations of reloading after cyclic preloading

Figure 11 Continuous creep curves after tension-compression loading

(242MPa) of the cyclic preloading. After that the creep test is conducted for the duration of 20000 seconds at the peak stress (242MPa). The stress-strain relation first shows cyclic hardening over a few cycles and then stabilizes. After stabilization the stress-strain relations follow the same trajectory for a number of cycles, without hardening or softening.

Figure 9 shows the relationship between the peak stress (maximum and minimum stresses) and cycle number during the cyclic preloading. The peak stresses are reached 242MPa after a few cycles, and then it kept its value for additional numbers of cycle. Namely, all continuous creep tests were conducted at the same stress level of 242MPa.

Figures 10 shows the stress-strain relations of the reloading to the stress levels of the creep test after the cyclic preloading, corresponding to the broken curve in Fig. 8. The stress-strain relations are similar irrespective of preloading cycle numbers. This shows that the subsequent plastic deformation is not affected by the number of preloading cycles after stabilization, allowing the conclusion that back stress at the maximum stress does not change with additional cyclic loading after the stabilization.

Figures 11 shows the relationship between the continuous creep strain and time in seconds. In Fig. 11, the solid circles show the results of original specimens, which are not subjected to preloading. The different continuous creep curves obtained after different cycle numbers in preloading. The continuous creep strain decreases with increases in cycle numbers in the preloading.

4.2 CONSTITUTIVE MODEL AND SIMULATIONS

In constitutive models of both plasticity and viscoplasticity, the back stress at the peak stress saturates when the stress-strain relation of the cyclic loading stabilizes. The saturated value of the back stress at the peak stress does not change after the additional cyclic loading until strain softening. The test results in Fig. 10 support the validity of the hardening laws for back stress incorporated in the constitutive models. However, the difference of the continuous creep curves due to additional cyclic loading cannot be simulated by the conventional constitutive model given by Eq.(1). Then, it is necessary to employ other internal variables, which are based on micromechanism such as dislocation etc., to explain the continuous creep curves. Estrin et al. (1996) proposed a constitutive model based on the dislocation density, and the model employs some internal variables related to dislocation mechanism. There is possibility to clarify the difference of the continuous creep curves from the dislocation mechanism if the continuous creep curves could be simulated by the constitutive model. In this paper, a constitutive model based on the dislocation density was proposed, and internal variables explaining the difference of the continuous creep curves were discussed.

4.2.1 Constitutive Model Based on Dislocation Density

The summary of the constitutive model based on dislocation density is as follows: Considering the dislocation density model proposed by Estrin et al. (1996), we assume that the viscoplastic potential is written as

$$F = \xi \frac{\sigma_0 \sqrt{Y - Z}}{m + 1} \left(\frac{\bar{\sigma}}{\sigma_0 \sqrt{Y - Z}} \right)^{m+1} \tag{2}$$

where ξ and m are material constants, and σ_0 is the drag stress. Y is the ratio of the immobile dislocation density to the mobile dislocation density. $\bar{\sigma}$ is the equivalent stress defined by $\bar{\sigma} = \left\{ (3/2) C_{ijkl} \left(\sigma_{ij} - \alpha_{ij} \right) \left(\sigma_{kl} - \alpha_{kl} \right) \right\}^{1/2}$ with 4th rank anisotropy coefficient tensor C_{ijkl} and the back stress α_{ij}. From the normality rule, we can have the viscoplastic flow rule as

$$\dot{\varepsilon}^v_{ij} = \frac{3}{2} \cdot \frac{1}{\bar{\sigma}} \xi \left(\frac{\bar{\sigma}}{\sigma_0 \sqrt{Y - Z}} \right)^m C_{ijkl} \left(\sigma_{kl} - \alpha_{kl} \right) \tag{3}$$

The evolution equation of Y is written as

$$\frac{\partial Y}{\partial t} = \left\{ A + A_1 \sqrt{Y} - A_2 \left(\frac{\dot{\varepsilon}^v}{\dot{\varepsilon}_0^v} \right)^{-1/n} Y \right\} \dot{\varepsilon}^v \tag{4}$$

The ratio of recoverable dislocation density Z is defined as the ratio of the immobile dislocation density at time zero to the recoverable dislocation density at time t, and its evolution equation is written as

$$\frac{\partial Z}{\partial t} = \left\{ A_5 \sqrt{Y} - A_2 \left(\frac{\dot{\varepsilon}^v}{\dot{\varepsilon}_0^v} \right)^{-1/n} Z \right\} \dot{\varepsilon}^v - A_6 Z \tag{5}$$

In Eqns. (4) and (5), A, A_1, A_2, A_5, A_6, $\dot{\varepsilon}_0^v$ and n are material constants.
    The evolution equation of the back stress is given by the following equation.

$$\frac{\partial \alpha_{ij}}{\partial \varepsilon_{ij}} = A_3 \sqrt{Y - Z} - A_4 \alpha_{ij} \tag{6}$$

where A_3 and A_4 are material constants.
    In the uniaxial case and for isotropic materials, i.e. $C_{1111} = 2/3$, the viscoplastic strain and the back stress can be reduced from equation (3) and (6) as

$$\dot{\varepsilon}^v = \xi \left(\frac{\overline{\sigma}}{\sigma_0 \sqrt{Y - Z}} \right)^m \tag{7}$$

$$\frac{\partial \alpha}{\partial \varepsilon^v} = A_3 \sqrt{Y - Z} - A_4 \alpha \tag{8}$$

where $\dot{\varepsilon}^v$ is the viscoplastic strain rate and α is the back stress in axial direction.

4.2.2 Simulations

Simulations of the cyclic tension-compression preloading were conducted to obtain the back stresses after the hysteresis loop stabilized. Figure 12 shows the simulation of the cyclic tension-compression preloading with strain amplitude of 0.25%. In Fig. 12, only stabilized loops are shown. The solid, broken and dash-dot lines show the experiment, simulation and back stress, respectively. The simulation has a good agreement with the experiment. The back stress after the hysteresis loop stabilizes is 55MPa. The material constants for cyclic loading are shown in Table 2.
    Considering the equivalence of the back stress during plasticity and creep, and

Table 2 Material constants for cyclic loading

A	A_1	A_2	A_3	A_4	A_5	A_6	ξ	m	n	Y_0	Z_0	$\dot{\varepsilon}_0^v$	σ_0 (MPa)
160	30	30	40000	250	5	0	1	12	18	1	0	1	250

Table 3 Material constants for creep

A	A_1	A_2	A_3	A_4	A_5	A_6	ξ	m	n	Y_0	Z_0	$\dot{\varepsilon}_0^v$	σ_0 (MPa)
2800	8000	10	3647	10	7995	0	1	7	18	1	0	1	400

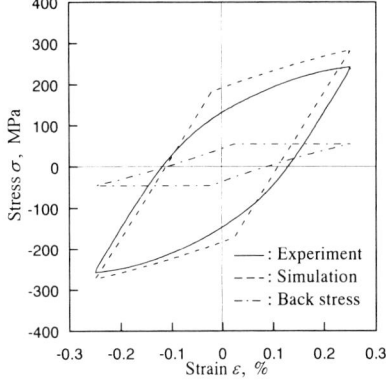

Figure 12 Cyclic tension-compression loading (experiments and simulation)

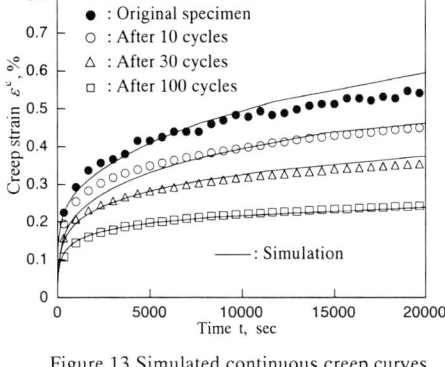

Figure 13 Simulated continuous creep curves

using the value of the back stress 55MPa after the cyclic loading, the simulation of the continuous creep curves were carried out and the results are shown in Fig. 13. The continuous creep curves are simulated well by the constitutive model. The values of the material constants for the creep curves of original specimen, which is indicated by solid circles, are shown in Table 3. The material constants for the continuous creep curves are the same as that for the original specimen except for A_3. The constant A_3 changes the value due to accumulated

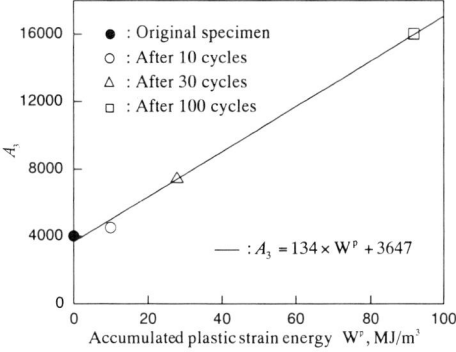

Figure 14 A_3 versus accumulated plastic strain energy W^p

plastic strain energy spent by the cyclic preloading, and is expressed by a formula of accumulated plastic strain energy as shown in solid line in Fig. 14.

The constant A_3 explains the effect of the immobile dislocation on the back stress during creep deformation, then the occurrence of the different continuous creep curves after cyclic preloading are caused by the change of the effect of the immobile dislocation density on the evolution of the back stress. However, the change of the immobile dislocation mechanism does not effect on the stress-strain relation of the reloading as shown in Fig. 10. This suggests that latent hardening should be considered during creep deformation. The latent hardening increases with increase in number of cycle of cyclic preloading, and it leads to the different continuous creep curves.

5. Conclusions

The equivalence of the back stress due to plasticity and creep was discussed. Considering the equivalence the simulations of the continuous creep curves after cyclic preloading were carried out by the dislocation density based constitutive model. As a result, the following conclusions were obtained:

1. The equivalence of the back stress due to plasticity and creep was discussed by the simple experiments such as the intermittent creep during tension.

2. The different continuous creep curves due to number of cycle were observed after the hysteresis loop of the cyclic tension-compression loading stabilized. The difference does not come from the change of the effective stress due to number of cycle because of the equivalence of the back stress.

3. The dislocation density based constitutive model well simulates the difference of the continuous creep curves due to the number of cycle of the preloading.

4. The different continuous creep curves after cyclic preloading are caused by the change of the effect of the immobile dislocation density on the evolution of the back stress due to the number of cycle of the preloading.

5. The latent hardening should be considered in the evolution equation of the back stress due to creep deformation.

References

Blum, W. Vogler, S. and Biberger, M. (1989) Stress Dependence of the Creep Rate at Constant Dislocation Structure, *Materials Science and Engineering*, **A112**, 93-106.

Davies, C. K. Older, A. G. and Stevens, R. N. (1992) Dislocation Configurations and Internal Stresses in the Creep Nimonic 91, *Journal of Materials Science*, **27**, 5365-5374.

Estrin, Y. Braasch, H. and Brechet, Y. (1996), A Dislocation Density Based Constitutive Model for Cyclic Deformation, *ASME Journal of Engineering Materials and Technology*, **118**, 441-447.

Freed, A. D. and Walker, K. P. (1993) Viscoplasticity with Creep and Plasticity Bounds, *International Journal of Plasticity*, **9**, 213-242.

Ikegami, K. and Niitsu, Y. (1985), Effect of Creep Prestrain on Subsequent Plastic Deformation, *International Journal of Plasticity*, **1**, 331-345.

Ishikawa, H. and Sasaki, K. (1994), Creep, Stress Relaxation and Biaxial Ratchetting of Type 304 Stainless Steel After Cyclic Preloading, *ASME Journal of Engineering Materials and Technology*, **116**, 133-141.

Kawai, M. (1994), A Constitutive Model for Anisotropic Creep Deformation, *ASME Journal of Engineering Materials and Technology*, **116**, 142-147.

Krempl, E. (1987), Models of Viscoplasticity, *Acta Mechanica*, **69**, 25-42.

Kujawski, D. Kallianpur, V. and Krempl, E. (1980), An Experimental Study of Uniaxial Creep, Cyclic Creep and Relaxation of AISI Type 304 Stainless Steel at Room Temperature, *Journal of Mechanics and Physics of Solids*, **28**, 129-148.

Malygin, G. A. (1993), Dislocation Mechanism of the Transient Creep, *Physics of Metals and Metallography*, **75**, 468-473.

Ohno, N. Kawabata, M. and Naganuma, J. (1990), Aging Effects on Monotonic, Stress-Paused, and Alternating Creep of Type 304 Stainless Steel, *International Journal of Plasticity*, **6**, 315-327.

Orlova (1993), An Estimation of Internal Stress Evolution in the Primary Creep Based on the Composite Model of Dislocation Structure, *Material Science and Engineering*, **A163**, 61-66.

Ruggles, M. and Krempl, E. (1989), The Influence of Test Temperature on the Ratchetting Behavior of Type 304 Stainless Steel, *ASME Journal of Engineering Materials and Technology*, **111**, 378-383.

Sasaki, K. and Ishikawa, H. (1995), Experimental Observation on Viscoplastic Behavior of SUS304, *JSME International Journal, Series A*, **38**, 265-272.

Valanis, K. C. (1975), On the Foundation of the Endochronic Theory of Plasticity, Archives of Mechanics, Vol. 27, pp. 857-868.

Wu, H.-C. and Ho, C.-C. (1993), Strain Hardening of Annealed 304 Stainless Steel by Creep, *ASME Journal of Engineering Materials and Technology*, **115**, 345-350.

Wu, H.-C. and Ho, C.-C. (1995), An Investigation of Transient Creep by Means of Endochronic Viscoplasticity and Experiment, *ASME Journal of Engineering Materials and Technology*, **117**, 260-268.

ASSESSMENT OF THE MULTIAXIAL CREEP DATA BASED ON THE ISOCHRONOUS CREEP SURFACE CONCEPT

Z.L. KOWALEWSKI
Institute of Fundamental Technological Research
Department for Strength of Materials
00-049 Warsaw, ul. Świętokrzyska 21
POLAND

Abstract

The paper presents a new methodology for determination of the same time to rupture surfaces on the basis of creep tests carried out on pure copper under plane stress state at elevated temperature (523K). It is proposed to adopt this methodology in order to determine the surfaces of the same time to obtain minimum creep rate and surfaces of the same time to tertiary creep period. Analysis of all types of the isochronous surfaces proved that the damage evolution in copper develops proportionally until the creep rupture is achieved.

1. Introduction

Creep data are normally presented in the form of creep strain versus time taken at constant load (or stress) and temperature. Such plots allow for a visual identification of the primary, secondary and tertiary creep regimes in addition to providing some quantitative indications of the effect of stress on the strain-time behaviour. It is also important to note that these plots serve as the source of various other graphical presentations which provide interpretations not readily available in the strain-time plots. Since a description of material creep behaviour should reflect the essential interrelations among stress, strain, and time, many graphical methods of the creep data presentation have been elaborated. One of the well known methods depicting these interrelation-ships is through isochronous stress-strain curves obtainable from the standard creep curves. The usual form of isochronous curve plots stress on the ordinate and total strain on the abscissa for constant temperature. It presents locus of total strains accumulated when different constant stresses are applied for a fixed time. The isochronous method of creep data presentation was first suggested by McVetty [1]. Later, Shanley [2] has shown how this technique can be of benefit to the stress analysis in certain types of design applications, where limitation of strain may be an important factor in preserving the structural integrity or functional capability of a component. The isochronous stress-strain method is very simple in use for presentation of the experimental data from

S. Murakami and N. Ohno (eds.), IUTAM Symposium on Creep in Structures, 401–410.
© 2001 *Kluwer Academic Publishers. Printed in the Netherlands.*

uniaxial creep tests. Its application to the graphical presentation of the results from multiaxial tests is much more complicated. Since at the beginning of the new millennium an increasing demand is observed on multiaxial creep testing, many efforts are focused on the development of more effective creep data analysis methods than that previously used. It has been found that multi-axial creep rupture results are conveniently plotted in terms of isochronous surfaces [3-9] being loci of constant rupture time in a stress space. This approach especially simplifies theoretical creep results analysis giving clear graphical representation of material lifetime. However, the accurate experimental determination of the shape of these surfaces requires a large number of creep rupture data from tests carried out under complex loading. The paper presents a methodology of determination of the same time to rupture surfaces on the basis of creep tests carried out on pure copper under plane stress state at elevated temperature. It is proposed to adopt this technique in order to determine the surfaces of the same time to obtain minimum creep rate (also called the surface of the same duration of primary creep) and surfaces of the same time to tertiary creep period.

2. Experimental Procedure

Creep investigations were carried out on thin-walled tubular testpieces with the use of the biaxial creep testing machine enabling to realise plane stress conditions by simultaneous loading of the testpieces by an axial force and twisting moment at elevated temperature (523K). The material tested was an electrolytic copper of 99.9% purity. According to Polish Standards it is denoted as M1E.

The experimental programme comprised creep tests up to rupture for copper specimens subjected to biaxial stress state obtained by various combinations of tensile and torsional stresses: $(\sigma_{12}/\sigma_{11}) = 0$, $(\sigma_{12}/\sigma_{11}) = \sqrt{3}/3$, $(\sigma_{12}/\sigma_{11}) = \infty$. Tests were carried out for three equivalent stress levels: 70.0; 72.5; and 75.0 MPa. The programme of creep tests is presented in Figure 1.

3. Experimental Data

Creep characteristics of copper are presented in Figures 2-4. The results prove that the process is stress state sensitive, in spite of the initial isotropy of the material in the sense of such parameters obtained from monotonic tension investigations as Young's modulus, yield limit or ultimate tensile stress. The variations of lifetime obtained for uni-axial tension and pure torsion tests exceeded one order of magnitude. In all cases the shortest lifetimes and moreover the lowest ductility were achieved for the tensioned testpieces. The minimum creep strain rate characterises the creep process in its secondary period. For all stress levels considered, the greatest value of this parameter was always obtained for the tension test, but the lowest - for pure torsion test. More thorough discussions of these results are reported in [10].

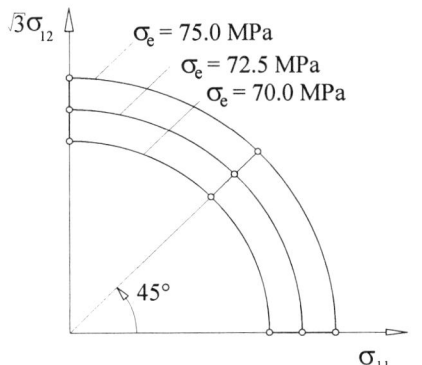

Figure 1. Experimental programme.

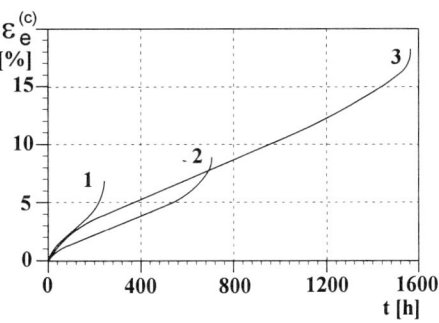

Figure 2. Creep curves for copper at 523K under stress equal to 70.0 MPa; $1 - (\sigma_{12}/\sigma_{11}) = 0$;

$2 - (\sigma_{12}/\sigma_{11}) = \sqrt{3}/3$; $3 - (\sigma_{12}/\sigma_{11}) = \infty$.

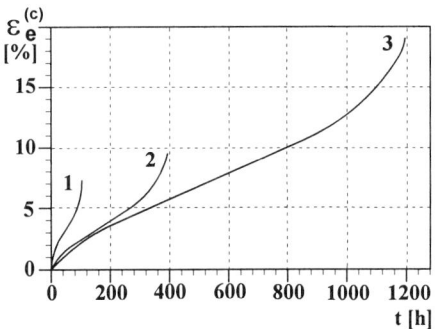

Figure 3. Creep curves for copper at 523K under stress equal to 72.5 MPa; $1 - (\sigma_{12}/\sigma_{11}) = 0$;

$2 - (\sigma_{12}/\sigma_{11}) = \sqrt{3}/3$; $3 - (\sigma_{12}/\sigma_{11}) = \infty$.

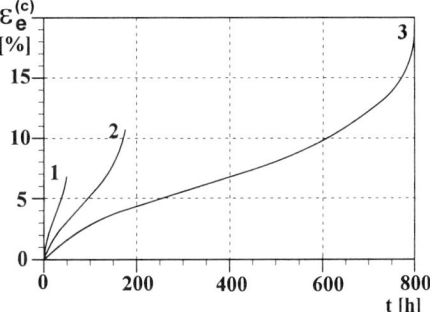

Figure 4. Creep curves for copper at 523K under stress equal to 75.0 MPa; $1 - (\sigma_{12}/\sigma_{11}) = 0$;

$2 - (\sigma_{12}/\sigma_{11}) = \sqrt{3}/3$; $3 - (\sigma_{12}/\sigma_{11}) = \infty$.

4. Experimental Data Analysis

As mentioned earlier, the creep process of copper is a stress state sensitive. Such material behaviour can be interpreted, on the one hand, as the material deformation due to different deformation mechanisms, activation of which is connected with the stress state type, on the other hand however, it is known [4, 5, 15] that the majority of microcracks appearing on the grain boundaries are observed on those grain boundaries which are perpendicular to the maximum principal stress. It seems that for the copper, the last conclusion can be confirmed by the shapes of the specimen cross-section in places where rupture has occurred. In the case of creep tension tests the failure line was perpendicular to the main specimen axis, in the case of complex stress states this line was inclined at angles equal typically to 15-20°, which were measured with respect to the line perpendicular to the main specimen axis, whereas for testpieces subjected to pure torsion these angles were approximately equal to 30-45°, Figure 5. Thus, it can be concluded that for copper, the basic deformation mechanism seems to be the same for all stress state types considered, while the resulting variations in lifetimes for the same effective stress follow from differences in the value of the maximum principal stress.

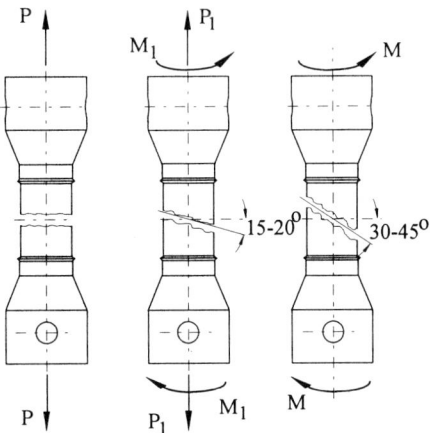

Figure 5. Failure lines for different types of stress states.

4.1. CREEP RUPTURE ANALYSIS THROUGH THE CONCEPT OF ISOCHRO-NOUS SURFACES

4.1.1. *Theoretical Approach*
In order to present creep data in the form of isochronous surfaces the multiaxial creep constitutive equations are required. Since it was not possible to obtain isochronous creep rupture locus on the basis of the uniaxial creep theory proposed by Kachanov [11] and Rabotnov [12], Leckie and Hayhurst [5] have attempted to generalise the single state Rabotnov-Kachanov equations to multiaxial states of stress. The generalisation of the uniaxial equations for multiaxial stresses has been achieved by making the assumption that the influence of continuum damage on the deformation rate process is scalar in character, and by the introduction of the homogeneous stress function which reflects the stress state effects on the time to rupture. Equations reflecting multiaxial conditions can be written in the following form

$$\frac{\dot{\varepsilon}_{ij}}{\dot{\varepsilon}_0} = \frac{3}{2}\left(\frac{\sigma_e}{\sigma_0}\right)^{n-1}\left(\frac{S_{ij}}{\sigma_0}\right)\left(\frac{1}{(1-\omega)^n}\right), \tag{1}$$

$$\frac{\dot{\omega}}{\dot{\omega}_0} = \Delta^{\nu}\frac{1}{(1+\eta)(1-\omega)^{\eta}}, \tag{2}$$

where n, η, ν, $\dot{\omega}_0$, σ_0, $\dot{\varepsilon}_0$ - constants, σ_e - effective stress, S_{ij} - stress deviator tensor,

$\dot{\varepsilon}_{ij}$ - strain rate tensor, ω - damage variable, $\varDelta$ - rupture criterion ($\varDelta = \varDelta(\sigma_{ij}/\sigma_0) = \sigma_{max}/\sigma_0$ for copper, and $\varDelta = \varDelta(\sigma_{ij}/\sigma_0) = \sigma_e/\sigma_0$ for aluminium alloys, where σ_{ij} - stress tensor, σ_{max} - maximum principal stress). Integration of the damage evolution equation (2) for the following boundary conditions: $\omega{=}0$, $t{=}0$ and $\omega{=}1$, $t{=}t_R$, yields, after normalisation, the relation describing time to rupture in the form:

$$\frac{t_R}{t_0} = \frac{1}{\varDelta^{\nu}}. \tag{3}$$

Substitution of $t_R = t_0$ in (3) gives the equation of isochronous surface.

According to Johnson [13, 14], the rupture criteria for aluminium alloy and pure copper represent the extremes of material behaviour, since the isochronous surface for many metals lies somewhere between these criteria. In spite of the fact that these observations have been made on the basis of relatively limited amount of the experimental data, and in certain cases they did not give precise description of rupture, they are still influencing the process of developing new creep damage models. Typical examples of such situation are: the rupture criterion applied by Dyson and McLean [15] in which they assumed that the creep damage of Nimonic 80A tested at 1023K was governed by the criterion being a product of the effective stress and maximum principal stress criteria, and the creep rupture criterion proposed by Sdobyrev [16] in the form reflecting the relation between time to rupture, effective stress (σ_e), and maximum principal stress (σ_{max})

$$t_R^{(Sdob)} = A\{\beta\sigma_{max} + (1-\beta)\sigma_e\}^{-\nu}, \tag{4}$$

where A, ν - material constants, and β - experimentally determined coefficient.

On the basis of the experimental results obtained from tests made on cruciform specimens [4] and Johnson's investigations [13, 14] on thin-walled tubular testpieces, Hayhurst [4] in attempts to describe the multi-axial behaviour of several materials suggested a general relationship

$$t_R^{(Hay)} = A\{a\sigma_{max} + bJ_1 + cJ_2^{\frac{1}{2}}\}^{-\nu}, \tag{5}$$

being a linear combination of the maximum principal tensile stress, the first invariant of the stress tensor (J_1), and the second invariant of the stress deviator (J_2'). In expression (5) a, b, c are constants with $a+b+c{=}1$ what limits the influence of particular terms in the criterion (5) on the time to rupture. By selecting appropriate values of these constants, the well known simple rupture criteria can be represented.

4.1.2. Experimental Approach

The isochronous surfaces can be determined on the basis of experimental data obtained from tests carried out under complex stress states. These tests should be realised at various stress states and for several stress levels. Having all creep curves a procedure to the creep data presentation in the form of isochronous surfaces requires only two steps. Firstly, a diagram representing the relationship $\log(\sigma_e){=}f[\log(t_R)]$, Figure 6, has to be prepared, and secondly, on the basis of this diagram, the isochronous surfaces should be determined.

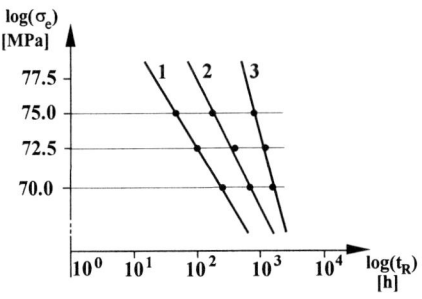

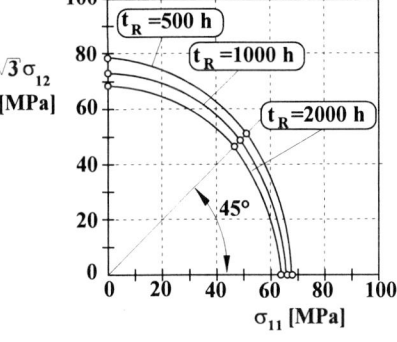

Figure 6. Logarithmic diagram of stress versus rupture ; 1 - $(\sigma_{12}/\sigma_{11}) = 0$;

2 - $(\sigma_{12}/\sigma_{11}) = \sqrt{3}/3$; 3 - $(\sigma_{12}/\sigma_{11}) = \infty$.

Figure 7. Experimental isochronous creep rupture surfaces.

In Figure 6 the results for the three types of stress states are presented. Data points for the chosen type of stress state with relatively high accuracy are located on the straight lines which have different mutual location. These lines have been obtained with the use of the least squares technique. Taking into account, as a reference point, the line representing pure torsion results the remaining straight lines are shifted and rotated. Analysis of the mutual location of these lines indicates that the material subjected to creep process under uni-axial tension is significantly more sensitive to stress variations, in comparison to the same material tested either at pure torsion or a combination of tension and torsion. On the basis of the diagram shown in Figure 6, drawing straight lines which are parallel to the stress axis and intersect the approximation lines representing the experimental data, it is easy to determine the points connecting the same times to rupture. The points for the rupture time taken into account provide the values of effective stresses necessary to achieve the rupture at the particular type of stress state considered. These values can be approximated on the stress plane $(\sigma_{11}, \sqrt{3}\sigma_{12})$ giving, as a consequence, the isochronous curves. Surfaces of the same lifetime determined in this way are shown in Figure 7 for the rupture time equal to 500, 1000, and 2000 [h]. The surface corresponding to the rupture time equal 500 [h] has been selected for further comparative studies presented in the next subsection.

4.1.3. Comparison of experimental isochronous creep rupture surface with surfaces predicted by various rupture criteria

The curve of the same time to rupture determined on the basis of experimental programme is compared with theoretical predictions of the three well known creep rupture hypotheses:

• the maximum principal stress rupture criterion

$$\sigma_R = \sigma_{max} = \frac{1}{2}\left(\sigma_{11} + \sqrt{\sigma_{11}^2 + 4\sigma_{12}^2}\right), \tag{6}$$

- the Huber-Mises effective stress rupture criterion

$$\sigma_R = \sigma_e = \sqrt{\sigma_{11}^2 + 3\sigma_{12}^2} , \qquad (7)$$

- the Sdobyrev creep rupture criterion [16]

$$\sigma_R = \beta\sigma_{max} + (1 - \beta)\sigma_e . \qquad (8)$$

The isochronous surfaces resulting from these rupture criteria are compared with the surface determined on the basis of experimental results, Figure 8.

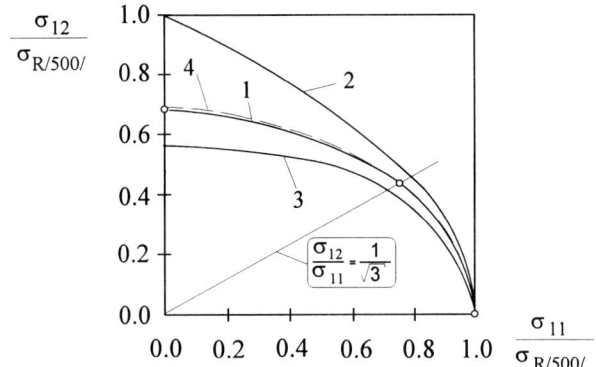

Figure 8. Comparison of the isochronous creep rupture surfaces determined for $t_R = 500$ [h] on the basis of experimental results (1) and theoretical calculations using the following rupture criteria; 2 – maximum principal stress criterion; 3 – effective stress criterion; 4 – Sdobyrev criterion (broken line).

All curves presented in the normalised co-ordinate system are referred to the rupture time equal to 500 [h]. Tensile stress corresponding to the lifetime of 500 [h] has been selected as the normalisation value ($\sigma_{R/500/}$). As it is clearly seen, the best description of the experimental data has been achieved for the Sdobyrev creep rupture criterion taken with the coefficient $\beta=0.4$, calculated on the basis of creep tests carried out. The value of β indicates that the damage mechanism governed by the effective stress as well as the maximum principal stress played a considerable role in the creep rupture of the copper tested.

4.2. ISOCHRONOUS LOCI REPRESENTING TYPICAL POINTS OF THE CREEP PROCESS

In the previous section it has been shown how to present clearly the experimental creep rupture results. Since the creep damage process already appears at the primary creep [17], it seems to be reasonable to adopt the isochronous surface concept at other characteristic points of the creep curve, such as the end of the primary or of secondary creep periods. Obviously, it is possible only then if the conditions of the process and the material tested ensure the typical creep curve, i.e. a curve having all creep stages. In practice, the loading conditions and materials of structural components satisfy these conditions.

4.2.1. *Isochronous Curves of the End of Primary Creep Period*

Determination procedure of the curves representing the same duration of primary creep is analogous to that used to obtain surfaces of the same time to rupture. Thus, in the first step, the diagrams $\log(\sigma_e)=f[\log(t_1)]$ have been prepared for all stress states types considered in the plane $(\sigma_{11}, \sqrt{3}\sigma_{12})$. Similarly to the rupture times considerations, data points for the chosen type of stress state, with relatively high accuracy, are located on the straight lines which have different mutual positions, Figure 9. On the basis of diagrams shown in Figure 9, it is easy to determine points connecting the same times to achieve stabilisation of the creep strain rate, Figure 10. It can be seen that the shape and dimensions of the isochronous surfaces representing the same time to rupture and the same duration of primary creep period are similar. However, in order to compare them accurately it is necessary to provide clearly defined reference point. In order to assess mutual correlation of both types of isochronous surfaces, the tension stress equal to 67.9 [MPa] has been selected as a reference point, which corresponds to the lifetime of 500 [h]. Taking into account the reference tension stress level and using diagrams $\log(\sigma_e)=f[\log(t_1)]$, shown in Figure 9, it is easy to find the time to creep strain rate stabilisation ($t_1=79$ [h]). Knowing this value, all stress levels for the considered stress states which correspond to the same duration of primary creep can be determined, Figure 9, giving the data necessary to obtain the isochronous surface. Comparison of the surfaces representing the same lifetimes and the same duration of primary creep period is presented in Figure 13. It is easy to see a great similarity of these surfaces.

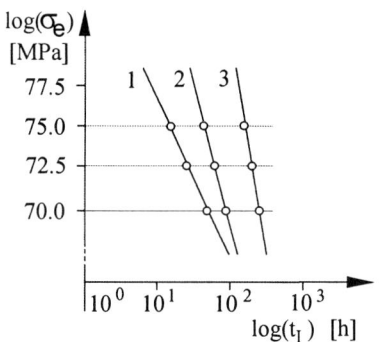

Figure 9. Logarithmic diagram of stress versus time to the end of primary creep ; 1 - $(\sigma_{12}/\sigma_{11}) = 0$; 2 - $(\sigma_{12}/\sigma_{11}) = \sqrt{3}/3$; 3 - $(\sigma_{12}/\sigma_{11}) = \infty$.

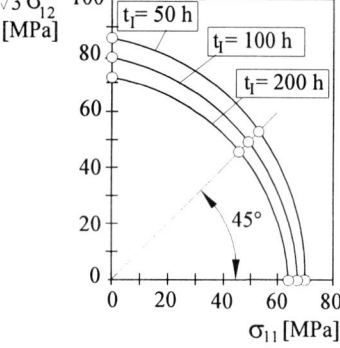

Figure 10. Experimental isochronous surfaces of the end of primary creep.

Thus, it can be stated for the tested copper that, on the basis of the degree of creep process development at the end of primary creep, represented by the surface of the same duration of primary creep, we can deduce the shape of isochronous surface describing the same time to rupture. In other words, the degree of damage evolution in copper for each loading type applied in the experimental programme maintained a constant value. In order to confirm this suggestion, the other characteristic point of the creep curve, i.e. time to tertiary creep, will be discussed in the next subsection.

4.2.2. *Isochronous Curves of the End of Secondary Creep Period*

The curves representing the same time to tertiary creep are determined with the use of the same method as that used to obtain surfaces of the same time to rupture and the same duration of primary creep, Figures 11 and 12.

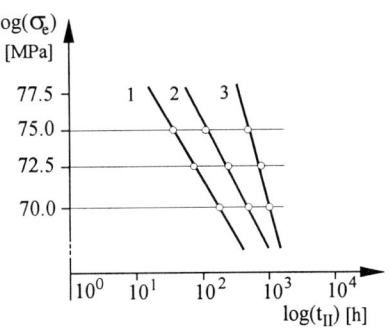

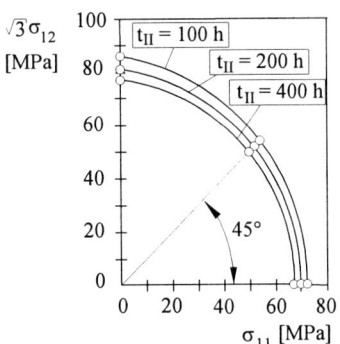

Figure 11. Logarithmic diagram of stress versus time to the end of secondary creep ; 1 - $(\sigma_{12}/\sigma_{11}) = 0$; 2 - $(\sigma_{12}/\sigma_{11}) = \sqrt{3}/3$; 3 - $(\sigma_{12}/\sigma_{11}) = \infty$.

Figure 12. Experimental isochronous creep of the end of secondary creep.

The shapes of the isochronous surfaces representing the same time to tertiary creep, the same duration of primary creep period, and the same time to rupture are similar [18]. These surfaces are compared in Figure 13 for the reference stress equal to 67.9 [MPa].

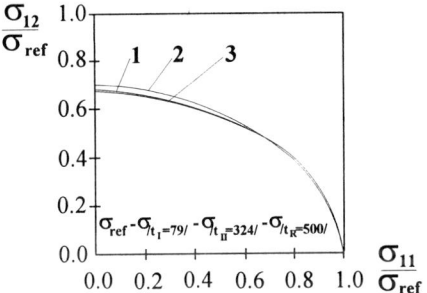

Figure 13. Comparison of the isochronous surfaces; (1) surface of the same time to rupture, (2) surface of the same duration of primary creep period, and (3) surface of the same time to tertiary creep period.

The reference stress level used as a normalisation factor at uni-axial tension corresponds to the uni-axial tensile stress giving the time to rupture of 500 [h], duration of primary creep of 79 [h], and time to tertiary creep of 324 [h]. The mutual location of the surfaces confirms the previously suggested thesis that the damage evolution in copper, measured as a ratio of the effective stress and the selected reference stress, develops proportionally for all directions considered in the two-dimensional stress plane.

5. Conclusions

For the copper such creep parameters as: duration of primary creep period, minimum creep rate, time to rupture and ductility were the functions of the type of stress state.

The procedure of determination of the isochronous creep rupture surfaces is proposed. Although this method can be applied to relatively limited range of stress levels, i.e. stress levels close to those used in experiments, it gives a promising tool in multiaxial data analysis. The method can be also applied to determine surfaces representing the same time to stabilise creep rate and surfaces of the same time to attain tertiary creep period. Analysis of all types of the isochronous surfaces for copper proved that the damage evolution develops proportionally until the creep rupture is achieved for different stress types taken into account in the $(\sigma_{11}, \sqrt{3}\sigma_{12})$ stress space.

Verification of the fundamental creep rupture criteria has shown that the Sdobyrev criterion gives the most accurate description of the material damage evolution during the long-term constant loading conditions at elevated temperature.

6. References

1.   McVetty, P.G.: Working stresses for high temperature service, *Mech. Eng.* **56**, 3 (1934), p 149.

2.   Shanley, F.R.: *Weight-Strength Analysis of Aircraft Structure*, McGraw-Hill, 1952.

3.   Piechnik, S. and Chrzanowski, M.: Time of total creep rupture of a beam under combined tension bending, *Inst. J. Solids Structures* **6** (1970), 453-477.

4.   Hayhurst, D.R.: Creep rupture under multi-axial states of stress, *J. Mech.Phys.Solids* **20** (1972), 381-390.

5.   Leckie, F.A. and Hayhurst, D.R.: Constitutive equations for creep rupture, *Acta Metallurgica* **25** (1977), 1059-1070.

6.   Chrzanowski, M. & Madej, J.: The construction of failure limit curves by means of a damage (in Polish), *Theoretical and Applied Mechanics*, **4**, 18 (1980), 587-601.

7.   Hayhurst, D.R., Trąmpczyński, W.A., and Leckie, F.A.: Creep rupture under non-proportional loading, *Acta Metall.*, **28** (1980), 1171-1183.

8.   Litewka, A. and Hult, J.: One parameter CDM model for creep rupture prediction, *Eur.J.Mech., A/Solids*, **8**, 3 (1989), 185-200.

9.   Kowalewski, Z.L., Hayhurst, D.R., and Dyson, B.F.: Mechanisms-based creep constitutive equations for an aluminium alloy, *J. Strain Analysis*, **29** (1994), 309-316.

10. Kowalewski, Z.L.: Experimental evaluation of the influence of stress state type on creep characteristics of copper at 523K, *Arch. Mech.*, **47** (1995), 13-26.

11. Kachanov, L.M.: *The Theory of Creep* (English translation edited by Kennedy A.J.), 1958, National Lending Library, Boston Spa.

12. Rabotnov, Y.N.: *Creep problems in structural members*, 1969, North Holland Publishing Company, Amsterdam.

13. Johnson, A.E., Henderson, J., and Mathur, V.D.: Combined stress fracture of commercial copper at 250 C, *The Engineer*, **202** (1956), 261.

14. Johnson, A.E., Henderson, J., and Khan, B.: *Complex-stress creep, relaxation and fracture of metallic alloys*, H.M.S.O., Edinburgh, 1962.

15. Dyson, B.F., and McLean, D.: Creep of Nimonic 80A in torsion and tension, *Metal Sci.*, **2**, 11 (1977), 37-45.

16. Sdobyrev, V.P.: Creep criterion for some high-temperature alloys in complex stress state (in Russian), *Izv. AN SSSR. Mekh. and Mashinostr.*, **6** (1959), 12-19.

17. Boettner, R.C., and Robertson, W.D.: Study of growth of voids in copper during creep process by measurement of accompanying change in density, *Trans. Metall. Soc. A.I.M.E.*, **221** (1961), 613.

18. Kowalewski, Z.L.: Biaxial creep study of copper on the basis of isochronous creep surfaces, *Arch. Mech.*, **48** (1996), 89-109.

BIAXIAL AND TRIAXIAL CREEP TESTING OF TYPE 304 STAINLESS STEEL AT 923 K

M. SAKANE
Department of Mechanical Engineering, Ritsumeikan University
1-1-1 Nojihigashi Kusatsu-shi Shiga, 525-8577, Japan
T. HOSOKAWA
Department of Mechanical Engineering, Ritsumeikan University
1-1-1 Nojihigashi Kusatsu-shi Shiga, 525-8577, Japan

Abstract

This paper describes the biaxial and triaxial creep damage evaluation experimentally. A biaxial and a triaxial creep machines were developed to evaluate the creep damage under multiaxial stress states. Multiaxial creep tests were carried out using the two creep machines and the suitability of multiaxial creep parameters was discussed. The maximum principal stress was a suitable parameter for the multiaxial creep damage evaluation but Mises stress was not. Microstructural observation was also made for discussing the grain boundary damage under a triaxial stress state.

1. Introduction

Multiaxial creep damage evaluation is one of the great issues for designing high temperature components. Recently Type IV cracking in fine grain regions in heat affected zones was reported and the stress multiaxiality is being suspected to accelerate the creep damage. However, there is little experimental research carried out on biaxial and triaxial creep fracture. For the biaxial creep research, one of the authors has developed a biaxial creep machine and discussed the suitability of the creep stress parameter[1] but there is no study for triaxial creep except for the early study made by Hayhurst[2]. He developed a triaxial creep machine and carried out triaxial creep tests using an aluminum alloy but he could not obtain definite conclusions for the parameter describing the multiaxial creep damage. Since multiaxial creep testing is significantly difficult technically and costly no one can succeed to obtain definite conclusions using practical heat resistant materials whereas many multiaxial creep damage models were proposed so far [3]. A reliable multiaxial creep damage model can not be developed without confirming with the experimental study.

The objective of this paper is to investigate the effect of stress multiaxiality on creep and creep rupture properties. Development of a triaxial creep machine is firstly presented and a new triaxial creep testing method utilizing a miniature specimen is

S. Murakami and N. Ohno (eds.), IUTAM Symposium on Creep in Structures, 411–418.

Figure 1. Photograph of triaxial creep specimen.

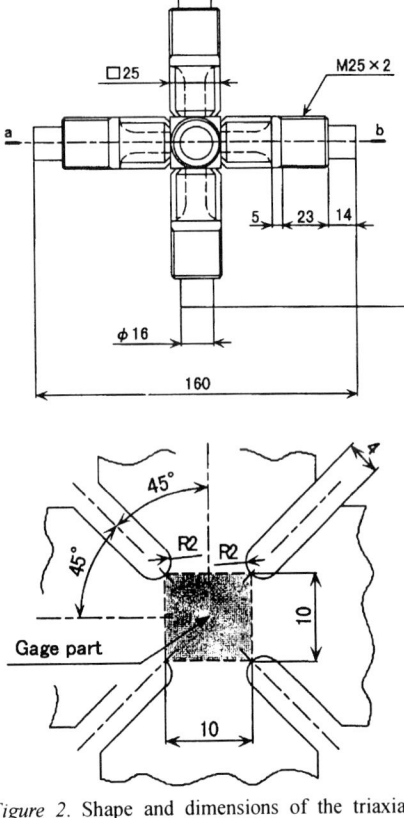

Figure 2. Shape and dimensions of the triaxial creep specimen.

introduced. Triaxial creep and creep rupture tests were carried out using the machine and the suitability of multiaxial stress parameters is discussed. Creep voids were observed by a scanning electron microscope and the microstructural creep damage is discussed in relation with the multiaxial damage parameters.

2. Triaxial Creep Testing

A new triaxial creep machine was developed for making triaxial creep tests because the machine is not commercially available. Figure 1 shows the photograph of triaxial creep specimen and Fig.2 illustrates the shape and dimensions of the specimen. The specimen has six arms for loading and the center cubic part with dimensions of 10 mm x 10 mm x 10 mm is the gage part. Deep grooves with 2 mm width were cut to near the gage part to achieve a uniform triaxial stress state in the gage part [4]. Making these grooves is essential for the uniform triaxial stress distribution in the gage part but this specimen shape has a fatal disadvantage that the specimen fails at grooves.

A new triaxial creep test method was developed utilizing a miniature creep specimen as shown in Fig.3. In the new creep testing, triaxial creep load was initially applied using the triaxial creep specimen shown in Fig.2 for a predetermined time and the damage under triaxial stress was induced in the gage part. A miniature uniaxial creep specimen was cut from the gage part of predamaged triaxial specimen. The diameter of the miniature creep specimen was 2 mm and the gage length 10 mm. The stress applied to the miniature specimen was set to the same as that used in triaxial creep tests. The triaxial creep rupture time was calculated based on the triaxial loading time and the miniature rupture time which will be described later in detail.

TABLE 1.   Chemical composition of the material tested (wt%).

C	Si	Mn	P	S	Cr	Ni
0.05	0.41	1.49	0.031	0.027	18.09	8.14

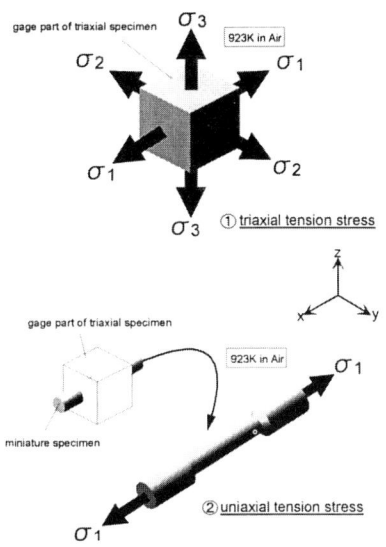

Figure 3. Triaxial creep testing utilizing a miniature creep specimen.

Figure 4 shows the photograph and schematic of the triaxial creep machine.   The machine has six electric-hydraulic servo systems including six hydraulic actuators. The machine is capable to perform a load controlled test as well as a stroke controlled test.   The maximum load capacity of the machine is 49 KN.   Specimens were heated by an electric resistant heater with an electric capacity of 3KW.

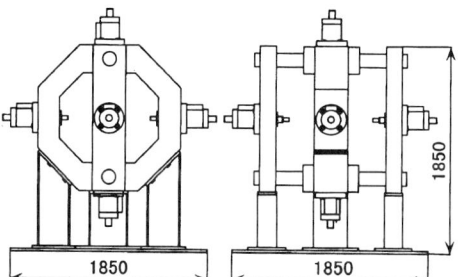

Figure 4. Photograph and schematic of triaxial creep apparatus.

The maximum temperature of the furnace was 923K.

Figure 5 shows the extensometry of the gage part in triaxial creep testing.   Three pairs of linear elongation gages were used.   Quartz rods of the extensometers were inserted into the grooves of specimens.   The deformation of the gage part was calculated by decomposing the output of the six extensometers into z, y and z elongation.

Test material used was a Type 304 stainless steel solution treated at 1343K with the chemical composition listed in Table 1.   Triaxial and biaxial creep tests were performed at 923 K under a constant load condition.   In triaxial tests, only equi-triaxial tests were performed where the three principal stresses have the same amplitude.

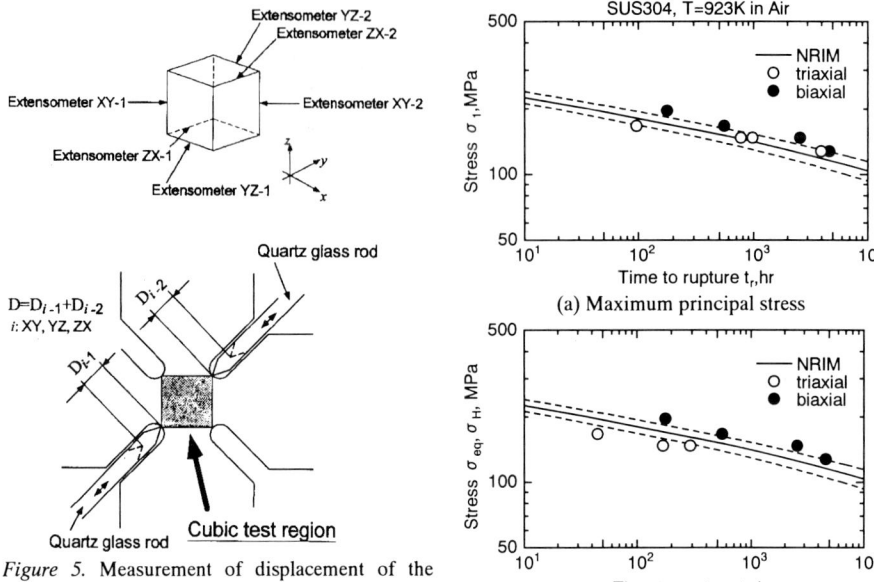

Figure 5. Measurement of displacement of the cubic test region.

(a) Maximum principal stress

(b) Mises' and Huddleston stress

Figure 7. Correlation of rupture time with the maximum principal and Mises stresses.

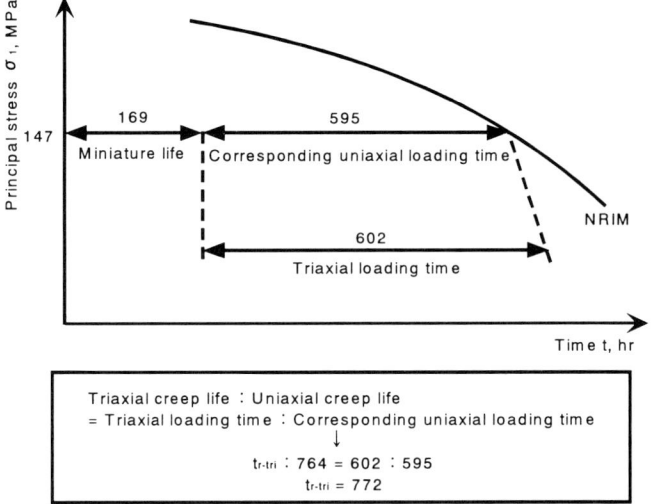

Figure 6. Calculation method of triaxial creep rupture life.

3. Experimental results and Discussion

3.1   TRIAXIAL AND BIAXIAL CREEP RUPTURE PARAMETERS

Figure 6 shows the calculation method of triaxial creep rupture lives, where the figure shows the relationship between the maximum principal stress and time.   A solid line is

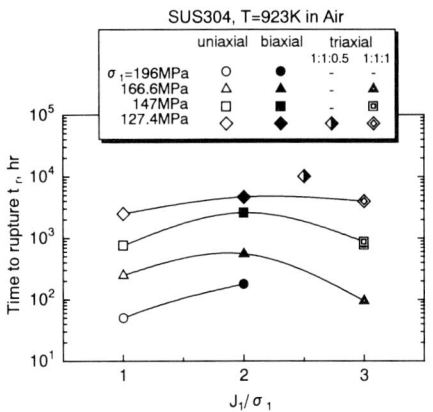

Figure 8. Relationship between rupture time and J_1/σ_1.

the creep rupture data of National Research Institute for Metals [5] in uniaxial tests. In this example, the rupture time of a miniature specimen at 147MPa, predamaged under a triaxial stress state for 602 hrs, was 169 hrs. The damage accumulated under the triaxial stress state for 602 hrs corresponds with that in uniaxial stress state for 595 hrs which is obtained by subtracting the rupture time of the miniature specimen from that of NRIM uniaxial data. From this procedure, the triaxial rupture time was obtained by using the proportional relationship in Fig.7 as 772 hrs, based on the uniaxial rupture time of 764 hrs.

Figure 7 correlates the creep rupture time between uniaxial, biaxial and triaxial stress states. Biaxial creep tests were carried out using the same Type 304 stainless steel at 923 K but the detailed description is omitted here because of the space of the paper. The detail of testing procedure was already published elsewhere [1].

In the correlation of the rupture times with the maximum principal stress in Fig.7 (a), the biaxial data locate slightly outside of the dashed line of a factor of two band, giving a conservative estimate against the uniaxial data. The maximum principal stress makes an unconservative estimate for the triaxial data at higher stress range but the estimate shifts to conservative estimate as the stress level lowers. The overall fitting using the maximum principal stress is appropriate but it does not take account of the effect of the second and third principal stresses in the correlation. The correlation may change if the second and third principal stresses vary so that the effect of these stresses on the correlation must be examined before giving a final conclusion of the appropriateness of the maximum principal stress.

Mises equivalent stress has a zero value in the equi-triaxial stress state used here since there exists no shear stress component in the stress state. The zero Mises stress means no damage accumulation in the correlation with Mises stress in the triaxial creep tests so that only the rupture times of miniature specimens were plotted in the figure. The triaxial data are correlated unconservatively in comparison with the uniaxial data and are slightly fall out of a factor of two band. Mises stress seems to be a suitable parameter for the correlation but essentially it has a discrepancy for the correlation. As mentioned earlier, rupture times of miniature specimens were plotted in Fig.8 (b) since Mises stress estimates no creep damage under the triaxial stress condition. If the loading time would increase under the triaxial stress condition, the rupture times of the miniature specimens may become shorter and then the triaxial data plotted in Fig.7 (b) move in shorter rupture time direction. Therefore, the correlation shown in Fig.7 (b) depends on the triaxial loading time. In this sense, Mises stress is not a appropriate parameter for the multiaxial data correlation, which is attributed to that the Mises stress becomes zero and estimates no creep damage under the equi-triaxial stress condition.

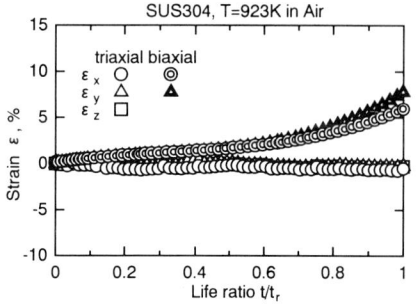 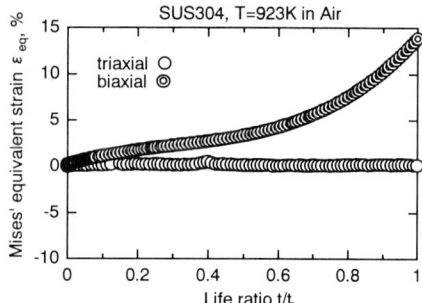

Figure 9. Creep curves in biaxial and triaxial test at $\sigma_1 = 127.4 MPa$.

Figure 10. Relationship between life ratio and Mises equivalent strain.

For the biaxial testing, Mises stress has the same value as the maximum principal stress, so the correlation of the biaxial data with Mises stress is the same as that with the maximum principal stress.

Figure 8 shows the variation of rupture time against J_1/σ_1, where J_1 is the first invariant of the principal stresses, i.e., $J_1 = \sigma_1 + \sigma_2 + \sigma_3$. The parameter expresses the severity of stress triaxiality. In maximum principal stress controlled tests, creep lives have almost constant lives and the stress multiaxiality have almost no influence on rupture time.

3.2 CREEP CURVES IN BIAXIAL ANS TRIAXIAL TESTS

Figure 9 plots the creep curves in biaxial and triaxial creep tests. The x directional creep strain is nearly the same as the y directional creep strain in biaxial test. In triaxial test, almost no creep strain was accumulated in the three directions but small contraction occurred in the gage part. The contraction of the gage part agrees with the finite element results.

Figure 10 shows Mises strain accumulation in biaxial and triaxial tests. A commonly seen creep curve was found in biaxial test consisting of primary, secondary and tertial stages. Fracture strain in biaxial test was around 15% that is smaller than that in uniaxial test. Creep ductility may be reduced by the stress biaxiality. Creep strain in triaxial test is nearly zero. The experimental results shown in Fig.10 clearly prove that no Mises strain occurred under zero Mises stress. Since the creep curve in biaxial stress state at $\sigma_{eq} = \sigma_1 = 127.4 MPa$ precisely agrees with the uniaxial creep curves in Fig.10, the creep deformation is determined by the Mises potential, considering that no creep strain arises under zero Mises stress. This consideration is consistent with the classical theory of creep deformation.

Figure 11 relates the minimum Mises strain rate and rupture time in uniaxial, biaxial and miniature tests. The relationship of the three tests is uniquely correlated by a line with unity slope. The equation shown by a dashed line can be expressed by the following equation.

$$t_r \cdot \dot{\varepsilon}_{eq} = 11.1 \tag{1}$$

It should be noted that the data of the miniature specimens triaxially predamaged are on

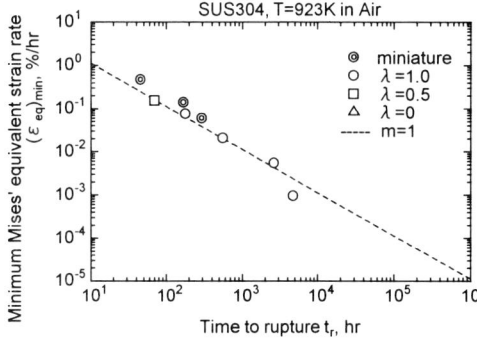

Figure 11. Relationship between the minimum Mises equivalent strain rate and time to rupture.

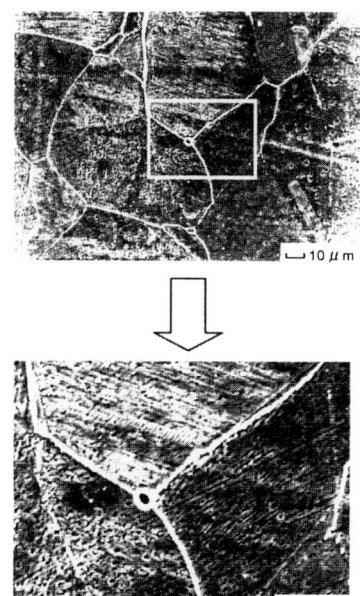

Figure 12. SEM photographs of triaxially damaged specimen at $\sigma_1 = \sigma_2 = \sigma_3 = 127.4$ $1MPa$ for 1500 hrs.

the line of no predamaged specimens.   This means that the minimum strain rate-rupture time relationship of the damaged specimen agrees with that of nondamaged specimens.

3.3 MICRO STRUCTURE OBSERVATIONS OF TRIAXIALLY DAMGED SPECIMEN

Figure 12 is the photographs of the triaxially crept specimen for at $\sigma_1 = \sigma_2 = \sigma_3 = 127.4 MPa$ for 1500 hrs.   Clear voids formed at grain boundaries specially at the triple junction.   The void formation at grain boundaries indicates that the grain boundary damage progressed under zero Mises stressing condition.

Considering the void observations and the results shown in Figs.10 and 12, the following two creep damage mechanisms are concluded to occur.   One is the deformation controlled creep fracture mechanism in which the case is high stress, short fracture time and transgranular fracture.   In this case, the relationship shown in eq. (1) holds as shown in Fig.11, so the rupture time is determined by the minimum Mises strain rate.   Since the minimum Mises strain rate is controlled by Mises stress as shown in Fig.10, the Mises stress expresses well the accumulation of creep damage in the case of the deformation controlled creep fracture, where the mean stress has almost no influence on creep fracture.

Second damage mechanism is the grain boundary void formation.   The results in Fig.12 indicates that the creep damage, grain boundary void formation, progresses under zero Mises stress, where zero Mises stress does not mean no stresses are applied but the three equal principal stresses are applied. This fracture mechanism is not deformation controlled but is a brittle fracture by the grain boundary void formation. If this fracture occurs, the minimum strain rate-rupture time relationship deviates significantly from that represented by the dotted line shown in Fig.11. In this case, mean stress has a significant effect on the creep fracture. This fracture type may occur in long term and low stressing condition, which corresponds with the actual damage condition of practical plants. A new equivalent stress that correctly expresses the damage accumulation in this case is still an open question and must be developed.

4. Conclusions

1. Triaxial creep test method was developed utilizing a miniature uniaxial creep testing.
2. The maximum principal stress was a suitable parameter for assessing biaxial and triaxial creep damage. However, further research must be made to examine the σ_2 and σ_3 effect on multiaxial creep damage.
3. Mises equivalent stress was not a suitable parameter for assessing multiaxial creep damage but was suitable for describing creep deformation.
4. Creep voids were observed under triaxial creep loading, which indicates that the creep damage was progressing under zero Mises stress.
5. Two creep fracture mechanisms were discussed. One was the deformation controlled creep fracture where Mises stress was effective for the damage assessment while the other was the void controlled fracture at grain boundaries where the mean stress may play a dominant role.

5. References

1. Mukai, S. et al., (1996) Development of Multiaxial Creep Testing Machine Using Cruciform Specimen, *J. Society Material Science Japan* **45**, 559-565.
2. Hayhurst, D. R. and Felce, D. (1986) Creep Rupture Under Tri-axial Tension, *Engineering Fracture Mechanics* **25**, 645-664.
3. Huddleston, R. L. (1985) An Improved Multiaxial Creep-Rupture Strength Criterion, *Trans. ASME, J. Pressure Vessel Technology* **107**, 421-429.
4. Mukai, S. et al., (1997) Development of Triaxial Tension Creep Test Machine, *J. Society Material Science Japan* **46**, 1374-1380.
5. NRIM Creep Data Sheet, (1986) No.4B, National Research Institute for Metals, Japan.

UNIAXIAL/MULTIAXIAL CREEP-RATCHETTING OF SEVERAL TYPES OF STEELS AND ITS CONSTITUTIVE MODELLING

Fusahito YOSHIDA
Department of Mechanical Engineering, Hiroshima University
1-4-1, Kagamiyama, Higashi-Hiroshima, 739-8527, Japan

1. Introduction

In the last two decades, ratchetting (or cyclic creep, i.e. cyclic stress-induced strain accumulation) has been intensively studied, and several interesting phenomena have been reported, such as the effects of maximum stress and mean stress in uniaxial ratchetting [1-6]; the effect of the primary stress and its variation in multiaxial ratchetting [6-12]; biaxial strain accumulation [13-17]; time-dependent behavior and creep effect [2, 7, 10, 15-17]; and the effects of temperature and its variation on the material behavior [10, 15-18]. If we look at carefully these experimental data, we can find that the ratchetting behavior strongly depends on types of materials. About this point, Hassan & Kyriakides [19] mentioned that cyclic hardening/softening nature has a significant effect on uniaxial/multiaxial ratchetting, and they also presented constitutive models to describe the different ratchetting behavior appearing for several types of steels. However, they restricted the discussion within the framework of rate-independent plasticity, even though the viscoplasticity of materials (rate-dependent ratchetting or creep effect) sometimes plays an important role in ratchetting.

This paper demonstrates how different deformation characteristics appear in uniaxial and multiaxial ratchetting for various types of steels. This includes a dicussion of creep-ratchetting interaction appearing at elevated temperature. Moreover, constitutive models of cyclic plasticity and viscoplasticity are presented that describe the difference in ratchetting behavior caused by the materials' cyclic hardening/softening (or non-hardening) natures, as well as their plastic (rate-independent) or viscoplastic (rate-dependent) deformation characteristics.

2. Experimental Observations

2.1 UNIAXIAL RATCHETTING

The ratchetting behavior is strongly affected by the cyclic hardening/softening nature of materials. As an example, Fig. 1 shows experimental data on the ratchetting strain accumulation vs. number of stress cycles registered in stress-controlled uniaxial ratchetting experiments on (a) 304 SS [7, 17] and 316 SS (cyclic hardening materials) tested at

S. Murakami and N. Ohno (eds.), IUTAM Symposium on Creep in Structures, 419–428.

room temperature (RT) and at 650 °C; (b) cyclic stabilized CS 1020 and CS 1026 steel (cyclic non-hardening [stabilized] materials) at room temperature [6]; and (c) virgin CS 1020 steel (a cyclic softening material) at room temperature [6]. For 304 SS and 316 SS, the rate of strain accumulation decreases with increasing number of stess cycles for the first about ten cycles and approaches its small steady value, in contrast, stabilized CS1020 and CS1026 exhibit almost constant rate of ratchetting in the whole stress cycles, and furthermore, virgin CS 1020 shows the acceleration of the rate of strain accumulation. These are the typical examples of the uniaxial ratchetting behavior for cyclic hardening, non-hardening (stabilized) and softening materials.

The uniaxial ratchetting is induced by two different material natures, i.e., a small strain difference in each stress cycle (rate-independent ratchetting), and viscoplasticity of materials (rate-dependent ratchetting or creep effect). It is worth noting that in the above cases of 304 SS and 316 SS, the viscoplasticity is mainly responsible for the ratchetting, rather than the rate-independent ratchetting, as seen in stress-strain response during the ratchetting (refer to [17]). In contrast to this, the stress-strain response of CS 1020 and CS 1026 show typical rate-independent ratchetting (refer to [6]). The decrease in the rate of

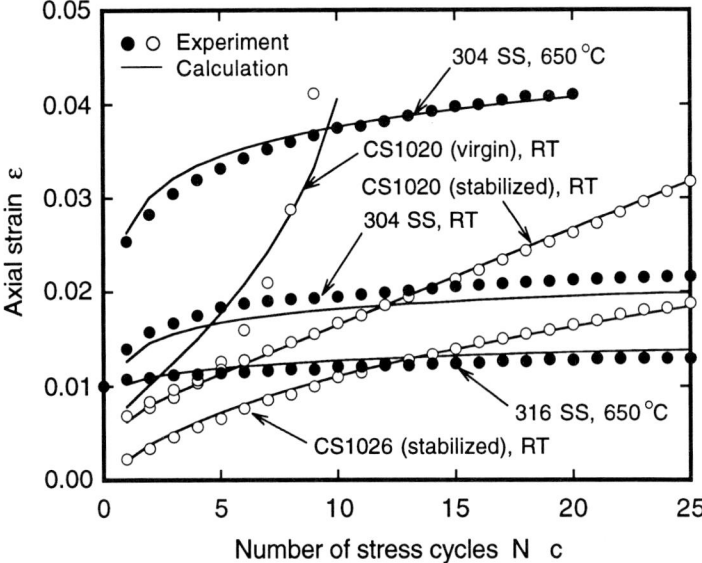

Fig. 1 Strain accumulations in uniaxial ratchetting for several types of steels under the conditions:
- 304 SS at 650 °C, σ_{max} = 150 MPa, σ_{min} = - 120 MPa, $\dot\sigma$ = 0.1 MPa s^{-1}
- 304 SS at RT, σ_{max} = 250 MPa, σ_{min} = - 100 MPa, $\dot\sigma$ = 30 MPa s^{-1};
- 316 SS at 650 °C, σ_{max} = 120 MPa, σ_{min} = - 96 MPa, $\dot\sigma$ = 0.5 MPa s^{-1};
- stabilized CS 1020 at RT, σ_{max} = 424 MPa, σ_{min} = - 297 MPa,
- stabilized CS 1026 at RT, σ_{max} = 269 MPa, σ_{min} = - 179 MPa,
- virgin CS 1020 at RT, σ_{max} = 536 MPa, σ_{min} = - 257 MPa.

ratchetting with increasing number of stress cycles appearing in the data on 304 SS and 316 SS is due to strain hardening induced during the ratchet-strain accumulation.

In order to investigate the effects of high-temperature creep, a ratchetting experiment with peak-stress hold was performed on 316 SS. Figure 2 shows the stress-strain response on 316 SS at 650°C during cyclic stressing with peak-stress hold (maximum stress σ_{max} = 120 MPa, minimum stress σ_{min} = - 96 MPa, stress rate $\dot{\sigma}$ = 0.5 MPa s^{-1}, 30 minute stress-hold at each stress cycle). Strain accumulation in the creep-ratchetting was found to be almost the same as that in the static creep of the same hold stress (120 MPa). The same phenomenon was reported in the creep-ratchetting of 304 SS at room temperature [5, 7] and at 650°C [17], as well as 316 FR stainless steel at 650°C [20].

2.2 MULTIAXIAL RATCHETTING

Figure 3 shows typical experimental data on the ratchetting strain accumulation vs. number of strain cycles registered in multiaxial ratchetting tests for a cyclic non-hardening steel (CS 1026 at RT [6]) and a cyclic hardening steels (304 SS tested at RT [7] and at 650°C [17]; and 316 SS at 300°C), respectively. These experiments were perfomed by subjecting thin-walled tubular specimens to combined axail load, internal pressure and torsion. The experimental conditions were the followings:
   - for CS 1026 at RT, steady internal pressure (circumferential stress σ_θ = 70 MPa) and cyclic axial straininig (strain range $\Delta\varepsilon_z$ = 0.01) were imposed, and the circumferential strain ε_θ accumulated,
   - for 304 SS at 650°C, steady axial load (axial stress σ_z = 70 MPa) and cyclic torsional straininig (shear strain range $\Delta\gamma_{z\theta} / \sqrt{3}$ = 0.015 at strain rate $\dot{\gamma}_{z\theta} / \sqrt{3}$ = 10^{-5} s^{-1}), and the axial strain ε_z accumulated,

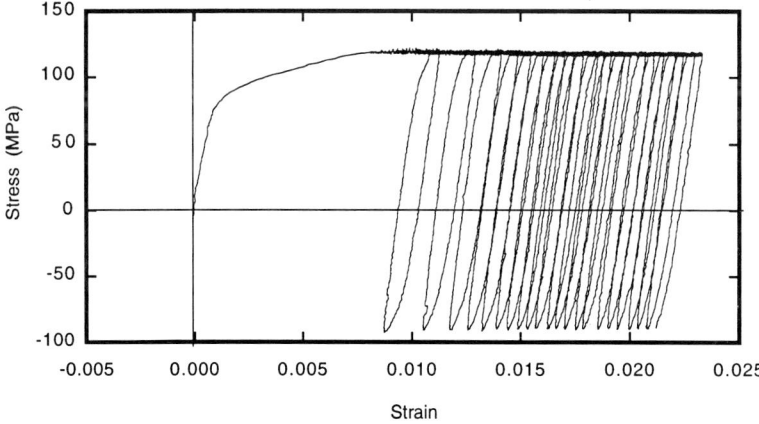

Fig. 2 Stress-strain response during uniaxial creep-ratchetting for 316 SS at 650 °C (σ_{max} = 120 MPa, σ_{min} = - 96 MPa, $\dot{\sigma}$ = 0.5 MPa s^{-1}, 30 minute stress-hold at each stress cycle).

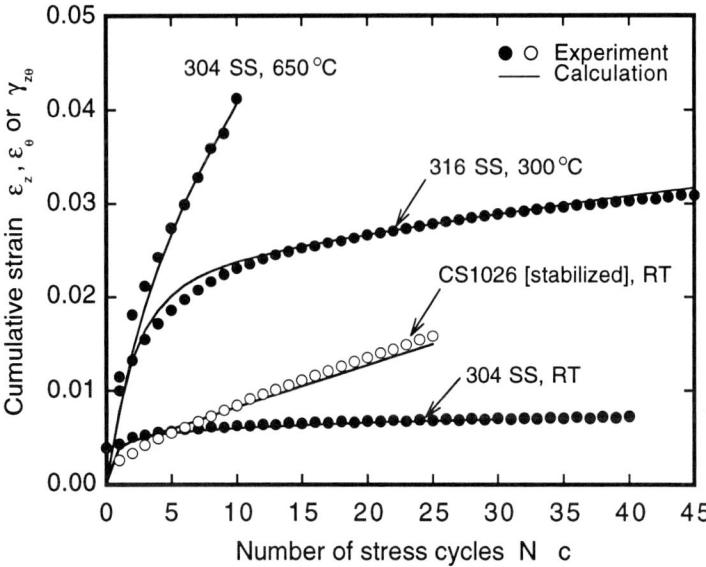

Fig. 3 Strain accumulations in multiaxial ratchetting for several types of steels.

- for 304 SS at RT, steady internal pressure (circumferential stress $\sigma_\theta = 40$ MPa) and cyclic axial straininig (strain range $\Delta\varepsilon_z = 0.01$ at strain rate $\dot{\varepsilon}_z = 10^{-3}$ s^{-1}), and the circumferential strain ε_θ accumulated,
- for 316 SS at 300°C, steady torsion (shear stress $\sqrt{3}\tau_{z\theta} = 50$ MPa) and cyclic axial straininig (axial strain range $\Delta\varepsilon_z = 0.01$ at strain rate $\dot{\varepsilon}_z = 5 \times 10^{-5}$ s^{-1}), and the shear strain $\gamma_{z\theta}$ accumulated.

From these figures, it is found that the multiaxial ratchetting behavior strongly depends on types of materials. The rate of ratchet-strain accumulation for 304 SS and 316 SS (cyclic hardening steels) decreases with increasing number of strain cycles for the first about ten cycles and approaches its small steady value, in contrast, for the cyclic non-hardening steels it remains almost constant in the whole strain cycles.

3. Constitutive Models

In order to describe the above-mentioned deformation characteristics in uniaxial/multiaxial ratchetting, constitutive models of cyclic plasticity and viscoplasticity are discussed in this section.

3.1 RATE-INDEPENDENT PLASTICITY MODEL

For a rate-independent constitutive model, the yield function and the associated flow rule is given by the equations:

$$f = \frac{3}{2}(s - \alpha) : (s - \alpha) - (Y_o + R)^2, \quad \dot{\varepsilon}^p = \frac{\partial f}{\partial s}\dot{\lambda},$$ (1)

where s and α denote the stress deviator and the back stress deviator, and Y_o and R stand for the initial radius of the yield surface and the isotropic hardening stress, respectively.

The description of ratchetting in terms of constitutive models is mainly related to the kinematic hardening rules. Chaboche-Rosselier [27] has given the back stress evolution equation as a summation of M components of the back stress of the Armstrong-Frederick [21] type, α_i $(i=1, 2, ..., M)$ as:

$$\alpha = \sum_{i=1}^{M} \alpha_i, \quad \dot{\alpha}_i = C_i\left(\frac{2}{3}a_i\dot{\varepsilon}^p - \alpha_i\dot{\bar{\varepsilon}}\right),$$ (2)

where $\dot{\varepsilon}^p$ and $\dot{\bar{\varepsilon}}$ are the plastic strain rate and the effective plastic strain rate, respectively, and C_i and a_i are material constants. Hereafter, we call the above nonlinear kinematic hardening rule 'NLK-rule'.

It is already well known that the use of the NLK rule leads to the overprediction of the uniaxial/multiaxial ratchetting strain accumulation because the recall (or the dynamic recovery) term is too active. Therefore, in addition to the NLK rule, in the present paper is used the nonlinear kinematic hardening rule with a threshold ('NLK-T' rule) which has been proposed by Chaboche [22] as:

$$\dot{\alpha}_T = C_T\left(\frac{2}{3}a_T\dot{\varepsilon}^p - \left\langle 1 - \frac{\alpha_{th}}{\overline{\alpha}_T}\right\rangle\alpha_T\dot{\bar{\varepsilon}}\right), \quad \overline{\alpha}_T = \sqrt{\frac{3}{2}\alpha_T : \alpha_T},$$ (3)

where $<>$ is the MacCauley bracket, and α_{th} is the threshold for the back stress α_T. In this equation, the recall term is active only when $\overline{\alpha}_T > \alpha_{th}$. Ohno-Wang [23, 24] also proposed another type of threshold for the back stress evolution, which may be used instead of Eq. (3).

Moreover, in order to improve the description of multiaxial ratchetting, Burlet-Cailletaud [25] has proposed an evolution equation with the radial evanecence of the recall term in the back stress evolution ('NLK-RE' rule) as:

$$\dot{\alpha}_{RE} = C_{RE}\left[\frac{2}{3}a_{RE} - (\alpha_{RE} : n)\right]\dot{\varepsilon}^p,$$ (4)

where n is the outward normal to the yield surface at the current stress point. This model shows exactly the same stress-strain response as the NLK model for uniaxial ratchetting, but it predicts the dramatic decrease of ratchetting rate with increasing number of cycles for the case of multiaxial ratchetting. One of the advantages in using this model, combined with the NLK and NLK-T, is that we can have an additional degree of freedom to describe the multiaxial ratchetting behavior independently of the uniaxial ratchetting.

In the present paper, the back stress evolution equations are given as a summation of the above-mentioned three components of NLK, a NLK-T and a NLK-RE. The details of the equations will be discussed in the next section.

The isotropic hardening is described by the evolution of the non-isotropic hardening region of which concept has been presented by Chaboche *et al.* [26] and generalized by Ohno [27]. The cyclic softening was described by the contraction of yield surface ($\dot{R}$ <0) during the cyclic plastic deformation.

3.2 VISCOPLASTICITY MODEL

A viscoplasticity constitutive model based on the overstress concept are given by the equations:

$$\dot{\varepsilon}^p = \frac{3(s - \alpha)}{2\bar{\sigma}}\dot{\bar{\varepsilon}}, \qquad \dot{\bar{\varepsilon}} = \left\langle \frac{\bar{\sigma} - (Y_o + R)}{D} \right\rangle^n,$$

$$\bar{\sigma} = \sqrt{\frac{3}{2}(s - \alpha):(s - \alpha)},$$

(5)

where D stands for the drag stress and n the stress-rate sensitivity exponent. The temperature dependence of the flow stress is expressed such a way that the back stress α and the radius of the yield surface, $(Y_o + R)$, are functions of temperature T.

4. Comparison between Experimental and Predicted Results

The performance of the above-mentioned constitutive models in predicting the uniaxial/multiaxial ratchetting is evaluated by comparing the predictions with the corresponding experimental results.

4.1 UNIAXIAL RATCHETTING

For uniaxial ratchetting on virgin CS 1020, stabilized CS 1020 and CS 1026, the rate-independent plasticity model was employed, where the back stress evolution was given as a summation of three NLK components, together with a NLK-T rule ($\alpha = \alpha_1 + \alpha_2 + \alpha_3 + \alpha_T$). In this evolution equation, the NLK-back stress component α_2 with a large value of C_2 was introduced to describe the smooth change in the tangent modulus $d\sigma/d\varepsilon$ of the stress-strain curve in the vicinity of the reyielding point, and α_3 with a small value of C_3 for the description of small value of steady ratchetting rate. The isotropic softening was taken into account for the virgin CS 1020, but not for CS 1026. For 304 SS and 316 SS, the isotropic hardening was introduced into the model. For the numerical simulation of uniaxial ratchetting on 304 SS and 316 SS, the viscoplasticity models were employed. The back stress evolution for the viscoplasticity models was expressed by the same rules as those for the rate-independent plasticity models.

In Fig. 1 is illustrated the predictions of uniaxial ratchetting for several types of steels, together with the corresponding experimental results. It is worth noting that the above results of numerical simulation well describe the difference in ratchetting behavior appearing between these steels, i.e., the rate of strain accumulation decreases with increasing number of stess cycles in the first about ten cycles and approaches its small steady value for 304 SS and 316 SS (cyclic hardening materials), in contrast, the rate of ratchetting

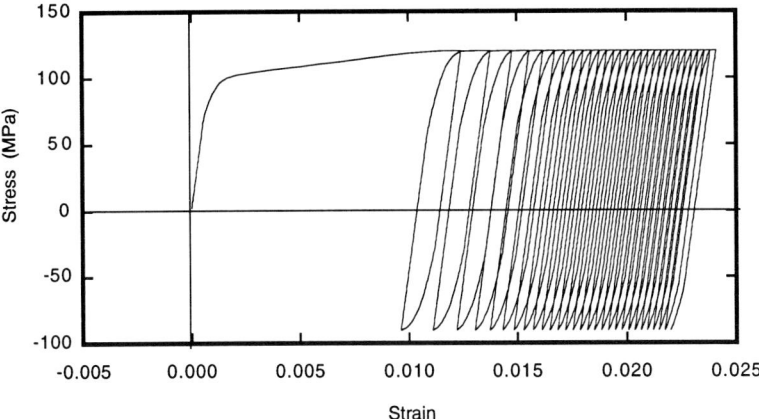

Fig. 4 Stress-strain response for 316 SS at 650°C during uniaxial creep-ratchetting calculated by the viscoplasticity model (the corresponding experimental result is shown in Fig. 2).

remains almost constant in the whole cycles for stabilized CS 1020 and CS 1026 (cyclic non-hardening materials). Furthermore, the use of this model with the isotropic softening allow us to predict the accelerationof the rate of ratchetting for a cyclic softening material, as shown in Fig. 1 for virgin CS 1020 steel. The result of numerical simulation of stress-strain response in the ratchetting with peak-stress hold, which corresponds to the experimental result shown in Fig. 2, is illustrated in Fig. 4. The calculated result well reproduces the creep-ratchetting behavior.

4.2. MULTIAXIAL RATCHETTING

Numerical simulation of multiaxial ratchetting was performed using viscoplasticity models for the ratchetting on 304 SS and 316 SS; and a rate-independent plasticity model for the ratchetting on CS 1020 and CS1026.

Two different types of back stress evolution equations, the following *models A and B*, consisting of a NLK-RE rule superposed on NLK components as well as a NLK-T rule, are evaluated:

$$(Model\ A) \qquad \alpha = \mu_A \alpha_1 + (1 - \mu_A)\alpha_{RE} + \alpha_2 + \alpha_3 + \alpha_T, \qquad (6)$$

where $a_{RE} = a_1$ and $C_{RE} = C_1$, and

$$(Model\ B) \qquad \alpha = \alpha_1^* + \alpha_2 + \alpha_3 + \alpha_T,$$

$$\dot{\alpha}_1^* = C_1 \left[\frac{2}{3} a_1 \dot{\varepsilon}^p - \mu_B \alpha_1^* \dot{\bar{\varepsilon}} - \sqrt{\frac{2}{3}}(1 - \mu_B)(\alpha_1^* : n)\dot{\varepsilon}^p \right], \qquad (7)$$

where, μ_A and μ_B are material parameters controlling the contribution of the NLK-RE rule. The *model A* is identical with the Burlet-Cailletaud model [25] , and *model B* is suggested by Chaboche [4].

In Fig. 3 are illustrated the predicted results of strain accumulation vs. number of strain cycles in the multiaxial ratchetting, together with the corresponding experimental results. For the simulation, the *model A* was used for the ratchetting on 304 SS and 316 SS, and *model B* for CS1026, respectively. The predictions are in good agreement with the experimental results.

5. Concluding Remarks

The deformation characteristics of uniaxial/multiaxial creep-ratchetting for several types of steels observed in the experiments are summarized as follows:

- *Uniaxial ratchetting* — for cyclic hardening materials (304 SS and 316 SS), the rate of strain accumulation decreases with increasing number of stess cycles for the first about ten cycles and approached its small steady value, in contrast, cyclic non-hardening (stabilized) materials (stabilized CS1020 and CS1026 steels) exhibit almost steady rate of ratchetting in whole stress cycles. The acceleration of ratchet-strain accumulation is found for a cyclic softening material (virgin CS1020 steel) .
- *Multiaxial ratchetting* — the behavior is quite similar to the uniaxial case, i.e., for cyclic hardening steels (304 SS and 316 SS), the rate of strain accumulation decreases with increasing number of strain cycles for the first about ten cycles and approaches its small steady value, in contrast, cyclic non-hardening steels (CS1020 and CS1026 steels) show almost steady rate of ratchetting in the whole strain cycles.
- *Creep effect* — some of these steels, i.e., 304 SS and 316 SS at RT and elevated temperatures, the ratchetting is mainly caused by the viscoplasticity of the materials, while the carbon steels (CS1020 and CS1026) show rate-independent ratchetting. In the uniaxial ratchetting on 316 SS with peak-stress hold at 650 °C, the strain accumulation was found to be almost the same as that in the static creep of the same hold stress.

The above-mentioned ratchetting behavior is well predicted if we make an appropriate choice of constitutive models of cyclic plasticity and viscoplasticity. The performance of the models in predicting the ratchetting is indicated as follows:

- The rate-dependent plasticity model with NLK+NLK-T can predict quantitatively well the steady rate of uniaxial ratchetting.
- The viscoplasticity NLK+NLK-T model satisfactorily calculate the viscoplastic uniaxail ratchetting including the creep effect. Furthermore, if the model is combined with the evolution of isotropic hardening, it allows us to describe the decrease in the rate of ratchetting with increasing number of cycles for the first about ten cycles.
- Two types of NLK+NLK-RE models, *models A* and *B*, were evaluated. The *model A* describes the behavior of the rapid decrease of ratchetting rate appearing for the first about ten cycles followed by the small steady rate, but in contrast, the *model B*

predicts almost steady rate of ratchetting in the whole strain cycles. The predictions of multiaxial ratchetting by *models A* and *B* correspond to the ratchetting behavior of cyclic hardening and non-hardening materials, respectively.

It is thus concluded that the several combinations of the above nonlinear kinematic hardening rules together with the isotropic hardening/softening rule, as well as the expression of material's viscous effect (viscoplasticity) based on the overstress concept, have a high degree of freedom of describing both uniaxial and multiaxial ratchetting.

References

[1] Yoshida, F., Murata, K. and Shiratori, E., 1980, Constitutive Equation of Cyclic Creep under Increasing-Stress Condition, *Bulletin of JSME*, Vol. 23, pp. 337-344.

[2] Kujawski, D., Kallianpur, V. and Krempl, E., 1980, An Experimental Study of Uniaxial Creep, Cyclic Creep and Relaxation of AISI Type 304 Stainless Steel at Room Temperature, *Journal of Mechanics and Physics of Solids*, Vol. 28, pp. 129-148.

[3] Yoshida, F., Ohta, I. and Shiratori, E., 1981, Effect of Change of Maximum Stress and Stress Ratio on the Cyclic-Creep Behavior, *Bulletin of JSME*, Vol. 24, pp. 507-514.

[4] Chaboche, J. L. and Nouailhas, D., 1989, Constitutive Modeling of Ratchetting Effects, Part I and II, *ASME Journal of Engineering Materials Technology*, Vol. 111, 384-392.

[5] Yoshida, F., Kondo, J. and Kikuchi, Y., 1989, Viscoplastic Behavior of SUS304 Stainless Steel under Cyclic Loading at Room Temperature, *JSME International Journal*, Ser. I, Vol. 32, 136-141.

[6] Hassan, T. Corona E. and Kyriakides, S., 1992, Ratchetting in Cyclic Plasticity, Part I: Uniaxial Behavior, *International Journal of Plasticity*, Vol. 8, pp. 91-116.

[7] Yoshida F.,1990, Uniaxial and Biaxial Creep-ratcheting Behavior of SUS304 Stainless Steel at Room Temperature, *Internatonal Journal of Pressure Vessels and Piping*, Vol. 44, pp. 207-223.

[8] Yoshida, F., Tajima, N., Ikegami, K. and Shiratori, E., 1978, Plastic Theory of Mechanical Ratcheting, *Bulletin of JSME*, Vol. 24, pp. 507-514.

[9] Shiratori, E., Ikegami, K. and Yoshida, F., 1979, Analysis of Stress-Strain Relations by Use of an Anisotropic Hardening Plastic Potential, *Journal of Mechanics and Physics of Solids*, **27**, 213-229.

[10] Inoue, T., Yoshida, F., Ohno, N., Kawai, M. Niitsu, Y. and Imatani, S., 1991, Evaluation of Inelastic Constitutive Models under Plasticity-Creep Interaction in Multiaxial Stress State., *Nuclear Engineering & Design*, Vol. 126, pp.1-11.

[11] Jiang, Y. and Sehitoglu, H., 1994, Cyclic Ratchetting of 1070 Steel under Multiaxial Stress State, *International Journal of Plasticity*, Vol. 11, pp. 579.

[12] Jiang, Y. and Sehitoglu, H., 1994, Multiaxial Cyclic Ratchetting under Multiple Step Loading, *International Journal of Plasticity*, Vol. 11, pp. 849-.

[13] Yoshida, F., Yamamoto, S., Itoh, M. and Ohmori, M., 1984, Biaxial Strain Accumulation in the Mechanical Ratchettng, Bulletin of JSME, Vol. 27, pp. 2100-2106.

[14] Hassan, T. Corona E. and Kyriakides, S., 1992, Ratchetting in Cyclic Plasticity, Part II: Multiaxial Behavior, *International Journal of Plasticity*, Vol. 8, pp. 117-146.

[15] Inoue, T., Yoshida, F., Ohno, N., Kawai, M. Niitsu, Y. and Imatani, S., 1994, Inelastic Stress-Strain Response of 2 1/4 Cr-1Mo Steel under Combined Tension-Torsion at 600 °C., *Nuclear Engineering & Design*, Vol. 150, pp. 107-118.

[16] Delobelle, P., Robinet, P. and Bocher, L., 1995, International Journal of Plasticity, Vol. 11, pp. 295-330.

[17] Yoshida, F., 1995, Ratchetting Behaviour of 304 Stainless Steel at 650 °C under Multiaxially Strain-Controlled and Uniaxially/Multiaxially Stress-Controlled Conditions, *European Journal of Mechanics, A/Solids*, Vol. 14, pp. 97-117.

[18] Ruggles, M. B. and Krempl, E., 1989, The Influence of Test Temperature on the Ratcheting Behavior of Type 304 Stainless Steel, *ASME Journal of Engineering Materials Technology*, Vol. 111, 378-383.

[19] Hassan, T. and Kyriakides, S., 1994, Ratcheting in Cyclically Hardening and Softening Materials: I. Uniaxial Behavior, II. Multiaxial Behavior, *International Journal of Plasticity*, Vol. 10, pp. 149-212.

[20] Yoshida, F., Kobayashi, M., Tsukimori, K., Uno, T., Fukuda, Y., Igari, T. and Inoue, T., Inelastic Analysis of Material Ratchetting of 316FR at Varing Temperature - Results of Benchmark Project (B) by JSMS, *SMiRT-14, Div. L*, Paper No. LW/3.

[21] Armstrong, P.J. and Frederick, C.O. 1966, A Mathematical Representation of the Multiaxial Bauschinger Effect., *GEGB Report* RD/B/N731, Berkeley Nuclear Laboratories.

[22] Chaboche, J. L., 1991, On Some Modifications of Kinematic Hardening to Improve the Description of Ratchetting Effects, *International Journal of Plasticity*, Vol. 7, pp. 661-678.

[23] Ohno, N. and Wang, J.-D., 1993, Kinematic Hardening Rules with Critical State of Dynamic Recovery, Part I: Formation and Basic Features for Ratchetting Behavior, Part II: Application to Experiments of Ratchetting Behavior, *International Journal of Plasticity*, Vol. 9, pp. 375-403.

[24] Ohno, N. and Wang, J.-D., 1994, Kinematic Hardening Rules for Simulation of Ratcheting Behavior, *European J. Mechanics, A/Solids* Vol. 13, pp.519-531.

[25] Burlet, H. and Cailletetaud, G., 1987, Modeling of Cyclic Plasticity in Finite Element Codes, in *Proceedings of 2nd Int. Conf. on Constitutive Laws for Engineering Materials, Theory and Application*, pp. 1157-1165.

[26] Chaboche, J. L., Dang-Van, K. and Cordier, G., 1979, Modelization of the Strain Memory Effect on the Cyclic Hardening of 316 Stainless Steels, *SMiRT-5, Div. L*, Paper No. L. 11/3.

[27] Ohno, N., 1982, A constitutive Model of Cyclic Plasticity with a Non-hardening Strain Range, *ASME Journal of Applied Mechanics*, 49, pp. 721-727.

TEMPERATURE MEASUREMENT AND LIFETIME PREDICTION OF A HIGH-PRESSURE TURBINE ROTOR

Y. SUGITA, N. SHINOHARA and K.SUGIYAMA
Chubu Electric Power Co., Inc.
20-1 Kitasekiyama Ohdaka Midori-ku Nagoya, Japan

1. Introduction

With an increase in number of fossil power plants operated for long term, it is becoming important to predict time to replace a high temperature component as precisely as possible. In a steam turbine, an analytical method is normally used to predict lifetime, since it's difficult to get creep specimens out of a part where damage is supposed to be the most severe. However, the analytical method has several problems.

The first one is that input data affect the predicted lifetime remarkably. For example, the lifetime may change by a factor of two with a 10°C difference in temperature. Although the temperature on the bore surface of a turbine rotor was already measured [1], the inner shell temperature has been referred to as the temperature on the outer surface, that is, the steam side surface, of the rotor, since it is difficult to measure it directly.

The second problem is how to take into account creep relaxation at a part where stress is concentrated. An elastic analysis without creep relaxation may underestimate the lifetime. Since a 10MPa difference in stress may change the predicted lifetime by a factor of two as well, it is necessary to calculate the amount of creep relaxation precisely.

The third problem is that there are few examples to compare the predicted lifetime with creep test results of actual turbine components.

In this study, a measurement of temperature on the outer surface of a high-pressure turbine rotor was performed during operation and the analytical method was improved by using results together with data in the paper [1]. This paper describes the results of a comparison of an analytically predicted lifetime with an experimental lifetime of two rotors.

2. Temperature Measurement of a Turbine Rotor during Operation

2.1. MEASUREMENT METHOD

The measurement was carried out with a GE type high-pressure turbine rotor of a

S. Murakami and N. Ohno (eds.), IUTAM Symposium on Creep in Structures, 429–438.

220MW unit. The unit had been in service since 1960 until 1997, and the measurement was performed during the last service period of the unit. The pressure and temperature of the main steam are 16.6MPa and 566°C, respectively.

Temperatures on the control-stage dovetail and the outer surface of wheel and rotor were measured as shown in Figure 1. Thermocouples and leads from the thermocouples were embedded in grooves machined on the blade dovetail and the surface of the wheel. The leads were then extended from the root of the wheel, through the tapered groove on the rotor surface, to the root of the thrust collar. Finally, they were connected to FM telemetry transmitters mounted in the groove that was machined on the rear face of the thrust collar. The inner shell temperature was also measured as a reference temperature. Details are shown in the paper [2].

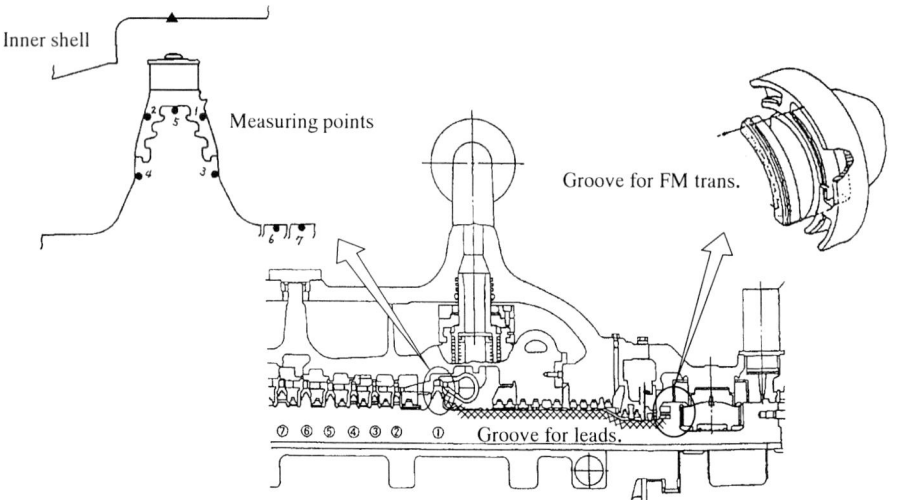

Figure 1.   Temperature measuring points and the method of signal transmission

2.2 MEASURMENT RESULTS

As shown in Figure 2, the temperature on the inlet side of the blade dovetail and the wheel is higher than that on the outlet side and that on the inner shell at the rated load of 220MW. The reason for this is considered as follows. A leakage flow with higher enthalpy from the nozzle exit is directed inward in a radial manner to the wheel root. This leakage flow warms the inlet side of the blade dovetail and the wheel, and is mixed with the flow with lower enthalpy from the blade outlet and is finally directed to the gland.

Practically, the inner shell temperature has been used as a reference temperature to evaluate the life of dovetail portion against creep rupture, and the softening of the material. From the measured results, the difference in temperature of the inner shell and

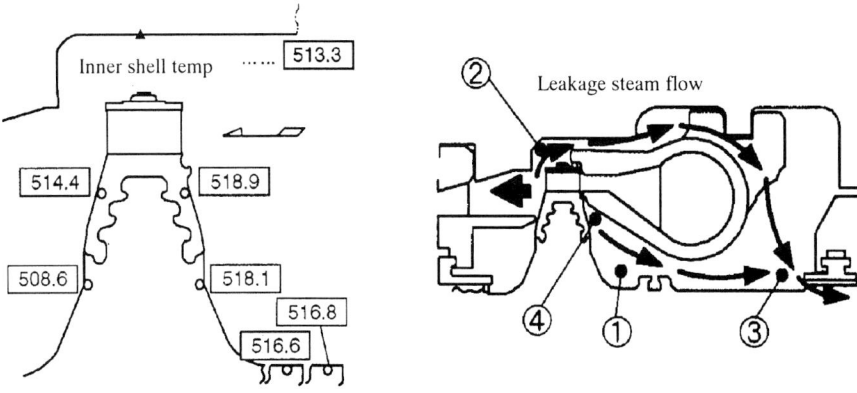

Figure 2.   Temperature distribution at the rated load of 220MW and leakage flow of steam

the blade dovetail is less than 7°C at every load. Therefore it is considered reasonable to use the temperature of the inner shell as a reference for evaluation.

3. Analytical Prediction of Lifetime

3.1 CONDITIONS

Two rotors were chosen for lifetime prediction. One is the rotor that is same as the rotor in chapter 2 (called as 220MW HP rotor here). The other is a high- and intermediate-pressure turbine rotor of a 375MW unit that is different from the rotor in the paper [1] but the same type of rotor. The high-pressure part of this rotor (called as 375MW HIP(HP) rotor here) was investigated. Steam conditions and operating hours of both rotors are shown in TABLE 1.

TABLE 1.   Steam conditions and operating hours of rotors

	220MW HP rotor	375MW HIP rotor
Pressure of main steam	16.6 MPa	16.6 MPa
Temperature of main steam	566 °C	566 °C
Total operating hours	183,000 hrs	200,000 hrs

The temperature and the stress were calculated by using FEM software, MARC. The mechanical model of the rotor was constructed with axis-symmetrical elements. Taking stress concentration into account, the part around the bore was divided into small elements. The temperature of the steam around the control stage was assumed to be a little bit higher than the inner shell temperature. At other stages, the temperature of steam was approximated by linear interpolation between the temperatures at the control stage and at the outlet of the HP turbine. The pressure distribution of the steam was also estimated in the same way as the temperature.

3.2 CALCULATED TEMPERATURE

Figure 3 shows the comparison of the calculated temperatures of the 220MW HP rotor during load up with the measurement data. The calculated temperature agrees well with the measured temperature on the control-stage outer surface. On the other hand, the temperature on the bore surface was verified by using the 375MW HIP(HP) rotor, since the temperature was already measured in this type of rotor [1]. Figure 4 shows the results. The measured data is quoted from the paper [1]. It is known that the analytical method can predict the temperature on the bore surface as well.

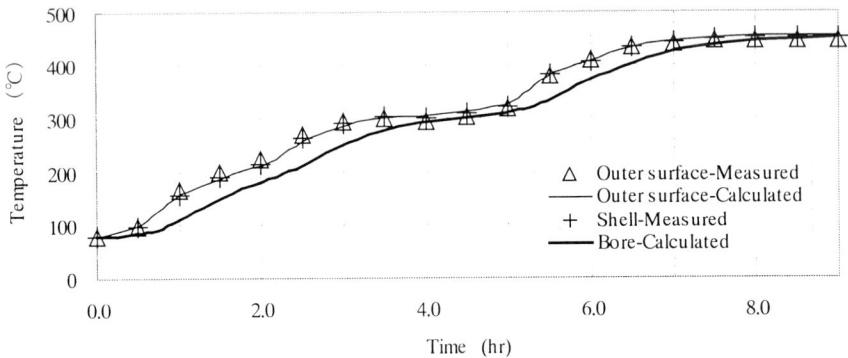

Figure 3. A comparison of the calculated temperatures with the measurement data in the 220MW HP rotor during load up

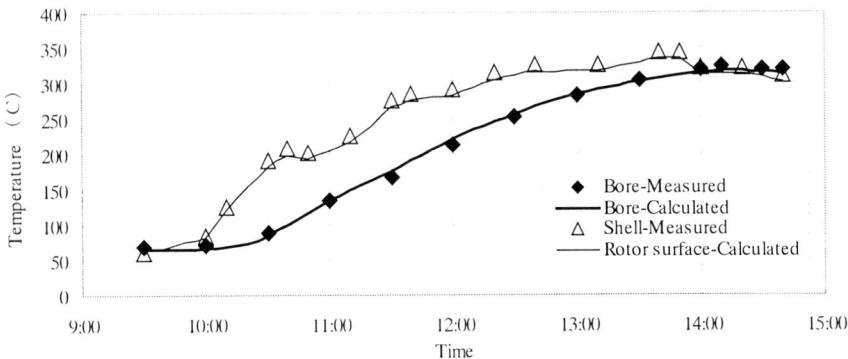

Figure 4. A comparison of the calculated temperatures with the measurement data [1] in the 375MW HIP(HP) rotor during load up

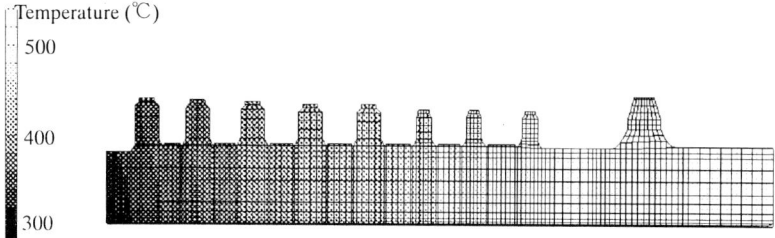

Figure 5. Temperature distribution of the HP rotor at the rated load of 220MW

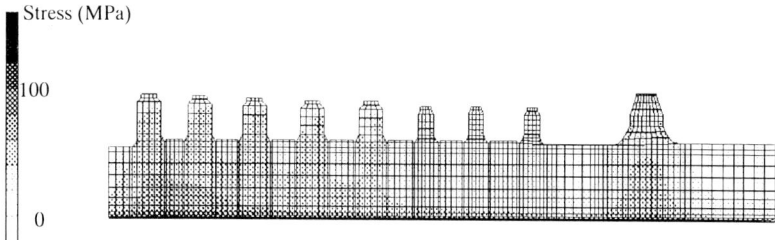

Figure 6. Elastic stress distribution of the HP rotor at the rated load of 220MW

Figure 5 shows the temperature distribution of the HP rotor at the rated load of 220MW. The difference in temperature is observed to occur mainly along the rotor axis. The temperature on the control-stage bore surface where creep damage is supposed to be the largest is slightly lower than that on the control-stage outer surface. The result of the 375MW HIP(HP) rotor is almost same as that of the 220MW HP rotor.

3.3 CALCULATED STRESS

Figure 6 shows the elastic stress distribution of the HP rotor at the rated load of 220MW, where the stress is Mises' equivalent stress. The thermal stress caused by the temperature distribution shown in Figure 5, the inertia stress due to the rotation of 3,600 rpm, and the stress caused by the steam pressure are included. It is found that the stress is high near the bore surface.

The creep analysis was performed to estimate stress relaxation during the long-term operation. Only the minimum creep strain rate was considered for the constitutive equation of creep. The minimum creep strain rate of the rotor material, the forged CrMoV steel, is published from National Research Institute for Metals (NRIM). Since all of them are obtained from accelerated tests, the minimum creep strain rate under the condition experienced in actual use must be extrapolated. In this study, the minimum creep strain rate was estimated from Monkman-Grant relation and the creep rupture data as follows.

Monkman-Grant relation is known to give the following equation between the minimum creep strain rate $\dot{\varepsilon}_c$ and the creep rupture hour t_f [3].

$$\dot{\varepsilon}_c \cdot t_f^{\ k} = C \tag{1}$$

Here C and k are constants. On the other hand, the creep rupture data are normally described as the relation between Larson-Millar parameter LM and the stress σ, which was simplified as a following quadratic equation in this study.

$$LM = T(\log t_f + 20) = l(\log \sigma)^2 + m(\log \sigma) + n \tag{2}$$

Here T is the absolute temperature, and l, m and n are constants. Equation (1) and (2) give the following equation.

$$\log(\dot{\varepsilon}_c) = \log C + k(20 - LM / T)$$

$$\dot{\varepsilon}_c = C \cdot 10^{k[20-\{l(\log \sigma)^2 + m(\log \sigma) + n\}/T]} \tag{3}$$

Equation (3) is used as the creep law.

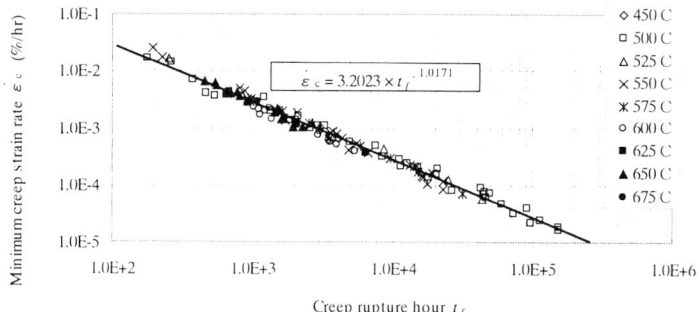

Figure 7. A relation between minimum creep strain rate and creep rupture time of forged CrMoV steel.

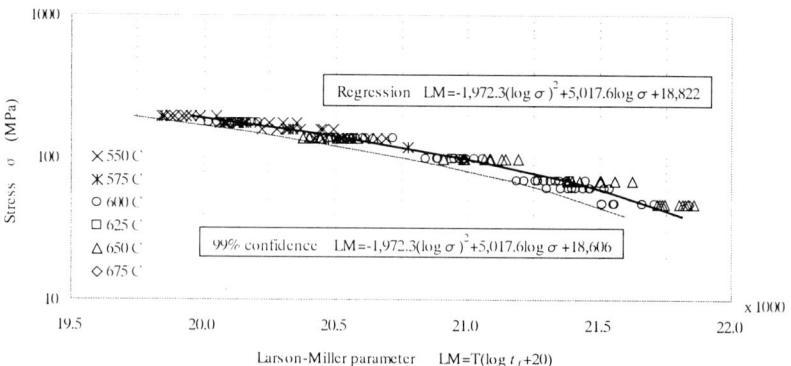

Figure 8. A relation between stress and Larson-Miller parameter of forged CrMoV steel

Figure 7 shows the relation between the minimum creep strain rate and the creep rupture time of the forged CrMoV steel, both of which are quoted from NRIM database [4]. As shown in Figure 7, Monkman-Grant relation is satisfied with constants C and k of 3.2023 and 1.0171 respectively.

Figure 8 shows the creep rupture properties of the forged CrMoV steel as the relation between the stress and Larson-Miller parameter *LM*. The values of *l*, *m* and *n* in equation (3) are respectively determined as −1,972.3, 5,017.6 and 18,822 from the regression curve that is obtained from the creep data under the lower stress condition experienced in actual use. The 99% confidence lower curve is used later for the life prediction.

Figure 9 shows the stress relaxation during total operating hours, 183,000hrs, near the control-stage bore surface of the 220MW HP rotor, while Figure 10 shows the variation of creep strain that causes the stress relaxation. In this study, the stress relaxation is assumed to continue during total operation, since the stress does not exceed the yield strength during a cycle from start to stop. As shown in Figure 9 and Figure 10, it is found that the stress decreases gradually from 92MPa to 66MPa with small creep strain of 0.04% that is much smaller than the ductility of the material.

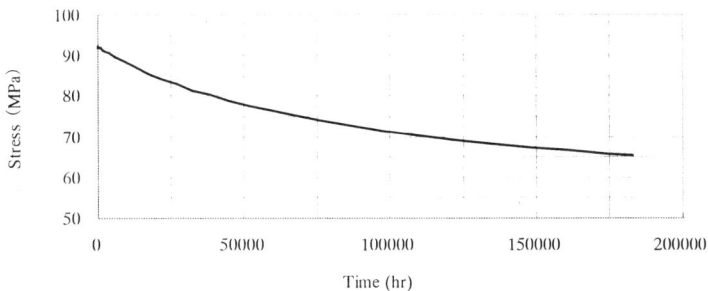

Figure 9.   Stress relaxation near the control-stage bore surface of the 220MW HP rotor

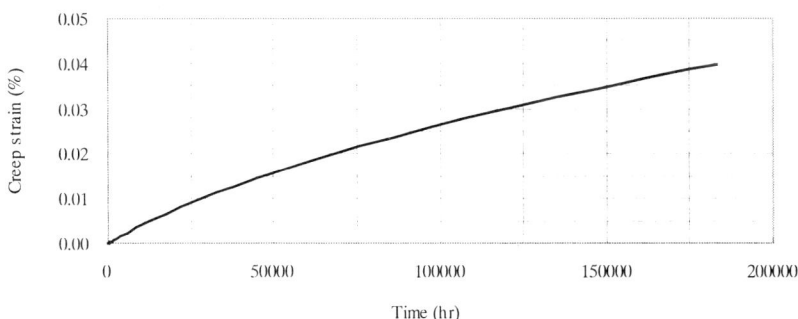

Figure 10.   Variation of creep strain near the control-stage bore surface of the 220MW HP rotor

3.4 LIFE PREDICTION

The creep rupture time changes sensitively with the magnitude of stress as shown in

Figure 8. In this study, the stress relaxation curve was divided into many steps with small duration of time Δt_i , and the creep rupture hour $(t_f)_i$ corresponding to the stress in each step was obtained from Figure 8. Following that, the creep damage fraction ϕ_c was calculated by the linear damage fraction rule.

$$\phi_c = \sum_i \frac{\Delta t_i}{(t_f)_i} \tag{4}$$

TABLE 2 shows the predicted creep damage fractions of the 220MW HP rotor and the 375MW HIP(HP) rotor during their long-term operations as shown in TABLE 1, where both the regression curve and the 99% confidence lower curve in Figure 8 were used as the creep rupture properties. The creep damage fraction predicted only by the elastic stress is also shown in this TABLE. The damage fraction was calculated on the control-stage bore surface and at the same stage mean radius point that is shown later in Figure 11. TABLE 2 indicates that the creep damage fraction is influenced significantly by the stress relaxation and the creep rupture curve. The higher the stress is, the larger the effect of the stress relaxation in the creep analysis is observed. For example, creep damage fraction decreases to one-fifth in the 375MW HIP(HP) rotor with the stress relaxation taken into account. If the 99% confidence lower curve is used instead of the regression curve, the fraction becomes almost double. Compared with the fraction at the mean radius point where the stress is low, the fraction on the bore surface is large. The difference is remarkable in the 375MW HIP(HP) rotor.

TABLE 2.   Predicted creep damage fraction of the 220MW HP rotor and 375MW HP(HP) rotor

Analytical method	Creep rupture curve	220MW HP rotor		375MW HIP(HP) rotor	
		bore surface	mean radius point	Bore surface	mean radius point
Creep Analysis	Regression	2 %	<1 %	8 %	3 %
	99% confidence	3 %	<1 %	15 %	6 %
Elastic Analysis	Regression	3 %	<1 %	39 %	4 %
	99% confidence	6 %	<1 %	73 %	7 %

Both the bore surface and the mean radius point are in the control-stage

4. Creep Tests

4.1 SPECIMEN

Figure 11 shows the locations in the 220MW HP rotor to get specimens out of. At first, two large blocks were cut out of both the control-stage and the journal, and then specimens were produced from the parts near the bore surface and the mean radius point of two blocks. The specimens from the journal were used to obtain the reference creep data, since that part was not under high temperature. The specimens from the 375MW HIP(HP) rotor were produced in the same way. The 6mm diameter of specimen was used to investigate the creep properties near the bore surface.

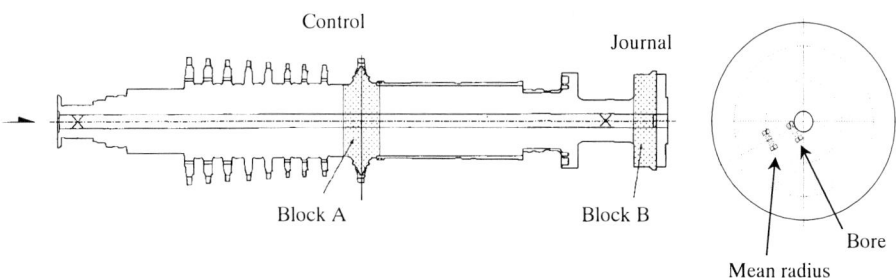

Figure 11.   Locations of specimens

4.2 TEST RESULTS

As shown in Figure 12, the creep strength of the control-stage is lower than that of the journal that is almost on the regression curve for virgin materials. This difference becomes small with the decrease of stress. The reason is considered to be that the deterioration by heating is accumulated even in virgin materials during creep tests under the condition of low stress, high temperature and long rupture time, which is the same condition as experienced in actual use. Accordingly the influence of the heat deterioration appears largely under the test condition of high stress. Although a little difference between creep rupture data of the rotors and the virgin materials is observed even under the condition of low stress, there is no significant difference between creep strengths near the control-stage bore surface and at the same stage mean radius point. Consequently, it is found that both rotors were deteriorated by long term heating, while the creep damages were small.

5. Discussion

The result predicted by creep analysis in TABLE 2 shows a good approximation of the experimental result that the creep damage of both rotors was small. In this study, only the minimum creep rate was used as the constitutive equation of creep. If the transient creep is added, the stress relaxation will become more rapid and large. Therefore, the calculation based on the minimum strain rate gives the conservative prediction of lifetime, while the elastic analysis is too conservative.

Although the creep damage was small, the deterioration by long term heating was observed. Then, the lifetime prediction based on the creep rupture curve of long-term heated materials is recommended.

In this study, the material of the rotor is assumed to be homogeneous. If large inclusions or impurities exist around the bore surface, damage concentration due to the different relation between stress and strain must be considered.

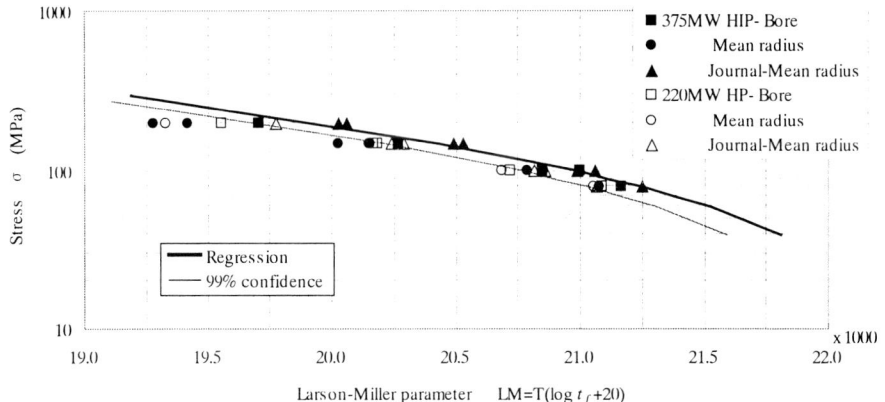

Figure 12.   Results of creep rupture tests

6. Conclusions

In this study, the measurement of temperature on the outer surface of a high-pressure turbine rotor, the improvement of the analytical life prediction, and the verification by creep tests of specimens from two rotors were performed. The results are as follows.

(1) It is reasonable to use the temperature of the inner shell as a reference temperature for the evaluation of that on the control-stage outer surface of a rotor.
(2) The creep analysis with the minimum creep strain rate shows a good approximation of the creep test results, while the elastic analysis is too conservative.
(3) Both the 220MW HP rotor and the 375MW HIP rotor were deteriorated by long term heating, but their creep damage was small.

Acknowledgement

The measurement of temperature was performed in cooperation with Toshiba Corporation. The authors wish to thank Mr. Toshihiro Matsuura, Mr. Hiroshi Uchida and coworkers in Toshiba Corporation for the achievement of this difficult work. The cooperation of personnel in Shin-Nagoya power station is also greatly appreciated.

References

1. Nakahira, T.; Non-steady thermal stress of a steam turbine rotor, J. JSME **74** (1971), 402-409 in Japanese.
2. Matsuura, T. et.al., Vibration and temperature measurement of a high-pressure steam turbine control-stage blade, Proc. 1999 Int. Joint Power Gen. Conf., **2** (1999) 573-580
3. Viswanathan, R.; Damage Mechanisms and Life Assessment of High-temperature Components, ASM International, Ohio,   (1989)
4. National Research Institute for Metals, NRIM creep data sheet, No.9B, (1990), in Japanese

EFFECTS OF MATRIX CREEP ON FIBER STRESS PROFILES IN UNIDIRECTIONAL COMPOSITES: MESOSCOPIC ANALYSIS BASED ON A VARIATIONAL METHOD

N. OHNO[1], T. ANDO[1], T. MIYAKE[2] and S. BIWA[1]
[1]*Department of Micro System Engineering, Nagoya University,*
Furo-cho, Chikusa-ku, Nagoya 464-8603, Japan
[2]*Nagoya Municipal Industrial Research Institute,*
Rokuban, Atsuta-ku, Nagoya 456-0058, Japan

1. Introduction

Unidirectional composites reinforced with long brittle fibers may suffer from fiber breaks. Fiber stress in broken fibers changes from zero to that of intact fibers as axial distance increases from each fiber break. The distance for such a transient change of fiber stress is called the stress recovery, or transfer, length and is important for evaluating the longitudinal tensile strength of composites.

Matrix creep may occur in metal matrix and polymer matrix composites. Then, the profiles of fiber stress in broken fibers change with time, even if stress applied to composites is constant This time-dependent change causes the stress transfer length to increase with time, so that the carrying load of broken fibers reduces; then, since the carrying load of intact fibers increases, further fiber breaks may occur to induce eventually the creep rupture of composites.

Hence, the effect of matrix creep on the profiles of fiber stress is a fundamental subject, but analytical solutions have been obtained only in a few cases: Lifshitz and Rotem (1970) derived approximately a linear viscoelastic solution for an axisymmetric cell containing a broken fiber by assuming the perfect bonding at fiber/matrix interface and by utilizing the Laplace transformation. Mason et al. (1992) succeeded in getting an exact solution for a 2D plate model by assuming the power-law creep of matrix and the perfect bonding at interface, though they ignored elastic strain of matrix. Iyengar and Curtin (1997) estimated analytically the time-dependent extension of the stress transfer length on the basis of the numerical work of Du and McMeeking (1995), in which an axisymmetric model was analyzed by taking account of interfacial initial slip in addition to nonlinear matrix creep.

The solutions mentioned above were obtained by solving differential equations based on the shear lag assumption. It is, however, possible to utilize a variational method if a functional is known for the problem considered. Variational methods are effective especially if the forms of solutions can be assumed appropriately. The methods enable us to obtain approximate solutions most accurately within the

S. Murakami and N. Ohno (eds.), IUTAM Symposium on Creep in Structures, 439–452.

assumed forms of solutions, since the unknown coefficients in the assumed forms of solutions are determined rationally by getting the functionals stationary. Variational methods, however, have not been utilized for elastic-creep problems of unidirectional fiber composites so far. Incidentally, Ohno and Miyake (1999) considered an energy approach to evaluate analytically the time-dependent change of fiber stress profiles in broken fibers.

In this work, the time-dependent increase of a stress transfer length is evaluated analytically by developing a variational method for two kinds of axisymmetric models and by taking account of both elastic and creep strains of matrix, in which an elastic fiber is embedded. First, a functional based on incremental complementary energy is demonstrated for the axisymmetric models, and it is shown that the functional has a stationary function, which is the solution of a shear lag differential equation. Second, by supposing bilinear profiles of fiber stress and the power-law creep of matrix, an approximate solution is derived for the time-dependent increase of the stress transfer length. Third, the solution is compared with finite-difference computations of the shear lag equation. Finally, the present solution is applied to one of the experiments done by Miyake et al. (1998, 2000), in which the time-dependent change of fiber stress profiles in single fiber model composites was measured using laser Raman spectroscopy.

2.   Axisymmetric Shear Lag Models

In the present work, we consider two kinds of shear lag models, i.e., a fiber breakage model and a fiber pull-out model, which are axisymmetric. To begin with, we describe the models briefly.

In the fiber breakage model, a broken fiber of a radius r_f is embedded in matrix in a cell, which has an outer radius R and an axial length 2ℓ (Fig. 1). Let us denote radial and axial coordinates by r and z, respectively. We suppose that the cell is subjected to uniform, axial strain $\varepsilon_0(t)$ at the outer peripheral and at the axial ends:

$$\partial u/\partial z = \varepsilon_0(t) \quad \text{at} \quad r = R, \tag{1}$$

$$\partial u/\partial z = \varepsilon_0(t) \quad \text{at} \quad z = \pm\ell, \tag{2}$$

where t indicates time, and u axial displacement.

Employing the shear lag assumption (Cox, 1952), we regard the fiber as a 1D bar and consider only the shear deformation in matrix to be responsible for fiber stress profiles. Then, if Hooke's law is assumed for the broken fiber, we have

$$\frac{\partial u_f}{\partial z} = \frac{\sigma_f}{E_f}, \tag{3}$$

where u_f and σ_f denote the axial displacement and axial stress in fibers, respectively, and E_f Young's modulus of fibers. The fiber stress σ_f is related with the shear stress acting on the fiber/matrix interface, τ_i, as

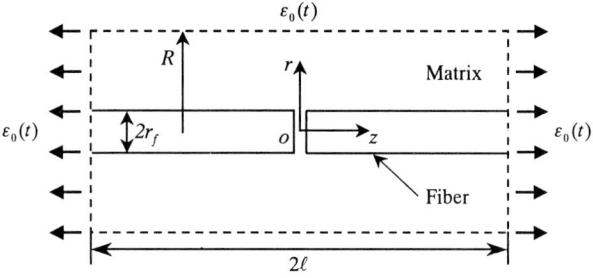

Figure 1. Axisymmetric cell consisting of broken fiber and elastic-creeping matrix subjected to overall strain $\varepsilon_0(t)$.

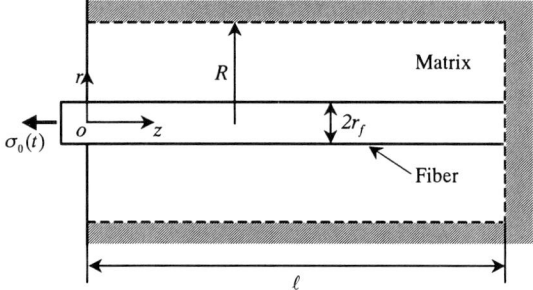

Figure 2. Axisymmetric model for pull-out of fiber embedded in elastic-creeping matrix.

$$\tau_i = -\frac{r_f}{2}\frac{\partial \sigma_f}{\partial z}. \tag{4}$$

If the matrix exhibits creep in addition to elastic deformation, the shear strain rate in matrix, $\dot{\gamma}_m$, has an expression

$$\dot{\gamma}_m = \frac{\dot{\tau}_m}{G_m} + \dot{\gamma}_m^c, \tag{5}$$

where τ_m and $\dot{\gamma}_m^c$ represent the shear stress and shear creep rate in matrix, respectively, and G_m the shear rigidity of matrix. The superposed dot stands for the differentiation with respect to time t. In the model, $\dot{\gamma}_m$ may satisfy accurately $\dot{\gamma}_m = \partial \dot{u}/\partial r$, so that using the above equation we can have

$$\dot{u}_R - \dot{u}_f = \int_{r_f}^{R}(\dot{\tau}_m/G_m + \dot{\gamma}_m^c)\,dr, \tag{6}$$

where u_R denotes u at $r = R$.

In the present work, since the difference between R and r_f cannot be small, we take into account the radial variation of τ_m in matrix by assuming the following self-equilibrium equation of τ_m in matrix (Clyne and Withers, 1993; Li and Grubb, 1994):

$$r\tau_m = r_f\tau_i .$$ (7)

Then, (6) becomes

$$\dot{u}_R - \dot{u}_f = \ln\frac{R}{r_f} \cdot \frac{r_f\dot{\tau}_i}{G_m} + \int_{r_f}^R \dot{\gamma}_m^c dr .$$ (8)

Differentiating the above equation with respect to z, and using (1) and (3), we derive

$$\frac{r_f^2}{2G_m}\ln\frac{R}{r_f} \cdot \frac{\partial^2\dot{\sigma}_f}{\partial z^2} - \frac{\dot{\sigma}_f}{E_f} - \int_{r_f}^R \frac{\partial\dot{\gamma}_m^c}{\partial z}dr + \dot{\varepsilon}_0(t) = 0 ,$$ (9)

which is a differential equation for $\dot{\sigma}_f(z,t)$ subjected to boundary conditions

$$\sigma_f(0,\ t) = 0 ,$$ (10)

$$\sigma_f(\pm\ell,\ t) = E_f\varepsilon_0(t) .$$ (11)

It is obvious that equation (9) is applicable to the fiber pull-out model shown in Fig. 2. For this model, we assume that the hatched part in the figure has no strain, so that in (9)

$$\varepsilon_0(t) = 0 .$$ (12)

Moreover, the boundary conditions are

$$\sigma_f(0,\ t) = \sigma_0(t) ,$$ (13)

$$\sigma_f(\ell,\ t) = 0 .$$ (14)

Incidentally, if only elastic deformation takes place in matrix at $t = 0$, the following shear lag equation is applicable to the initial states in the two modes mentioned above:

$$\frac{r_f^2}{2G_m}\ln\frac{R}{r_f} \cdot \frac{d^2\sigma_f}{dz^2} - \frac{\sigma_f}{E_f} + \varepsilon_0 = 0 .$$ (15)

3. Functional and Stationary Function

For the two kinds of axisymmetric models described in the preceding section, it has been assumed that fibers are regarded as 1D bars deforming uniaxially, and that only shear deformation in matrix has influence on the axial distribution of fiber stress. In accordance with such shear lag assumptions, now let us consider an incremental complementary energy function

$$U = \pi r_f^2 \int_0^\ell \frac{\dot{\sigma}_f^2}{2E_f}dz + 2\pi\int_0^\ell\int_{r_f}^R \left(\frac{\dot{\tau}_m^2}{2G_m} + \dot{\tau}_m\dot{\gamma}_m^c\right)rdrdz - 2\pi R\int_0^\ell \dot{\tau}_R\dot{\varepsilon}_0 zdz ,$$ (16)

where τ_R denotes τ_m at $r = R$. In the right hand side of the above equation, the first term represents the incremental elastic energy in the fiber embedded in matrix, the second term the incremental elastic energy and energy dissipation induced by the shear deformation in matrix, and the third term the variation of incremental complementary energy due to the axial strain rate $\dot{\varepsilon}_0(t)$ at $r = R$.

Using (4) and (7), we obtain

$$\tau_m(r,z,t) = -\frac{r_f^2}{2r}\frac{\partial \dot{\sigma}_f(z,t)}{\partial z}. \tag{17}$$

Substitution of the above equation into (16) allows the incremental complementary energy U to become a functional of $\dot{\sigma}_f(z,t)$:

$$U[\dot{\sigma}_f(z,t)] = \pi r_f^2 \int_0^\ell F(z,\dot{\sigma}_f,\dot{\sigma}_f')dz, \tag{18}$$

where $\sigma_f' = \partial \sigma_f / \partial z$, and

$$F = \frac{\dot{\sigma}_f^2}{2E_f} + \frac{r_f^2}{4G_m}\ln\frac{R}{r_f}\cdot\dot{\sigma}_f'^2 - \dot{\sigma}_f'\int_{r_f}^R \dot{\gamma}_m^c dr + \dot{\sigma}_f'\dot{\varepsilon}_0 z. \tag{19}$$

The boundary conditions for $\dot{\sigma}_f(z,t)$ in (18) are

$$\dot{\sigma}_f(0,t) = \dot{\sigma}_0(t), \tag{20}$$

$$\dot{\sigma}_f(\ell,t) = E_f\dot{\varepsilon}_0(t). \tag{21}$$

Here it is noted that $\sigma_0(t) = 0$ in the fiber breakage model, and that $\varepsilon_0(t) = 0$ in the fiber pull-out model.

Stationary functions of the above functional U satisfy Euler's equation

$$\frac{\partial F}{\partial \dot{\sigma}_f} - \frac{d}{dz}\left(\frac{\partial F}{\partial \dot{\sigma}_f'}\right) = 0. \tag{22}$$

Substitution of (19) into (22) yields

$$\frac{r_f^2}{2G_m}\ln\frac{R}{r_f}\cdot\frac{\partial^2\dot{\sigma}_f}{\partial z^2} - \frac{\dot{\sigma}_f}{E_f} - \int_{r_f}^R \frac{\partial\dot{\gamma}_m^c}{\partial z}dr + \dot{\varepsilon}_0(t) = 0. \tag{23}$$

This is just equation (9), i.e., the shear lag differential equation given in the preceding section. Therefore, solutions of equation (9) can be obtained equivalently by finding stationary functions of the functional U given by (18) and (19).

If only elastic deformation takes place in matrix, we can consider a complementary energy function

$$U = \pi r_f^2 \int_0^\ell \frac{\sigma_f^2}{2E_f}dz + 2\pi\int_0^\ell\int_{r_f}^R \frac{\tau_m^2}{2G_m}rdrdz - 2\pi R\int_0^\ell \tau_R\varepsilon_0 zdz. \tag{24}$$

Substitution of (17) into the above gives a functional

$$U[\sigma_f(z,t)] = \pi r_f^2 \int_0^\ell F(z,\sigma_f,\sigma_f')dz, \tag{25}$$

where

$$F = \frac{\sigma_f^2}{2E_f} + \frac{r_f^2}{4G_m} \ln\frac{R}{r_f} \cdot \sigma_f'^2 + \sigma_f' \varepsilon_0 z \, . \tag{26}$$

Then, we can derive equation (15) by substituting (26) into the following Euler's equation relevant to (25):

$$\frac{\partial F}{\partial \sigma_f} - \frac{d}{dz}\left(\frac{\partial F}{\partial \sigma_f'}\right) = 0 \, . \tag{27}$$

4. Analyticall Solutions Based on Bilinear Profiles of Fiber Stress

Figures 3(a) and (b) schematically illustrate the profiles of fiber stress in the fiber breakage and fiber pull-out models, respectively. It may be most simple to approximate bilinearly the profiles, as indicated by the dashed lines in the figures. In the present section, adopting such a bilinear approximation, we derive analytical solutions on the basis of the functional U demonstrated in the preceding section.

Here we consider the fiber pull-out model first, since it is simpler. Let us suppose that the stress σ_0 applied to the fiber is constant, and let us approximate bilinearly $\sigma_f(z,t)$ in the fiber pull-out model as

$$\sigma_f(z,\, t) = \begin{cases} \sigma_0[1 - z/a(t)] \, , & 0 \le z \le a, \\ 0, & a \le z \le \ell, \end{cases} \tag{28}$$

where $a(t)$ denotes a stress transfer length (Fig. 3(b)). Then, $\dot{\sigma}_f(z,t)$ has an expression

$$\dot{\sigma}_f(z,\, t) = \begin{cases} \sigma_0 z \dot{a}(t)/a^2 \, , & 0 \le z \le a, \\ 0, & a \le z \le \ell. \end{cases} \tag{29}$$

Substituting (29) into the functional U given by (18) and (19), we obtain

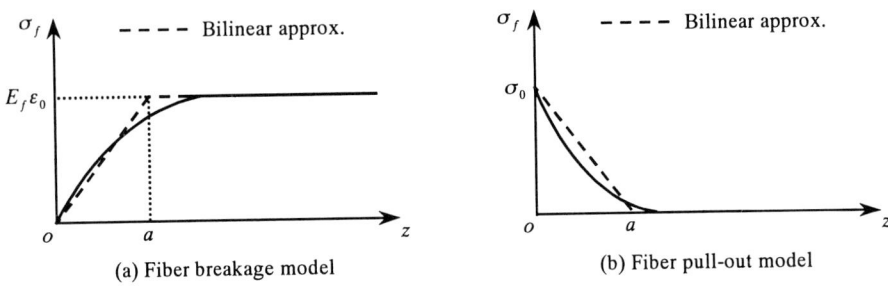

(a) Fiber breakage model   (b) Fiber pull-out model

Figure 3. Schematic illustration of fiber stress profile; (a) fiber breakage model, (b) fiber pull-out model.

$$U = \frac{\pi r_f^2 \sigma_0}{a}\left[\sigma_0 \dot{a}^2\left(\frac{1}{6E_f} + \frac{r_f^2}{4G_m a^2}\ln\frac{R}{r_f}\right) - \frac{\dot{a}}{a}\int_0^a\int_{r_f}^R \dot{\gamma}_m^c\, dr\, dz\right].$$
(30)

As was represented in (18), formulation of the present variational method is incremental. Accordingly, the above functional U should be taken to be stationary with respect to $\dot{a}$ by supposing that the current value of a is known. Thus, using

$$\partial U/\partial\dot{a} = 0,$$
(31)

we obtain

$$\sigma_0\left(\frac{1}{3E_f} + \frac{r_f^2}{2G_m a^2}\ln\frac{R}{r_f}\right)\dot{a} - \frac{1}{a}\int_0^a\int_{r_f}^R \dot{\gamma}_m^c\, dr = 0.$$
(32)

In the fiber pull-out model, τ_m is the dominant stress in matrix. Thus, if we assume for simplicity Norton's law for $\dot{\gamma}_m^c$, we can have

$$\dot{\gamma}_m^c = B\left|\tau_m\right|^{n-1}\tau_m,$$
(33)

where B and n are material constants. Substitution of (17) and (28) into (33) gives

$$\dot{\gamma}_m^c = B\left(\frac{r_f^2\sigma_0}{2ar}\right)^n.$$
(34)

Then, since (32) results in

$$\left(\frac{a^n}{3E_f} + \frac{r_f^2 a^{n-2}}{2G_m}\ln\frac{R}{r_f}\right)\dot{a} = \frac{Br_f^{2n}\sigma_0^{n-1}}{2^n}\int_{r_f}^R \frac{dr}{r^n},$$
(35)

we obtain an analytical solution

$$\frac{a(t)^2 - a(0)^2}{3E_f} + \frac{r_f^2}{G_m}\ln\frac{R}{r_f}\ln\frac{a(t)}{a(0)} = Br_f^2\ln\frac{R}{r_f}\cdot t, \quad n = 1,$$
(36a)

$$\frac{a(t)^{n+1} - a(0)^{n+1}}{3(n+1)E_f} + \frac{r_f^2\ln(R/r_f)[a(t)^{n-1} - a(0)^{n-1}]}{2(n-1)G_m} = \frac{Br_f^{n+1}\sigma_0^{n-1}}{2^n(n-1)}\left[1 - \left(\frac{r_f}{R}\right)^{n-1}\right]t, \quad n \neq 1.$$

(36b)

The initial value $a(0)$ in the above solution can be evaluated using the elastic functional (25) as follows: Substitution of (28) and $\varepsilon_0 = 0$ into (25) with (26) provides

$$U = \pi r_f^2\sigma_0^2\left(\frac{a}{6E_f} + \frac{r_f^2}{4G_m a}\ln\frac{R}{r_f}\right).$$
(37)

Then, using

$$dU/da = 0, \tag{38}$$

we obtain

$$a(0) = r_f \sqrt{\frac{3E_f}{2G_m} \ln \frac{R}{r_f}} . \tag{39}$$

An analytical solution for the fiber breakage model can be derived as well, if overall strain ε_0 is constant. For this model, $\sigma_f(z,t)$ can be approximated bilinearly as

$$\sigma_f(z,t) = \begin{cases} E_f \varepsilon_0 z / a(t), & 0 \le z \le a, \\ E_f \varepsilon_0, & a \le z \le \ell. \end{cases} \tag{40}$$

Then, (18) and (19) give

$$U = \frac{\pi r_f^2 E_f \varepsilon_0}{a} \left[E_f \varepsilon_0 \dot{a}^2 \left(\frac{1}{6E_f} + \frac{r_f^2}{4G_m a^2} \ln \frac{R}{r_f} \right) + \frac{\dot{a}}{a} \int_0^a \int_{r_f}^R \dot{\gamma}_m^c dr dz \right]. \tag{41}$$

Hence, using (31), we obtain

$$E_f \varepsilon_0 \left(\frac{1}{3E_f} + \frac{r_f^2}{2G_m a^2} \ln \frac{R}{r_f} \right) \dot{a} + \frac{1}{a} \int_0^a \int_{r_f}^R \dot{\gamma}_m^c dr = 0 . \tag{42}$$

In the fiber breakage model, the axial normal stress in matrix is as dominant as matrix shear stress τ_m in the initial state, in contrast to the fiber pull-out model. The axial normal stress in matrix, however, tends to relax much more quickly than τ_m as matrix shear creep proceeds (Du and McMeeking, 1995; Ohno and Miyake, 1999). Thus, we may assume (33) to express $\dot{\gamma}_m^c$. Equation (42) then can be integrated analytically as

$$\frac{a(t)^2 - a(0)^2}{3E_f} + \frac{r_f^2}{G_m} \ln \frac{R}{r_f} \ln \frac{a(t)}{a(0)} = Br_f^2 \ln \frac{R}{r_f} \cdot t , \quad n = 1 , \tag{43a}$$

$$\frac{a(t)^{n+1} - a(0)^{n+1}}{3(n+1)E_f} + \frac{r_f^2 \ln(R/r_f)[a(t)^{n-1} - a(0)^{n-1}]}{2(n-1)G_m} = \frac{Br_f^{n+1} (E_f \varepsilon_0)^{n-1}}{2^n (n-1)} \left[1 - \left(\frac{r_f}{R} \right)^{n-1} \right] t ,$$

$$n \ne 1 . \tag{43b}$$

For the fiber breakage model, equation (25) with (26) is reduced to

$$U = \pi r_f^2 (E_f \varepsilon_0)^2 \left(\frac{a}{6E_f} + \frac{r_f^2}{4G_m a} \ln \frac{R}{r_f} + \frac{\ell}{2E_f} \right). \tag{44}$$

Hence, applying (38) to (44), we can obtain again equation (39), which gives the initial value $a(0)$ in the solution (43).

5. Discussion

We have derived the analytical solutions (36) and (43) by approximating bilinearly the profiles of fiber stress. It is seen that the solution (36) with σ_0 replaced by $E_f \varepsilon_0$ is identical to (43). In this section, to discuss the validity of the solutions, we compare (36) with finite difference computations and an experiment based on Raman spectroscopy.

5.1. COMPARISON OF PRESENT SOLUTION AND COMPUTATIONS

Figures 4(a) and (b) compare the solution (36) and finite difference computations of equations (9) and (15). The computations were done in the same way as in the work of Du and McMeeking (1995), who employed nondimensional quantities

$$\hat{r} = \frac{r}{r_f}, \quad \hat{z} = \frac{z}{r_f}, \quad \hat{t} = BE_f\sigma_0^{n-1}t, \quad \hat{\sigma}_f = \frac{\sigma_f}{\sigma_0}. \tag{45}$$

Figures 4(a) and (b) deal with linear and nonlinear cases of matrix creep, respectively;

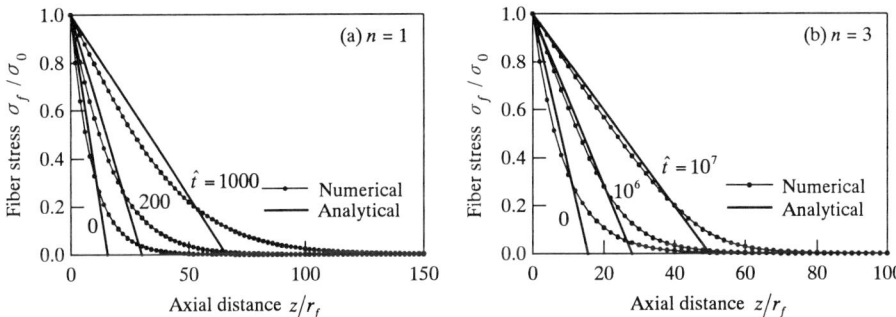

Figure 4. Time-dependent profiles of fiber stress by analytical solution and finite difference computation ($G_m/E_f = 0.01$, $R/r_f = 5$, $\ell/r_f = 200$); (a) $n = 1$, (b) $n = 3$.

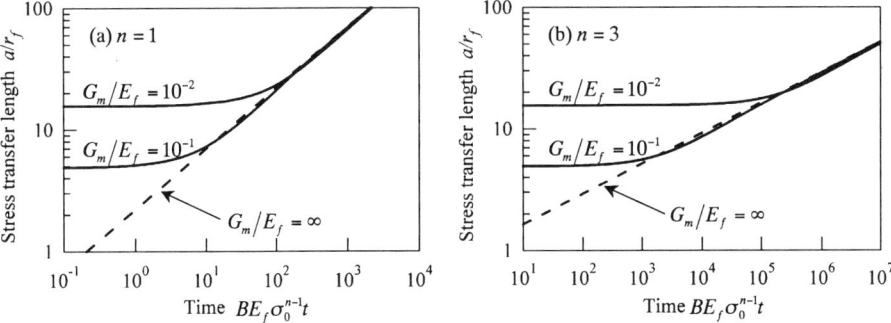

Figure 5. Influence of matrix shear rigidity on time-dependent increase of stress transfer length; (a) $n = 1$, $R/r_f = 5$, (b) $n = 3$, $R/r_f = 5$.

the material parameters employed are given in the figure caption. It is seen from the figures that, within the bilinear approximation of fiber stress profiles, the solution (36) agrees well with the computations in the two cases. Especially when matrix creep is nonlinear, the solution (36) works better, as seen in Fig. 4(b). This is because the nonlinearity of matrix shear creep gets fiber stress to distribute more bilinearly, as was found by Du and McMeeking (1995).

5.2. EFFECT OF ELASTIC SHEAR DEFORMATION OF MATRIX

Before moving on to the comparison with an experiment, we discuss the effect of the elastic shear deformation of matrix, which has been taken into account in the present work.

When $G_m = \infty$, we have $a(0) = 0$ from (39), and consequently the solution (36) is reduced to

$$a(t) = r_f \sqrt{3BE_f \ln(R/r_f) \cdot t} \ , \quad n = 1, \tag{46a}$$

$$a(t) = r_f \left\{ \frac{3(n+1)BE_f \sigma_0^{n-1}}{2^n(n-1)} \left[1 - \left(\frac{r_f}{R} \right)^{n-1} \right] t \right\}^{1/(n+1)} , \quad n \neq 1. \tag{46b}$$

Hence, when $G_m = \infty$, the stress transfer length a develops from zero in proportion to $t^{1/(n+1)}$. This is the finding in the exact solution of Mason et al. (1992), who ignored the elastic shear strain in matrix in deriving their solution.

Figures 5(a) and (b) illustrate the effect of the elastic shear strain of matrix on the time-dependent increase of a in the two cases of $n = 1$ and $n = 3$, respectively. As seen from the figures, $a(t)$ is larger just after loading if G_m/E_f is smaller, though the dependence of $a(t)$ on G_m/E_f fades with time. It is noticed that the solution with $G_m = \infty$, equation (46), becomes valid if this solution gets providing such a as comparable with $a(0)$ given by (39). Hence, we can say that the effect of the elastic shear strain of matrix cannot be ignored until matrix shear creep develops to be comparable with the initial elastic shear strain in matrix.

5.3. COMPARISON OF PRESENT SOLUTION AND EXPERIMENT

Now let us apply the present solution to one of the fiber pull-out experiments done by Miyake et al. (2000). In the experiment, a single carbon fiber/acrylic model composite was employed, and one end of the carbon fiber was subjected to a constant stress of $\sigma_0 = 2.7$ GPa for 500 hours. Figure 6 shows the fiber stress profiles measured using laser Raman spectroscopy in the experiment.

A notice is necessary for applying the present solution to the experiment mentioned above. It was concluded that the sliding at fiber/matrix interface occurred under the loading to $\sigma_0 = 2.7$ GPa, since the maximum value of interfacial shear stress determined from the fiber stress profiles in Fig. 6 was only about one tenth of the theoretical estimation based on the perfect bonding at interface (Miyake et al.,

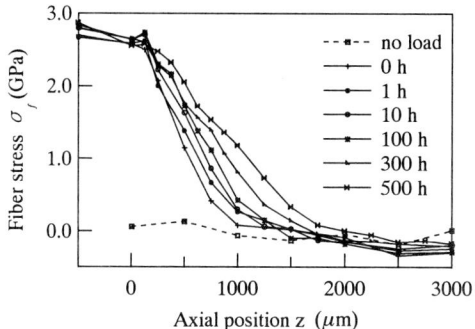

Figure 6. Time-dependent change of fiber stress profile in single carbon fiber/acrylic model composite tested using laser Raman spectroscopy (Miyake et al., 2000).

2000). The sliding has not been taken into account in the preceding sections. Nevertheless, the functional (18) and the resulting solution (36) remain valid, if the sliding took place *only initially* in the experiment, and if the initial value of a is taken to be equal to a slip length a_s, i.e.,

$$a(0) = a_s. \tag{47}$$

Using (4), a_s can be expressed in terms of interfacial slip stress τ_s as

$$a_s = \frac{r_f \sigma_0}{2\tau_s}. \tag{48}$$

Figure 7 compares the experiment and the solution (36) with respect to $a(t)$. It is seen that the solution (36) with the initial slip mentioned above agrees well with the experimental relation indicated by solid circles, which were determined by approximating bilinearly the fiber stress profiles in Fig. 6. The material constants used for the solution are given in Table 1. The value of R/r_f in the table was shown

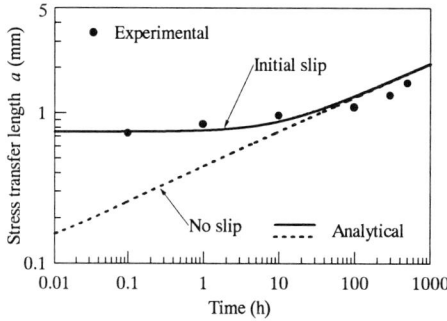

Figure 7. Comparison of experiment and analytical solution based on creep constants determined by steady-state fitting of creep curves.

Table 1 Material constants

Elastic constants *etc.*		$E_f = 490\text{MPa}, \quad G_m = 570\text{MPa}$ $r_f = 2.7\mu\text{m}, \quad R/r_f = 4$
Matrix creep constants	No hardening	$B = 1.25 \times 10^{-4}\,\text{MPa}^{-n}\text{h}^{-1}, \quad n = 3.5$
	Time-hardening	$B^* = 1.25 \times 10^{-4}\,\text{MPa}^{-n}\text{h}^{-1}, \quad n = 3.5,$ $t_0 = 50\text{h}, \quad k = 0.65$
Interfacial slip stress		$\tau_s = 4.9\text{MPa}$

to be appropriate for a fiber embedded in infinite matrix by Li and Grubb (1994), and the creep constants were determined by fitting the creep tests of acrylic (Appendix). Incidentally, R/r_f is ascertained to have negligible influence in (36b) on simulating the experiment, since R/r_f does not appear in $a(0)$ given by (47) and (48).

The dashed line in Fig. 7 illustrates the prediction based on the perfect bonding at interface, i.e., the solution (36) with $a(0)$ prescribed as (39). It is seen that this prediction fails to agree with the experiment up to $t \approx 10$ h. We are thus led to confirm that the assumption of the perfect bonding at interface is not appropriate for simulating the experiment.

The solution (36) is based on Norton's law expressed as (33). As a consequence, the transient creep behavior of the matrix material is disregarded in the predictions shown in Fig. 7. Now, just for discussing the effect of such creep behavior, let us assume simply the time-hardening of matrix shear creep

$$\frac{\partial \gamma_m^c}{\partial t^*} = B^* \left| \tau_m \right|^{n-1} \tau_m, \quad t^* = t_0 \left(\frac{t}{t_0} \right)^k, \tag{49}$$

where B^*, n, t_0 and k are material constants (see Table 1 and Appendix). Then, (36a) and (36b) are extended to

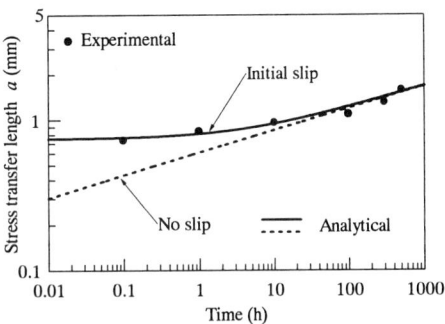

Figure 8 Comparison of experiment and analytical solution with time-hardening of matrix creep.

$$\frac{a(t^*)^2 - a(0)^2}{3E_f} + \frac{r_f^2}{G_m}\ln\frac{R}{r_f}\ln\frac{a(t^*)}{a(0)} = B^* r_f^2 \ln\frac{R}{r_f}\cdot t^*, \quad n=1, \tag{50a}$$

$$\frac{a(t^*)^{n+1} - a(0)^{n+1}}{3(n+1)E_f} + \frac{r_f^2 \ln(R/r_f)[a(t^*)^{n-1} - a(0)^{n-1}]}{2(n-1)G_m} = \frac{B^* r_f^{2n}\sigma_0^{n-1}}{2^n(n-1)}\left(\frac{1}{r_f^{n-1}} - \frac{1}{R^{n-1}}\right)t^*,$$

$$n \neq 1. \tag{50b}$$

The above solution gives the predictions shown in Fig. 8. It is seen that the predictions in this figure are a little better than those in Fig. 7. It is, however, not necessary to change the conclusions, which we have obtained by discussing the results in Fig. 7.

6. Conclusions

The present work dealt with a variational method to analyze the time-dependent change of fiber stress profiles in axisymmetric models for unidirectional fibrous composites. A functional based on incremental complementary energy was demonstrated and proved to have the stationary function satisfying a differential equation based on the shear lag assumption. Then, an analytical solution was derived by assuming bilinear profiles of fiber stress and by getting the functional stationary. A power law of matrix shear creep was employed to derive the analytical solution. It was shown that the solution agrees well with finite-difference computations of the differential equation, and that the elastic deformation of matrix has significant influence on the time-dependent increase of the stress transfer length until time elapses sufficiently long. It was also shown that the present solution simulates well the experiment performed for a carbon fiber/acrylic model composite if the initial slip at fiber/matrix interface is taken into account.

7. Appendix

Figure 9 shows tensile creep curves of acrylic (Miyake et al., 2000). The dashed lines in the figure indicate the simulation based on

$$\varepsilon = \sigma/E_m + A\sigma^n t, \tag{51}$$

where $A = 5.53 \times 10^{-6}$ (MPa^{-n}h^{-1}) and $n = 3.5$. The solid lines, on the other hand, are based on

$$\varepsilon = \sigma/E_m + A^*\sigma^n t^*, \quad t^* = t_0(t/t_0)^k, \tag{52}$$

where $A^* = 5.53 \times 10^{-6}$ (MPa^{-n}h^{-1}), $n = 3.5$, $t_0 = 50$ (h) and $k = 0.65$.

The equivalence between tension and torsion ascertained for inelastic deformation of metals tends to hold for polymers as well (Kitagawa et al., 1992; Qiu and Kitagawa, 1993). Hence, if we assume simply the Tresca equivalence, B in (33) and B^* in (49) satisfy $B = 2^{n+1}A$ and $B^* = 2^{n+1}A^*$, respectively, so that we have the creep constants given in Table 1.

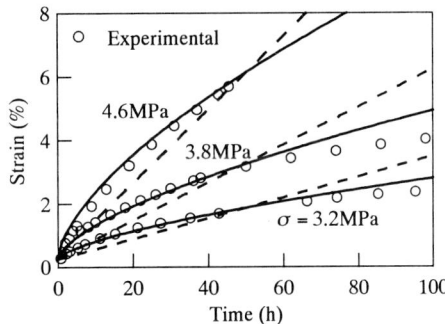

Figure 9. Tensile creep curves of acrylic at constant stresses; dashed and solid lines fitted with equations (51) and (52), respectively.

8. Acknowledgement

This work was supported in part by the Ministry of Education under a Grant-in-Aid for Scientific Research C (No. 11650086).

9. References

Clyne, T.W. and Withers, P.J. (1993) *An Introduction to Metal Matrix Composites*, Cambridge Univ. Press.

Cox, H.L. (1952) The elasticity and strength of paper and other fibrous materials, *Brit. J. Appl. Phys.* **3**, 72-79.

Du, Z.-Z. and McMeeking, R.M. (1995) Creep models for metal matrix composites with long brittle fibers, *J. Mech. Phys. Solids* **43**, 701-726.

Iyengar, N and Curtin, W.A. (1997) Time-dependent failure in fiber-reinforced composites by means of matrix and interface shear creep, *Acta Mater.* **45**, 3419-3429.

Kitagawa, M., Onoda, T., and Mizutani, K. (1992) Stress-strain behavior at finite strains for various strain paths in polyethylene, *J. Mater. Sci.* **27**, 13-23.

Li, Z.-F. and Grubb, D.T. (1994) Single-fibre polymer composites, Part I: interfacial shear strength and stress distribution in the pull-out test, *J. Mater. Sci.* **29**, 189-202.

Lifshitz, J.M. and Rotem, A. (1970) Time-dependent longitudinal strength of unidirectional fibrous composites, *Fibre Sci. Tech.* **3**, 1-20.

Mason, D.D., Hui, C.-Y., and Phoenix, S.L. (1992) Stress profiles around a fiber break in a composite with a nonlinear, power law creeping matrix, *Int. J. Solids Struct.* **29**, 2829-2854.

Miyake, T., Yamakawa, T., and Ohno, N. (1998) Measurement of stress relaxation in broken fibers embedded in epoxy using Raman spectroscopy, *J. Mater. Sci.* **33**, 5177-5183.

Miyake, T., Kokawa, S., and Ohno, N. (2000) Evaluation of time-dependent change of fiber stress profiles in constant load single-fiber pull-out tests using Raman spectroscopy, *Trans. Jpn. Soc. Mech. Eng., Ser. A*, **66**, 464-471, (in Japanese).

Ohno, N. and Miyake, T. (1999) Stress relaxation in broken fibers in unidirectional composites: modeling and application to creep rupture analysis, *Int. J. Plasticity* **15**, 167-189.

Qiu, J. and Kitagawa, M. (1993) Cyclic stress-strain curves of polyoximethylene under combined tension-torsion, *Trans. Jpn. Soc. Mech. Eng., Ser. A*, **59**, 2926-2933, (in Japanese).

MICROMECHANICS BASED CREEP DAMAGE ANALYSIS OF UNIDIRECTIONAL METAL MATRIX COMPOSITES

S. KRUCH, J.L. CHABOCHE, N. CARRERE
ONERA
29, Av. Division Leclerc
B.P. 72
92322 Chatillon Cedex

1. Introduction

The introduction of composite materials to reinforce industrial components submitted to cyclic thermo-mechanical loads including hold times, requires new developments in non-linear constitutive and damage modeling. This is particularly true with unidirectional MMCs where strong non-linearities, governed by damage and viscoplasticity, are coupled with the material anisotropy. A particular emphasis is devoted to the analysis of creep in order to take into account the stress redistributions inside the composite during the application of a hold time. Many studies have been carried out on the longitudinal creep behavior of MMCs showing the importance of the fibers resistance [1, 2]. Ohno and co-workers [3] performed creep tests on several directions of a SCS6/Ti-15-3 composite and analyzed the results on the basis of the stress relaxation in the matrix.

In this analysis, the longitudinal direction (i.e. the reinforced direction) is supposed strong enough to develop only plasticity, when both damage and plasticity dictate the transverse behavior. In-situ micrographic observations performed during single tension tests have shown the importance of damage, located at the interface between the matrix and the fibers, evolving at the very beginning of the test. The numerical analysis of the composite under creep loads applied in the transverse direction, is based on a micromechanical approach performed on a hexagonal unit cell representative of the internal microstructure. The regular distribution of fibers inside the matrix allows the use of the homogenization of periodic media technique to perform the micro-macro analysis and to impose the adequate boundary conditions on the unit cell. Fibers are supposed to remain elastic isotropic, the titanium matrix is characterized with an elasto-viscoplastic behavior and the interface is described with a debond law based on Tvergaard's model. An important

S. Murakami and N. Ohno (eds.), IUTAM Symposium on Creep in Structures, 453–462.

effort was devoted to the determination of material parameters, particularly for the constitutive equations of the matrix [4], and for the interface which was identified from micromechanical tests (Push out and single tension with in situ observations).

The first part of this paper presents the experimental procedure and the main results obtained from several tests. The numerical micromechanical analysis, performed with the finite element method, is detailed in the second part. Finally, the third part is devoted to the discussion and the correlation between the experimental results and the numerical ones.

2. Experimental results

The material analyzed in this study is a titanium matrix (Ti6242) reinforced with long unidirectional SiC (SCS6) fibers with a volume fraction of 35%. The specimen used for the experimental analysis are 8 ply plate manufactured by Snecma using the fiber-foil technique. The behavior of the titanium matrix is elasto-visco-plastic, the fibers are supposed to remain elastic and the protecting coat around the fibers (to prevent a reaction between the titanium and the SiC) is a brittle zone where the decohesion and damage in the transverse direction of the composite will take place.

Creep tests were performed at 500°C, under vacuum to avoid environmental problems, for several loads (150, 200, 225 and 250 MPa) [5]. Experimental results show the classical creep behavior with the three characteristic states: primary, secondary and tertiary as shown on figure 1. The primary state is short and it is induced by the initial loading. The secondary state is the longest one and it is characterized by a constant creep rate. Finally, the tertiary state is very short and always precedes the fracture of the specimen.

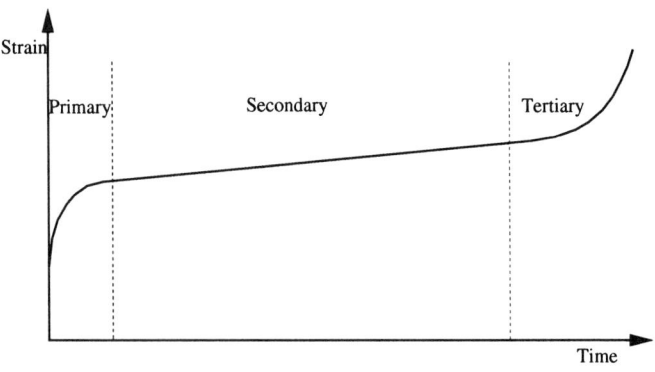

Figure 1. schematical representation of the creep evolution

Results can be summarized as follows:

- Under a critical stress, estimated here around 225 MPa, the creep rate increases gradually and the life time is long (there is no rupture even for long creep time (> 1500h)).
- For applied stresses over 225 MPa, the creep rate increases drastically leading to short failure times.

Figure 2 shows the evolution of creep rate as a function of the applied load. The modes described above can be characterized by two straight lines of a slope ratio around 100 showing that a new phenomena takes place when applying a stress over the critical one.

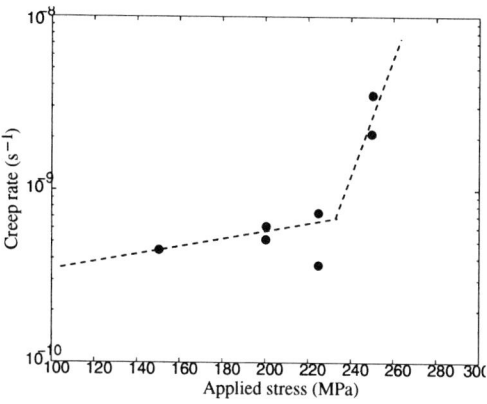

Figure 2. creep rate evolution as a function of the applied stress

Micrographic observations performed on the tested specimen and a micromechanical analysis with the finite element method using sophisticated constitutive equations have been carried out in order to investigate and understand local mechanisms involved in the creep process. It's an evidence that the matrix controls the global deformation of the composite, but the load carried by the matrix is controlled by the strength of the interface. For high stress levels (near the critical stress), the load is totally carried by the matrix (the interface being completely broken), leading to the failure of the composite when a crack initiates and propagates in the matrix. For lower stress levels (around 150 MPa), the interfacial damage is not very important and the load is partly carried by the fibers and the matrix leading to long lifetimes.

3. Micromechanical analysis

The regular position of the fibers inside the matrix allows to isolate a unit cell representative of the reference volume element (RVE) of the composite. The macroscopic behavior can be obtain applying periodic boundary conditions and averaging the local stresses and strains over this reference volume. As mentioned before, the matrix is elasto-viscoplastic and the parameters of the constitutive equations have been identified for several temperature [4], fibers will remain elastic and an interface debond law has been developed, based on Tvergaard's works [6], to describe the damage process. The parameters of the debond law were determined from micromechanical tests: Push-out for the shear behavior and single tension for the normal mode, with in situ observations [7].

Figure 3 presents a micrography of the microstructure and the mesh of the hexagonal unit cell used to perform the numerical analysis.

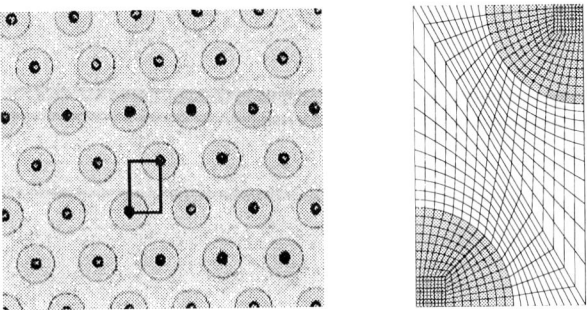

Figure 3. SiC/Ti microstructure (Snecma) and unit cell

3.1. INTERFACE DAMAGE ANALYSIS

As mentioned before, the initial damage, which will control the behavior of the specimen under the creep load in the transverse direction, is located at the interface between the matrix and the fiber. Micrographic observations have shown that the interfacial zone is constituted with different layers of carbon and with a complex reaction zone. It is not the purpose here to model all those elements, but to introduce a phenomenological model, through an interfacial debond law, giving an averaged behavior of the local mechanical fields, as it is schematically presented in the figure 4.

The phenomenological model characterizing the interface behavior is based on the works carried out by Needelman [8] and Tvergaard [6] and describes the progressive decohesion of two surfaces (in 3D). This local constitutive law relies two initially superposed nodes of the matrix and the fiber. Two relative displacements are then defined, U_n and U_t, the normal and tangential ones respectively, giving the evolution of a damage variable λ which cumulates the tension and shear damages as follows:

$$\lambda = \sqrt{\left(\frac{U_n}{\delta_n}\right)^2 + \left(\frac{U_t}{\delta_t}\right)^2} \tag{1}$$

Adjustable parameters δ_n and δ_t represent a characteristic length of the interface. The damage variable varies from 0 (local safe state) to 1 (local broken state). A

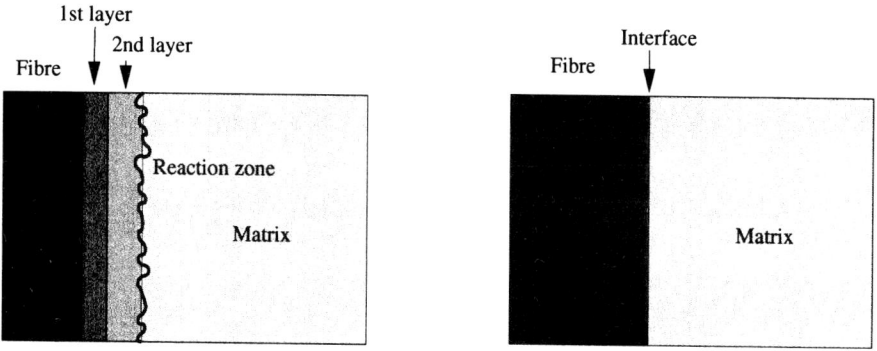

Figure 4. schematical representation of interphases and the modeling approach

schematic evolution of the pure normal and the pure tangential behaviors is represented on figure 5, showing the main differences between both damages. For

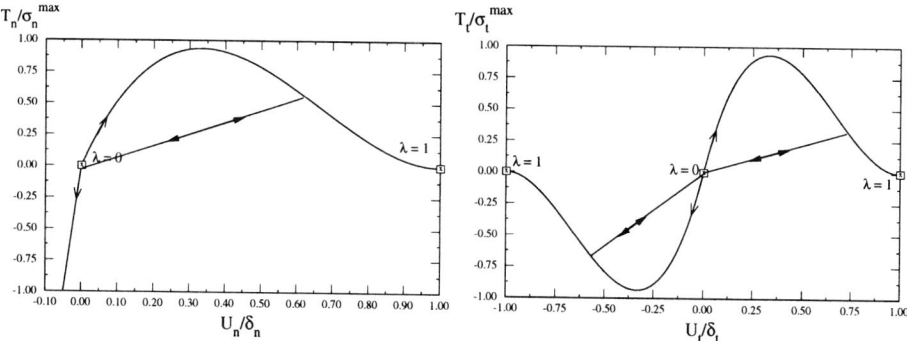

Figure 5. Normal and tangential behavior of the interface law

negative values of U_n, this is the case when the fiber is binded by the matrix due to the residual stresses, the normal component of the stress doesn't induce any damage and this one is only influenced by the shear stress state. For positive values of U_n, both damages are active, increasing the evolution of the interface decohesion. As it can be seen on figure 5 the unloading/reloading path characterizes the irreversibility of the damaging process.

Figure 6 presents the macroscopic response of a square unit cell loaded in tension in the transverse direction at 20^0C. This test was performed and simulated to identify the normal parameters of the debond law. The transverse curve gives a lot of information concerning the degradation of the composite. After the elastic domain, the damaging process takes place without plasticity, this is directly linked to the beginning of the interface degradation. During the first unloading path, the influence of damage is clearly observed with the change in the initial elastic slope. If the unloading is carried on, the damage deactivation process begins until the complete recovery of the initial stiffness, it can be observed that the residual strain is almost equal to 0 pointing out that, at this load level, the behavior is restricted to elasticity coupled with damage. As the load increases, plasticity becomes active and interacts with damage which is visible on the hysteresis loops and the residual strain. With this kind of analysis, once the interface is partly broken, the damage deactivation is directly taken into account with the unilateral contact condition between the matrix and the fiber. Figure 7 presents the contour of the shear plastic strain for a long crack around the interface inducing an important stress redistribution.

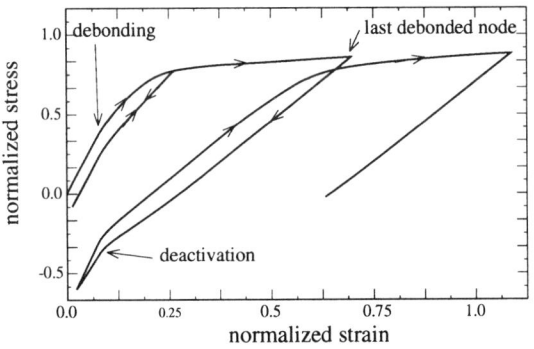

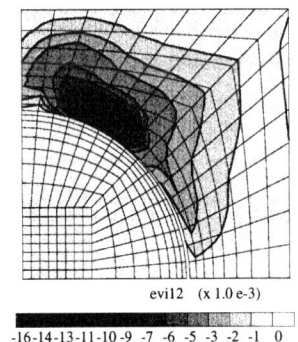

Figure 6. Averaged tensile curve   *Figure 7.* Shear plastic strain contour

The same kind of analysis was carried out to determine the tangential parameters of the debond law. Figure 8 presents the micro-mechanical simulation of the push-out test performed on an axisymetrical geometry. The manufacturing process of the sample is completely simulated in order to have the accurate residual state of the stresses at the interface as shown on this figure representing the contour of the shear stress before and during the test. The binding of the fibers by the matrix induces a stable crack propagation along the interface where the friction between both materials will have an important influence on the macroscopic response. The interface model, presented above, has been improved to take into account the influence of friction during the damaging process [9].

3.2. TITANIUM VISCOPLASTICITY

The amount of cumulated damage at the interface will affect the evolution of the creep rate, as mentioned before, but this one is directly linked to the viscoplasticity of the Ti matrix. An important experimental program was devoted to identify the parameters of the constitutive equations of the titanium matrix. Single

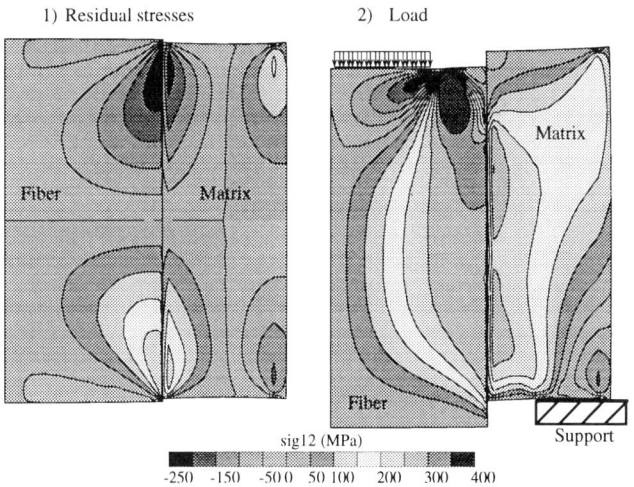

Figure 8. Push-out test: residual shear stresses and their evolution during the test

tension/compression tests, performed for several temperature from 20°C up to 870°C, have shown a classical isotropic elasto-viscoplastic behavior which has been characterized with one non-linear isotropic hardening and one non-linear kinematic hardening [4]. The effects of viscosity observed for high temperatures were described with a classical Norton's law identified from relaxation [10] and creep tests [5].

4. Numerical results and discussion

The numerical simulation performed on the unit cell presented on figure 3 has followed as close as possible the experimental procedure. The main steps of the calculation are summarized following:

- The first step consisted to determine the residual stresses induced by the manufacturing process, applying the load/temperature history undergone by the composite (referred as "manufacturing" in the following curves).
- The creep load is imposed, giving the response of the primary state (referred as "loading"),
- finally, the creep test itself is simulated (secondary state).

The third state, directly linked to the final failure of the specimen, is not modeled because, in this analysis, the crack initiation and propagation in the matrix are not taken into account.

Simulations have been performed for each experimental load, but in this paper only the results obtained for two loads are presented: the first one, 200 MPa, under the critical stress, and the second one, 270 MPa, over this stress. Figure 9 presents the evolution of the damage parameter around the interface between the fiber and the matrix, at different output times corresponding to the steps described before. It can be seen that the manufacturing process doesn't induce damage, both the applied pressure and the cooling process binding the fibers. Once the load is applied, damage starts and increases progressively. The comparison between the damage evolution induced by both loads shows that for low levels, damage increases slowly (and sometimes saturates) during the creep without breaking the interface, for higher loads, the interface begins to break and this process continues until the complete failure.

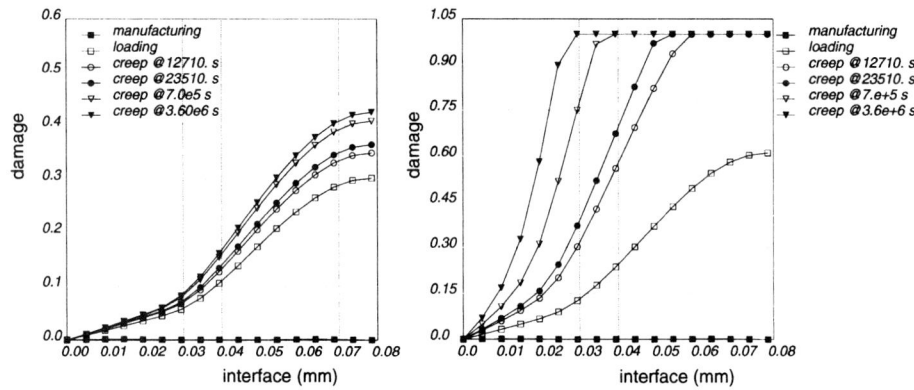

Figure 9.   Damage evolution around the interface for two loads

Figure 10 presents a comparison of the creep rates for both loads. When the damaging process is important the creep rate is much more higher, as it was observed in the experimental results. The numerical creep rates are 1.4410^{-9} and

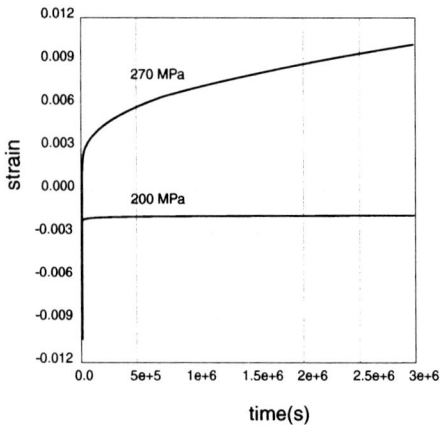

Figure 10.   Creep rate evolution

3.410^{-11} for the high and low load levels respectively. The comparison of these values with respect to the experimental ones presented in figure 2, shows a good estimate of the creep rate for the high load level but a bad one for the low level. Many unknown parameters could explain this discrepancy, as the real behavior of the matrix inside the composite. The titanium matrix was identified with specimen coming from a standard monolithic material which microstructure could be different from the one of the foil material used to manufacture the composite. The probable anisotropic texture of the matrix is right now analyzed at Onera using the EBSD technique. Also, the parameters of the interface debond law were determined from room temperature tests, with the hypothesis that the behavior evolves with the temperature through the relaxation of the residual stresses. Additional micromechanical observations should be performed to analyze the evolution of the interfacial reaction zone as a function of the temperature. An other important point to investigate concerns the influence of the strain or stress rate on the interface behavior, actually, the micromechanical tests were performed for just one strain rate, the model being written without any viscous effect.

5. Conclusion

Experimental results shown the classical creep curves with the primary, secondary and tertiary phases. Both the linear secondary creep and the tertiary one are dependent on the amount of damage initiated at the interface during the primary creep. Two main evolutions could be arose: the first one, for loads under a given threshold, leads to long life tests (some of them without tertiary creep or stopping during the secondary creep), and the second one, for loads over the threshold, inducing an important damage at the beginning of the secondary creep, leads to the tertiary creep and the failure of the specimen. The numerical micromechanical analysis, using the finite element method, performed to simulate the creep tests, helped to understand the influence of the damaging process, evolving at the interface, on the macroscopic response. In order to improve the lifetime under creep loads, it is important that the interface remains safe as long as possible in order to share the load between the fiber and the matrix. It has been shown that once the interface begins to break, the crack continues to propagate until the complete decohesion, then, a macroscopic crack leads to the failure of the specimen, the load being only carried by the unreinforced matrix. These are just qualitative observations, and an effort is devoted to improve the modeling in order to be able to reproduce quantitatively the experimental results.

Acknowledgment

The authors are grateful to Onera's people who performed and analyzed the experimental tests and Snecma for manufacturing the composite plates.

References

1. Jeng S. M., Yang J. M., "Creep behavior and damage mechanisms of SiC-fiber-reinforced titanium matrix composite", Materials Science and Enginneering, Vol. A171, pp. 65-75, 1993.
2. Faucon A., "Etude de quelques mécanismes d'endommagement au sein d'un matériau composite à matrice d'alliage de titane renforcée par des filaments de carbure de silicium", PhD Thesis, Université de Bordeaux, 1999.
3. Ohno N., Toyoda K., Okamoto N., Miyake T., Nishide S., "Creep behavior of a unidirectional SCS6/Ti-15-3 metal matrix composite at 450°C, Transactions of the ASME, Vol. 116, pp. 208-214, April 1994.
4. Baroumes L., Vinçon I., "Identification du comportement thermo-élastoviscoplastique de l'alliage Ti-6242", Int. Rep, LMT-Cachan, 1995.
5. Carrere N., Kruch S., Valle R., Vassel A., "Transverse creep behaviour of a unidirectional SCS-6/Ti-6242 composite", to be presented at ECCM 9, Brighton (U.K.), June 2000.
6. Tvergaard V., "Micromechanical modelling of fibre debonding in metal reinforced by short fibres", Inelastic deformation of composites materials, G. Dvorak Ed., IUTAM symposium, Troy (USA), 1990.
7. Guichet B., "Identification de la loi de comportement interfaciale d'un composite SiC/Ti", PhD Thesis, Ecole Centrale de Lyon, 1998.
8. Needleman A., "A continuum model for void nucleation by inclusion debonding", J. Applied Mech., 54, pp. 525-531, 1987.
9. Chaboche J.L., "On the interface debonding models", Int. J. of damage mechanics, vol. 6, pp. 250-257, 1997.
10. Malon S., "Caractérisation des mécanismes d'endommagement dans les composites à matrice métallique de type SiC/Ti", PhD Thesis, ENS Cachan, 2000.

MICROMECHANICAL MODELS FOR CREEP IN THE CONSOLIDATION OF COMPOSITES

J. CARMAI, F.P.E. DUNNE
Department of Engineering Science, University of Oxford,
Parks Road, Oxford, OX1 3PJ, England

Abstract

This work addresses the development of physically-based constitutive equations for the consolidation of fibre-matrix-void systems typically arising in the manufacture of matrix-coated fibre metal matrix composite materials. The analyses consider square array packing of the coated fibres under symmetrical in-plane compressive load and take into account the power-law creep of the matrix. The model is based on an energy approach in which assumed velocity fields in the deforming matrix are considered and are expressed in terms of an unknown parameter. In this way, the dependence of the deformation rate on volume fraction of voids and fibres is derived through the use of Hill's minimum principle for velocities. A micromechanical finite element model has also been developed against which the physically-based model has been tested.

1. Introduction

Continuous fibre reinforced metal matrix composite materials can be produced by consolidating bundles of Ti-6Al-4V coated SiC fibres. The consolidation process is important in determining the quality of the final composite. Modelling of the process allows predictions of the optimum temperature, pressure and time required to obtain fully dense composites with satisfactory engineering properties. The present work addresses the development of physically-based constitutive equations for consolidation of fibre-matrix-void systems.

2. Model development

2.1 PHYSICALLY-BASED MODEL

The model has been developed based on a micro-mechanical approach in which the constitutive behaviour of a representative volume element (a unit cell) of composite is

S. Murakami and N. Ohno (eds.), IUTAM Symposium on Creep in Structures, 463–468.

modelled and then used to predict the macroscopic response. In the present work, matrix-coated fibres are assumed to form a symmetrical, periodic array, repeating in two directions. The coated fibres are assumed to be perfect circular cylinders of infinite length so plane strain conditions hold on planes perpendicular to the fibre axis. SiC fibres are assumed not to undergo plastic deformation at the temperature of processing. The present analyses consider the matrix material to be incompressible and to deform by power-law creep. The analyses assume perfect interfaces between fibres and matrix and sticking friction at the matrix-matrix interface. Figure 1 shows schematically the unit cell during consolidation.

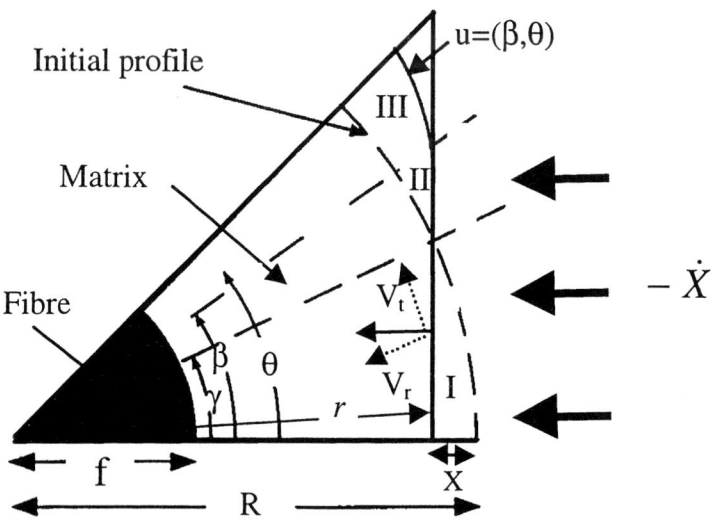

Figure 1 Schematic diagram showing the unit cell during consolidation.

One of the pioneering constitutive models for porous materials, which was developed based on micro-mechanical analysis and using the potential method is Cocks' model (Cocks, 1989) in which a lower bound expression for the potential of a porous creeping material was derived in terms of an assumed deformation rate field. Duva and Crow (1994) also developed a predictive model based on the potential energy method for the consolidation of fibres surrounded by porous matrix material. The deformation rate of the cell is estimated through the use of Hill's minimum principle for velocities (Hill, 1956). In the present work, it has also been possible to develop constitutive equations for consolidation of matrix coated fibre systems by assuming velocity fields in the deforming matrix, and using a repeating cell approach together with Hill's minimum principle for velocities. The principle states that of all kinematically admissible velocity fields (v_i), the actual field minimises the functional

$$F = \int_V W \, dV - \int_s T_i v_i dS \tag{1}$$

F denotes the total potential energy rate of the body which consists of the strain rate potential (W) and the potential energy rate due to external pressure (T_i). V is the volume of the deformed matrix and S is the outer surface where the pressure T_i is prescribed. In order to use Hill's minimum principle in the present analysis, a single free parameter, λ, is introduced into the imposed horizontal velocity. The assumed velocity fields in regions I, II and III shown in figure 1 then depend on the free parameter λ.

The matrix boundary in regions I and II has horizontal velocity, $-\lambda \dot{X}$, imposed upon it. The radial and tangential velocity components at the matrix boundary can be determined by resolving $-\lambda \dot{X}$ into the radial and the tangential directions. The radial velocity of the matrix boundary in region III is obtained by application of the requirement for incompressibility of the matrix.

An expression for λ in terms of applied pressure, the creep parameters, relative density and volume fraction of fibres was determined by applying Hill's minimum principle for velocities. The total potential energy rate for the analysed problem is given by

$$F = \sum_{i=1}^{3} (\int_{V_i} W(\dot{\varepsilon}_{e_i}) \, dV_i) - \int_s T \lambda \dot{X} \, dS \tag{2}$$

where $W(\dot{\varepsilon}_{e_i})$ denotes the strain rate energy in region i and $i =$ I, II and III.

Minimising the functional F with respect to λ gives

$$\lambda = (T\dot{X}(R - X))^n \times A \times \left(\int_0^\gamma \int_f^r P(\dot{\varepsilon}_{e_1}) r dr d\theta + \int_\gamma^\beta \int_f^r P(\dot{\varepsilon}_{e_2}) r dr d\theta + \int_\beta^{\frac{\pi}{4}} \int_f^r P(\dot{\varepsilon}_{e_3}) r dr d\theta \right)^{-n} \tag{3}$$

where

$$P(\dot{\varepsilon}_{e_i}) = \frac{n+1}{n} W(\dot{\varepsilon}_{e_i}) \frac{1}{\lambda^{\frac{n+1}{n}}} \tag{4}$$

The relative density, which is defined as the ratio of current density to the density at the fully dense state, can be obtained as

$$D = \frac{\left(\dfrac{m_{fibre} + m_{matrix}}{V_{fibre} + V_{matrix} + V_{void}} \right)}{\left(\dfrac{m_{fibre} + m_{matrix}}{V_{fibre} + V_{matrix}} \right)} = \frac{V_{fibre} + V_{matrix}}{V_{fibre} + V_{matrix} + V_{void}} \tag{5}$$

It can be shown that the overall dilatation rate for the cell is given by

$$\dot{\varepsilon}_{kk} = \frac{-2\lambda \dot{X}}{R \cos \gamma} \tag{6}$$

The overall densification rate for the cell is given by

$$\dot{D} = -D\dot{\varepsilon}_{kk} \qquad\qquad (7)$$

The physically-based constitutive equations for densification of fibre-matrix-void systems (equations (3), (6) and (7)) have been derived using a variational method. It explicitly accounts for fibre radius, matrix properties, and the voids through the parameter λ.

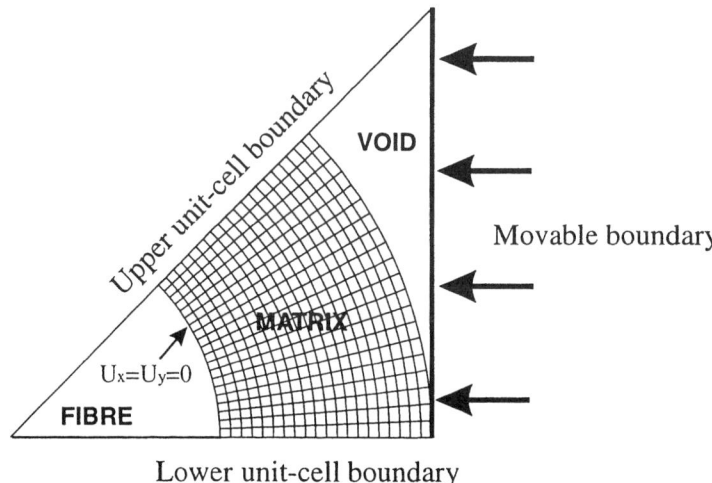

Figure 2 Schematic Diagram showing the micro-mechanical finite element model.

2.2   MICROMECHANICAL FINITE ELEMENT MODEL

An explicit micro-mechanical finite element model has been developed for square array packing using ABAQUS which is shown schematically in figure 2. Since the fibre is assumed to be rigid, the finite element mesh is developed only for the matrix which consists of two dimensional plane strain, 4-noded elements. All nodes lying on the fibre-matrix boundary are fixed. A movable boundary is used for the application of the distributed load, simulating the hydrostatic pressure state during the hot isostatic pressing process. The movement of this boundary is restricted to be along the x-axis. The multiaxial power-law creep equation was implemented into ABAQUS using a UMAT subroutine.

3. Results

Figures 3(a) and (b) show comparisons of density evolution over time for applied pressures of 20, 30, 50 and 70 MPa obtained from the energy method and the finite element model respectively for consolidation at 900°C for a volume fraction of fibres of 25%.

It can be seen that the majority of the composite consolidation occurs within the initial stage and that increasing the applied pressure leads to higher densification rate. The relative density evolution curves predicted by the energy method show good agreement with those predicted by the finite element model for the initial stage of consolidation.   At relatively high densities some differences can be seen.The prolongation in time

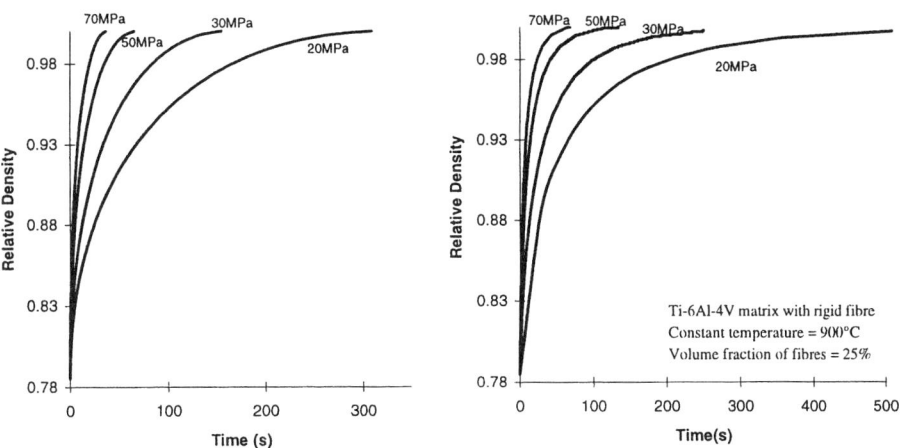

Figure 3 Graphs showing the variation of relative density with time at a constant temperature of 900 °C for a range of constant pressures predicted (a) by the energy method and (b) by the finite element model.

predicted by the finite element model at high relative densities is caused by the locking behaviour of elements in which the elements become too "stiff" to deform. Locking behaviour occurs in elements when material is almost or fully incompressible (Hibbitt et al., 1998).

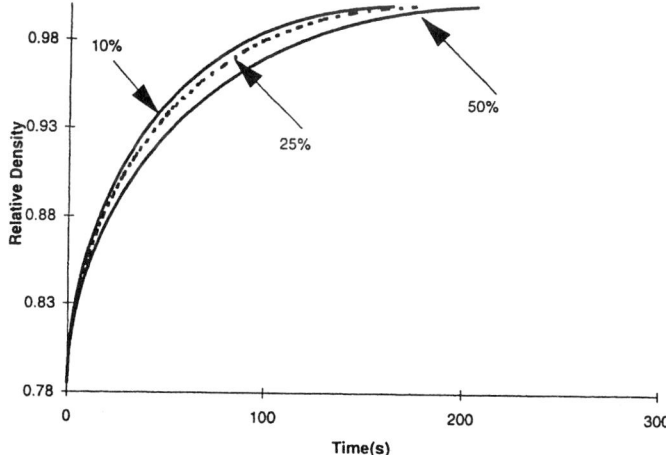

Figure 4 Graphs showing the variation of relative density with time predicted by the energy method for a range of volume fractions of fibres.

Figure 4 shows the density evolution curves for fibre volume fractions of 10%, 25% and 50% obtained from the energy method. It can be seen that as volume fraction of fibres increases, so the time at which the fully dense stage is reached increases, i.e. the densification rate decreases. The higher the volume fraction of fibres the lower the densification rate. This conclusion is supported by the results of simulations in Duva and Crow (1994); Kunze (1996); Elzey et al. (1998); Kunze and Wadley (1998) and the experimental evidence in Besson and Evans (1992).

4. Conclusions

A physically-based model has been developed for composite consolidation based on a variational method. The model was developed by consideration of an assumed velocity field in a repeating unit cell which was expressed in terms of unknown parameters. The deformation rate of the unit cell was estimated through the use of Hill's minimum principle for velocity. In this way, the dependence of the deformation rate on the volume fraction of fibres was derived. The results of the model have been compared with explicit finite element representations of the consolidating composite, and good agreement demonstrated. The predictions obtained from the models show that high applied pressures enhance densification rate. It was also found that volume fraction of fibres has an effect on the densification. Increasing the volume fraction of fibres leads to a reduction in densification rate.

5. References

Besson, J., Evans, A. G., (1992) The effect of reinforcements on the densification of a metal powder. Acta. Metallurgica et Materialia 40(9), 2247-2255.

Cocks, A.C.F., (1989) Inelastic deformation of porous materials. Journal of the Mechanics and Physics of Solids 37(6), 693-715.

Duva, J.M., Crow, P.D., (1994) Analysis of consolidation of reinforced materials by power-law creep. Mechanics of Materials 17, 25-32.

Elzey, D.M., Gampala, R., Wadley, H.N.G., (1998) Inelastic contact deformation of metal coated fibers. Acta Materialia 46(1), 193-205.

Hibbitt, Karlsson, Sorensen, Inc., (1998) ABAQUS/Standard User's Manual Version

Hill, R., (1956) New horizons in the mechanics of solids. Journal of the Mechanics and Physics of Solids 5, 66-74.

Kunze, J.M., Wadley, H.N.G., (1998) The vacuum hot pressing behavior of siliconcarbide fibers coated with nanocrystalline Ti-6Al-4V. Materials Science and Engineering A, 138-144.

Kunze, J.M., (1996) Mechanisms and models of metal matrix coated fiber densification. PhD Dissertation, University of Virginia.

OFF-AXIS CREEP BEHAVIOR OF UNIDIRECTIONAL POLYMER MATRIX COMPOSITES AT HIGH TEMPERATURE

M. KAWAI

Institute of Engineering Mechanics and Systems,
University of Tsukuba, Tsukuba 305-8573, JAPAN

Abstract

A short-term anisotropic creep behavior of a unidirectional carbon fiber-reinforced polymer matrix composite T800H/Epoxy under constant and variable loading conditions at 100°C was examined. Constant-stress creep tests in tension were first performed at three different stress levels for five hours on off-axis plain coupon specimens with various fiber orientation angles: $\theta = 0$, 10, 30, 45 and 90°. Then, creep deformations of off-axis plain coupon specimens subjected to a single step change, either increase or decrease, in stress were measured to observe effects of load histories on the off-axis tensile creep for the unidirectional composite. Off-axis creep recovery after completely removing the creep stress for each of the tensile creep tests was also observed for five hours at the same temperature. To evaluate effects of the physical aging of matrix resin, moreover, tensile flow curves of off-axis specimens preconditioned in the same test environment for different periods of time before testing were compared. On the basis of these experimental results, capabilities required by a desired inelastic constitutive model for describing a matrix-dominated anisotropic creep behavior of unidirectional polymer matrix composites were elucidated. Finally, an attempt to develop an anisotropic constitutive model that fulfilled some of the requirements was made.

1. Introduction

Advanced aerospace structures demand high performance polymer matrix composites that can be used at elevated temperatures, practically lower than 300°C [1, 2]. In general, polymers exhibit a larger elongation and a more significant rate- and time-dependent behavior at a higher temperature [3]. The structure of polymers that typically consists of long molecular chains changes with time to reach a more stable configuration at a given temperature. Accordingly, physical and mechanical properties of polymers vary with time [4]. Such a phenomenon is called physical aging, and it becomes more significant at high temperatures below the glass transition temperature of polymer. Deformations of composites and laminates are usually governed by shear responses of

469

matrix material along reinforcing fibers and between stacked plies.   Therefore, inelastic deformation, as well as damage and fracture, for polymer matrix composites significantly depend on temperature.   Moreover, they are governed by complicated interactions between viscoelastic and viscoplastic responses of polymer matrix and by changes in mechanical properties of polymer matrix due to physical aging, especially in a high temperature range.   Such a matrix-dominated behavior is essential to the high-temperature durability of polymer matrix composites.   To understand properties for polymer matrix composites and to correctly evaluate their performance and reliability under service stress conditions at elevated temperatures, therefore, it is one of the most important issues to elucidate by experiment their time- and rate-dependent nonlinear behavior and their history dependence.   Furthermore, a realistic and rational modeling of time- and rate-dependent nonlinear behavior of polymer matrix composites is prerequisite to making reliable evaluations of the ultimate capacity of high-temperature composite structures and to establishing accurate design methods based on local stress analysis.

The present study primarily aims to examine a short-term creep behavior of unidirectional carbon fiber-reinforced polymeric composite T800H/Epoxy under off-axis constant and variable stress conditions at elevated temperature.   Constant-stress creep tests in tension are first performed at three stress levels on plain coupon specimens with various fiber-orientation angles.   Stress-dependence and directional nature of the off-axis tensile creep behavior are elucidated from these constant-stress creep tests.   Then, creep responses of off-axis specimens subjected to either a stepped increase or a stepped decrease in stress are observed to manifest characteristics of a matrix-dominated creep hardening of composite.   To evaluate the effects of physical aging of the epoxy resin, moreover, tensile flow curves of off-axis specimens preconditioned in the creep test environment for different periods of time are compared.   Finally, a phenomenological approach to an inelastic constitutive modeling is adopted for the anisotropic inelastic behavior of unidirectional polymer matrix composites.

2. Experimental

2.1. MATERIALS AND SPECIMENS

The material used in this study is a unidirectional fiber composite fabricated using the autoclave method from the prepreg (P2212-15, TORAY) which consists of the carbon fiber (T800H) and the thermosetting epoxy resin (No. 3631).   The lay-up of the virgin laminate is $[0]_{12}$.   The T800H/Epoxy prepreg was cured at 180°C for two hours.   The glass transition temperature of the epoxy resin is 215°C.

Five kinds of plain coupon specimens with different off-axis angles ($\theta = 0$, 10, 30, 45, 90°) were cut from 400 mm by 400 mm unidirectional laminate panels.   The shape and dimensions of the off-axis specimens are based on the testing standards JIS K 7087 [5] and ASTM D2990 [6]; the specimen length $L = 200$ mm and the thickness $t = 1.7$ mm.   The specimen width is 10 mm for $\theta = 0°$, and 20 mm for other off-axis angles. Rectangular-shaped aluminum-alloy tabs were attached on both ends of the specimens using epoxy adhesive (ARALDITE); the thickness of the end-tabs is 1.0 mm.

2.2. TEST PROCEDURE

Off-axis constant and variable stress creep tests for the unidirectional T800H/Epoxy composite were performed using a closed-loop hydraulic MTS-810 testing machine of the dual servovalve type; the flow capacities of two servovalves are 4 L/min and 19 L/min, respectively. For raising the temperature of specimens, a heating chamber with a precise digital control capability was employed. The variation of the specimen temperature in time from the prescribed value was less than 1.0°C. The specimens were clamped in the heating chamber by the high temperature hydraulic wedge grips fitted on the testing machine.

Off-axis constant-stress creep tests were carried out at 100°C for five hours. Three different stress levels were selected for each of the off-axis creep tests; the middle level was specified to be 60 % of the static tensile strength at test temperature. Then, off-axis variable-stress creep tests that include either a single stepped increase or decrease in stress between two levels adopted for the constant-stress creep tests were conducted at the same temperature. The duration of the creep test at each stress level was specified to be five hours. Off-axis creep recovery after complete unloading of the prior creep stress was also measured for five hours. The loading and unloading of the creep stress were controlled at a constant stroke rate of 1.0 mm/min. The off-axis specimens were heated up to 100°C in air keeping load at zero, and they were preconditioned in the test environment for 60 minutes prior to being creep tested.

To determine the level of creep stress at 100°C and evaluate the effect of aging of the epoxy matrix, static tension tests were carried out on the off-axis specimens preconditioned in the creep test environment (100°C) for different periods of time: 1, 25, 50 and 100 hours. The off-axis static tension tests were carried out with stroke control; the crosshead speed was specified to be 1.0 mm/min on the basis of the testing standards: JIS K7087 [5] and ASTM D2990 [6]. Longitudinal and lateral strains of each off-axis specimen were monitored with two-element L-type rosette strain gauges. These strain gauges were mounted back to back at the center of each specimen.

3. Experimental Results and Discussion

3.1. OFF-AXIS CREEP UNDER CONSTANT STRESS CONDITIONS

Off-axis constant-stress creep curves are shown in *Figures 1 a-d* for representative off-axis angles $\theta = 0, 10, 45$ and 90°, respectively. From these figures, we can see that the creep deformation of the unidirectional T800H/Epoxy composite clearly appears for all of these off-axis angles, including the fiber direction ($\theta = 0°$). As expected, the magnitude of creep strain significantly depends on the fiber-orientation angle. For each off-axis angle, the creep strain becomes larger at a higher creep stress and the creep rate rapidly decreases with time in an early stage of the creep response. Such stress-dependent and transient creep responses of composite are similar to those for

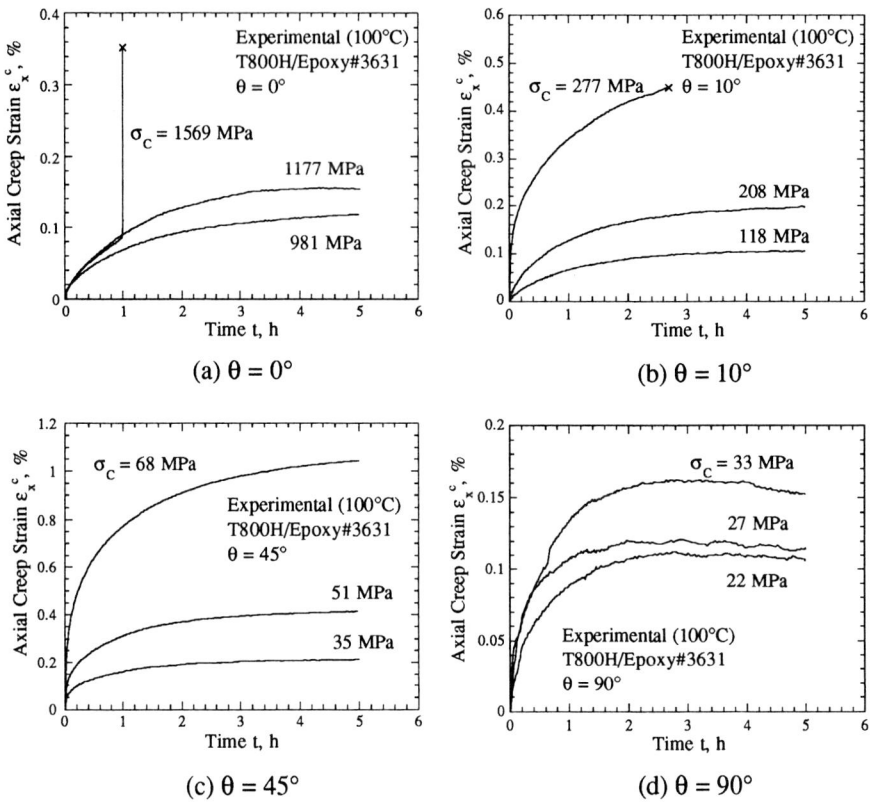

Figure 1. Off-axis creep curves for unidirectional T800H/Epoxy at 100°C.

conventional metallic materials [7].   Unlike a typical creep response of metals, however, the composite creep rate almost vanishes in a short time as the creep strain increases, regardless of the sustained stress level and of the off-axis angle.   Accordingly, the creep deformation of composite is eventually terminated.   Namely, the steady-state creep response at a constant strain rate in a classical sense is not substantially followed with the load duration.

Such a decay of the composite creep rate with the off-axis load duration indicates that the polymer matrix hardens to apparently suppress time-dependent structural changes of molecular chains under the applied constant stress condition; i.e., the driving force for the creep of polymer matrix is diminished.   This mechanism is consistent with the creep response in the fiber direction ($\theta = 0°$); in this case, the disappearance of the composite creep rate is due to the stress relaxation of matrix up to a certain level that depends on the stress applied to the composite.   From a phenomenological point of view, a decay of the creep rate under a constant stress condition may be described using a viscoplastic model based on the overstress concept [8].

In *Figure 1d*, it is observed that the creep strain in the transverse direction ($\theta = 90°$) tends to slightly decrease after a certain amount of the normal creep deformation proceeds.   Such an anomalous creep response indicates that a contraction of composite specimen takes place with time in the loading direction.   We speculate that this phenomenon is ascribed to a volume contraction (i.e., a volume relaxation) of the epoxy matrix due to the physical aging [4].   It may be brought about by the Poisson effect due to a delayed creep deformation of the matrix in the fiber direction.   A similar anomaly was observed also on the stress relaxation behavior in the transverse direction for the same unidirectional composite at the same test temperature [9].

3.2. OFF-AXIS CREEP UNDER A STEPPED INCREASE IN STRESS

A creep curve for the off-axis specimen subjected to a single stepped increase in stress is shown in *Figure 2a* for $\theta = 45°$.   This figure includes the constant-stress creep curves for respective stress levels.   From this figure, we can see that the creep rate of composite is enhanced by a stepped increase in stress and the creep rate just after the stress change is as high as the initial rate of the prior creep at a lower stress level. Such a temporal creep softening is similar to that for conventional metals.   With the subsequent load duration at a higher stress level, however, the creep rate rapidly decreases to vanish, as observed in the constant-stress creep response.   The results for other off-axis angles have the same tendency.   It seems that the creep strain of composite developing after the stepped change in stress becomes smaller than that under a constant stress of the same intensity, although we need more data to characterize effects of stress histories on the off-axis creep of unidirectional composites.

As an initial step, let us examine the classical phenomenological approaches [7] to a description of the off-axis creep of composite subjected to a stepped increase in stress.

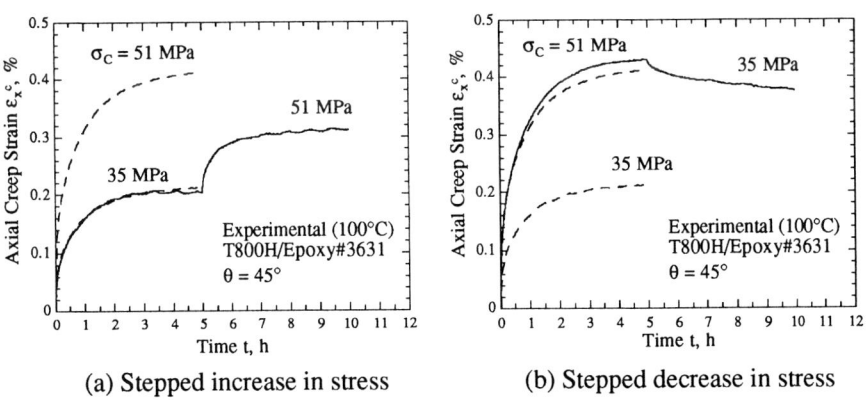

(a) Stepped increase in stress   (b) Stepped decrease in stress

Figure 2. Off-axis creep curves for unidirectional T800H/Epoxy subjected to stepped changes in stress at 100°C ($\theta = 45°$).

It is obvious that the time-hardening theory significantly underestimates the subsequent creep deformation; the subsequent creep curve shifted in the upper direction is completely different from the associated part of the constant-stress creep curve for the same stress level.   To check the strain-hardening theory, we assume a creep curve which is obtained by shifting the subsequent creep curve to the left in such a way that the starting point of the subsequent creep curve is located on the constant-stress creep curve for the same stress level.   It is easy to see that the shifted subsequent creep curve is lower than the compared constant-stress creep curve.   This comparison indicates that the strain-hardening theory overestimates the creep deformation of composite after a stepped increase in stress.   The actual creep response of composite, therefore, lies between these predictions given by these classical hardening rules.   It appears that the strain-hardening assumption is slightly better than the time-hardening one.   In reality, a combined strain and time hardening rule allows a physically better description of the short-term creep, as manifested later on.

3.3. OFF-AXIS CREEP UNDER A STEPPED DECREASE IN STRESS

An off-axis creep response to a stepped decrease in stress is shown in *Figure 2b* for $\theta = 45°$.   This figure also includes the constant-stress creep curves for respective stress levels.   We can observe that the creep strain gradually decreases with time after the stepped decrease in stress.   Such a creep recovery after partial unloading reveals the existence of a driving force acting in the opposite direction of the applied stress.   A mechanism of the creep recovery is usually explained by an internal stress that develops with inelastic deformation.   The creep recovery rate gradually decreases with the load duration and it tends to diminish, as in the constant-stress creep.   This indicates that the driving force disappears with time and the associated internal stress changes accordingly with subsequent creep deformation.   It is noted that the subsequent creep

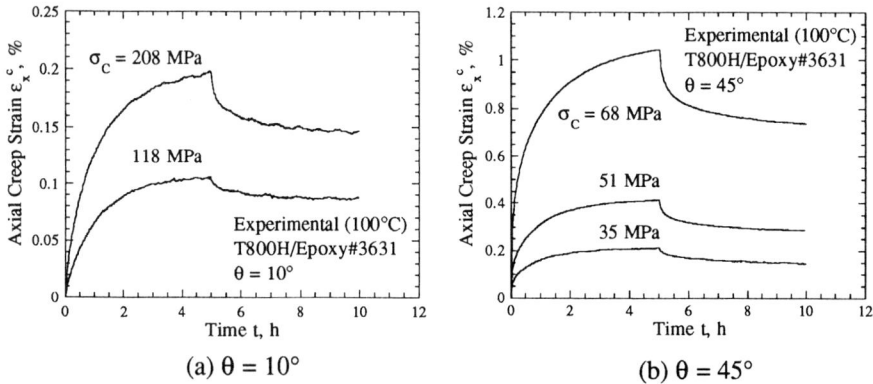

Figure 3. Off-axis creep recovery curves for unidirectional T800H/Epoxy after complete unloading at 100°C.

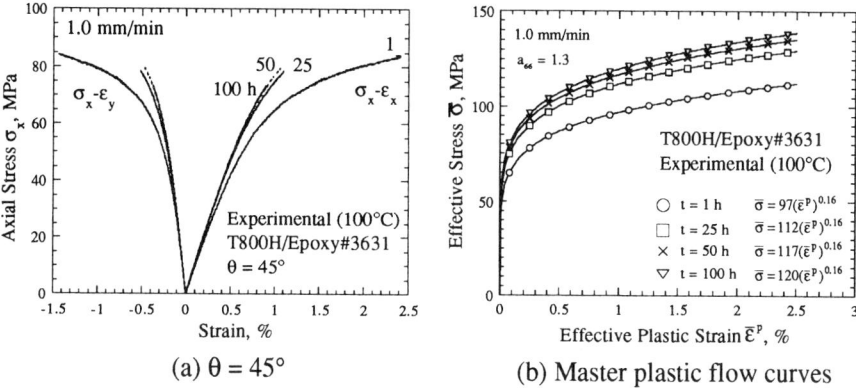

Figure 4. Effects of aging on off-axis tensile stress-strain curves.

strain does not recover to the level of the constant-stress creep curve for the same lower stress. This indicates that irrecoverable strain is involved by the off-axis creep deformation of composite.

Off-axis creep recovery curves after completely unloading the creep stresses are shown in *Figures 3a* and *b* for $\theta = 10°$ and $45°$, respectively. Irrespective of the creep stress level and of the off-axis angle, a certain amount of creep strain rapidly recovers in a short time after unloading. The creep recovery rate decreases to vanish. A certain amount of permanent strain eventually remains for every off-axis specimen. These features of the off-axis creep recovery are the same as those observed in the creep response after partial unloading. Again, this observation reveals that the off-axis creep behavior of the unidirectional T800H/Epoxy composite subjected to a relatively high stress at elevated temperature is significantly influenced by a viscoplastic property of the polymer matrix.

3.4. OFF-AXIS AGING

Tensile stress-strain curves for specimens subjected to stress-free aging at 100°C for 1, 25, 50 and 100 h are shown in *Figure 4a* for the off-axis angle $\theta = 45°$. This figure also displays a change in the transverse strain with an increasing longitudinal stress. From this figure, we can see that the flow stress of off-axis specimen increases with an increasing aging time prior to tensile testing. This observation proves that the thermosetting epoxy resin hardens by the physical aging even if it is free of stress and this matrix hardening enhances the flow strength of the unidirectional T800H/Epoxy composite. This also reveals that in reality the creep hardening of composite has a nature influenced by both strain hardening and time hardening. It is expected that the vanishing creep rate of off-axis specimen under a constant stress condition would be attributed to the age hardening of the epoxy resin as well as an intrinsic characteristic of polymers for creep.

The relationships between stress and plastic strain for all off-axis specimens subjected to the above-prescribed stress-free aging are plotted in *Figure 4b*, using the effective stress and effective plastic strain [10]. The increase in the flow stress of composite due to aging becomes much clearer in this figure. We also see that the effect of aging, i.e. an increase in the flow stress, is more significant for a shorter duration. It is noted that the directional nature of the off-axis stress-strain response of unidirectional composite is not influenced by aging; the effect of aging is characterized by the increase in the initial yield stress as well as the elastic modulus of composite.

4. Anisotropic Constitutive Models for Unidirectional Composites

An anisotropic constitutive model for the viscoplastic behavior of unidirectional polymer matrix composites has been developed on the basis of the irreversible thermodynamic formalism [11, 12]. This attempt is briefly summarized as follows.

4.1. A PURE KINEMATIC HARDENING MODEL

Some features of the directional creep and creep recovery of unidirectional composites can be described using a kinematic hardening model for transversely isotropic continua [13, 14, 15]:

$$\dot{\boldsymbol{\varepsilon}}^{p} = \frac{3}{2}\dot{P}\frac{\mathbf{u}}{U}, \tag{1}$$

$$\dot{\mathbf{p}} = \frac{2}{3}H\dot{\boldsymbol{\varepsilon}}^{p} - \zeta_{D}L\dot{P}\mathbf{u} - \zeta_{S}MV^{\ell}\frac{\mathbf{v}}{V}, \tag{2}$$

where

$$\dot{P} = KU^{m}, \qquad U = \sqrt{\frac{3}{2}\mathbf{s}^{*}:\mathbf{A}\mathbf{s}^{*}}, \qquad V = \sqrt{\frac{3}{2}\mathbf{p}:\mathbf{A}\mathbf{p}}, \tag{3a}$$

$$\mathbf{u} = \mathbf{A}\mathbf{s}^{*}, \qquad \mathbf{v} = \mathbf{A}\mathbf{p}, \qquad \mathbf{s}^{*} = \mathbf{s} - \mathbf{p}. \tag{3b}$$

In the above expressions, **A** represents the transversely isotropic tensor of the fourth rank [16], **s** is the deviator of Cauchy stress and **p** denotes a kinematic hardening variable. The coefficients H, K, L, M, ℓ, m, ζ_{D} and ζ_{S} are material constants. This is an extension of the Malinin-Khadjinsky model [17], and it is similar to the anisotropic constitutive models developed by Robinson [18] and Nouailhas and Freed [19].

4.2. A COMBINED KINEMATIC AND ISOTROPIC HARDENING MODEL

Capabilities to describe a partial creep recovery after a stepped decrease in stress and an age hardening can be considered by modifying the above constitutive model as follows:

$$\dot{\varepsilon}^p = \frac{3}{2} K \left\langle \sqrt{\frac{3}{2} \frac{s-a}{J(s-a)} : A \frac{s-a}{J(s-a)}} \left(J(s\text{-}a) - r - r^* - r_0\right) \right\rangle^m \frac{A(s-a)}{\sqrt{\frac{3}{2}(s-a):A(s-a)}}, \quad (4)$$

$$\dot{a} = \beta \dot{p}, \quad (5)$$

$$\dot{r} = \frac{3}{2}(1 - \beta)J(\dot{p}), \quad (6)$$

$$\dot{r}^* = (C - Dr^*)|\dot{f}^*|, \quad (7)$$

$$\dot{f}^* = -\zeta_T \frac{f^{*2}}{t}, \quad (8)$$

where $<\cdot>$ denotes the Macauley bracket, and the function $J(\cdot)$ returns a scalar similar to the von Mises equivalent stress. The coefficient β is a combined hardening parameter and it is taken in the range: $0 \leq \beta \leq 1$. This model assumes that the age hardening r^* is isotropic and the growth rate is in proportion to the rate of the free volume fraction f^* [4]. The variable t in Eq. (8) represents time and other coefficients C, D and ζ_T are material constants.

Assuming $\beta = 1$, $\zeta_T = 0$, and $r_0 = 0$ in this model, we can reproduce the pure kinematic hardening model expressed by Eqs. (1) and (2). A pure isotropic hardening model is derived by setting $\beta = 0$, $r_0 = 0$ and $\zeta_D = 0$; it was successfully applied to a description of off-axis creep for a unidirectional metal matrix composite [20].

5. Conclusion

Anisotropic short-term creep and creep recovery behavior for the unidirectional carbon fiber-reinforced composite T800H/Epoxy under off-axis constant and variable stress conditions at 100°C were examined. From the present study, it was elucidated that the creep deformation of unidirectional polymer matrix composites is characterized by a temporal softening due to stress changes, a rapid hardening to diminish creep rate under a continuously applied stress, a development of an internal stress and a hardening due to aging.

These salient features of the directional creep of unidirectional composite may be moderately described using a viscoplastic constitutive model furnished with an adequate combination of isotropic and kinematic hardening. An attempt to model the short-term creep of composite as such was made on the basis of the irreversible thermodynamic formalism.

Further study is needed to explain the anomalous behavior of composite under transverse loading conditions and to verify the validity of the combined isotropic and kinematic hardening model developed for describing the viscoplastic behavior of unidirectional polymer matrix composites.

Acknowledgment

This study was supported in part by the Ministry of Education, Science and Culture of Japan under a Grant-in-Aid for Scientific Research (C) (2) (No. 10650072).

References

1. Morita, W.H. and Graves, S.R.: Graphite/Polyimide technology overview and space shuttle orbiter application, *14th National SAMPE Technical Conference*, Oct. 12-14, 1982, 387-401.
2. Jones, J.S.: Celion/Larc-160 Graphite/Polyimide composite processing technique and properties, *14th National SAMPE Technical Conference*, Oct. 12-14, 1982, 402-417.
3. Nielsen, L.E.: *Mechanical Properties of Polymers and Composites*, Marcel Dekker, New York, 1975.
4. Matuoka, S.: *Relaxation Phenomena in Polymers*, Carl Hanser Verlag, 1992.
5. JIS K 7087: Testing methods for tensile creep of carbon fiber reinforced plastics, in *Japanese Industrial Standard*, Japanese Industrial Association, (1996), 1-18.
6. ASTM D2990: Standard test methods for tensile, compressive and flexural creep and creep-rupture of plastics, in *ASTM Standards and Literature References for Composite Materials*, ASTM, (1987), 445-455.
7. Odqvist, F.K.G. and J. Hult: *Kriechfestigkeit Metallischer Werkstoffe*, Springer-Verlag, Berlin, 1962.
8. Perzyna, P.: Thermodynamic theory of viscoplasticity, *Advances in Applied Mechanics* 11 (1971), 313-354.
9. Kawai, M., Kazama, T., Shinbo, S., and Masuko, Y.: Off-axis stress relaxation behavior of unidirectional CFRP at high temperature, *Proceedings of the 29th FRP Symposium*, JSMS, (2000), 199-200.
10. Sun, C.T. and Chen, J.L.: A simple flow rule for characterizing nonlinear behavior of fiber composites," *J. Composite Materials* 23 (1989), 1009-1020.
11. Chaboche, J.L.: Viscoplastic constitutive equations for the description of cyclic and anisotropic behavior of metals, *Bulletin de L'Academie Polonaise des Science, Série de Science Techniques* 25- 1 (1977), 33-42.
12. Lemaitre, J. and Chaboche, J.L.: *Mechanics of Solid Materials*, Cambridge University Press, Cambridge, 1985.
13. Kawai, M.: A phenomenological constitutive model for metal matrix composites, in M. Tokuda, B. Xu, and M. Senoo (eds.), *Macro/Micro/Meso Mechanical Properties of Materials*, Mie Academic Press, (1993), 599-606.
14. Kawai, M.: Formulation of inelastic constitutive model for fiber-reinforced composite materials, *Transactions of the Japan Society of Mechanical Engineers* 60-578A (1994), 2334-2341.
15. Kawai, M.: Constitutive model for nonlinear behavior of unidirectional fiber reinforced metal matrix composites, in B. Xu and W. Yang (eds.), *Proceedings of Asia-Pacific Symposium on Advances in Engineering Plasticity and Its Application*, International Academic Publishers, (1994), 95-100.
16. Boehler, J.P. (ed.): *Applications of Tensor Functions in Solid Mechanics*, CISM No. 292, Springer-Verlag, 1987.
17. Malinin, N.N. and Khadjinsky, G.M.: Theory of creep with anisotropic hardening, *Int. J. Mech. Sci.* 14 (1972), 235-246.
18. Robinson, D.N.: Constitutive relationships for anisotropic high-temperature alloys, *Nuclear Eng. Design* 83 (1984), 389-396.
19. Nouailhas, D. and Freed, A.D.: A viscoplastic theory for anisotropic materials, *Trans. ASME J. of Eng. Mater. and Technol.* 114 (1992), 97-104.
20. Kawai, M.: Coupled inelasticity and damage model for metal matrix composites, *International Journal of Damage Mechanics* 6-4 (1997), 453-478.

CREEP OF ICE AND MICROSTRUCTURAL CHANGES UNDER CONFINING PRESSURE

P.D. BARRETTE AND I.J. JORDAAN
Ocean Engineering Research Centre
Memorial University of Newfoundland
St. John's, Newfoundland
Canada

1. Introduction

Glaciers are one of the most striking examples of a material undergoing creep. The flow of these large bodies of ice is driven by self-weight and occurs as a result of the very high homologous temperature at which ice exists under atmospheric conditions. Deformation-related motions are in the order of centimetres to metres per year and the hydrostatic pressure at the base of the largest ice caps may be up to 40 MPa. Numerous investigations have been conducted to unravel the physical processes taking place in what may be considered the longest lived, *in situ* creep experiments available to material scientists.

This is in sharp contrast to the dynamically more vigorous events the ice engineering community is confronted with when attempting to estimate the strength and behaviour of ice during its interaction with man-made structures. Examples are: the stresses exerted by an ice sheet on the hull of a ship, the interaction of icebergs with offshore oil production rigs, and loading of bridge piers by river jams during spring thaws. These events may involve significant load fluctuations and hydrostatic pressures exceeding 70 MPa. Displacement rates are in the order of millimetres to metres per second (40 m/sec in the case of ice/propeller interactions). In addition to gravity, water currents and winds acting on floating ice as well as ship inertia are the main driving forces.

Mechanical testing and medium-scale field investigations are commonly carried out, either as a complement or as an alternative to the real scale event, to promote a better understanding of the processes involved in the ice during these interactions. This paper is an example of investigations conducted for that purpose. It reports on the creep and microstructural response of laboratory-made polycrystalline ice in compression, at various levels of hydrostatic pressures and up to large strains. These are conditions for which, prior to our testing program, little information was provided in the open literature. An analysis of a brittle and a ductile failure planes produced at low and high confinement is also presented. The implications of these results for the high-speed compressive loading of natural ice by a flat indentor are discussed.

S. Murakami and N. Ohno (eds.), IUTAM Symposium on Creep in Structures, 479–488.

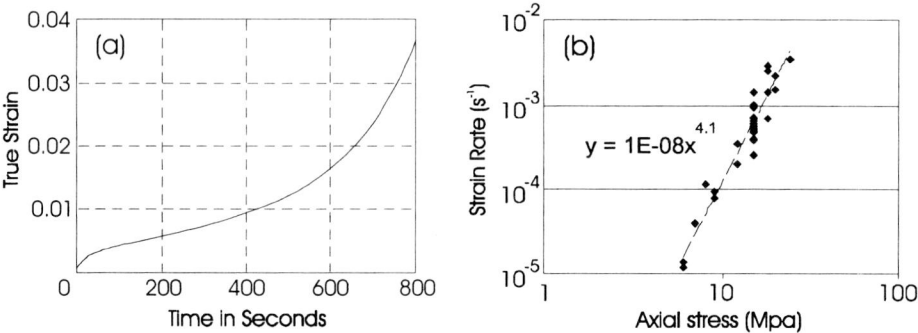

Figure 1. a) Typical result of a creep test for a confinement of 53 MPa and axial stress of 6 MPa.
b) Minimum strain rate as a function of axial stress for all tests.

2. Laboratory procedures

Machining and testing of the ice were done in a cold room at -10°C. Cylindrical specimens of polycrystalline ice measuring 70 mm in diameter and 155 mm in length were made following a procedure described in [1]. The crystals were in the order of 2 to 3 mm in size. They were equiaxed in shape and had a random crystallographic orientation. The specimens were pressurised in a triaxial cell to levels ranging from 5 to 60 MPa, and submitted to an axial compressive load up to 96 kN (25 MPa), to true strains reaching 44% [2]. The temperature of the ice during testing was between -9 and -10°C (corresponding to a homologous temperature of 0.96). A total of 34 tests are considered in this paper.

Post-testing analysis included the 'thin sectioning' of slices cut from each specimen at the desired location and orientation, following a procedure described in [3]. The end products were sections 0.5 mm in thickness welded on a glass plate that could be observed and photographed through cross-polarised transmitted light (see [4] for instance). Reflected light was also used to bring out microfractures.

3. Results

3.1. PHENOMENOLOGY

Figure 1a displays a plot of strain versus time for a typical creep test. The response of the ice is similar to that observed with common metals at high homologous temperatures. The data is compatible with Norton's power law (also referred to as Glen's law in ice engineering):

$$\frac{d\varepsilon}{dt} = A\sigma^n \tag{1}$$

where $d\varepsilon/dt$ is the strain rate and σ is the deviatoric stress, A and n are constants reflecting the response of the ice to the applied load. Figure 1b is a plot of strain rate against deviatoric stress on logarithmic scales for all tests. A stress exponent of 4.1 is consistent with earlier reports. When these constants are determined for different ranges of hydrostatic pressure (P), the material displays a lower compliance between 20 and 50 MPa (Table 1).

TABLE 1. A comparison of compliance
for three different ranges of hydrostatic pressure.

P (MPa)	A (MPa^{-n} s^{-1})	n
10	1E10^{-8}	4.1
20-50	3E10^{-8}	3.7
50-65	7E10^{-9}	4.2

A previous assessment of the data for higher levels of strain has indicated that the relationship represented by eq. 1 varies with the amount of deformation [2]. This is not unexpected as the nature of the material was shown to be in constant evolution [1,2,5]. A new formulation, with parameters that reflect the internal structural changes in the ice, was thus proposed [6]. It infers that the viscoelastic behaviour of the ice is a function of state variables. Hence,

$$\mu \propto g^{-1}(S_{\sigma 1}, S_{\sigma 2}) s^{-(n-1)} \qquad (2)$$

where μ is an apparent viscosity, s is the Von Mises stress, n is a constant, and $S_{\sigma i}$ expresses the 'damage' incurred by the ice at any given time and for a low and a high pressure regime ($i=1,2$). Since

$$\frac{d\varepsilon}{dt} = \frac{s^n}{\mu} \qquad (3)$$

which is an alternate form of eq. 1, then

$$\frac{d\varepsilon}{dt} \propto g(S_{\sigma 1}, S_{\sigma 2}) s^n \qquad (4)$$

This formulation was used to generate empirical fits to the data collected during this research program. It lends support, but for high strain levels, to the phenomenon outlined in Table 1, whereby a minimum value for the compliance of the ice is seen at mid-pressure range [6,7].

3.2. THE DEVELOPMENT OF SHEAR PLANES

A new set of experiments was conducted to investigate the development of shear planes in ice specimens upon rupture. In order to promote the development of a single yielding

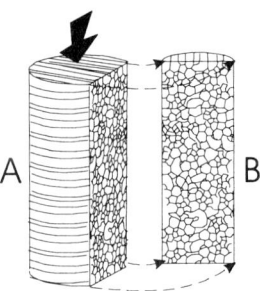

Figure 2. Columnar-grained ice. For the high pressure test, a slab (B) of the specimen (A), was removed before testing. B was thus a mirror image of A. The arrow represents the compression axis.

plane (as opposed to a uniform strain distribution in the bulk of the specimen), testing was done with columnar-grained ice (Fig. 2). The columns were oriented normal to the loading direction, thus allowing a fracture plane to make its way across the specimen with less interference from the grain boundaries. High axial stresses were also used for that purpose.

Figure 3a displays a photograph of a thin section cut perpendicular to the columns in an undeformed specimen. It should be noted that this plane is parallel to the crystallographic c-axis of the crystals, which are randomly oriented within that plane [8]. Figure 3b is a specimen deformed to a true strain of 8.5% under a confinement pressure of 10 MPa and an initial axial stress of 15 MPa (corresponding to a hydrostatic pressure of 15 MPa). The specimen failed along a plane oriented at about 45 deg. with respect to the loading axis. Figures 3c,d are a closer view of the shear plane. The presence of tilt wall boundaries in some crystals shows that at least part of the deformation was accommodated by dislocation glide. The shear plane is an intensely fractured area varying in width from less than one mm to 10 mm. Intracrystalline fracturing is noticeable in the vicinity of the failure plane and becomes almost absent away from it.

Specimen preparation for the test done at high hydrostatic pressure differed from that used for the low-pressure test just described. Prior to testing, a section of the specimen was removed and stored away (Fig. 2). This was done to preserve the undeformed state of the grain structure. The specimen was then loaded with a confinement pressure of 69 MPa and an axial stress of 30 MPa (for a hydrostatic pressure of 79 MPa). The total strain achieved after specimen rupture was 22%. The failure plane in this case was also oriented at about 45 deg. to the load axis but occurred in the lower portion of the specimen (Fig. 4). Microcracking is observed throughout the specimen, but more so near the failure plane (Fig. 4d). It is, however, substantially less developed than in the specimen deformed under low confinement. Instead, the deformation event led to the nucleation of new crystals, both inside the shear surface and throughout the specimen (Fig. 5). The result is the overall reduction in grain size of the ice, a feature that is not apparent in the low-pressure test (compare Figs. 3 and 4), where grain refinement occurs only along the failure plane and generally results from pulverisation.

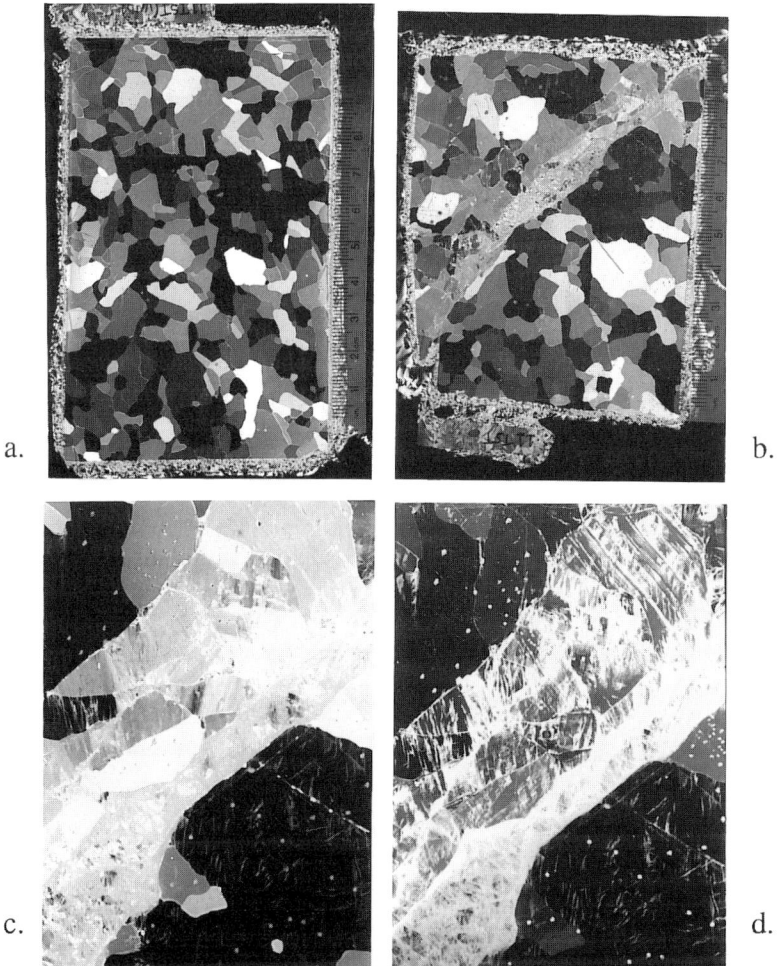

Figure 3. a) Thin section across an undeformed columnar-grained ice specimen. b) Thin section across a specimen deformed with a confinement pressure of 10 MPa and an axial stress of 15 MPa. c) Close-up view of the shear zone in (b). d) Same as (c) with side lighting. Scale in mm.

The strain-time relationships for these tests, shown in Fig. 6, are divided into three segments. Segment A is the deformation recorded by the specimen as it was being loaded to the desired level. Segment B is the creep response of the ice. It displays a slight acceleration in both cases. Segment C corresponds to specimen rupture. This last segment records the strain rate during that event: 0.28 s^{-1} and 0.53 s^{-1} for the test done at low and high hydrostatic pressure, respectively. These correspond to the value for the bulk of the specimen (and are therefore higher near the failure surfaces). The shear plane that formed at lower hydrostatic pressure displays a degree of microfracturing that is consistent with these rates. The zone that developed at high pressure, however, seems

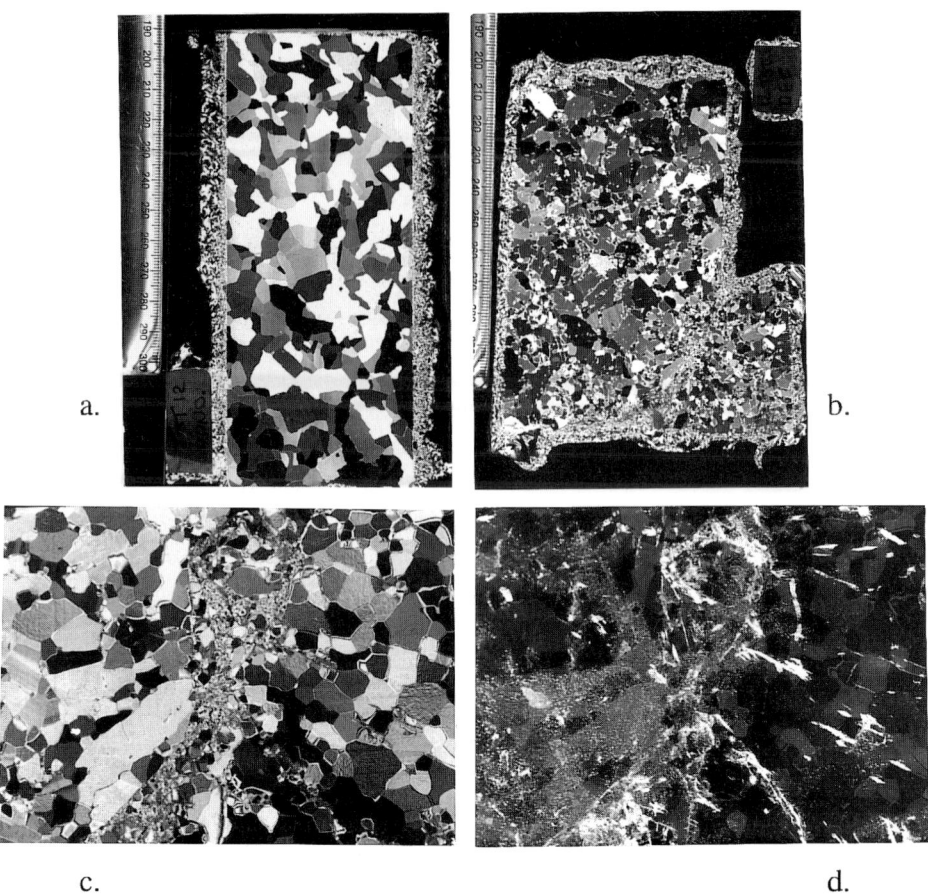

Figure 4. a) Thin section across the undeformed section of columnar-grained ice specimen. b) Thin section across a specimen deformed with a confinement pressure of 69 MPa and an axial stress of 30 MPa. c) Close-up view of the shear zone shown in (b). d) Same as (c) with side lighting. Scale in mm.

to have accommodated the displacement mostly through recrystallisation processes (in places, a line can be drawn across the shear zone without encountering a single fracture surface).

4. Discussion

Jones [9] and Jones and Chew [10] reported a peak in strength and a decrease in minimum strain rate up to 20 to 30 MPa confinements, followed by a reverse in trend upon further increase in confinement. The results of our testing program extend these

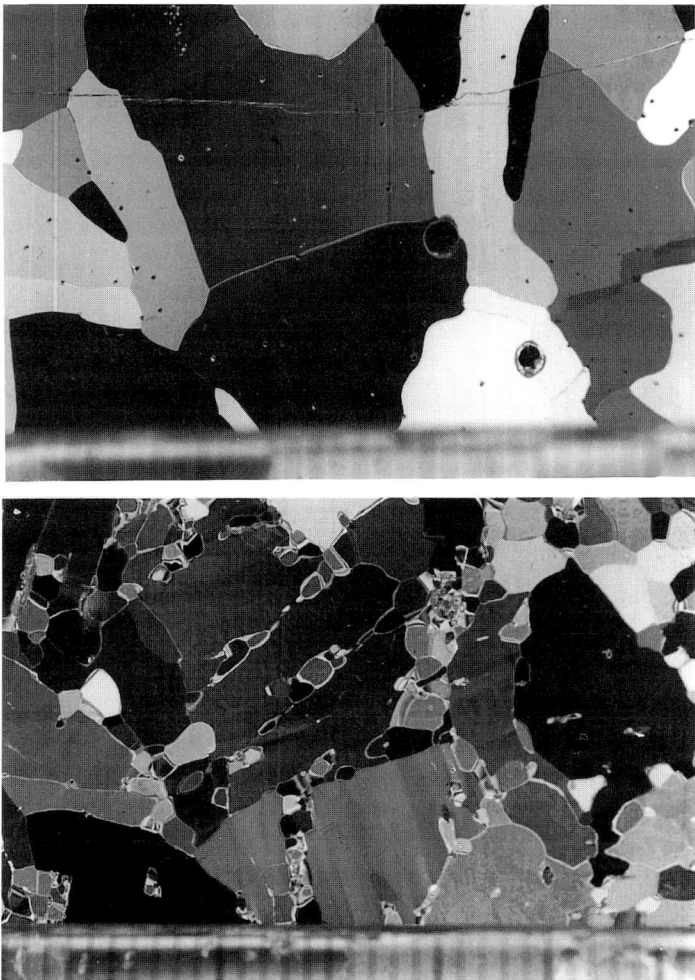

Figure 5. Columnar ice tested at high pressure from the undeformed portion of the specimen (above), and from the deformed portion (below). Note growth of new crystals at the expense of old ones. Scale in mm.

observations to higher rates of deformation and higher strain levels. We further document the deformation mechanisms taking place in the ice at various confinement levels. The contrast in the morphology of the shear zones described in this paper is noteworthy: it indicates that dynamic recrystallisation is an effective strain-accommodating mechanism even at very high strain rates.

Contrary to most crystalline substances, the melting point of ice decreases with hydrostatic pressure (see [11] for instance). This has to do with the fact that ice expands upon freezing. For the temperature at which our tests were conducted, this pressure is

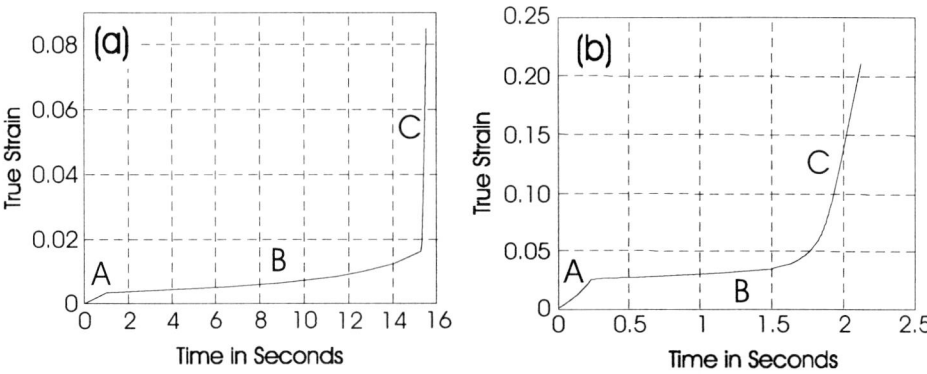

Figure 6. Deformation traces for the low pressure test (a) and the high pressure test (b) with columnar-grained ice. See text for discussion.

approximately 110 MPa [12]. A decrease in strength (or increase in minimum strain rate) at high confinement levels has thus been attributed to the generation of liquid at grain boundaries [6,9]. No direct evidence of this phenomenon could be obtained during our study.

5. Implications for ice/structure interactions

A medium-scale field testing program was carried out during which a natural ice cover was subjected to high-speed compressive loading by a flat indentor [13,14]. For the purpose of these tests, the indentor and its actuator were located in a trench that was dug into the ice (Fig. 7). The forces applied by the indentor, actuator displacement, indentor strains and local pressures were amongst the parameters measured during the interaction events [14,15]. Figure 8 displays an example of the load response. This data was recently attributed to cyclic processes taking place within a 'damaged' layer that developed along the ice-indentor interface [6]. The processes involve the occurrence of high pressure zones in localised areas along the ice structure interface, where the ice appears to be strongly recrystallised. This is ascribed to their position at the interface where the hydrostatic pressure builds up as a result of the interaction while they sustain the bulk of the load [6].

A finite element simulation with the physical basis described above was able to reproduce the force-time response of the ice and the development and propagation of a damaged layer [6]. Our earlier investigations [1,2,5] further provide a useful framework to explain the mechanical and microstructural response of the ice. In this paper we report a style of deformation that contrasts with the level of confinement. The occurrence of a brittle shear surface under low hydrostatic pressure and leading to intense fracturing may be akin to what is occurring in the ice away from high pressure zones. A high confinement, on the other hand, led to the production of a ductile shear

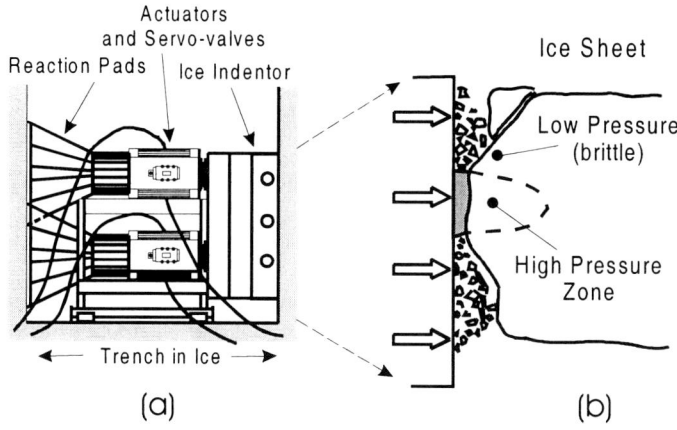

Figure 7. a): Field set-up for the medium-scale indentation event. b) Ice failure near the ice/structure interface, showing an example of a high pressure zone.

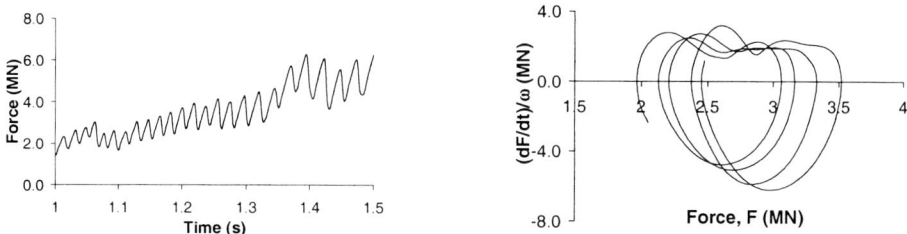

Figure 8. Left: Load response of the ice with time. Right: Force against normalised rate of change of force, emphasising the repetitive nature of the load cycles.

surface, where recrystallisation is the main strain-accommodating mechanism. It was generated in an ice specimen while the displacement rate between the platens was 90 mm/sec, which is comparable to the range of indentor displacement rates used during the medium-scale interaction events [14,15]. This feature thus provides additional evidence that recrystallisation can operate even at such high deformation rates and may reproduce the nature of the damage in the ice during a real interaction where the hydrostatic pressure is high.

Additional work is being done to further investigate this subject as well as to attempt to obtain direct evidence of pressure melting. Mechanical testing will also be conducted to enlarge our phenomenological database and to look at the effects of other important parameters such as ice temperature.

6. Acknowledgements

Financial support for this research was provided by the Natural Sciences and Engineering Research Council of Canada, the Government of Canada through the Program on Energy Research and Development, and the Canada-Newfoundland Offshore Development Fund.

7. References

1. Melanson, P., Meglis, I., Jordaan, I.J., and Stone, B.M. (2000) Microstructural change in ice: I. Constant deformation-rate tests under triaxial stress conditions, *J. Glaciology*, in press.

2. Meglis, I., Melanson, P. and Jordaan I.J. (2000) Microstructural change in ice: II. Creep behavior under triaxial stress conditions, *J. Glaciology*, in press.

3. Sinha, N.K. (1977) Dislocations in ice as revealed by etching, *Philosophical Mag.* **36**, 1385-1404.

4. Bloss, F.D. (1961) An introduction to the methods of optical crystallography. Holt, Rinehart and Winston, Toronto.

5. Stone, B.M., Jordaan, I.J., Xiao, J. and Jones, S.J. (1997) Experiments on the damage process in ice under compressive states of stress, *J. Glaciology*, **43**, 11-25.

6. Jordaan, I.J., Matskevitch, D.G. and Meglis, I.L. (1999) Disintegration of ice under fast compressive loading, *Int. J. Fracture*, **97**, 279-300.

7. Melanson, P.M., Jordaan, I.J. and Meglis, I.L. (1998) Modelling of damage in ice, Proceedings, 14th IAHR Symposium on Ice, Potsdam, New York.

8. Perey, F.G.J. and Pounder, E.R. (1958) Crystal orientation in ice sheets, *Can. J. Physics*, **36**, 494-502.

9. Jones, S.J. (1982) The confined compressive strength of polycrystalline ice, *J. Glaciology*, **28**, 171-177.

10. Jones, S.J. and Chew, H.A.M. (1983) Creep of ice as a function of hydrostatic pressure, *J. Physical Chemistry*, **87**, 4064-4066.

11. Bridgman, P.W. (1952) The physics of high pressure. G.Bell and Sons Ltd, London.

12. Nordell, B. (1990) Measurement of P-T coexistence curve for ice-water mixture. *Cold Regions Science Technology*, **19**, 83-88.

13. Frederking, R.M.W., Jordaan, I.J. and McCallum, J.S. (1990) Field tests of ice indentation at medium scale: Hobson's choice ice island, 1989. Proceedings, 10th IAHR Symposium on Ice, Espoo, Finland, 931-944.

14. Masterson, D.M., Frederking, R.M.W., Jordaan, I.J. and Spencer, P.A. (1993) Description of multi-year ice indentation tests at Hobson's Choice Ice Island, Proceedings, 12th Int. Conference OMAE, Glasgow, Scotland, 145-155.

15. Gagnon, R.E. (1998) Analysis of visual data from medium scale indentation experiments at Hobson's Choice Island. *Cold Regions Science Technology*, **28**, 45-58.

NUMERICAL AND EXPERIMENTAL CREEP BENDING BEHAVIOR OF POLYETHYLENE BEAMS

T. HIROE, H. MATSUO, K. FUJIWARA AND F. OHASHI
Faculty of Engineering, Kumamoto University
Kurokami 2-39-1, Kumamoto 860-8555, Japan

Abstract

The monotonic compressive loading tests previously performed for high-density solid polymers including polyethylene (PE) revealed that the inelastic deformation behavior of these engineering plastics is remarkably dependent on loading rate and temperature under the condition of normal use. The viscoplastic constitutive theory based on overstress (VBO) had reproduced such monotonic uniaxial deformation of the polymers successfully. In this study the VBO model is applied to analyze the creep behavior of PE beams subjected to a linearly increasing moment which is subsequently held constant. The analysis based on the Bernoulli-Euler beam theory shows an analogy between the curvature-moment and uniaxial stress-strain relations, and the existence of two possible states of equilibrium: termination of primary creep or secondary creep.

In the numerical analysis, the basic equations are transformed into nonlinear ordinary differential equations by the use of Gaussian quadrature for the spatial integrals, and the numerical solution is performed using material functions; equilibrium, viscosity and viscosity-control functions for PE at 25℃ determined in the compressive loading tests. The numerical experiments of three kinds of loading cases show the significant loading rate effects, illustrating that the curvature increase and the spatial distribution of stresses at the end of the moment increase depend on the moment rate. During the creep period the curvatures increase and reach the equilibrium moment curve. The rate effects on stresses disappear with time when stresses are redistributed and reach the equilibrium stress curve.

In order to verify the numerical results, four-point bending tests are performed for PE at 25℃ using a hydraulic servo-controlled testing machine within a rather small deformation range of loading cases. The observed deformation of lattice-ruled beam surfaces indicates that the Bernoulli-Euler hypothesis is valid and the applied bending moment needs some modifications. The time-histories of curvatures measured with use of strain gages show a fairly good correspondence with the numerical results.

1. Introduction

Recent growing use of solid polymers as primary structural members in modern society

489

S. Murakami and N. Ohno (eds.), IUTAM Symposium on Creep in Structures, 489–498.

has been desirous of reliable deformation analysis as well as metals. Most constitutive models in the integral representation had not succeeded in reproduction of nonlinear and unique rate-dependent behavior observed in polymeric materials loaded in a servo-controlled testing system. Presently a differential model, the viscoplastic theory based on overstress (VBO) developed by Krempl (1982) has been adopted in various version forms. An early version of the VBO model was applied successfully only for monotonic loaded deformation of solid polymers by Kitagawa et al. (1992) and Ariyama et al. (1992). The updated version of the VBO model has shown modeling capabilities of responses including for cyclic loading in Krempl (1998) and Krempl and Bordonaro (1998), but the model has not simulated the structural deformation of polymeric materials.

An early version of the VBO model modified by Hiroe and Igari (1985) has well reproduced monotonic deformation behavior under time-varying strain-rates and temperatures effectively on the concept of time-temperature equivalence for high-density polymers of polyethylene (PE), polypropylene (PP) and polycarbonate (PC) in Hiroe et al. (1995, 1998, 1999). In this study, the model is applied to calculate the creep bending behavior of PE beams similarly in the previous studies of Hiroe and Krempl (1982, 1984) and Delph (1981) with use of another unified model for metals. Bending creep tests are also performed for PE beams at 25℃ in addition to creep bending analyses and demonstrative numerical experiments.

2.   Creep bending analysis and equilibrium solutions

2.1   CONSTITUTIVE EQUATION AND BEAM BENDING MODEL

The uniaxial form of the modified early version of the VBO model used in this study for PE is represented as a relation of the engineering stress σ and strain ε :

$$\dot{\varepsilon} = \frac{\dot{\sigma}}{E} + \frac{X}{E \cdot K[X, \beta[\varepsilon]]} \, , \quad X = \sigma - g[\varepsilon] \tag{1}$$

where g: equilibrium stress function, X: overstress, K: viscosity function、 β: viscosity control function and E: modulus of elasticity. A superposed dot and square brackets denote differentiation with respect to time t and "function of" respectively.

In this study, pure bending of a beam with a symmetrical cross section is taken into consideration. According to the Bernoulli-Euler beam theory, the longitudinal total strain ε in the beam is given by

$$\varepsilon[y, t] = y \cdot \gamma[t] \tag{2}$$

where γ: curvature and y: a distance from the neutral axis. The axial force and moment balance yield

$$\int_{-y_2}^{y_1} \sigma[y, t] b[y] dy = 0 \tag{3}$$

$$M[t] = \int_{-y_2}^{y_1} y \, \sigma[y, t] b[y] dy \tag{4}$$

where b: width of the section and y_1, y_2: distances from the neutral axis to the outermost fibers. Equation (1) and differentiating of Eqs. (2) and (4) yield

$$\dot{\gamma} = \frac{1}{EI} \int_{-y_2}^{y_1} y \frac{X}{K[X, \beta[\varepsilon]]} b \, dy + \frac{\dot{M}}{EI} \tag{5}$$

where I is the area moment of inertia. Differentiating Eq. (2) and substitution of (1) and (5) yield

$$\dot{\sigma} = \frac{y}{I} \int_{-y_2}^{y_1} y \frac{X}{K[X, \beta[\varepsilon]]} b \, dy + \frac{\dot{M}}{I} y - \frac{X}{K[X, \beta[\varepsilon]]} \tag{6}$$

Eqs. (3), (5) and (6) describe the behavior of the beam. (Eq. (3) is used for determining the neutral axis.) The initial and loading conditions for monotonic creep bending are

$$t = 0 \quad : \quad \gamma = \sigma = 0 \quad , \quad t_0 > t > 0 \quad : \quad \dot{M} = const$$

$$t \geq t_0 \quad : \quad \dot{M} = 0 \, (M = M_0)$$

Furthermore when the cross section has symmetrical neutral axis (height: $2h$), Eq. (5) can be rewritten using integration by parts as follows

$$\dot{\gamma} = \frac{\dot{M}}{EI} + \frac{1}{EI} \frac{M - G}{\overline{K}} \tag{7}$$

where

$$\frac{1}{\overline{K}} = \frac{2}{K[\sigma[h] - g[\gamma h] \beta[\gamma h]]} \tag{8.a}$$

$$G = Me + \overline{K}F \tag{8.b}$$

$$F = 2 \int_0^h \left\{ \int_0^y y(\sigma - g[\varepsilon]) b \, dy \right\} \frac{d}{dy} \left(\frac{1}{K[\sigma - g[\varepsilon] \beta[\varepsilon]]} \right) dy \tag{8.c}$$

$$Me = 2 \int_0^h y \, g[\varepsilon] b \, dy \tag{8.d}$$

The moment Me is denoted as the equilibrium moment and G approaches Me as σ approaches g everywhere. A comparison of (7) and (1) shows an analogy between the two equations, and the creep behavior of beam bending is predicted to be similar to the uniaxial creep behavior as noticed in the previous works.

2.2   EQUILIBRIUMS OF CREEP DEFORMATION

In case of the uniaxial creep ($\sigma = \sigma_0$), two equilibrium solutions are obtained from Eq. (1) under the conditions of Eqs. (9) and (10).

$$\sigma = g[\varepsilon] \tag{9}$$

$$\sigma_0 - g[\varepsilon] = const \tag{10.a}$$

$$\beta[\varepsilon] = const \tag{10.b}$$

For condition (9), the overstress vanishes and all the rates are zero, and for conditions (10) beyond some strain, the other equilibrium: pseudo-equilibrium (constant creep rate) is realized where $g[\varepsilon] = const$, ($dg/d\varepsilon = g' = 0$) $d\beta / d\varepsilon = \beta' = 0$ beyond some strain.

In case of the beam bending creep, similar equilibrium solutions are considered. If condition (9) holds everywhere in the beam and if $\dot{\gamma} = \dot{M} = 0$, Eqs. (7) and (8) yield

$$M = G = Me \tag{11}$$

so that the first equilibrium is possible in the beam if the applied moment is equal to the equilibrium moment. During the creep, Eq. (7) is rewritten as the following integral form, which shows that infinite time is required theoretically to reach this equilibrium.

$$EI \int_{\gamma_0}^{\gamma} \frac{d\gamma \, \overline{K}}{M_0 - G} = t - t_0 \tag{12}$$

To show that secondary creep: $\dot{\gamma} = const$ is a possible solution of Eq. (7), it is supposed that Eq. (10) is true for every fiber of the beam. This assumption is equivalent to postulating $g' = 0$, $\beta' = 0$ beyond some strain. In this case, the value of Me approaches a constant value M_∞ and when $M_0 > M_\infty$, this equilibrium is also realized theoretically after the infinite elapsed time. PE used in this study does not possess such characteristics expressed in Eq. (10).

3. Numerical simulations

3.1 NUMERICAL SCHEME

In numerical calculation of creep bending, a beam with a rectangular cross-section (height: $2h$, width: b, $b/h = 1$) is considered for simplicity. Every parameter is non-dimensionalized as follows: $\overline{y} = y/h$, $\overline{b} = b/h = 1$, $\overline{\gamma} = \gamma h$ ($= \varepsilon_{max}$), $\overline{\gamma}_y = \sigma_y / E = \gamma_y h$, $\overline{I} = I/h^4 = 2\overline{b}/3 = 2/3$, $\overline{M} = \dot{M}/M_y$, where $M_y = \sigma_y I/h$, σ_y is the zero-strain intercept when the slop line of $g[\varepsilon]$ curve is back extrapolated. Equation (5) is rewritten as

$$\dot{\overline{\gamma}} = \frac{2}{E\overline{I}} \int_0^1 \overline{y} f[\sigma, \overline{y}\overline{\gamma}] \overline{b} \, d\overline{y} + \frac{\dot{\overline{M}} \sigma_y}{E} \tag{13}$$

where $f[\ ] = X/K[\ ]$. Substituting Eq. (13) into Eq. (6) leads to

$$\dot{\sigma} = \frac{2}{\overline{I}} \overline{y} \int_0^1 \overline{y} f[\sigma, \overline{y}\overline{\gamma}] \overline{b} \, d\overline{y} + \dot{\overline{M}} \sigma_y \overline{y} - f[\sigma, \overline{y}\overline{\gamma}] \tag{14}$$

In order to approximate the spatial integrals in Eqs. (13) and (14), a Gaussian quadrature was adopted as

$$\int_0^1 \overline{y} f[\sigma, \overline{y}\overline{\gamma}] d\overline{y} = \sum_{i=1}^{Ng} \omega_i \left(\frac{1}{2}\right) \overline{y}_i \, f[\sigma_i, \overline{y}_i \overline{\gamma}] + R$$

$$\cong \sum_{i=1}^{Ng} \omega_i \left(\frac{1}{2}\right) \overline{y}_i \, f[\sigma_i, \overline{y}_i \overline{\gamma}]$$

$$\tag{15}$$

where $\overline{y}_i = \frac{1}{2}(\tilde{y}_i + 1)$, $i = 1 \sim Ng$: point number of Gaussian quadrature, ω_i: the weight factor of the i-th Gaussian point, y_i: the position of i-th point, R: remnant. As the results, Eqs. (13) and (14) are written as a system of ($Ng +1$) nonlinear ordinary differential equations of (16) and (17).

$$\dot{\overline{\gamma}} = \frac{1.5}{E} \sum_{i=1}^{Ng} \omega_i \, \overline{y}_i \, f[\sigma_i, \overline{y}_i \overline{\gamma}] + \frac{\dot{\overline{M}} \sigma_y}{E} \tag{16}$$

$$\dot{\sigma}_i = 1.5\bar{y}_i \sum_{i=1}^{Ng} \omega_i\, \bar{y}_i\, f\!\left[\sigma_i, \bar{y}_i\bar{\gamma}\right] + \dot{M}\,\sigma_y\,\bar{y}_i - f\!\left[\sigma_i, \bar{y}_i\bar{\gamma}\right] \quad , i = 1\sim Ng \tag{17}$$

Numerical solutions are obtained by computing above equations with time-histories of applied moment.

3.2 NUMERICAL CONDITIONS FOR HIGH-DENSITY POLYETHYLENE BEAMS

3.2.1 *Material functions and constants in the VBO model for PE*
High-density PE (0.95g/cc) used in this study was purchased in extruded rods of 30 or 60mm diameter. In the preliminary study, they were machined in the same longitudinal direction to column specimens of $20^\phi \times 30^l$mm, and monotonic compressive loading tests at three strain-rates and a 10^4 s relaxation test at the maximum strain are performed at 25°C to determine the material functions and constants in the VBO model. True stress based on volume-constant-hypothesis and logarithmic strain is employed. Figure 1 shows the observed stress-strain curves and estimated g function, and the stress σ_y used for non-dimensionalization is indicated as a symbol on the stress axis. The material functions are derived from the rewritten expression of Eq. (1) using these data as described in Hiroe et al. (1995). The computed stress-strain diagrams are also illustrated for comparison in the figure.

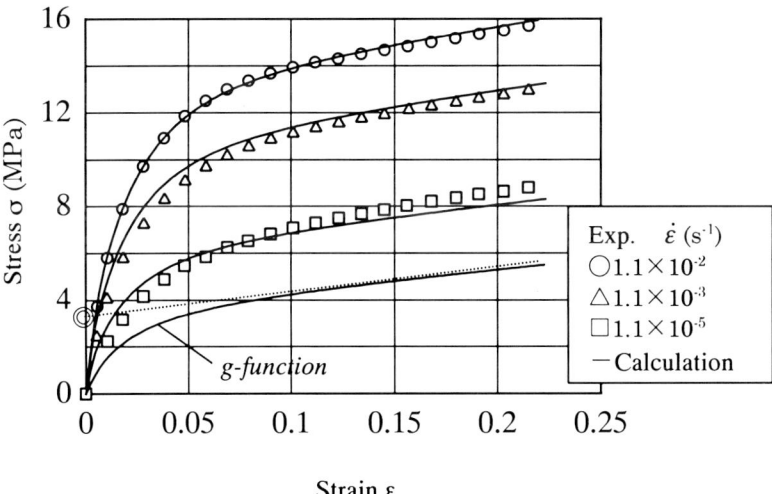

Fig. 1   Stress-strain curves of compressive loading tests
and numerical results with g-function

3.2.2 *Numerical test cases*

In this study, ten-point Gaussian quadrature with weight factors of 30 digits was used for spatial integrals, and eleven nonlinear ordinary differential equations were solved with the relative error tolerance of 10^{-7} using the method in Bulrish-Stoer (1966). The locations of Gauss points Si (i =1-10) are given in Table 1. Table 2 shows the details of three cases studied for PE. Cases 1 and 2 differ by a hundred times of the loading rate and case 3 does from others by the applied creep moment and the loading rate. The hold time for loaded bending moment is 10^6 s.

Table 1   Positions of Gauss points

G. P	S1	S2	S3	S4	S5	S6	S7	S8	S9	S10
y_i/h	.013	.067	.160	.283	.426	.574	.717	.840	.933	.987

Table 2   Loading conditions of numerical experiments

Case	$\dot{M}/My$ (s^{-1})	M/My	Hold Time (s)
1	2.0×10^{-4}	2.0	10^6
2	2.0×10^{-2}	2.0	10^6
3	2.0×10^{-3}	2.5	10^6

3.3   RESULTS AND DISCUSSIONS

Figure 2 shows a plot of normalized moment M/M_y vs. normalized curvature γ/γ_y. The equilibrium moment M_e is also shown. For Cases 1 and 2, an effect of loading rate is demonstrated where the curvature at a hundredth of loading rate increases 2.7 times. During the creep period the curvatures grow four- to twelve-fold and almost reach the equilibrium moment curve. The redistribution of the stresses σ_i at the Gauss points Si is depicted for three cases on a stress-strain diagram with g-function in Fig. 3. The difference of stress-distribution due to loading rate between Cases 1 and 2 disappears during the creep deformation. In general the outer fiber stresses drop and inner ones increase approaching the g-curve, but in Case 3 the stresses are redistributed rather non-monotonically. The time history of such stress-redistribution and the curvature are graphed for Cases 1 and 2 in Fig. 4. It is known that the initial effect of loading rate disappears after 10^5 s (25 hours) and the equilibrium solution for practical purposes is reached after 10^6 s (12 days).

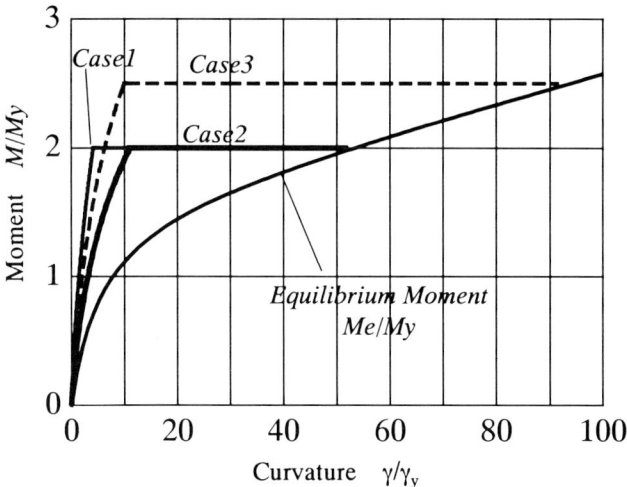

Fig. 2   Numerical moment ratio vs. curvature ratio
          for Cases 1,2 and 3

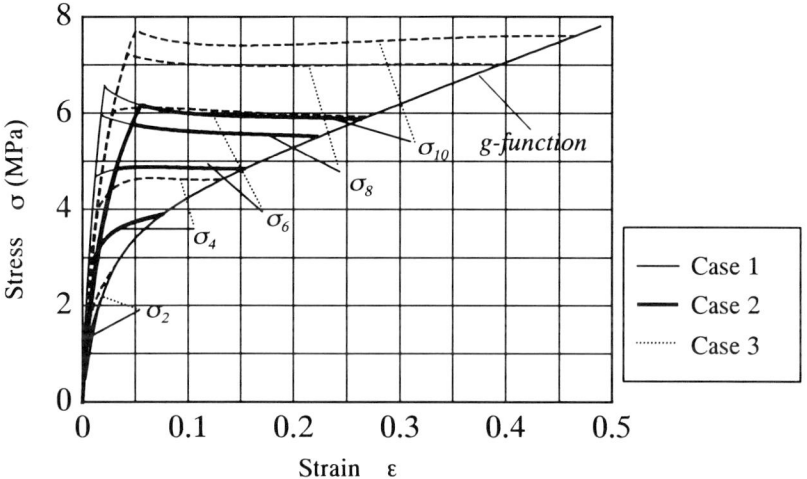

Fig. 3   Numerical stress vs. strain with g-function
          for Cases 1,2 and 3

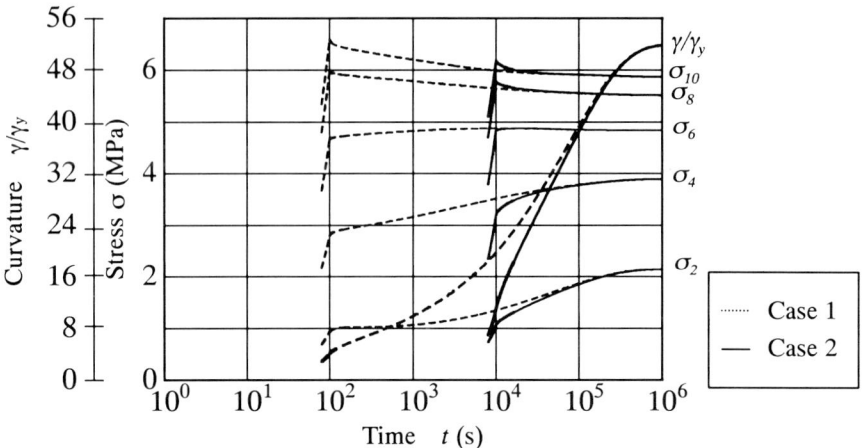

Fig. 4   Numerical time-histories of stresses and curvature
for Cases 1 and 2

4. Experiments and discussions

Four point bending tests are performed for PE beams to verify the numerical study. The specimens are machined from the same extruded PE rods as those used in the preliminary study to the prismatic beams of a rectangular section ($b \times 2h$: 20×40mm) and a length of 300mm. They are loaded at 25℃ in a temperature-controlled water chamber using a full-digital servo-hydraulic testing system. The distances between external forces and supports are set to 80 mm ($= L_o$) and 200mm ($= 2L + L_o$) respectively. Time histories of curvature are observed measuring the strains at the top and bottom surfaces with use of high-elongation foil strain gages type KEF-5-120 C1 (max. strain: 15%) and a data-logger & analyzer.

Table 3   Loading conditions of four-point bending tests

Case	$\dot{M}/My\ (s^{-1})$	M/My	Hold time (s)
A	3.6×10^{-2}	3.6	500
B	7.2×10^{-2}	3.6	500
C	4.5×10^{-2}	4.5	350

The test conditions for creep bending are given in Table 3. The specimens are subjected to rather similar loads for the restrictions in strain measurement. Figure 5 shows a typical side view of a bent PE beam.   Lattice lines ruled on the side surface of

Fig. 5   A typical side view of a bent PE beam
(Lattice lines are ruled on the side surface)

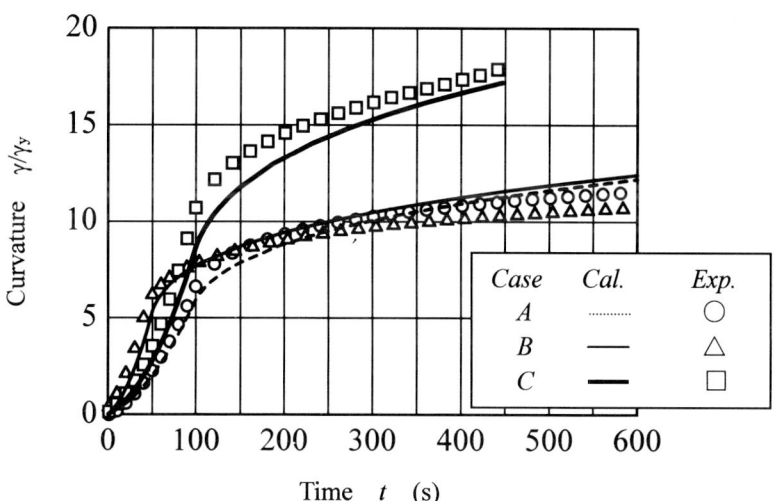

Fig. 6   Experimental and numerical time-histories of curvature ratio
for Cases A, B and C

the beam indicate that the Bernoulli-Euler hypothesis is valid within such a flexure range. This photo also suggests that some modifications to M $(= P \cdot L/2$, P: the external force) are required to calculate an actual bending moment M_{exp} taking deflection and friction into consideration approximately as follows:

$$M_{exp} = Q \cdot (L' - 2\mu h) \tag{18}$$

where the perpendicular force $Q = P/(2 \cdot \cos\theta)$; the deflection angle $\theta = \gamma \cdot L_o/2$; $L' = (L - 2h \cdot \sin\theta)/\cos\theta$. The coefficient of quasi-static friction is assumed as $\mu = 0.34$ referring to Nielson (1975). Therefore in this experiment the applied moment is not strictly held constant while the external force is sustained.

The beam bending tests is numerically simulated in cooperation with Eq. (18) and the results are compared with experiments as shown in Fig. 6. There exists some amount of discrepancy but it seems that the computational results have reproduced the experiments practically within such deformation range.

The VBO model used in this study is valid only for monotonic loading and the model is basically a small strain theory. Therefore further studies on update modeling for cyclic loading, large deformation and multiaxial loading are necessary in order to simulate the deformation of solid polymers as structural members precisely.

References

Krempl, E. (1982) The role of servo-controlled testing in the development of the theory of viscoplastic based on total strain and overstress, *ASTM STP* **765**, 5-23.

Kitagawa, M., Qui, J. and Mizutani, K. (1992) Stress-strain curves for polyethylene and polypropylene at finite strain under combined tension and torsion, *Trans. JSME Series A* **59**-548, 621-626.

Ariyama, T., Sakuma, M. and Kaneko, K. (1992) Viscoelastic-plastic deformation characteristics of polypropylene after cyclic preloading, *Trans. JSME Series A* **58**-556, 113-118.

Krempl, E. (1998) Some general properties of solid polymer inelastic deformation behavior and their application to a class of clock model, *J. Rheology* **42**, 713-725.

Krempl, E. and Bordonaro, C. M. (1998) Non-proportional loading of Nylon 66 at room temperature, *Int. J. Plasticity* **14**-(1-3),245-258.

Hiroe, T. and Igari, T. (1985) An application of the viscoplasticity to the inelastic analysis at elevated temperature, *Trans. JSME Series A* **51**-461, 248-253.

Hiroe, T., Matsuo, H., Fujiwara, K., Miyata, M., Shibata, Y., and Sakai, K., (1995) Viscoplastic deformation of high-density polyethylene , *Trans. JSME Series A* **61**-584, 53-59.

Hiroe, T., Matsuo, H., Fujiwara, K. and Tsuda, Y. (1998) Time-temperature equivalence for the stress-strain behavior of high density solid polymers, *Trans. JSME Series A* **64**-624, 69-74.

Hiroe, T., Matsuo, H., Fujiwara, K. and Ohashi, F. (1999) Experimental and numerical deformation behavior under time-varying strain-rates and temperatures for high-density polymers, *Proc. of 4th Int. Conf. on Constitutive Laws for Engineering Materials*, 449-452.

Hiroe, T. and Krempl, E. (1982) Creep behavior of beams using the viscoplasticity theory based on total strain and overstress, *ASME PVP* **60**, 31-42.

Hiroe, T. and Krempl, E. (1984) The viscoplasticity theory based on overstress applied to the analysis of beams under monotonic and cyclic creep loading, *Nuclear Engineering and Design* **77**,139-148.

Delph, T.J.(1981) Creep, relaxation and cyclic behavior of a beam using a state-variable constitutive model, *Nuclear Engineering and Design* **65**, 411-421.

Buirish, R. and Stoer, J. (1966) Numerical treatment of ordinary differential equations by extrapolation methods, *Numerishe Mathematik* **8**, 1-13.

Nielson, L. E. （1975） *Mechanical properties of polymers and composites*, Marcel Dekker, Inc..

A NUMERICAL ELASTIC-VISCOPLASTIC COLLAPSE ANALYSIS OF CIRCULAR CYLINDRICAL SHELLS UNDER AXIAL COMPRESSION

L.P. MIKKELSEN
Assistant Professor
Aalborg University,
Dept. of Building Technology and Structural Engineering,
Sohngaardsholmsvej 57, DK-9000 Aalborg, Denmark

1. Introduction

In offshore, aerospace and car industries, circular cylindrical shells are frequently used as structural components. In many applications, this is due to a requirement for a high axial compressive strength-to-weight ratio and an ability to absorb energy during complete structural collapse. On the other hand, the high imperfection sensitivity of circular cylindrical shells under axial compression requires a thorough knowledge of the occurrence of geometric imperfections and their influence on the collapse behaviour of the structure. In addition to the geometric imperfection sensitivity, a rate dependence of the material behaviour may have a significant influence on the collapse behaviour.

The transition from a dominating non-axisymmetric (diamond) collapse mode for thin-walled tubes to an axisymmetric (concertina) mode for more thick-walled tubes, has been well known to experimentalists for decades (Batterman, 1965; Lee, 1962). A transition which to a great extent is dependent on the material behaviour.

A time-independent plastic bifurcation analysis shows that the collapse behaviour of a circular cylindrical shell is governed by an axisymmetric bifurcation point with a buckling mode in the shape of a sinusoidal function in the axial direction. In the post-buckling region this bifurcation point is followed by a secondary bifurcation into a non-axisymmetric sinusoidal mode with an axial wavelength given by the double of the wavelength of the axisymmetric mode. At the maximum loading point the initial periodic buckling mode will localize into one or a few buckles in the axial

499

S. Murakami and N. Ohno (eds.), IUTAM Symposium on Creep in Structures, 499–508.

direction, a localization initiated by e.g. a boundary condition or a geo-metrical localized imperfection. A plastic bifurcation analysis shows that a low hardening material or a low yielding material tends to favour the axisymmetric buckling mode, a trend which can be specified as the am-plitude of the axisymmetric collapse mode at the point for occurrence of the secondary non-axisymmetric bifurcation point (Tvergaard, 1983a). The collapse modes dependency of the material parameters agree nicely with experimental observations, show axisymmetric collapse modes for radius thickness ratio as low as $R/h = 150$ for low hardening mild steel, (Shridha-ran et al., 1981) and non-axisymmetric modes for $R/h > 4$ for rigid P.V.C. (polyvinylchloride) with a high initial yield stress, $\sigma_0/E = 0.02$, (Johnson et al., 1977).

For a time-dependent elastic-plastic material behaviour, the governing role played by the critical bifurcation point is replaced by a strong sensi-tivity to small initial imperfections, (Tvergaard, 1985). Only an unrealisti-cally high elastic bifurcation point exists for the viscoplastic case, (Obrecht, 1977). Therefore, an analysis of the elastic-viscoplastic collapse behaviour is performed by a full numerical analysis.

The present full numerical analysis is based on an incremental finite element discretization of a finite rotation shell theory allowing for arbitrar-ily large rotation and deflection of the shell wall, (Başar and Ding, 1990). A finite rotation shell theory is used in order to follow the initially devel-oped buckling mode far into the collapse region. The rate dependence of the material is modelled by a standard elastic-viscoplastic continuum de-scription, see e.g. (Tvergaard, 1985). Numerical examples of the influence of the prescribed deformation rate on the resulting collapse mode, on the maximum load-carrying capacity and on the energy absorption (the mean load-carrying capacity) of the circular cylindrical shell are presented.

2. Problem formulation

A moderately long elastic-viscoplastic circular cylindrical shell under ax-ial compression is analysed. The shell is kept sufficiently long so that the symmetric boundary condition will not affect the collapse behaviour, and sufficiently short so that the Euler column buckling mode will not turn out to be the most critical buckling mode. The shell has the middle surface radius R, and the thickness h . The middle surface coordinates x^1 and x^2 measure the axial and circumferential direction, respectively. The deforma-tion of the shell is assumed to satisfy the condition for a small strain theory, while a finite rotation shell theory used, see (Başar and Ding, 1990), is able to account for arbitrarily large rotations and deflection of the shell wall. In the present analysis the deformation of the shell is considered until the first contact between the shell walls is obtained, at which point the first collapse fold is assumed to be fully developed.

Two cases have been considered, one where the localization of the collapse mode in the axial direction is initiated by an geometrical localized imperfection and one where a clamped boundary condition initiates the localization of the collapse mode.

The case of an initialy localized geometrical imperfection, the collapse mode is initiated in the middle of the analysed part of the shell of the length $2L$ by an initial imperfection given by

$$\bar{w} = \frac{1}{2}\left(1 - \cos\frac{\pi x^1}{L}\right)\left(-\bar{\xi}_0 h \cos\frac{\pi x^1}{\ell_c} + \bar{\xi}_M h \cos\frac{\pi x^1}{2\ell_c}\cos\frac{M x^2}{R}\right) \quad (1)$$

where the term in the first pair of parentheses is the part which initiates the localized collapse mode similar to the one used by Tvergaard (1983a). The two terms in the second pair of parentheses are taken to be of the form of the critical bifurcation modes for the corresponding time-independent material for the first axisymmetric buckling mode with an amplitude $\bar{\xi}_0$ and the secondary non-axisymmetric buckling mode with an amplitude $\bar{\xi}_M$, respectively (Tvergaard, 1983b). The parameters ℓ_c and M are the critical half wavelength in the axial direction and the critical wave number in the circumferential direction, respectively, found by the time-independent bifurcation analysis, (Tvergaard, 1983b). Symmetric boundary condition is applied to both ends of the analysed part of the shell $x^1 = 0$ and $x^1 = 2L$ where L is given by an uneven number of the critical half wavelength ℓ_c .

The case with a clamped boundary condition at one end of the shell, the initial geometrical imperfection, is taken to be of the form

$$\bar{w} = \bar{\xi}_M h \cos\frac{\pi x^1}{2\ell_c}\cos\frac{M x^2}{R} \quad (2)$$

similar to the non-axisymmetric term in (1). The axisymmetric buckling mode and the localization behaviour are initiated by the clamped boundary condition. The analysed part of the shell has the length L equal to an even number of the critical half wavelength ℓ_c . The clamped boundary condition is applied to the end $x^1 = 0$, while a symmetric boundary condition is applied to the end $x^1 = L$.

Due to symmetry in the circumferential direction for both cases mentioned above, only one strip of the cylinder given by $0 < x^2 < \pi R/M$ is analysed applying the symmetric boundary condition to both boundaries $x^2 = 0$ and $x^2 = \pi R/M$.

A standard elastic-viscoplastic material model has been used, see e.g. (Tvergaard, 1985; Mikkelsen, 1993). In the time-independent limit the model will coincide with a J_2-flow theory. The effective viscoplastic strain rate is given by

$$\dot{\varepsilon}_e^P = \dot{\varepsilon}_0 \left(\frac{\sigma_e}{g(\varepsilon_e^P)} \right)^{1/m} \tag{3}$$

where m is the rate-hardening exponent and $\dot{\varepsilon}_0$ is a positive reference strain-rate. The term σ_e denotes the von Mises stress. The strain function $g(\varepsilon_e^P)$ represents the flow stress in a uniaxial tensile test performed at a strain rate corresponding to $\dot{\varepsilon}_e^P = \dot{\varepsilon}_0$, which in the present paper is modelled by a power law variation given by

$$\varepsilon_e^P = \frac{\sigma_0}{E} \left\{ \frac{1}{n} \left(\frac{g(\varepsilon_e^P)}{\sigma_0} \right)^n - \frac{1}{n} + 1 \right\} - \frac{g(\varepsilon_e^P)}{E} \tag{4}$$

with an initial value of $g(0)$ given by the reference stress σ_0 (initial yield stress). In (4), n is the strain hardening exponent, and E is Young's modulus.

The total strain rate $\dot{\eta}_{ij}$ is taken to be a sum of an elastic part and a viscoplastic part, where the elastic part is given by Hooke's law while the direction of the viscoplastic part is given by

$$\dot{\eta}_{ij}^P = \dot{\varepsilon}_e^P \left(\frac{3s_{ij}}{2\sigma_e} \right) \tag{5}$$

where s_{ij} is the stress deviator.

The numerical results are based on a finite element discretization of the virtual work on incremental form (Başar and Ding, 1990). Owing to the simple geometry, the shell is discretized into a rectangular mesh using a 4 node shell element in which the three displacement rates, $\dot{u}$, $\dot{v}$, and $\dot{w}$, are approximated by Hermitian cubics. The integration over the middle surface is performed by a 2×2 point Gauss quadrature, while the integration in the thickness direction is performed by a seven-point Simpson integration.

To allow for larger step size in the incremental finite element solution of the elastic-viscoplastic shell structure, a forward gradient method suggested by Peirce et al. (1984) with an interpolation parameter given by $\phi = 0.9$ is used.

3. Numerical results

The aim of the present numerical analysis is to study the effect of a material rate sensitivity on the collapse behaviour of circular cylindrical shells under axial compression. The study is focused on the influence of the material's rate-hardening exponent, m, and the prescribed average shortening rate $\dot{\varepsilon}_L$ on the first developed collapse fold. The collapse behaviour is followed until the first contact between the shell walls occurs, corresponding to the point where the curves in the figures stop. All the cases considered are given by a reference stress $\sigma/E = 0.001$, a strain hardening exponent $n = 10$ and Poisson's ratio $\nu = 0.3$. For a shell given with a radius-to-thickness ratio $R/h = 100$ this corresponds in the time-independent limit to a shell with a critical half wavelength $\ell_c \approx 0.19R$.

3.1. RATE HARDENING SENSITIVITY

In figure 1, circular cylindrical shells are compressed at a prescribed average shortening rate, $\dot{\varepsilon}_L$, given by the reference strain rate $\dot{\varepsilon}_L = \dot{\varepsilon}_0$. The geometry of the shells is given by the radius-to-thickness ratio, $R/h = 100$, with an initial localized imperfection of the form (1) where $M = 5$ and $\bar{\xi}_0 = \bar{\xi}_5 = 0.01$. The length of the analysed part is given by $L = 11\ell_c$.

A viscoplastic case, $m = 0.05$, ais compared with the corresponding time-independent J_2-flow theory. For a uniform deformation state, the two curves in Figure 1 would have to coincide, but due to the inhomogeneous deformation state a difference is observed. Similar to (Mikkelsen, 1995) a rate-hardening effect of the viscoplastic material increases the maximum load peak and the energy absorption during the initial buckling behaviour.

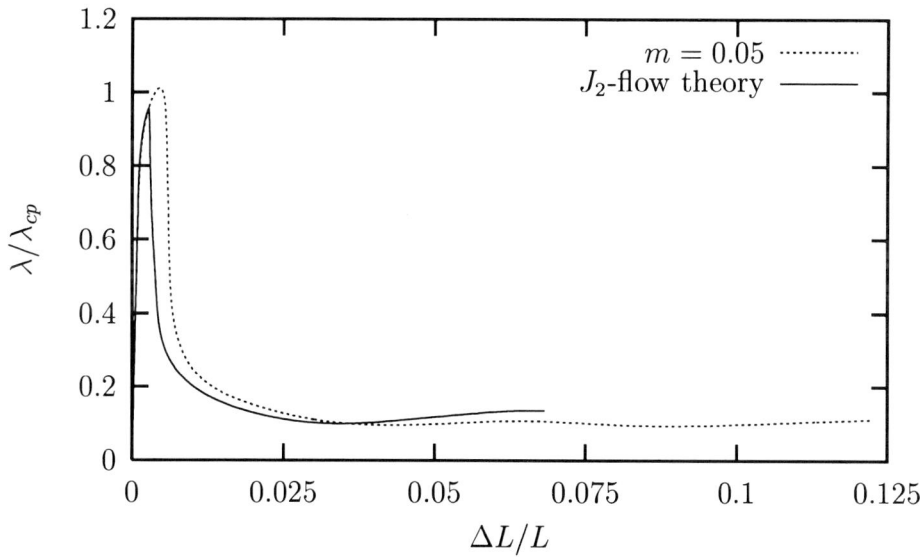

Figure 1.   Influence of the rate-hardening exponent m on the load versus shortening behaviour for a circular cylindrical shell with $R/h = 100$, $\sigma_0/E = 0.001$, $\nu = 0.3$, $n = 10$, $\bar{\xi}_0 = \bar{\xi}_5 = 0.01$ and $L = 11\ell_c$ compressed at a prescribed rate given by $\dot{\varepsilon}_L = \dot{\varepsilon}_0$.

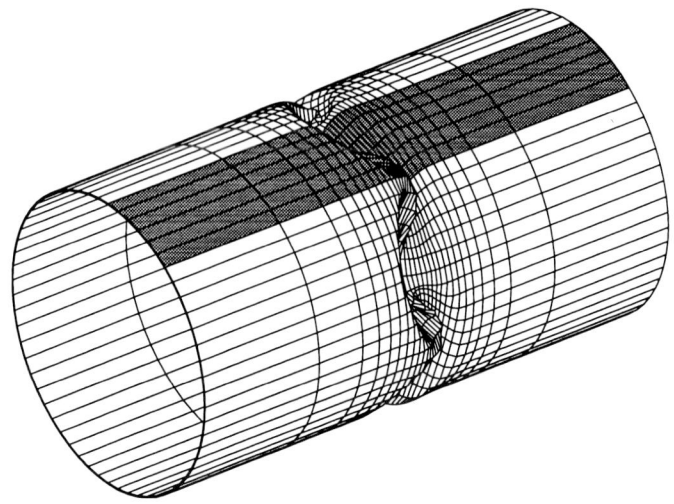

Figure 2. The deformed shell at the point where the first contact between the shell wall is obtained for a shell with $R/h = 100$, $\sigma_0/E = 0.001$, $\nu = 0.3$, $n = 10$, $m = 0.05$, $\dot{\varepsilon}_L = \dot{\varepsilon}_0$, $\bar{\xi}_0 = \bar{\xi}_5 = 0.01$ and $L = 11\ell_c$. The gray region indicates the analysed part of the shell.

On the other hand, the rate-hardening material behaviour has a delaying effect on the development of a contact point between the shell walls. An effect also observed for axisymmetric collapsing shell, see (Mikkelsen, 1999). Note that a higher material hardening effect in the viscoplastically deforming part of the shell results in a higher radius of curvature of the developed fold in the collapsed shell compared with the corresponding time-independent case, see (Mikkelsen, 1999). In Figure 1, the delayed contact point has the effect that even though the maximum load peak increases by approximately 6% for the case $m = 0.05$ compared with the time-independent case, the mean load-carrying capacity developing the first collapse fold decreases by 8 1/2%.

Figure 2 shows the deformed viscoplastic shell, $m = 0.05$, at the end point on the load versus deformation curve in Figure 1, where the first contact between the shell walls is obtained. The gray region shows the part of the shell which is analysed by 40×6 elements, which in the figure have been copied in the circumferential direction in order to generate the complete shell. The deformed mesh in Figure 2 (and latter in Figure 4) is constructed by drawing flat quadrangle on either side of the middle surface in order to represent the thickness of the shell.

3.2. DEFORMATION RATE SENSITIVITY

In order to study the sensitivity of the viscoplastic collapse behaviour to the prescribed shortening rate, the shell from the previous section is analysed using a clamped boundary condition. An initial imperfection of the form (2) with the amplitude $\bar{\xi}_3 = 0.01$ initiates the non-axisymmetric collapse mode, while the axisymmetric collapse mode and the localization of the initial periodic buckling pattern are initiated by the clamped boundary condition. The analysed part of the shell has the length $L = 32\ell_c$ and is analysed by 60×10 elements concentrated in the localization region.

Both the viscoplastic material model with a rate-hardening exponent $m = 0.05$ prescribed to deform at a prescribed average shortening rate $\dot{\varepsilon}_L = \dot{\varepsilon}_0$ and the corresponding time-independent J_2-flow theory results in a shell which buckles into a near axisymmetric (concertina) collapse mode similar to the one shown later in Figure 4c. Note that even though the first couple of collapse folds at a boundary are of a concertina form, the subsequent collapse folds can be of a non-axisymmetric form.

In Figure 3, the viscoplastic shell is compressed at a number of different rates in the region $10^{-6} < \dot{\varepsilon}_L/\dot{\varepsilon}_0 < 1000$. An increased shortening rate is seen to results in a higher stress level during the collapse behaviour. Each

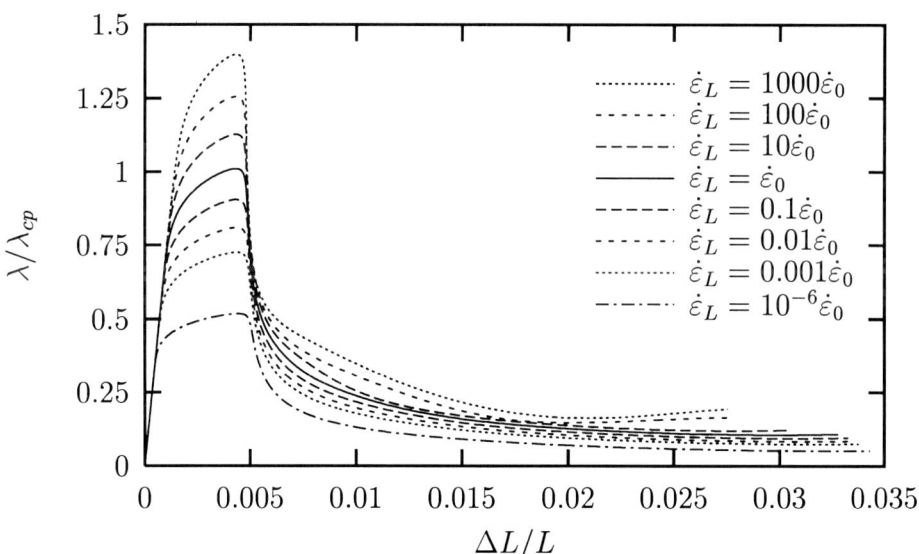

Figure 3. Influence of the prescribed compression rate, $\dot{\varepsilon}_L$ on the load versus shortening behaviour for a clamped imperfect shell with $R/h = 100$, $\sigma_0/E = 0.001$, $\nu = 0.3$, $n = 10$, $m = 0.05$, $\bar{\xi}_3 = 0.01$ and $L = 32\ell_c$.

time the average shortening rate is increased by a factor 10, the maximum load peak is found to increase by 11.4–11.7%. The increase in the mean load-carrying capacity is found to be between 10.5–10.6 when the shortening rate is in the region $10^{-6} < \dot{\varepsilon}_L/\dot{\varepsilon}_0 < 1$ while the increase is found to be significant larger, 14-18%, for higher shortening rates, $1 < \dot{\varepsilon}_L/\dot{\varepsilon}_0 < 1000$.

By looking at the shape of the collapse fold for the analysed cases, Figure 4(a–d), it can be observed that an increasing shortening rate favour a non-axisymmetric collapse mode. The larger growth in the mean load-carrying capacity for the high shortening rates may be sought in the collapse mode change into non-axisymmetric modes. Note that even though the non-axisymmetric mode is initiated by an initial imperfection given in the shape of the most critical buckling mode for the corresponding time-independent case, which for the present case is given by a circumferential wave-number $M = 3$, the shell in Figure 4a has actually collapsed into a mode with a circumferential wave-number equal to 6. Thus, the most critical buckling mode found for the corresponding time-independent case cannot any longer be the most critical imperfection for the viscoplastic case. Due to the symmetric boundary condition of the analysed part of the shell, the collapse mode is determined as an integral number of M .

With negligible elastic deformations the viscoplastic constitutive law (3) would predict an increasing stress level given by the factor $(\dot{\varepsilon}_L^b/\dot{\varepsilon}_L^a)^m$ if the prescribed deformation rate is changed from $\dot{\varepsilon}_L^a$ to $\dot{\varepsilon}_L^b$. If the prescribed shortening rate is increased by a factor 10, this corresponds to an increase of the stress level by 12%, in nice agreement with the observation shown above as long as the collapse mode stay in the concertina (near axisymmetric mode).

4. Concluding remarks

In the case of a time-independent plasticity theory a bifurcation theory shows a high influence of both the radius-to-thickness ratio and the material parameters on the collapse behaviour. On the other hand, all materials show some kind of rate sensitivity in which case the inelastic bifurcation point vanishes. A time-dependent study is restricted to a full numerical analysis. In order to get an overview of the rate sensitivity of the collapse behaviour of a circular cylindrical shell under axial compression, a viscoplastic full numerical analysis is performed.

For a fixed viscoplastic material, $m = 0.05$, the analysis shows that an increase in the prescribed average deformation rate results in an increase in the stress level and thereby in the maximum load peak and the mean load-carrying capacity. As long as the collapse mode is of a concertina form, the increase in the load level is in nice agreement with the prediction by

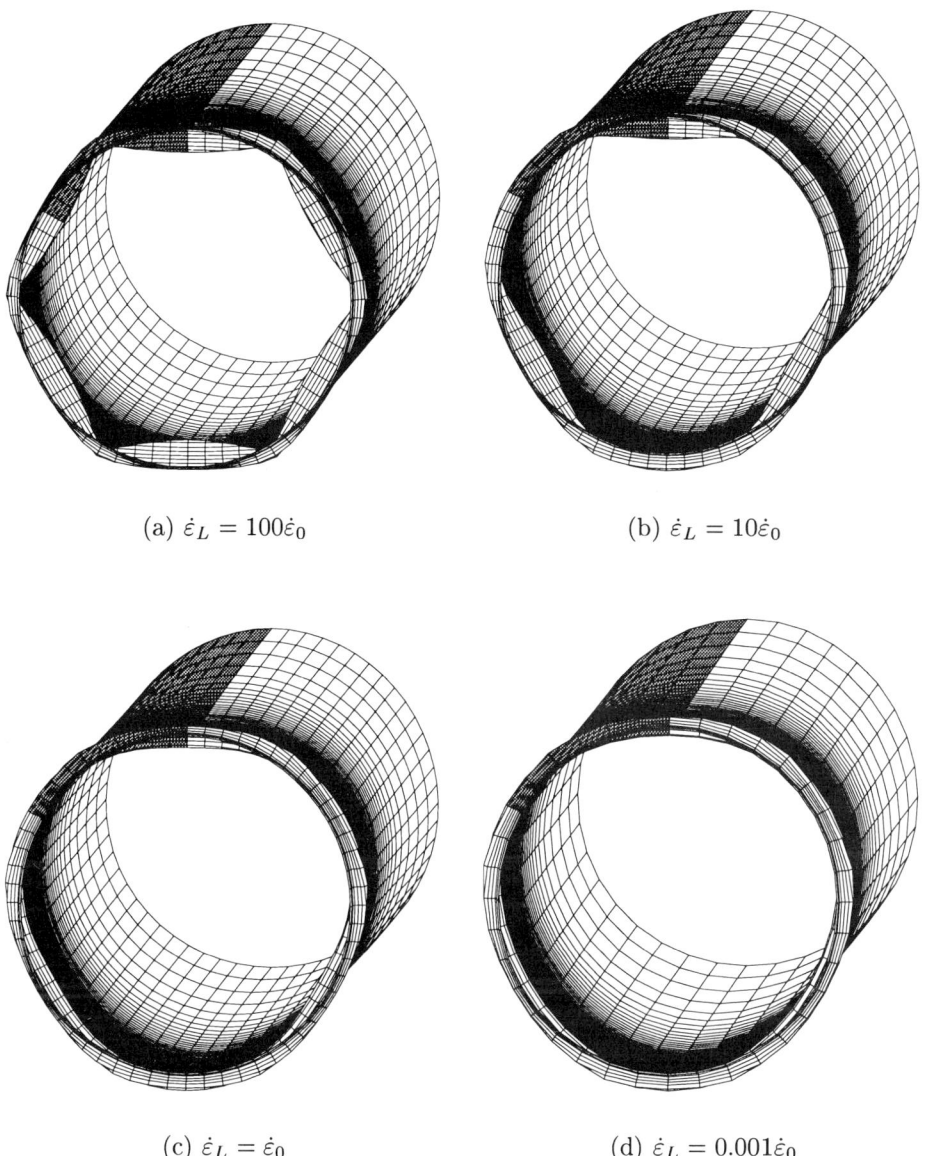

(a) $\dot{\varepsilon}_L = 100\dot{\varepsilon}_0$

(b) $\dot{\varepsilon}_L = 10\dot{\varepsilon}_0$

(c) $\dot{\varepsilon}_L = \dot{\varepsilon}_0$

(d) $\dot{\varepsilon}_L = 0.001\dot{\varepsilon}_0$

Figure 4. The deformed shells at the point where the first contact between the shell wall is obtained for a clamped shell with $R/h = 100$, $\sigma_0/E = 0.001$, $\nu = 0.3$, $n = 10$, $m = 0.05$, $\bar{\xi}_3 = 0.01$ and $L = 32\ell_c$. The gray region is the analysed part of the shell.

the viscoplastic constitutive law. An observation similar to the one found by the axisymmetric collapse analysis in (Mikkelsen, 1999). In (Mikkelsen, 1999), negligible change in the shape of the collapse fold was found for increasing shortening rates. This is not the case for the shell analysed in the present paper, see Figure 4, where a higher deformation rate is found to favour a non-axisymmetric collapse mode.

For a fixed prescribed moderate deformation rate, $\dot{\varepsilon}_L = \dot{\varepsilon}_0$, a high rate hardening material is found to increase the maximum load-carrying capacity of the shell structure. On the other hand, for the cases analysed, a delaying effect on the occurrence of contact between the shell walls results in a decreasing mean load for a high rate hardening material compared to a lower rate hardening material. Whether or not a similar effect is found for a total structural non-axisymmetric collapse requires a study of the collapse behaviour for more than one subsequent collapse fold. For the axisymmetric case (Mikkelsen, 1999) both the maximum and mean load-carrying capacity are found to be increased in a total structural collapse by a high rate hardening material.

References

Başar, Y. and Y. Ding: 1990, 'Finite-Rotation Elements for the Non-Linear Analysis of Thin Shell Structures'. *Int. J. of Solids and Structures* **26**, 83–97.

Batterman, S. C.: 1965, 'Plastic Buckling of Axially Compressed Cylindrical Shells'. *AIAA-J* **3**, 316–325.

Johnson, W., P. D. Soden, and S. T. S. Al-Hassani: 1977, 'Inextensional Collapse of Thin-Walled Tubes under Axial Compression'. *J. Strain Analysis* **12**, 317–330.

Lee, L. H. N.: 1962, 'Inelastic Buckling of Initially Imperfect Cylindrical Shells Subject to Axial Compression'. *J. Aerospace Sci.* **29**, 87–95.

Mikkelsen, L. P.: 1993, 'On the analysis of viscoplastic buckling'. *Int. J. Solids and Structures* **30**, 1461–1472.

Mikkelsen, L. P.: 1995, 'Elastic-viscoplastic buckling of circular cylindrical shells under axial compression'. *Eur. J. Mech. A/Solids* **14**, 901–920.

Mikkelsen, L. P.: 1999, 'A numerical axisymmetric collapse analysis of viscoplastic cylindrical shells under axial compression'. *Int. J. Solids and Structures* **36**, 643–668.

Obrecht, H.: 1977, 'Creep Buckling and Postbuckling of Circular Cylindrical Shells under Axial Compression'. *Int. J. solids and Structures* **13**, 337–355.

Peirce, D., C. F. Shih, and A. Needleman: 1984, 'A Tangent Modulus Method for Rate Dependent Solids'. *Computures and Structures* **18**, 875–887.

Shridharan, S., A. C. Walker, and A. Andronicou: 1981, 'Local Plastic Collapse of Ring-Stiffened Cylinders'. *Proc. Inst. Civ. Engres.* **71**, 341–367.

Tvergaard, V.: 1983a, 'On the transition from a diamond mode to an axisymmetric mode of collapse in cylindrical shells'. *Int. J. Solids and Structures* **19**, 845–856.

Tvergaard, V.: 1983b, 'Plastic buckling of axially compressed circular cylindrical shells'. *Thin Walled Structures* **1**, 139–163.

Tvergaard, V.: 1985, 'Rate-Sensitivity in Elastic-Plastic Panel Buckling'. In: D. J. Dawe, R. W. Horsington, A. G. Kamtekar, and G. H. Little (eds.): *Aspects of the Analysis of Plane Structures, A Volume in Honour of W.H. Wittrick*. Clarendon Press, pp. 293–308.

TIME-DEPENDENCE OF BUCKLING LOAD FOR A VISCOELASTIC PLATE UNDER CREEP CONDITION [*]

Ting-Qing Yang [1], Xiaochun Zhang [2], Qinguo Gang [3] and Qunli An [1]

[1] *Huazhong University of Science & Technology, Wuhan, 430074, China*
[2] *China University of Mining & Technology, Xuzhou, 221008, China*
[3] *Hebei University, Baoding, 071002, China*

Abstract This paper presents an investigation of the stability of a viscoelastic plate under creep condition. Of particular interest is the time-dependent characteristic of buckling load. Analyses show that creep buckling is a type of structural failure that may occur even when the stress remains constant. Under a certain in-plane compression loading, instability of viscoelastic structures arises after a time delay. This "delay buckling" event is more pronounced for a viscoelastic structure under creep condition. The critical load of instability is time-dependent and relies on function of material behaviors as well as the structure dimensions. Experimental studies are also conducted and the test results agree with the theoretical analyses.

1. Introduction

Creep buckling is a type of structural failure in many situations where compressive stresses are present. The investigation of these time-dependent structural behaviors is very important in numerous engineering applications. When the material creep behavior, mostly described by a power-law or Norton's law, is involved, the phenomenon of time-dependent structural instability occurs. A brief outline of some development in the study on structure creep buckling can be found in the literatures [1-4]. Tvergaard [2] considered the creep buckling behavior of simply supported plates under axial compression. He introduced the earlier researches performed by Hoff N. J., Zyczkowski M. et al. and reviewed many failure criteria for creep buckling, such as the prediction of infinite deflection at a finite critical time, the occurrence of infinite deflection rates at a finite time, allowable deflections or allowable strains, etc.

---

[*] This project is supported by the National Natural Science Foundation of China (19632030)

S. Murakami and N. Ohno (eds.), IUTAM Symposium on Creep in Structures, 509–548.

Minahen and Knauss [5] investigated the creep buckling of viscoelastic beam/column analytically and experimentally. As an example of a standard linear solid, the distinct nature of solutions for the structure behavior depends on the magnitude of compressive load and is related to deformation mode and material constant. The methodology can be generalized to apply to other structures. They also pointed out that further and more detailed investigations had to be performed. Drozdov and Kolmanovskii [6] addressed the stability problems for viscoelastic structures in details, especially when the material was subjected to aging. Recently, Huang [7] conducted a viscoelastic analysis for laminated plates under in-plane compression. The effect of various magnitudes of imperfections in the plates was discussed.

Since structures may collapse at a modest stress level due to an instability of the equilibrium, as a start, an investigation for the long-term behavior of a viscoelastic plate under quasi-static compression loading is necessary. From the engineering point of view, the main task in stability research is to evaluate the buckling load of structures. The present analysis emphasizes on the critical load and the loading duration. Time-dependent characteristics of buckling load under creep condition are studied by theoretically analyzing the stability of a viscoelastic thin plate. By means of integral form of constitutive relationship in linear viscoelasticity, a generalized buckling equation for a viscoelastic plate is formulated. A time-dependent buckling load function is derived from the buckling equation. The analyses show that the stability of the viscoelastic plate is dependent upon not only the structure dimensions but also the material viscoelastic behaviors. When the material is described by linearly viscoelastic solid model where the Poisson's ratio is a constant, a formula related the critical load to buckling time is obtained analytically. To demonstrate the developed formulation, experimental investigation is also conducted by testing thin plate specimens made from a high-density polyethylene (HDPE) material. The theoretical prediction of the plate response is in close agreement with the experimental curves.

2. Buckling Equation of A Viscoelastic Thin Plate

For a linear viscoelastic material, its integral form of relationship between stress σ_{ij}, strain ε_{ij} and time t can be written as [8]

$$\sigma_{ij} = \delta_{ij}\lambda(t) * d\varepsilon_{kk}(t) + 2G(t) * d\varepsilon_{ij}(t) \tag{1}$$

where $i, j = 1, 2, 3$; δ_{ij} is a Kronecker symbol; $G(t)$ and $\lambda(t)$ are material functions; and the asterisk "*" indicates Stieltjes convolution [9]. The Stieltjes convolution of two functions $\phi(t)$ and $\psi(t)$ is defined by $\phi * d\psi = \int_{-\infty}^{t} \phi(t-\tau)\, d\psi(\tau)$.

If deflection of the middle surface (x_1, x_2) for a viscoelastic thin plate with constant thickness h is denoted as $w(x_1, x_2, t)$, neglecting the influence of stress σ_{i3} ($i = 1,2,3$) on deformation, the strain displacement relations are

$$\varepsilon_{ij} = -x_3 w_{,ij} \qquad (i, j = 1,2) \tag{2}$$

From Eq.(2) and Eq.(1), stresses of the plate can be given by

$$\sigma_{11} = -x_3 [\lambda'(t) * d(w_{,11} + w_{,22}) + 2G'(t) * d(w_{,11})]$$

$$\sigma_{22} = -x_3 [\lambda'(t) * d(w_{,11} + w_{,22}) + 2G'(t) * d(w_{,22})] \tag{3}$$

$$\sigma_{12} = \sigma_{21} = -x_3 2G'(t) * d(w_{,12})$$

where $G'(t)$ and $\lambda'(t)$ are material parameters of the plate.

The resultant moments related to stresses in the plate can be written as

$$M_{ij} = \int_{-\frac{h}{2}}^{\frac{h}{2}} \sigma_{ij} x_3 \, dx_3 \tag{4}$$

By means of Eq.(3), one has

$$M_{11} = -\frac{h^3}{12}[\lambda'(t) * d(w_{,11} + w_{,22}) + 2G'(t) * d(w_{,11})]$$

$$M_{12} = -\frac{h^3}{6} G'(t) * d(w_{,12}) \tag{5}$$

$$M_{22} = -\frac{h^3}{12}[\lambda'(t) * d(w_{,11} + w_{,22}) + 2G'(t) * d(w_{,22})]$$

Considering an in-plane loading N_{ij}, from the equilibrium equation, the following equation can be derived

$$\frac{h^3}{12}[\lambda'(t) + 2G'(t)] * d(w_{,1111} + 2w_{,1122} + w_{,2222}) - N_{ij} w_{,ij} = 0 \tag{6}$$

Defining

$$F(t) = \frac{h^3}{12}[\lambda'(t) + 2G'(t)] \tag{7}$$

and

$$\nabla^4 w = w_{,1111} + 2w_{,1122} + w_{,2222}$$

Eq.(6) can be expressed as

$$F(t) * d(\nabla^4 w) - N_{ij} w_{,ij} = 0 \qquad (i, j = 1,2) \tag{8}$$

This is the buckling equation of the viscoelastic thin plate. $F(t)$ is a function of time-dependent material behaviors and the plate thickness.

3. Time-Dependent Buckling Load

The physical problem to be considered here is a viscoelastic rectangular plate with the length a and the width b subjected to in-plane compression loading. When the two sides with load application, $x_1=0$ and $x_1=a$, are simply supported and other two sides are load-free, one has

$$N_{11} = -p \qquad N_{22} = 0 \qquad N_{21} = 0$$

If the deflection $w(x_1, x_2, t)$ is assumed to be

$$w(x_1, x_2, t) = \sum_{m=1}^{\infty} w_m = \sum_{m=1}^{\infty} \eta_m(t) f_m(x_2) \sin\frac{m\pi x_1}{a} \tag{9}$$

Then Eq.(8) is taken as

$$F(t) * d[(\frac{m\pi}{a})^4 \eta_m(t) f_m - 2(\frac{m\pi}{a})^2 \eta_m(t) f_{m,22} + \eta_m(t) f_{m,2222}] - \eta_m(t)(\frac{m\pi}{a})^2 p f_m = 0 \tag{10}$$

To satisfy this equation, $f_m(x_2)$ and $\eta_m(t)$ fulfill, respectively,

$$\frac{f_{m,2222} - 2(\frac{m\pi}{a})^2 f_{m,22} + (\frac{m\pi}{a})^4 f_m}{(\frac{m\pi}{a})^2 f_m} = \beta_m \tag{11}$$

and

$$\frac{p\eta_m(t)}{F(t) * d\eta_m(t)} = \beta_m \tag{12}$$

where β_m is a constant.

The eigenvalues from Eq.(11) are given by

$$r_{1,2} = \pm\sqrt{\frac{m\pi}{a}\left(\frac{m\pi}{a} - \sqrt{\beta_m}\right)} \quad , \quad r_{3,4} = \pm\sqrt{\frac{m\pi}{a}\left(\frac{m\pi}{a} + \sqrt{\beta_m}\right)}$$

It can be seen that different boundary conditions at $x_2=0$ and $x_2=b$ will lead to different forms of shape function $f_m(x_2)$. Further analytical discussions need to be based on the details of the plate edges. The numerical methods can be used in the cases that analytical solutions are not readily available [10,11].

For simplicity, both ends of the plate at $x_2=0$ and $x_2=b$ are assumed to be free. The plate will bend like a cylindrical surface and $f_m(x_2)$ is independent upon x_2. Then, Eq.(11) yields

$$\beta_m = \left(\frac{m\pi}{a}\right)^2 \tag{13}$$

Eq.(12) gives

$$\beta_m \int_0^t F(t-\tau)\ddot{\eta}_m(\tau)d\tau = p\eta_m(t) \tag{14}$$

where $\beta_m = (m\pi)^2 / a^2$. For a symmetrical bending when $m=1$, one can find the minimum buckling load.

To find the nontrivial solution of Eq. (14) for $\eta_m(t)$, making the Laplace transform of both sides with zero initial condition provides

$$[p - \beta_m s \overline{F}(s)] \overline{\eta}_m(s) = 0 \tag{15}$$

where "-" represents the Laplace transform of the variable. Since $\overline{\eta}_m(s)$ is not always zero as the same as $\eta_m(t)$, one then has

$$\beta_m s \overline{F}(s) = p \tag{16}$$

This expression can be referred to a buckling equation in Laplace transform space for the viscoelastic thin plate. For a given compression load p_α, the corresponding buckling time s_α is determined by

$$s_\alpha = \overline{Q}^{-1}\left(\frac{p_\alpha}{\beta_m}\right) \tag{17}$$

where $\overline{Q}^{-1}$ is the inverse of $\overline{Q}(s)$, and $\overline{Q}(s) = s\overline{F}(s)$.

Considering $p = p_\alpha H(t)$, $H(t)$ is the Heaviside function, by taking the inverse Laplace transform of Eq. (16), one has

$$\beta_m F(t) = p_\alpha \tag{18}$$

For a given load p_α, there exists a buckling time t_α which can be written as

$$t_\alpha = F^{-1}\left(\frac{p_\alpha}{\beta_m}\right) \tag{19}$$

Eq.(19) indicates that the plate will buckle under the load p_α at the time t_α. This t_α is the buckling delay time of the viscoelastic plate under the loading p_α.

4. Evaluation of Buckling Time

For further the discussion of the buckling load and the evaluation of the buckling time, one can assume that the Poisson ratio of viscoelastic solids is a constant, i.e. $\mu(t) = \mu_0$. When the material functions in the Laplace transformation space are expressed as

$$\overline{G}'(s) = \overline{G}(s), \quad \overline{\lambda}'(s) = \frac{2\overline{G}(s)\overline{\lambda}(s)}{\overline{\lambda}(s) + 2\overline{G}(s)},$$

$$2\overline{G}(s) = \frac{\overline{E}(s)}{1 + \mu_0}, \quad \overline{\lambda}(s) = \frac{2\mu_0}{1 - 2\mu_0}\overline{G}(s).$$

Transforming back to the physical space, these equations yield

$$G'(t) = G(t), \quad 2G(t) = \frac{E(t)}{1 + \mu_0}, \quad \lambda'(t) = \frac{2\mu_0}{1 - \mu_0} G(t)$$

Therefore Eq.(7) can be written as

$$F(t) = \frac{h^3}{12} \frac{1}{1 - \mu_0} 2G(t)$$

If the relaxation function in shear is

$$2G(t) = g_0 + \sum_{i=1}^{n} g_i e^{-t/\tau_i}$$

where g_0, g_i and τ_i are material constants, one has

$$F(t) = \frac{h^3}{12} \frac{1}{1 - \mu_0} (g_0 + \sum_{i=1}^{n} g_i e^{-t/\tau_i}) = A_0 + \sum_{i=1}^{n} A_i e^{-t/\tau_i} \tag{20}$$

where A_0, A_i and τ_i are the constants dependent on the material coefficients and the plate thickness.

Then, Eq. (18) becomes

$$p_\alpha = \beta_m F(t_\alpha) = \beta_m (A_0 + \sum_{i=1}^{n} A_i e^{-t_\alpha/\tau_i}) \tag{21}$$

Eq. (21) does not mean the load decreases with time. It simply implies that under the compression load p_α, the plate might be unstable after a creep time t_α. Eq. (21) also shows two extremes: when $t_\alpha \to 0$, $p_\alpha^0 = \beta_m (A_0 + \sum_{i=1}^{n} A_i)$ is the instantaneous buckling load; when p_α is smaller than $p_\alpha^\infty = \beta_m A_0$, the plate is stable even $t_\alpha \to \infty$.

There are three distinct cases for the loading:

1) When $p_\alpha \geq p_\alpha^0$ i.e. the load is higher than the instantaneous elastic-buckling level, the thin plate buckles immediately.

2) When $p_\alpha \leq p_\alpha^\infty$ i.e. the load is lower than the long-term stable load, the plate remains stable.

3) In case of $p_\alpha^0 > p_\alpha > p_\alpha^\infty$, the plate will experience creep buckling after a certain time t_α.

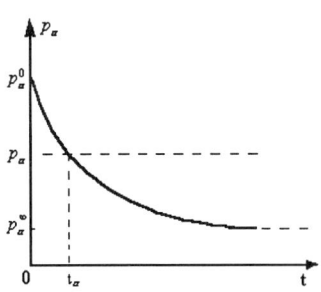

Figure 1. Buckling load vs Time

For a given load, there may exist a "delay buckling" for the structures under creep condition, i.e. structural instability may occur after a time delay. The value of critical load of instability is time-dependent and relies on the material functions and the plate geometry dimensions.

It can be found that the stability problem of viscoelastic beam/column [12] is a special case of the formulation discussed in sections 3 and 4.

5. Experimental Investigations

To demonstrate present analysis, an experimental investigation is also conducted. First, a group of high-density polyethylene (HDPE) specimens are tested in order to determine the description of the material viscoelastic behavior and the corresponding material coefficients. Then, experimental tests are performed for thin-plate specimens made of HDPE to validate the developed formulation.

5.1. MATERIAL VISCOELASTIC BEHAVIOR

A kind of HDPE is chosen as the plate material. The creep test at room temperature is conducted and the creep curves under different stresses are shown in Fig. 2. The experimental observations under a lower stress level condition illustrate that the HDPE is a linearly viscoelastic material. It will fracture under a higher stress level such as $\sigma = 9.26$MPa. When specimens are subjected to a lower stress such as $\sigma = 4.9$MPa, the material demonstrates a stable viscoelastic behavior. By means

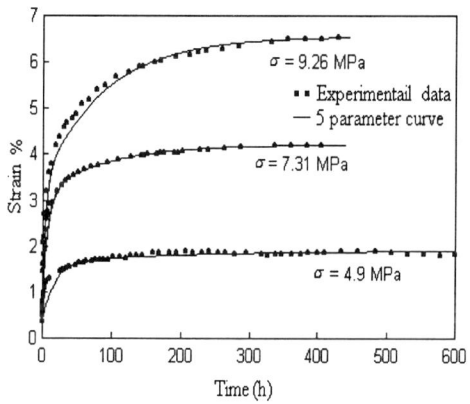

Figure 2. Creep curves

of a 5- parameter solid model, the creep function can be determined as

$$J(t) = 8.16 \times 10^{-4} + 2.45 \times 10^{-3} (1 - e^{-t/\tau_1}) + 6.12 \times 10^{-4} (1 - e^{-t/\tau_2}) \quad (1/MPa) \quad (22)$$

where $\tau_1 = 16.7\ h$, $\tau_2 = 166.7\ h$. From the relation between creep function and relaxation function [8], the relaxation function can be obtained as

$$E(t) = 257 + 857 \times e^{-t/4.13} + 110 \times e^{-t/136} \qquad (MPa) \qquad (23)$$

In case of creep buckling, considering a lower stress in the plate, in case of creep buckling, the relaxation function is modified as

$$E^*(t) = 385.5 + 1285.5 \times e^{-t/4.13} + 165 \times e^{-t/136} \qquad (MPa) \qquad (24)$$

According to the tests, the Poisson's ratio of the HDPE, μ, is around 0.44~0.46. Here μ=0.45 is used in the calculation of materials functions, the F(t), and then p_α.

When $\beta_m = \beta_1$, taking a=160 mm and h= 6 mm, the compression load versus time function can be obtained theoretically from Eq. (21). The $p(t)$ and $p^*(t)$ corresponding to the relaxation functions shown in Eqs.(23) and (24) can be derived as, respectively,

$$p(t_\alpha) = 9.94 + 33.14 \times e^{-t_\alpha/4.13} + 4.25 \times e^{-t_\alpha/136} \qquad (KN / m) \qquad (25)$$

and $\quad p^*(t_\alpha) = 13.91 + 46.39 \times e^{-t_\alpha/4.13} + 5.95 \times e^{-t_\alpha/136} \qquad (KN / m) \qquad (26)$

These equations can be applied to evaluate p_α^0 and p_α^∞.

5.2. CREEP BUCKLING EXPERIMENTS

Experimental tests are also conducted for several HDPE plates under compression to verify the accuracy of the formulation developed in the preceding sections. All plate specimens with dimensions: $160 \times 80 \times 6 \ mm^3$ are manufactured from panels made of the HDPE material. The test set-up is shown in Fig.3. Two opposite loading sides of the plate are simply supported and other two sides are free. The deflection at the center of plate varies with time and is measured by the test system as shown in Fig.3. A Dataaker-100 device performs data acquisition. Data collecting time is once every two minutes in the first hour and once every 10 minutes afterwards.

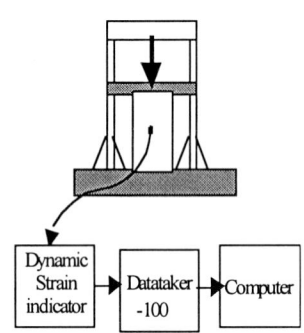

Figure 3. Sketch of buckling test

Experimental deflection-time curves are illustrated Fig. 4. It can be seen that when loaded at $p_\alpha^0 \geq p_\alpha \geq p_\alpha^\infty$, the specimen presents a creep behavior. If the buckling time is assumed to be at the beginning of the third stage in creep, the load versus buckling time curves can be obtained and are depicted in Fig. 5.

With the material functions determined from section 5.1, predictions of overall responses of HPDE plates are made by using Eq.(21). A comparison of the experimental results and theoretical prediction is also provided in Fig. 5. Two prediction curves are obtained from Eqs.(25) and (26), respectively. From this figure, a close agreement between the analysis and the test results can be observed. The discrepancy comes from the supposition that $f_m(x_2)$ is independent upon x_2 for obtaining Eq.(13), and some experimental factors such as the data processing of material parameters, the test equipment, and the measurement system.

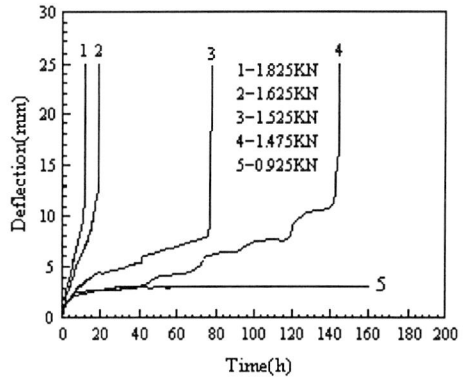

Figure 4.   Deflection-time curves of plate

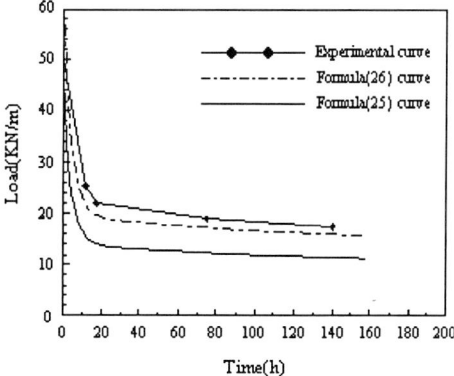

Figure 5.   Loads vs buckling times

6. Conclusions

The description of stability for a viscoelastic thin plate subjected to in-plane compression has been analytically established and can be employed to determine the critical load. Analysis shows that the creep buckling in viscoelastic structure is time-dependent under a certain load level. There are two distinct critical loading: instantaneous instability load p_α^0 and stable load p_α^∞. For a given plate structure, p_α^0 and p_α^∞ can be estimated from Eq. (21). In the case of $p_\alpha^0 > p_\alpha > p_\alpha^\infty$, there exists a delay buckling after time t_α. This buckling time, t_α, can be evaluated by means of an expression, such as Eq. (19).

To demonstrate the presented formulation, experimental investigations have also been performed first by testing a HDPE material. The responses of thin plate specimens made of the HDPE material subjected to in-plane compression have then been studied experimentally and analytically. The results show that the theoretical analyses agree with experimental observations.

References

1   Flügge W. *Viscoelasticity*. 2 nd Ed. Springer-Verlag, 1975

2   Tvergaard V. Creep Buckling of Rectangular Plates under Axial Compression, *Int. J. Solids Structures*, 1979, (15): 441-456

3   Schapery R A. Viscoelastic buckling of an Eular-type beam. *Texas A & M report*, RR-87-03

4   Paley M &Aboudi J. Creep buckling of inelastic plates. In: Zyczkowski M (Ed.) *Creep in Structure*, IUTAM Symposium Cracow/Poland, 1990; Springer-Verlag Berlin Heidelberg, 1991

5   Minahen T M, Knauss W G. Creep buckling of viscoelastic structures. *Int. J. Solids Structures*, 1993, 30(8): 1075-1092

6   Drozdov A. D., Kolmanovskii V.B. *Stability in Viscoelasticity*, North-Holland Series in Applied Mathematics and Mechanics, North-Holland, 1994

7   Huang N-N. Viscoelastic analysis of Von Karman laminated plates under in-plane compression with initial deflection. *Int. J. Non-linear Mechanics*, 1997, 32(6):1065-1075

8   Yang Ting-Qing. *Theory of viscoelasticity*. Wuhan: HUST Press, 1990 (in Chinese with English abstract)

9   Gurtin M E and Sternberg E. On the linear theory of viscoelasticity. *Archive Rat. Mech. Analysis*, 1962, 11:291-356

10   Yang T-Q, Wang R, and Yang Z-W. Dynamic response of a viscoelastic circular plate on a Viscoelastic half-space foundation. IUTAM Symp. 1990; in: Zyczkowski M (Ed). *Creep in Structures*, pp.685-692, Springer-Verlag, 1991.

11   Yang Ting-Qing, Qu Xuyan. Axisymmetric bending of viscoelastic circular plate with large deflection. *Acta Mechanica Solida Sinica*. 1990, 11(4):313-320 (in Chinese with English abstract)

12   Zhang Xiaochun, Yang Ting-Qing. A stability analysis for time-dependence of plate-beam structure of rock. *Journal of Wuhan Transportation Univ.*, 1999, 23(2):158-160

AUTHOR INDEX

Abe, F.	267	Fang, H. E.	131
Adam, L.	197	Fischer, F. D.	17
Ainsworth, R. A.	321	Fujiwara, K.	489
Altenbach, H.	141	Furtado, H. C.	241
An, Q.	509		
Ando, T.	439	Ganczarski, A.	207
		Gang, Q.	509
Barrette, P. D.	479	Golubovskiy, E.	231
Benallal, A.	151		
Bettinson, A. D.	95	Härkegård, G.	287
Biwa, S.	439	Hayhurst, D. R.	1, 175, 375
Bodnar, A.	189	Hellmich, Ch.	217
Boyle, J. T.	297	Hirano, T.	165
Brückner, U.	231	Hiroe, T.	489
Budden, P. J.	321	Ho, K.	361
Busso, E. P.	41	Hosokawa, T.	411
		Huth, H.-J.	287
Carmai, J.	463	Hyde, T. H.	85
Carrere, N.	453		
Chaboche, J.L.	453	Igari, T.	351
Charkaluk, E.	341	Ishikawa, H.	391
Chatterjee, A.	17		
Chow, C. L.	131	Jordaan, I. J.	479
Chrzanowski, M.	189		
Clemens, H.	17	Kablov, E.	231
Constantinescu, A.	341	Kawai, M.	469
		Kimmins, S. T.	257
Dang Van, K.	331	Kimura, K.	267
Dean, D. W.	321	Kitamura, T.	105
Dennis, R. J.	41	Knowles, D. M.	31
Desmorat, R.	125	Kowalewski, Z. L.	375, 401
Dunne, F. P. E.	463	Krempl, E.	361
Dyson, B. F.	3,375	Kruch, S.	453
		Kubo, S.	277
Epishin, A.	231	Kushima, H.	267

519

Lackner, R.	217	Seshadri, R.	297
Le May, I.	241	Shibutani, T.	105
Lechner, M.	217	Siad, L.	151
Lemaitre, J.	125	Sinohara, N.	429
Lin, J.	375	Smith, D. J.	257
Link, T.	231	Sugita, Y.	429
Liu, Y.	165	Sugiyama, K.	429
		Sun, W.	85
Macht, J.	217	Svetlov, I.	231
MacLachlan, D. W.	31		
Mang, H.A.	217	Tada, N.	75
Marketz, W. T.	17	Tagami, N.	277
Matsuo, H.	489	Takahashi, Y.	311
Mayama, T.	391	Tamura, E.	277
McLean, M.	3	Tokiyoshi, T.	351
Mikkelsen, L. P.	499		
Miyake, T.	439	Van Der Giessen, E.	51, 65
Miyazaki, N.	115	Verger, L.	341
Mizokami, Y.	351		
Murakami, S.	165	Walker, N. S.	257
		Webster, G. A.	95
Neilsen, M. K.	131	Wei, Y.	131
Nguyen, B.-N.	51, 65		
Nikbin, K.	95	Yagi, K.,	267
		Yang, T.-Q.	509
O'Dowd, N. P.	41, 95	Yoshida, F.	419
Ohashi, F.	489		
Ohji, K.	277	Zhang, X.	509
Ohno, N.	439		
Ohtani, R.	75		
Onck, P.	51,65		
Ponthot, J. P.	197		
Portella, P.	231		
Sakane, M.	411		
Sasaki, K.	391		
Sermage, J. P.	125		

Mechanics

SOLID MECHANICS AND ITS APPLICATIONS
Series Editor: G.M.L. Gladwell

49. J.R. Willis (ed.): *IUTAM Symposium on Nonlinear Analysis of Fracture*. Proceedings of the IUTAM Symposium held in Cambridge, U.K. 1997   ISBN 0-7923-4378-6

50. A. Preumont: *Vibration Control of Active Structures*. An Introduction. 1997   ISBN 0-7923-4392-1

51. G.P. Cherepanov: *Methods of Fracture Mechanics: Solid Matter Physics*. 1997   ISBN 0-7923-4408-1

52. D.H. van Campen (ed.): *IUTAM Symposium on Interaction between Dynamics and Control in Advanced Mechanical Systems*. Proceedings of the IUTAM Symposium held in Eindhoven, The Netherlands. 1997   ISBN 0-7923-4429-4

53. N.A. Fleck and A.C.F. Cocks (eds.): *IUTAM Symposium on Mechanics of Granular and Porous Materials*. Proceedings of the IUTAM Symposium held in Cambridge, U.K. 1997   ISBN 0-7923-4553-3

54. J. Roorda and N.K. Srivastava (eds.): *Trends in Structural Mechanics*. Theory, Practice, Education. 1997   ISBN 0-7923-4603-3

55. Yu.A. Mitropolskii and N. Van Dao: *Applied Asymptotic Methods in Nonlinear Oscillations*. 1997   ISBN 0-7923-4605-X

56. C. Guedes Soares (ed.): *Probabilistic Methods for Structural Design*. 1997   ISBN 0-7923-4670-X

57. D. François, A. Pineau and A. Zaoui: *Mechanical Behaviour of Materials*. Volume I: Elasticity and Plasticity. 1998   ISBN 0-7923-4894-X

58. D. François, A. Pineau and A. Zaoui: *Mechanical Behaviour of Materials*. Volume II: Viscoplasticity, Damage, Fracture and Contact Mechanics. 1998   ISBN 0-7923-4895-8

59. L.T. Tenek and J. Argyris: *Finite Element Analysis for Composite Structures*. 1998   ISBN 0-7923-4899-0

60. Y.A. Bahei-El-Din and G.J. Dvorak (eds.): *IUTAM Symposium on Transformation Problems in Composite and Active Materials*. Proceedings of the IUTAM Symposium held in Cairo, Egypt. 1998   ISBN 0-7923-5122-3

61. I.G. Goryacheva: *Contact Mechanics in Tribology*. 1998   ISBN 0-7923-5257-2

62. O.T. Bruhns and E. Stein (eds.): *IUTAM Symposium on Micro- and Macrostructural Aspects of Thermoplasticity*. Proceedings of the IUTAM Symposium held in Bochum, Germany. 1999   ISBN 0-7923-5265-3

63. F.C. Moon: *IUTAM Symposium on New Applications of Nonlinear and Chaotic Dynamics in Mechanics*. Proceedings of the IUTAM Symposium held in Ithaca, NY, USA. 1998   ISBN 0-7923-5276-9

64. R. Wang: *IUTAM Symposium on Rheology of Bodies with Defects*. Proceedings of the IUTAM Symposium held in Beijing, China. 1999   ISBN 0-7923-5297-1

65. Yu.I. Dimitrienko: *Thermomechanics of Composites under High Temperatures*. 1999   ISBN 0-7923-4899-0

66. P. Argoul, M. Frémond and Q.S. Nguyen (eds.): *IUTAM Symposium on Variations of Domains and Free-Boundary Problems in Solid Mechanics*. Proceedings of the IUTAM Symposium held in Paris, France. 1999   ISBN 0-7923-5450-8

67. F.J. Fahy and W.G. Price (eds.): *IUTAM Symposium on Statistical Energy Analysis*. Proceedings of the IUTAM Symposium held in Southampton, U.K. 1999   ISBN 0-7923-5457-5

68. H.A. Mang and F.G. Rammerstorfer (eds.): *IUTAM Symposium on Discretization Methods in Structural Mechanics*. Proceedings of the IUTAM Symposium held in Vienna, Austria. 1999   ISBN 0-7923-5591-1

Mechanics

SOLID MECHANICS AND ITS APPLICATIONS
Series Editor: G.M.L. Gladwell

69. P. Pedersen and M.P. Bendsøe (eds.): *IUTAM Symposium on Synthesis in Bio Solid Mechanics.* Proceedings of the IUTAM Symposium held in Copenhagen, Denmark. 1999
ISBN 0-7923-5615-2

70. S.K. Agrawal and B.C. Fabien: *Optimization of Dynamic Systems.* 1999
ISBN 0-7923-5681-0

71. A. Carpinteri: *Nonlinear Crack Models for Nonmetallic Materials.* 1999
ISBN 0-7923-5750-7

72. F. Pfeifer (ed.): *IUTAM Symposium on Unilateral Multibody Contacts.* Proceedings of the IUTAM Symposium held in Munich, Germany. 1999   ISBN 0-7923-6030-3

73. E. Lavendelis and M. Zakrzhevsky (eds.): *IUTAM/IFToMM Symposium on Synthesis of Non-linear Dynamical Systems.* Proceedings of the IUTAM/IFToMM Symposium held in Riga, Latvia. 2000   ISBN 0-7923-6106-7

74. J.-P. Merlet: *Parallel Robots.* 2000   ISBN 0-7923-6308-6

75. J.T. Pindera: *Techniques of Tomographic Isodyne Stress Analysis.* 2000 ISBN 0-7923-6388-4

76. G.A. Maugin, R. Drouot and F. Sidoroff (eds.): *Continuum Thermomechanics.* The Art and Science of Modelling Material Behaviour. 2000   ISBN 0-7923-6407-4

77. N. Van Dao and E.J. Kreuzer (eds.): *IUTAM Symposium on Recent Developments in Non-linear Oscillations of Mechanical Systems.* 2000   ISBN 0-7923-6470-8

78. S.D. Akbarov and A.N. Guz: *Mechanics of Curved Composites.* 2000   ISBN 0-7923-6477-5

79. M.B. Rubin: *Cosserat Theories: Shells, Rods and Points.* 2000   ISBN 0-7923-6489-9

80. S. Pellegrino and S.D. Guest (eds.): *IUTAM-IASS Symposium on Deployable Structures: Theory and Applications.* Proceedings of the IUTAM-IASS Symposium held in Cambridge, U.K., 6–9 September 1998. 2000   ISBN 0-7923-6516-X

81. A.D. Rosato and D.L. Blackmore (eds.): *IUTAM Symposium on Segregation in Granular Flows.* Proceedings of the IUTAM Symposium held in Cape May, NJ, U.S.A., June 5–10, 1999. 2000   ISBN 0-7923-6547-X

82. A. Lagarde (ed.): *IUTAM Symposium on Advanced Optical Methods and Applications in Solid Mechanics.* Proceedings of the IUTAM Symposium held in Futuroscope, Poitiers, France, August 31–September 4, 1998. 2000   ISBN 0-7923-6604-2

83. D. Weichert and G. Maier (eds.): *Inelastic Analysis of Structures under Variable Loads.* Theory and Engineering Applications. 2000   ISBN 0-7923-6645-X

84. T.-J. Chuang and J.W. Rudnicki (eds.): *Multiscale Deformation and Fracture in Materials and Structures.* The James R. Rice 60th Anniversary Volume. 2001   ISBN 0-7923-6718-9

85. S. Narayanan and R.N. Iyengar (eds.): *IUTAM Symposium on Nonlinearity and Stochastic Structural Dynamics.* Proceedings of the IUTAM Symposium held in Madras, Chennai, India, 4–8 January 1999   ISBN 0-7923-6733-2

Kluwer Academic Publishers – Dordrecht / Boston / London